Modification and Processing of Biodegradable Polymers

Modification and Processing of Biodegradable Polymers

Editor

Krzysztof Moraczewski

MDPI • Basel • Beijing • Wuhan • Barcelona • Belgrade • Manchester • Tokyo • Cluj • Tianjin

Editor
Krzysztof Moraczewski
Department of Materials Engineering
Kazimierz Wielki University
Bydgoszcz
Poland

Editorial Office
MDPI
St. Alban-Anlage 66
4052 Basel, Switzerland

This is a reprint of articles from the Special Issue published online in the open access journal *Materials* (ISSN 1996-1944) (available at: www.mdpi.com/journal/materials/special_issues/modification_process_biodegradable_polym).

For citation purposes, cite each article independently as indicated on the article page online and as indicated below:

LastName, A.A.; LastName, B.B.; LastName, C.C. Article Title. *Journal Name* **Year**, *Volume Number*, Page Range.

ISBN 978-3-0365-7373-1 (Hbk)
ISBN 978-3-0365-7372-4 (PDF)

© 2023 by the authors. Articles in this book are Open Access and distributed under the Creative Commons Attribution (CC BY) license, which allows users to download, copy and build upon published articles, as long as the author and publisher are properly credited, which ensures maximum dissemination and a wider impact of our publications.

The book as a whole is distributed by MDPI under the terms and conditions of the Creative Commons license CC BY-NC-ND.

Contents

About the Editor vii

Li-Han Lin, Hung-Pang Lee and Ming-Long Yeh
Characterization of a Sandwich PLGA-Gallic Acid-PLGA Coating on Mg Alloy ZK60 for Bioresorbable Coronary Artery Stents
Reprinted from: *Materials* 2020, 13, 5538, doi:10.3390/ma13235538 1

Ewa Olewnik-Kruszkowska, Magdalena Gierszewska, Agnieszka Richert, Sylwia Grabska-Zielińska, Anna Rudawska and Mohamed Bouaziz
Antibacterial Films Based on Polylactide with the Addition of Quercetin and Poly(Ethylene Glycol)
Reprinted from: *Materials* 2021, 14, 1643, doi:10.3390/ma14071643 17

Iwona Tarach, Ewa Olewnik-Kruszkowska, Agnieszka Richert, Magdalena Gierszewska and Anna Rudawska
Influence of Tea Tree Essential Oil and Poly(ethylene glycol) on Antibacterial and Physicochemical Properties of Polylactide-Based Films
Reprinted from: *Materials* 2020, 13, 4953, doi:10.3390/ma13214953 39

Sylwia Grabska-Zielińska, Magdalena Gierszewska, Ewa Olewnik-Kruszkowska and Mohamed Bouaziz
Polylactide Films with the Addition of Olive Leaf Extract—Physico-Chemical Characterization
Reprinted from: *Materials* 2021, 14, 7623, doi:10.3390/ma14247623 55

Tannaz Soltanolzakerin Sorkhabi, Mehrab Fallahi Samberan, Krzysztof Adam Ostrowski, Tomasz M. Majka, Marcin Piechaczek and Paulina Zajdel
Preparation and Characterization of Novel Microgels Containing Nano-SiO_2 and Copolymeric Hydrogel Based on Poly (Acrylamide) and Poly (Acrylic Acid): Morphological, Structural and Swelling Studies
Reprinted from: *Materials* 2022, 15, 4782, doi:10.3390/ma15144782 75

Arianna Pietrosanto, Paola Scarfato, Luciano Di Maio and Loredana Incarnato
Development of Eco-Sustainable PBAT-Based Blown Films and Performance Analysis for Food Packaging Applications
Reprinted from: *Materials* 2020, 13, 5395, doi:10.3390/ma13235395 91

Anna Czajka, Radosław Bulski, Anna Iuliano, Andrzej Plichta, Kamila Mizera and Joanna Ryszkowska
Grafted Lactic Acid Oligomers on Lignocellulosic Filler towards Biocomposites
Reprinted from: *Materials* 2022, 15, 314, doi:10.3390/ma15010314 107

Florentyna Markowicz and Agata Szymańska-Pulikowska
Assessment of the Decomposition of Oxo- and Biodegradable Packaging Using FTIR Spectroscopy
Reprinted from: *Materials* 2021, 14, 6449, doi:10.3390/ma14216449 125

Łukasz Łopusiewicz, Emilia Drozłowska, Paulina Trocer, Mateusz Kostek, Mariusz Śliwiński and Marta H. F. Henriques et al.
Whey Protein Concentrate/Isolate Biofunctional Films Modified with Melanin from Watermelon (*Citrullus lanatus*) Seeds
Reprinted from: *Materials* 2020, 13, 3876, doi:10.3390/ma13173876 141

Łukasz Łopusiewicz, Szymon Macieja, Mariusz Śliwiński, Artur Bartkowiak, Swarup Roy and Peter Sobolewski
Alginate Biofunctional Films Modified with Melanin from Watermelon Seeds and Zinc Oxide/Silver Nanoparticles
Reprinted from: *Materials* 2022, 15, 2381, doi:10.3390/ma15072381 161

Agnieszka Derewonko, Wojciech Fabianowski and Jerzy Siczek
Mechanical Testing of Epoxy Resin Modified with Eco-Additives
Reprinted from: *Materials* 2023, 16, 1854, doi:10.3390/ma16051854 185

Emil Sasimowski, Łukasz Majewski and Marta Grochowicz
Analysis of Selected Properties of Injection Moulded Sustainable Biocomposites from Poly(butylene succinate) and Wheat Bran
Reprinted from: *Materials* 2021, 14, 7049, doi:10.3390/ma14227049 199

Emil Sasimowski, Łukasz Majewski and Marta Grochowicz
Artificial Ageing, Chemical Resistance, and Biodegradation of Biocomposites from Poly(Butylene Succinate) and Wheat Bran
Reprinted from: *Materials* 2021, 14, 7580, doi:10.3390/ma14247580 231

Magdalena Tomanik, Magdalena Kobielarz, Jarosław Filipiak, Maria Szymonowicz, Agnieszka Rusak and Katarzyna Mroczkowska et al.
Laser Texturing as a Way of Influencing the Micromechanical and Biological Properties of the Poly(L-Lactide) Surface
Reprinted from: *Materials* 2020, 13, 3786, doi:10.3390/ma13173786 267

Rafał Malinowski, Krzysztof Moraczewski and Aneta Raszkowska-Kaczor
Studies on the Uncrosslinked Fraction of PLA/PBAT Blends Modified by Electron Radiation
Reprinted from: *Materials* 2020, 13, 1068, doi:10.3390/ma13051068 281

Karolina Wiszumirska, Dorota Czarnecka-Komorowska, Wojciech Kozak, Marta Biegańska, Patrycja Wojciechowska and Maciej Jarzebski et al.
Characterization of Biodegradable Food Contact Materials under Gamma-Radiation Treatment
Reprinted from: *Materials* 2023, 16, 859, doi:10.3390/ma16020859 297

About the Editor

Krzysztof Moraczewski

Dr hab. inż. Krzysztof Moraczewski is a professor at the Kazimierz Wielki University in Bydgoszcz, Poland. Currently, he is the Dean of the Faculty of Materials Engineering. He is the author or co-author of 100 scientific publications. He conducts research in the field of materials engineering, which is one of the dynamically developing areas of polymer science and methods of their modification. In his scientific work, he deals with issues related to the modification and metallization of the surface layer of polymers and processes related to the aging of polymer materials.

Article

Characterization of a Sandwich PLGA-Gallic Acid-PLGA Coating on Mg Alloy ZK60 for Bioresorbable Coronary Artery Stents

Li-Han Lin [1], Hung-Pang Lee [2] and Ming-Long Yeh [1,3,]*

1. Department of Biomedical Engineering, National Cheng Kung University, Tainan 701, Taiwan; d09522010@ntu.edu.tw
2. Biomedical Engineering, Dwight Look College of Engineering, Texas A&M University, College Station, TX 77843, USA; qer6322129@tamu.edu
3. Medical Device Innovation Center, National Cheng Kung University, Tainan 701, Taiwan
* Correspondence: mlyeh@mail.ncku.edu.tw; Tel.: +886-6275-7575 (ext. 63429); Fax: +886-6234-3270

Received: 8 November 2020; Accepted: 2 December 2020; Published: 4 December 2020

Abstract: Absorbable magnesium stents have become alternatives for treating restenosis owing to their better mechanical properties than those of bioabsorbable polymer stents. However, without modification, magnesium alloys cannot provide the proper degradation rate required to match the vascular reform speed. Gallic acid is a phenolic acid with attractive biological functions, including anti-inflammation, promotion of endothelial cell proliferation, and inhibition of smooth muscle cell growth. Thus, in the present work, a small-molecule eluting coating is designed using a sandwich-like configuration with a gallic acid layer enclosed between poly (D,L-lactide-co-glycolide) layers. This coating was deposited on ZK60 substrate, a magnesium alloy that is used to fabricate bioresorbable coronary artery stents. Electrochemical analysis showed that the corrosion rate of the specimen was ~2000 times lower than that of the bare counterpart. The released gallic acid molecules from sandwich coating inhibit oxidation by capturing free radicals, selectively promote the proliferation of endothelial cells, and inhibit smooth muscle cell growth. In a cell migration assay, sandwich coating delayed wound closure in smooth muscle cells. The sandwich coating not only improved the corrosion resistance but also promoted endothelialization, and it thus has great potential for the development of functional vascular stents that prevent late-stent restenosis.

Keywords: magnesium alloy; cardiovascular stents; callic acid; dip coating; endothelialization; anticorrosion

1. Introduction

According to reports from the Global Health Organization in 2016, cardiovascular diseases (CVDs), the prevalence of which increased by 15% in the past decade, are the leading cause of death (44%) among non-communicable diseases [1]. In clinical settings, a percutaneous coronary intervention (PCI) combined with balloon angioplasty and stent implantation is the gold standard for treating stenotic arteries. Currently, dual antiplatelet therapy is suggested for at least 12 months to alleviate mural thrombosis caused by PCI [2]. The traditional materials used in cardiovascular stents are 316 L stainless steels, cobalt-chromium alloys, and nickel-titanium alloys [3]. However, the permanent installation of devices inside the human body causes a chronic inflammatory local reaction and long-term endothelial dysfunction [4]. Biodegradable magnesium (Mg) and its alloys are ideal candidates for stent platforms because they have less neointimal formation and provide long-term inhibition of the growth of smooth muscle cells (SMCs), unlike stainless steel [5,6]. The Mg alloys also have twice the tensile strength of unmodified biodegradable poly-L-lactide (PLLA), which leads to flexible stenting deployment and better radial force [7–9].

To properly use a magnesium alloy as a biodegradable stent platform for physiological applications, the corrosion rate must be controlled. To date, the surfaces of degradable magnesium alloys have been modified in several ways, such as alkaline heat treatment (AHT), the sol-gel process, chemical conversion, and micro-arc oxidation (MAO) [10,11]. Although ceramic MAO processing may provide a consolidated surface [12–14] for clinical applications, a prior study showed that this process did not improve corrosion resistance after three months [15]. In contrast, a sol-gel coating can greatly improve the corrosion resistance with chemical modifications [16]. Polymeric processes create a physical barrier [17] that enhances the corrosion resistance of Mg-stents, and the barrier also serves as an absorbable drug-loading system [18,19]. A polymer coating of a specific thickness can prevent complete degradation of Mg alloys for periods of 1 to 12 months [20].

Phenolic molecules, such as gallic acids (GA), exert and induce specific levels of selective viability in human endothelial cells (ECs) and SMCs [21–24]. Furthermore, these extracted molecules have anti-oxidation characteristics that inhibit inflammation. Since cellular antioxidants prevent the formation of oxidized low-density lipoprotein (LDL) and antiplatelet activation, this is a remarkable method by which to treat atherosclerosis [25]. Although several articles have indicated that the phenolic conversion coating on a Mg-alloy is non-toxic and anti-corrosive, chemical conversion may compromise stent integrity through the phenolic-Mg conversion process [26–29].

Mg-Zn alloys are well-known for their good corrosion resistance and physiological safety [30]. The ZK60 Mg-6Zn-0.5Zr alloy has been used to make vascular stents due to its better biocompatibility than those of other Mg alloys [31]. In addition, ZK60 has a high tensile strength (315 MPa) that prolongs the radial force of Mg stents. Nevertheless, the rapid degradation rate of bare ZK60 represents a shortcoming of the alloy for its use to fabricate bioresorbable coronary artery stents [15]. In this paper, a coating comprising of GA sandwiched between two layers of Poly (D,L-lactide-co-glycolide) (PLGA), a novel sandwich coating, on Mg alloy ZK60 concept was proposed. It can serve not only as a protective barrier that increases corrosion resistance but also as a reservoir for controlling GA release. The surface properties of coated-Mg stent and corrosion behavior of a Mg alloy were determined. Furthermore, cell viability and cell migration tests were performed to investigate the cell regulation of released GA in endothelialization as well as the effects on the growth of smooth muscle cells.

2. Materials and Methods

2.1. Materials and Specimen Preparation

In this study, extruded commercial ZK60 alloy bars (Zn 5.48 wt.%, Zr 0.42 wt.% and balance Mg) comprised the starting material for the substrate [32]. PLGA in molar ratio of 85/15 and GA with purity above 97.5% were purchased from Sigma-Aldrich (St. Louis, MO, USA; Figure 1a). ZK60 disks cut from the bars were reduced to 12 mm in diameter and 5 mm in thickness (Figure 1c). These specimens were ground with silicone carbide sandpaper (150–5000 mesh) and polished. Before being dried in a stream of air, all specimens were rinsed ultrasonically with acetone, ethanol, and distilled water for 5 min, respectively. Mg-stent was machined by an INTAI Technology and CHONG HUAI laser (Taichung, Taiwan) for further evaluation.

2.2. Sandwich Coating Films Preparation

Mirror-polished ZK60 specimens were soaked in an alkaline solution composed of 20 wt.% NaOH. The solution was stirred at 240 rpm and then heated for 90 min at 60 °C to equilibrate the specimens. After 90 min, the specimens were drawn out of the alkaline solution and rinsed with deionized water. The rinsed specimens were incubated at 80 °C for 30 min until they were dried. Subsequently, heat treatment at 120 °C for another 30 min was conducted to stabilize the oxide film. To fabricate the sol-gel PLGA coating layer, PLGA was dissolved at a concentration of 4 wt.% in 10 mL of chloroform.

The PLGA film thickness depended on the drawing speed of 3 mm/s, and the film was dried in a stream of air to form a uniform dip-coating surface. Subsequently, different coating layers were prepared with or without GA solution. The samples were dipped into GA solution at a concentration of 1 wt.% and 10 mL acetone. Finally, the samples were ultrasonically rinsed with ethanol and air-dried. The coating steps were conducted on the Mg-stent for surface morphology observation.

2.3. Characterization of the Surface, the Cross-Section Structure, and the Elements Content

The crystallinity of ZK60 was analyzed using thin film X-ray diffraction (TF-XRD Bruker D8 Discover, Brucker, Karlsruhe, Germany) with Cu-Kα radiation. Diffraction patterns were acquired at 2θ values of 20–80°. The surface morphology and element distributions of the coating films were studied by scanning electron microscopy (SEM JSM-6700F, JEOL, Tokyo, Japan) under 10 kV acceleration voltage and energy dispersive spectrometry (EDS JSM-6700F, JEOL, Tokyo, Japan) under 0.2 keV acquisition energy, respectively. The structures of the films were recorded with a Fourier transform-infrared (FT-IR) spectrophotometer (FTIR-4600, Jasco, Tokyo, Japan) at a transmitter ratio (T%) and infra spectra resolution of 4 cm^{-1}. The spectra were collected in the range of 600–4000 cm^{-1}.

2.4. Electrochemical Corrosion and Hydrogen Evolution Tests

The electrochemical and hydrogen evolution tests were performed in revised simulated body fluid (r-SBF) solution (per liter, 0.072 g of NaSO$_4$, 0.182 g of K$_2$HPO$_4$, 0.225 g of KCl, 0.310 g of MgCl$_2$ 6H$_2$O, 0.736 g of NaHCO$_3$, 0.923 g of CaCl$_2$, 2.036 g Na$_2$CO$_3$, 5.403 g of NaCl, and 11.928 g of 4-(2-hydroxyethyl)-1-piperazineethanesulfonic avid (HEPES) dissolved in deionized water), respectively. In the electrochemical corrosion tests, a PARSTAT 2273 electrochemistry workstation (PARSTAT 2273, AMETEK, Berwyn, PA, USA) was operated at a scanning rate of 1 mV s^{-1} at −2.0 V to 1.0 V with a step height of 2.5 mV [12]. A reference saturated calomel electrode (SCE KCl) combined with a conventional three-electrode electrode cell and a platinum plate was used for the electrochemical analysis. The area of the working electrode exposed to the electrolyte was controlled by a polylactic acid (PLA) holder to within 1 cm^2. Each sample was tested for once and immersed in r-SBF solution at least three hours [12]. In hydrogen evolution test, all samples (n = 3) were freshly prepared and then placed in the r-SBF solution for 48 h at a pH of 7.4 and a temperature of 37 °C. The equipment then recorded the total volume of the hydrogen released from the magnesium alloys [32].

2.5. Cytocompatibility Evaluation

The cytocompatibility tests examined cytotoxicity, cell proliferation, and migration. A human umbilical vein cell line (EA. hy926) and a rat aortic smooth muscle cell line (RASMC) were provided by Wen-Tai Chiu (National Cheng Kung University, Tainan, Taiwan). The growth behaviors of the EA. hy926 and RASMC represented the cardiovascular ECs and SMCs, respectively. Before cell culturing, the sterile samples were immersed in free fetal bovine serum Dulbecco's Modified Eagle Medium (FBS DMEM) (approximately 1.25 cm^2/mL in DMEM) for 24-h extraction. The resulting filtered (0.2 µm filter) medium was then diluted to a 15 mL volume. The cells were seeded separately at a density of 4000 cells/well in 96-well plates. The 90% DMEM with 10% dimethyl sulfoxide (DMSO) was used as the positive control, and 90% DMEM with 10% FBS was used as the negative control. The culture medium was replaced with 90% extracted medium with 10% FBS overnight. Before collection of the optical density (OD) values at 450 nm, cell counting kit-8 (CCK8) solution was added into each well and the mixture was incubated for 2 h. Briefly, the cell proliferation was recorded on days 1, 4, and 7. Three replications were conducted and then the results were calculated using Equation (1) below.

Relative growth rate % = [(OD test − OD positive)/(OD negative − OD positive)] × 100% (1)

2.6. Hemolysis Tests

The hemolysis tests were performed according to the ISO 10993-4 standard for biomaterials [33]. Sodium citrate (4 wt.%) to the fresh blood samples in the ratio of 1:9 was taken 30 min before the tests. All specimens were immersed in centrifuge tubes containing 10 mL of normal saline and incubated for 30 min at 37 °C. Deionized water was used as the positive control, and normal saline was used as the negative control. After 30 min of incubating, 0.2 mL of the diluted blood, prepared with normal saline at a volume ratio of 4:5, was added into the centrifuge tubes, and all the tubes were incubated for 60 min at 37 °C. After 60 min, the tubes were centrifuged for 5 min at 2500 rpm, and the supernatant was collected and carefully transferred to a 96-well plate for spectroscopic measurement. The hemolysis data, read by ELISA, were calculated using Equation (2) below and were based on the average of three replications:

$$\text{Hemolysis \%} = [(\text{OD test} - \text{OD negative})/(\text{OD positive} - \text{OD negative})] \times 100\% \qquad (2)$$

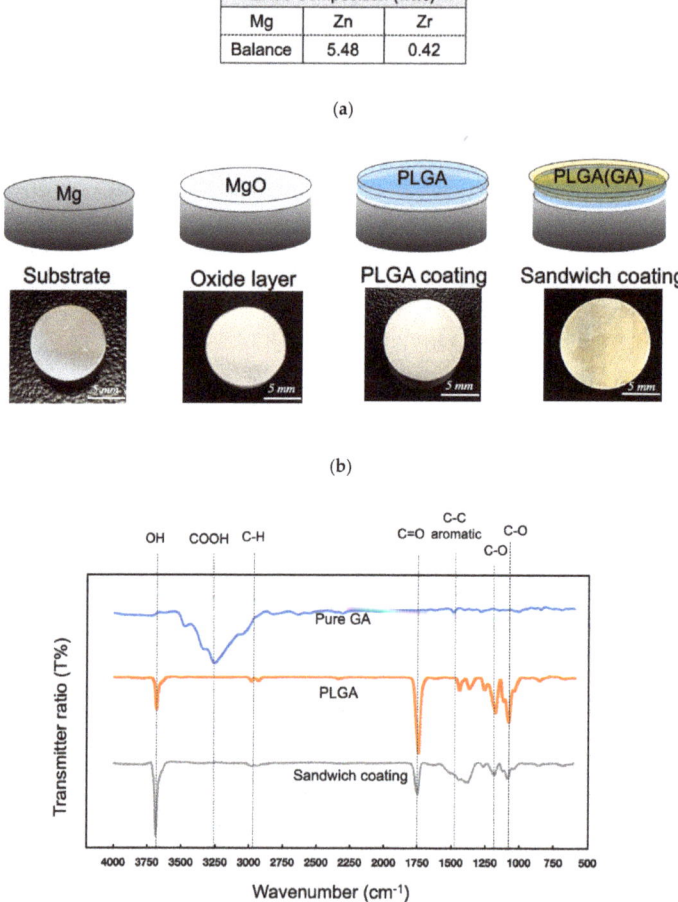

Figure 1. Composition of ZK60 (**a**) [34], the steps in the dip-coating procedure (**b**), and FT-IR spectra of GA, PLGA, and the sandwich coating (**c**).

2.7. Free Radical Activity Tests

The free radical activity tests, measured diphenyl-2-picrylhydrazyl (DPPH), were performed according to application on bleached teeth and several antioxidant tests on magnesium alloys [34]. The DPPH powder was diluted in 0.2 mmol/mL DPPH solution with 70% ethanol. All specimens were immersed in 2 mL of DPPH solution and incubated at 37 °C for 1 h in the dark. Similarly, DPPH solution was used as the control group. Then the absorbance of DPPH data, read by microplate reader, were calculated using Equation (3) below based on the average of three replications:

$$\text{Inhibition \%} = [(\text{OD control} - \text{OD test})/\text{OD control}] \times 100\% \qquad (3)$$

2.8. Statistical Analysis

In this study, all quantitative results are expressed as standard deviation (SD) unless indicated. Each assay was performed in at least three replicated tests, as described above. Measured experimental results from GraphPad Prism software (Prism 9.0, GraphPad, San Diego, CA, USA) were analyzed by a non-parametric test (Kruskal-Wallis Test) combined with Uncorrected Dunn's multiple comparisons test (each comparison stands alone), and a *p*-value < 0.05 was considered significant.

3. Results

3.1. Modification of the ZK60 Surface

An as-extruded ZK60 disk scanned by XRD indicated three peaks at 32.2°, 34.5°, and 36.8°, corresponding to Mg in Appendix A (Figure A1). As-extruded ZK60 cut disks were modified with AHT to develop a layer of magnesium hydroxide as Equation (4) below. The Mg(OH)$_2$ layer was then treated at high temperature to form a rough MgO layer as Equation (5) following by the PLGA dip coating. These chemical reactions produced a less-active ZK60 surface and a compact void-free oxide for the dip-coating process. The void-free structure prevented inner bulk erosion on the polymer layer and external pitting corrosion on the highly active magnesium alloy. In comparison to the bare ZK60, the MgO coating had higher corrosion resistance to delay magnesium ions from bursting out in Appendix A (Figure A2). The sandwich coating steps were schematized and are depicted in order in Figure 1b. As illustrated in that figure, the bare ZK60 substrate was first treated with AHT and then coated with sandwich coating layers. Since hydrophilic GA was unable to dissolve in the oil phase, the hydrophobic PLGA was able to remain compact in the sandwich layer coating process.

The FT-IR analysis confirmed the encapsulation of GA in the polymer sandwich layers (Figure 1c). The wavenumbers of the functional group in GA were sourced from a published FT-IR article [35]. The presence of the benzene signal at 1482 cm^{-1} allowed a decisive characterization of the phenolic GA compound even though PLGA and GA shared several similar functional groups. The GA group had benzene vibration peaks at 1482 cm^{-1}, and PLGA did not, while the interference of C=O (1750 cm^{-1}) and C-O (1250–1100 cm^{-1}) vibration peaks caused a broad benzene peak of the sandwich coating from 1600 cm^{-1} to 1300 cm^{-1}. Also, the O-H peaks, representing tri-hydroxyl groups at 3683 cm^{-1}, provided substantial evidence that the sandwich coating immobilized the GA in the sandwich structure.

$$Mg^{2+}(aq) + 2OH^-(aq) \rightleftharpoons Mg(OH)_2(s) \qquad (4)$$

$$Mg(OH)_2(s) \xrightarrow{\Delta} MgO(s) + H_2O(g) \uparrow \qquad (5)$$

3.2. Effects of PLGA Dip-Coating and Phenolic Layer on the Coating Morphology

The SEM images of the sandwich coating showed that the AHT-modified ZK60 surface became more uniform after a series of dip-coating processes (Figure 2). Initially, the morphology of the MgO after AHT was rough and coarse. After the PLGA dip-coating, a non-homogeneous microstructure formed on the MgO surface. After the second PLGA dip-coating, the top PLGA film layer smoothed the rough GA layer in the form of a sandwiched-layer structure. In addition, the SEM images indicated that PLGA and sandwich coatings had film thicknesses of 1.1 ± 0.4 µm and 2.1 ± 0.3 µm, respectively, in the cross-section view. The thickness of the coating's cross-sections was proportionate with the number of PLGA layers.

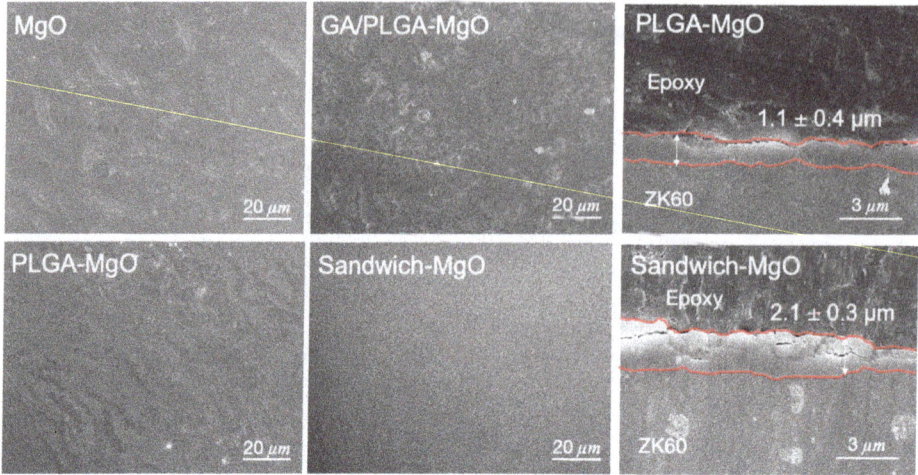

Figure 2. SEM images of the coating steps: MgO, PLGA-MgO, GA/PLGA-MgO, and Sandwich-MgO, the cross-section images of PLGA-MgO (1.1 ± 0.4 µm) and Sandwich-MgO (2.1 ± 0.3 µm).

EDS detection confirmed that the sandwich layers covered the bare ZK60, as shown in the elemental distribution in Figure 3a. In the MgO layer, magnesium and oxygen accounted for 36.9% and 63.1%, respectively, while in the final sandwich coating film, Mg and oxygen (O) accounted for 7.2% and 23.4%, respectively, with the balance being carbon (C). The decreases in the contents of Mg and O with the layers verified that the PLGA coating protected the MgO surface. Furthermore, the rising C contents of GA and PLGA indicated bare Mg substrate was covered by the coating. Next, an EDS line scan was applied to analyze the elemental changes in the sandwich coating (Figure 3b,c). Based on the changes in the ratios of the element weights, the highest content of Mg was in the ZK60 region. Due to the rich oxygen contents of MgO, GA, and PLGA, the increasing O and declining Mg contents indicated the MgO layer and sandwich coating. Furthermore, the rising carbon signal originated from the mounting epoxy resin. The film thickness observed in the SEM images was consistent with the EDS analysis measured from the first Mg–O crossing point to the second Mg–C crossing point in the PLGA and sandwich coating films.

To evaluate the dip-coating effect on the ZK60 stent platform, the coating steps mentioned above were repeated on the ZK60 stent prototype. Although the ZK60 stent developed irregular corrosive struts without the coating, the coated Mg-stent retained its stent integrity, and a uniform polymer coating formed on its surface (Figure 4).

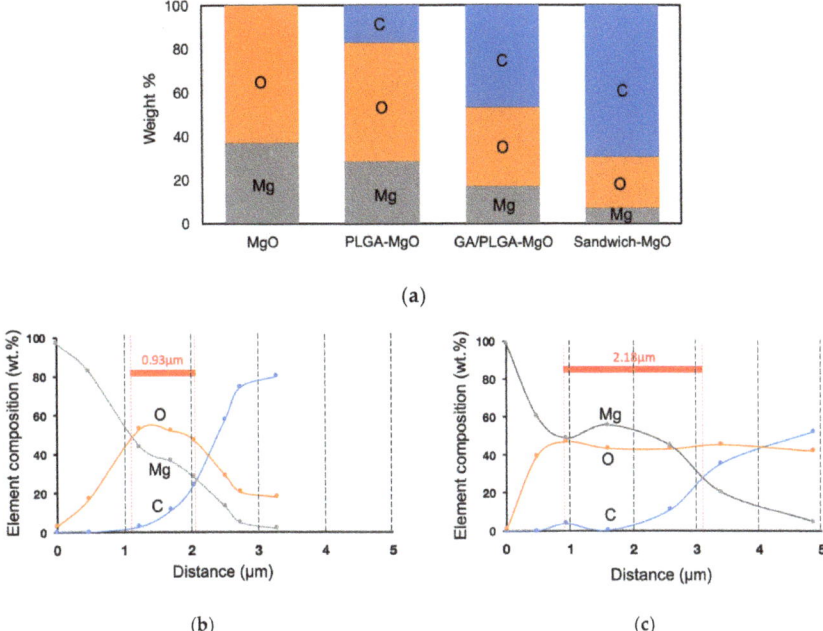

Figure 3. EDS composition analysis of (**a**) different layers and cross-sections of the (**b**) PLGA, and (**c**) sandwich coating films.

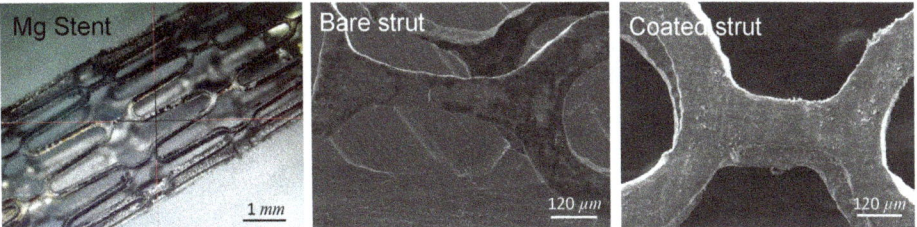

Figure 4. Mg-stent prototype, and the SEM images of the bare strut and sandwich-coated strut.

3.3. Anti-Corrosion Behavior

The hydrogen evolution was recorded for 48 h. The results indicated the highest H_2 volume for the untreated ZK60, and the lowest, for sandwich coating (Figure 5a). The electrochemical analysis showed results comparable to those for hydrogen release. The lowest corrosion density of the sandwich coating, acquired from the potentiodynamic polarization curves, suggested that the sandwich coating improved the corrosion resistance (Figure 5b). The unmodified ZK60 had the lowest corrosion potential, −1.6 V, and the highest corrosion current density, 20.51 μA/cm^2, which corresponded to the H_2 release, indicating that the bare surfaces did suffer a severe corrosion attack in the physiological environment (Table 1). Furthermore, the PLGA had a higher corrosion potential, −0.4 V, and a lower corrosion current density, 1.79 μA/cm^2, and the sandwich coating also exhibited a slightly increased corrosion potential of −0.2 V and a decreased corrosion current density of 0.01 μA/cm^2 than those of the bare ZK60. These findings suggested a significant difference between PLGA and sandwich coating in terms of electrochemical performance. This sandwich structure provided a better corrosion resistance.

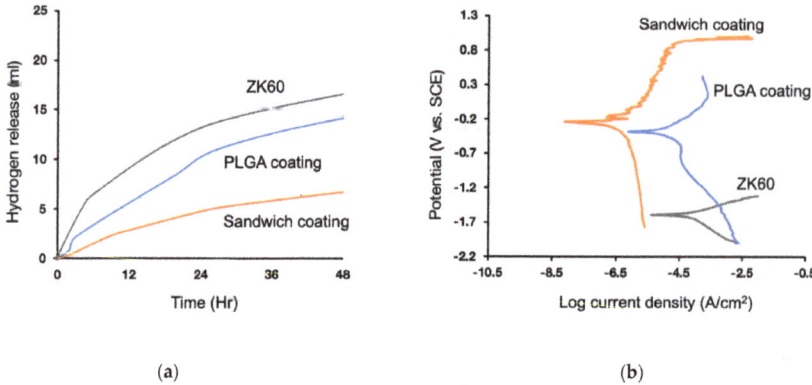

Figure 5. Corrosion resistance tests: (**a**) Hydrogen evolution curves of ZK60, PLGA coating and sandwich coating and (**b**) Potentiodynamic polarization curves of ZK60, PLGA coating, and sandwich coating in SBF solution.

Table 1. Fitting results for the potentiodynamic polarization curves related to Figure 5b. The inhibition efficiency was calculated as "$\eta\% = [1 - (Icorr. sample/Icorr. Bare ZK60)] \times 100\%$".

Specimen	Polarization Curves			
	Ecorr (V)	Log Icorr ($\mu A/cm^2$)	Icorr ($\mu A/cm^2$)	η (%)
Bare ZK60	−1.59	−4.69	20.51	0.00
PLGA coating	−0.40	−5.75	1.79	91.27
Sandwich coating	−0.24	−8.00	0.01	99.95

3.4. Effect of Phenolic Molecules on ECs and SMCs in TERMS of Cell Viability and Hemolysis

The sandwich coating layers promoted EC proliferation but inhibited the growth of SMCs (Figure 6). Changes in GA gradient resulted in varying EC vitality, demonstrating the relationship between EC tolerance and GA toxicity in Appendix A (Figure A3). The outcome demonstrated that the GA 1 wt.% group had better viability than did the other two groups, while the highly-concentrated GA suggested that direct over-exposure could cause apoptosis due to the high antioxidant activity. After comparison to different groups, including bare (ZK60), PLGA and sandwich coating, at 1, 4, and 7 days, sandwich coating showed no toxicity to ECs (Figure 6c).

Based on the cell viability of the ECs and SMCs, there were two notable points: the non-toxic growth rate of sandwich coating towards ECs, and the selective suppression of SMCs (Figure 6a,b). In comparison to bare ZK60, the PLGA and sandwich coatings exhibited significant viability in ECs. The SMCs exhibited proliferation in the PLGA group, while sandwich coating only had 90% viability after four days. Even though the proliferation of the ECs declined on Day 4, the decline could have been due to the high content of GA released in a static environment. The efficiency of GA, while not significant in the SMCs, was moderate in strength. These results indicated that the ECs had a robust proliferation growth rate of 150% compared to the SMC group, and sandwich coating inhibited SMC over-growth. None of the groups exhibited toxicity in the comparison tests; thus, the four-day test was sufficient to observe the growth trend, while a seven-day test would have soon reached 100% coverage and lost the trend in our relative growth study.

Sandwich coating presented a hemolysis ratio below 5%, corresponding to the clinical bio-safety standard (Figure 6d). The bare ZK60 had a high hemolysis ratio of more than 43%, while both sandwich coating and PLGA showed lower hemolysis rates of 2.1% and 3.8%, respectively. Overall, the sandwich demonstrated a promising ability to regulate ECs and SMCs with excellent hemocompatibility.

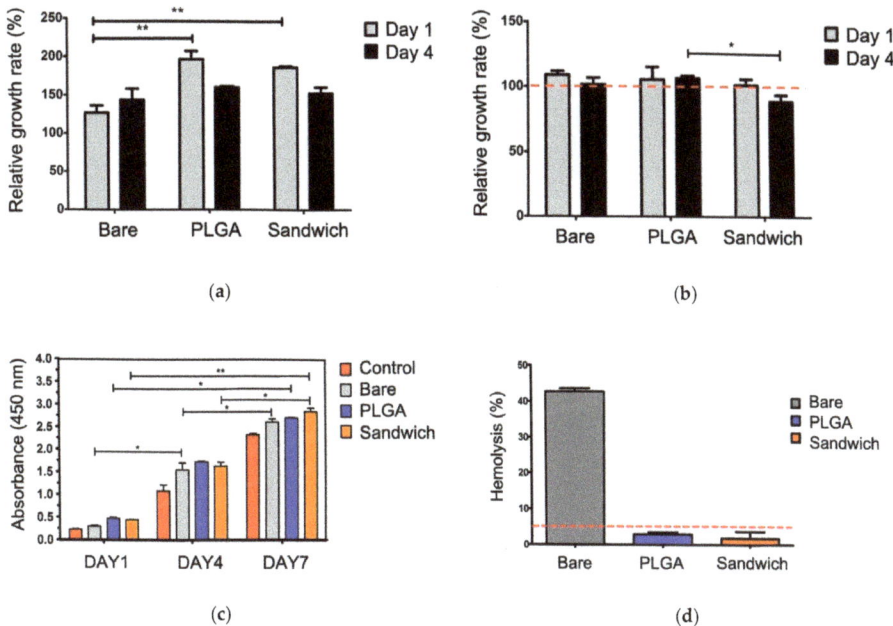

Figure 6. The cell viability evaluation of (**a**) ECs and (**b**) SMCs on day 1, day 4, (**c**) coatings versus EC proliferation (* $p < 0.05$ and ** $p < 0.01$, mean ± SD, N = 3) and (**d**) the hemolysis test.

3.5. Anti-Oxidation

The linear regression retrieved from a gradient GA content test defined a standard reference of GA released weight and concentration (Figure 7a). The sandwich coating film triggered GA weight release at 1, 2, 3, and 6 h, with a reduced speed after three hours (Figure 7b). The free radical scavenging analysis showed that the bare, PLGA, and sandwich coatings could eliminate oxidant stress with antioxidant capacities of approximately 28%, 36%, and 63%, respectively. These data verified that the GA released from the sandwich coating film promoted scavenging of free radicals and protected the vascular tissue due to its significant anti-oxidative ability, suggesting that the sandwich coating is a promising material for vascular stents (Figure 7c).

3.6. Cell Migration

The sandwiched-layer structure ensured that the GA had an anti-proliferative effect on the SMCs and a slight influence on EC migration (Figure 8). Re-endothelialization at the lesion site is a particularly important standard after PCI and requires healthy EC proliferation, migration, and spreading. Additionally, ideal regulation of SMCs and ECs could prevent penetration from SMCs as well as late-stent restenosis. As the previous viability results showed, GA has a specific sensitivity to SMCs and ECs at GA−1wt.% (~4 µg/mL) in the sandwich-structured film. The results indicated a robust EC migration without a significant difference in migration length among all groups (Figure 8a). In contrast, the SMCs had a larger migration distance in the PLGA group than in the other groups, while the sandwich and control groups were alike (Figure 8b).

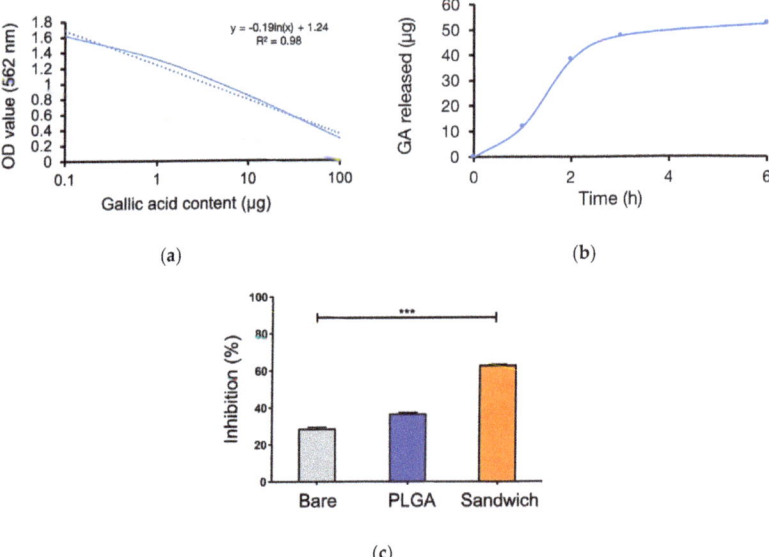

Figure 7. The anti-oxidation analysis of (**a**) the DPPH/GA efficiency curve, (**b**) the sandwich coating film release GA curve and (**c**) free radical scavenging activity (*** $p < 0.001$, mean ± SD, N = 6).

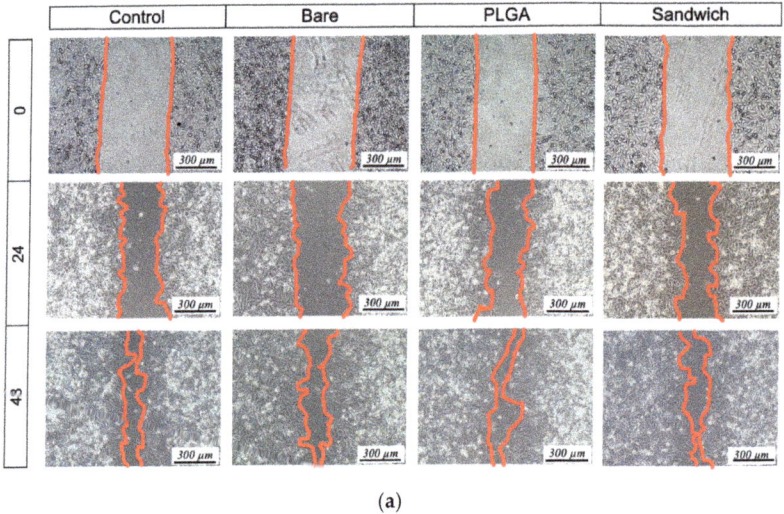

(**a**)

Figure 8. *Cont.*

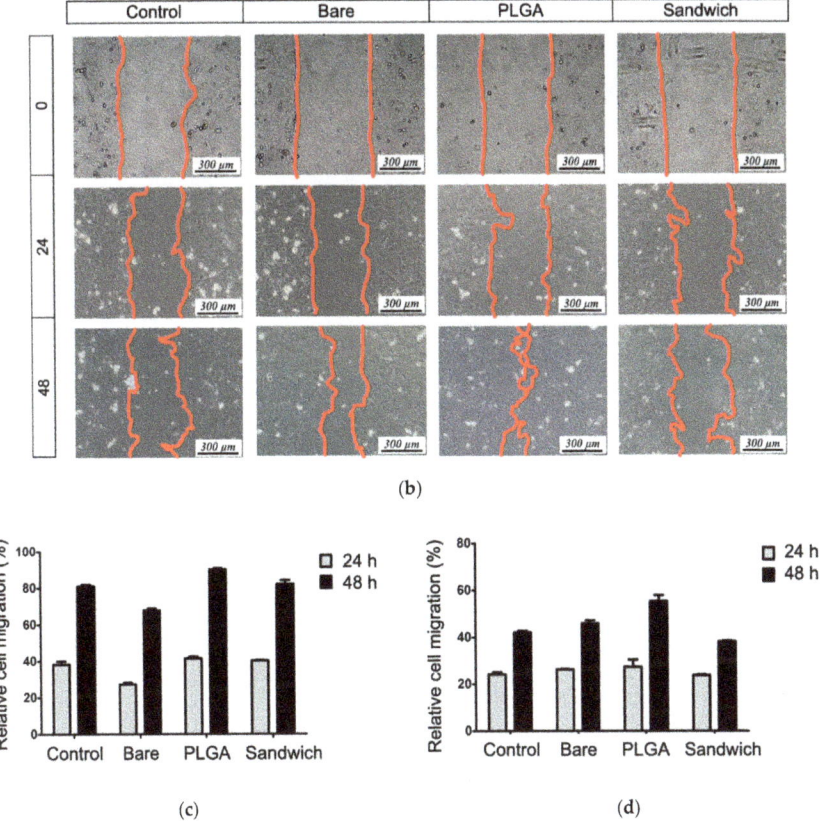

Figure 8. Cell migration of (**a**) EC, (**b**) SMC at 0, 24, and 48 h, and the fitting results for (**c**) EC and (**d**) SMC migration coverage (mean ± SEM, N = 3).

An overall analysis of the ECs and SMCs after 48 h showed 81% and 42% migration from sides to the middle wound closure, respectively, in the control group. In the PLGA group, both the ECs and SMCs exhibited migration of 90% and 55%, respectively (Figure 8c,d). In the bare group, the ECs exhibited mild movement of 68%, and the SMCs showed no significant difference. Furthermore, the sandwich coating, similarly to the control group, displayed migration rates of 82% for ECs and 38% for SMCs. The results were consistent with the cytocompatibility test, indicating that sandwich coating preserved the viability of ECs and inhibited the SMCs with the release of GA.

4. Discussion

One of the goals of surface modification of ZK60 is to enhance its corrosion resistance in simulated body fluid and hence improve its potential for use in fabricating bioresorbable coronary artery stents. $Mg(OH)_2$ forms a passive layer on the magnesium surface through AHT [36]. This process protects the Mg surface from ion attack and prevents the accelerated degradation of ZK60 magnesium, which is supported by the polarization test in Figure A2. Based on the limitations in our lab, discussing the LA:GA ratio of PLGA and the coating techniques was beyond the scope of the study. A comprehensive degradation study on PLGA (50:50) and (85:15) revealed that complete degradation of the PLGA (50:50) occurred after 102 days, whereas only about 60% of the PLGA (85:15) degraded within the same period [37]. Therefore, we simply discuss the weight percentage of PLGA (based on the 85:15 ratio) for its long-term degradation process and chose the best group as the control group in this study.

An oxide-passive layer on the Mg-disk led to a coarse surface until the second PLGA layer was deposited on it to fill the cracks (Figure 2), yet this passive layer prevented the initial hydrogen evolution from the ZK60 surface and prevented the acid degradation products of the PLGA layer from penetrating directly into the ZK60 substrate. This phenomenon explains why the NaOH passive layer is vital to make PLGA layer more completed and concrete on the coated surface. Although dip-coating might not be ideal for stent coating due to the complicated geometry, this technique can be simply conducted and qualifies for further bench testing, such as cytocompatibility and immersion test.

Many studies have investigated the use of phenolic molecules to regulate endothelialization based on their specific properties on SMCs and ECs [28,29]. These applications not only prevent the over-growth of SMCs from invading the EC wall but also assist in the formation of a uniform endothelial layer in the vascular system. Further, the goal of drug elution changed from anti-immunity to anti-proliferation since sirolimus (SR) replaced paclitaxel (PTX) as the main drug system for drug-eluting stent [38]. We believe our sandwiched GA concept can bring a novel direction to apply small-molecule on eluting stent instead of traditional SR or PTX. In this paper, the healing process can be nearly completed in 48 h, so a four-day viability test is consistent with the endothelial cell healing process (Figures 6 and 8). Endothelial cells are unique in their growth pattern, preferring to grow in flat structures, so the extreme growth rate without an appropriate dynamic environment, a laminar blood flow, tends to cause apoptosis in limited living space [39,40]. Nevertheless, the opposite behavior between ECs and SMCs in this study suggests that the PLGA films promoted the proliferation of SMCs, while sandwich coating notably delayed them with similar behavior to that of the control group. Many recent clinical trials have revealed that polymer-based stents [18] do not mitigate late-stent restenosis, the SMC migration results may explain the deficiencies related to the use of polymers and offer a possible solution.

The anti-inflammatory effect is another concern related to cardiovascular stents. Phenolic molecules exhibit extraordinary, inherent anti-oxidative abilities, thus helping to scavenge reactive oxygen species and protect vascular tissue. In addition, free radical scavenging decreases oxidants and inhibits the atherogenesis initiated by oxidation of LDL. These antioxidant mechanisms prevent late myocardial infarction and in-stent restenosis after PCI. Although the drug delivery strategy is beyond the scope of our study, the interaction of free-radical capture and the GA delivery trend are what we are interested in. Recently, a free radical scavenging analysis completed by DPPH was conducted to evaluate the antioxidant capacity of the stent platform [34]. Phenolic molecules, such as ECGC and TA [28–31], have been discussed in similar studies, yet the GA application under PLGA eluting stents has not been sufficiently investigated. In our previous study, a phenolic-modified ceramic coating on ZK60 was proven to be efficient on osteo-like cell activity [12]. Due to its ability to capture free radicals and cell selectivity, GA can potentially be applied in the modified polymer coating. However, the GA content is hard to characterize in the host because of the circulation system, which is also correlated to the cell tolerance and apoptosis occurring in a static environment.

Under the limitations of a coating strategy for the complicated geometry of stents and the in vivo environment, only the surface modification and the interaction of materials and cells can be discussed within our scope. To optimize the scale of this research, other coating techniques, such as spray coating, and the in vivo environment will be discussed in the near future.

5. Conclusions

PLGA dip-coatings with gallic acid (GA) were prepared on a ZK60 surface in a sandwiched-layer structure. The a close-packed sandwiched layer showed enhanced corrosion resistance and a homogeneous film surface. An in vitro assay demonstrated that the sandwich coating behaved selectively with bioactivity on ECs, significant suppression of SMC over-proliferation, anti-hemolysis ability, and anti-oxidation effects compared to PLGA. This simple technique with a sandwiched-layer structure is a promising surface treatment for a commercialized stent platform for treating atherosclerosis and preventing late-stent restenosis.

Author Contributions: Conceptualization, L.-H.L. and H.-P.L.; methodology, L.-H.L.; software, L.-H.L.; validation, L.-H.L.; formal analysis, L.-H.L.; investigation, L.-H.L.; resources, L.-H.L.; data curation, L.-H.L.; writing—original draft preparation, L.-H.L.; writing—review and editing, H.-P.L. and M.-L.Y.; visualization, L.-H.L. and H.-P.L.; supervision, M.-L.Y.; project administration, M.-L.Y.; funding acquisition, M.-L.Y. All authors have read and agreed to the published version of the manuscript.

Funding: This research was financially supported by Ministry of Science and Technology in Taiwan through Grants MOST-106-3114-E-006-012 and Medical Device Innovation Center (MDIC), National Cheng Kung University (NCKU) from the Featured Areas Research Center Program within the framework of the Higher Education Sprout Project by the Ministry of Education (MoE) in Taiwan.

Acknowledgments: This research was supported by the Nano Biomedical and Tissue Engineering LAB, Department of BME, National Cheng Kung University. We thank INTAI Technology and CHONG HUAI Laser for providing the Laser technique of Mg-stents.

Conflicts of Interest: The authors declare no conflict of interest. The funders had no role in the design of the study; in the collection, analyses, or interpretation of data; in the writing of the manuscript, or in the decision to publish the results.

Appendix A

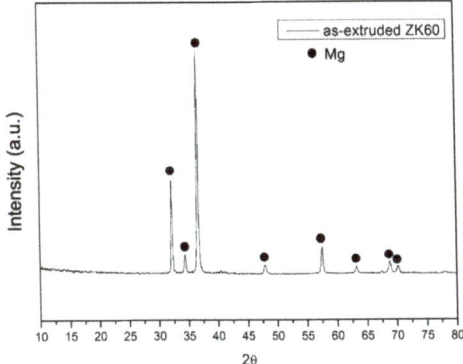

Figure A1. XRD patterns of as-extruded ZK60.

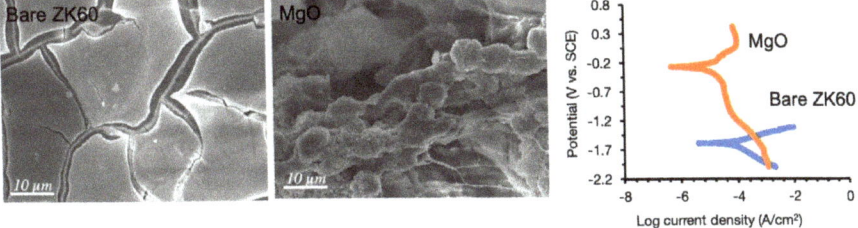

Figure A2. The MG63 attachment morphology of bare ZK60 and MgO and the potentiodynamic polarization curves of bare ZK60 and MgO.

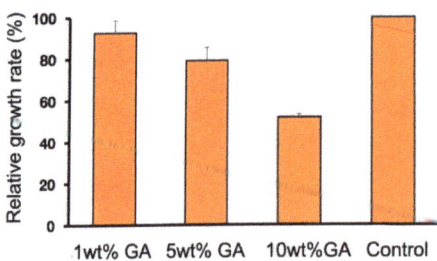

Figure A3. The cytocompatibility evaluation of GA content versus the relative growth rate (mean ± SD, N = 3).

References

1. Letchumanan, I.; Arshad, M.M.; Gopinath, S.C. Nanodiagnostic Attainments and Clinical Perspectives on C-Reactive Protein: Cardiovascular Disease Risks Assessment. *Curr. Med. Chem.* **2020**, *27*, 1. [CrossRef]
2. Stone, G.W.; Aronow, H.D. Long-term Care After Percutaneous Coronary Intervention: Focus on the Role of Antiplatelet Therapy. *Mayo Clin. Proc.* **2006**, *81*, 641–652. [CrossRef] [PubMed]
3. Hanawa, T. Materials for metallic stents. *J. Artif. Organs* **2009**, *12*, 73–79. [CrossRef] [PubMed]
4. Moravej, M.; Mantovani, D. Biodegradable Metals for Cardiovascular Stent Application: Interests and New Opportunities. *Int. J. Mol. Sci.* **2011**, *12*, 4250–4270. [CrossRef] [PubMed]
5. Waksman, R.; Pakala, R.; Eric, W.; Hartwig, S.; Harder, C.; Rohde, R.; Heublein, B.; Alex, H.; Andreae, A.; Waldman, K.H. Effect of magnesium alloy stents in porcine coronary arteries: Morphometric analysis of a long-term study. *J. Am. Coll. Cardiol.* **2006**, *47*, 23B.
6. Shi, Y.; Pei, J.; Zhang, L.; Lee, B.-K.; Yun, Y.; Zhang, J.; Li, Z.; Gu, S.; Park, K.; Yuan, G. Understanding the effect of magnesium degradation on drug release and anti-proliferation on smooth muscle cells for magnesium-based drug eluting stents. *Corros. Sci.* **2017**, *123*, 297–309. [CrossRef]
7. Ang, H.Y.; Huang, Y.Y.; Lim, S.T.; Wong, P.; Joner, M.; Foin, N. Mechanical behavior of polymer-based vs. metallic-based bioresorbable stents. *J. Thorac. Dis.* **2017**, *9*, S923–S934. [CrossRef]
8. Li, Y.; Wang, L.; Chen, S.; Yu, D.; Sun, W.; Xin, S. Biodegradable Magnesium Alloy Stents as a Treatment for Vein Graft Restenosis. *Yonsei Med. J.* **2019**, *60*, 429–439. [CrossRef]
9. Barkholt, T.Ø.; Webber, B.; Holm, N.R.; Ormiston, J.A. Mechanical properties of the drug-eluting bioresorbable magnesium scaffold compared with polymeric scaffolds and a permanent metallic drug-eluting stent. *Catheter. Cardiovasc. Interv.* **2019**, Available online: https://onlinelibrary.wiley.com/doi/full/10.1002/ccd.28545 (accessed on 11 November 2019). [CrossRef]
10. Kang, M.-H.; Jang, T.-S.; Kim, S.W.; Park, H.-S.; Song, J.; Kim, H.-E.; Jung, K.-H.; Jung, H.-D. MgF2-coated porous magnesium/alumina scaffolds with improved strength, corrosion resistance, and biological performance for biomedical applications. *Mater. Sci. Eng. C* **2016**, *62*, 634–642. [CrossRef]
11. Kim, S.-Y.; Kim, Y.K.; Ryu, M.-H.; Bae, T.-S.; Lee, M.-H. Corrosion resistance and bioactivity enhancement of MAO coated Mg alloy depending on the time of hydrothermal treatment in Ca-EDTA solution. *Sci. Rep.* **2017**, *7*, 1–11. [CrossRef] [PubMed]
12. Lee, H.-P.; Lin, D.-J.; Yeh, M.-L. Phenolic Modified Ceramic Coating on Biodegradable Mg Alloy: The Improved Corrosion Resistance and Osteoblast-Like Cell Activity. *Mater.* **2017**, *10*, 696. [CrossRef] [PubMed]
13. Lin, D.-J.; Hung, F.-Y.; Jakfar, S.; Yeh, M.-L. Tailored coating chemistry and interfacial properties for construction of bioactive ceramic coatings on magnesium biomaterial. *Mater. Des.* **2016**, *89*, 235–244. [CrossRef]
14. Lin, D.-J.; Hung, F.-Y.; Yeh, M.-L.; Lee, H.-P.; Lui, T.-S. Development of a novel micro-textured surface using duplex surface modification for biomedical Mg alloy applications. *Mater. Lett.* **2017**, *206*, 9–12. [CrossRef]
15. Lin, X.; Tan, L.; Wang, Q.; Zhang, G.; Zhang, B.; Yang, K. In vivo degradation and tissue compatibility of ZK60 magnesium alloy with micro-arc oxidation coating in a transcortical model. *Mater. Sci. Eng. C* **2013**, *33*, 3881–3888. [CrossRef]

16. Makkar, P.; Kang, H.J.; Padalhin, A.R.; Park, I.; Moon, B.-G.; Lee, B.T. Development and properties of duplex MgF2/PCL coatings on biodegradable magnesium alloy for biomedical applications. *PLoS ONE* **2018**, *13*, e0193927. [CrossRef]
17. Jiang, W.; Tian, Q.; Vuong, T.; Shashaty, M.; Gopez, C.; Sanders, T.; Liu, H. Comparison Study on Four Biodegradable Polymer Coatings for Controlling Magnesium Degradation and Human Endothelial Cell Adhesion and Spreading. *ACS Biomater. Sci. Eng.* **2017**, *3*, 936–950. [CrossRef]
18. Strohbach, A.; Busch, R. Polymers for Cardiovascular Stent Coatings. *Int. J. Polym. Sci.* **2015**, *2015*, 1–11. [CrossRef]
19. Lakalayeh, G.A.; Rahvar, M.; Haririan, E.; Karimi, R.; Ghanbari, H. Comparative study of different polymeric coatings for the next-generation magnesium-based biodegradable stents. *Artif. Cells Nanomed. Biotechnol.* **2017**, *46*, 1380–1389. [CrossRef]
20. Garcia-Garcia, H.M.; Wopperer, S.; Seleme, V.B.; Ribeiro, M.H.; Campos, C.M. The Development of Magnesium-Based Resorbable and Iron-Based Biocorrodible Metal Scaffold Technology and Biomedical Applications in Coronary Artery Disease Patients. *Appl. Sci.* **2019**, *9*, 3527.
21. He, Y.; Wang, J.; Yan, W.; Huang, N. Gallic acid and gallic acid-loaded coating involved in selective regulation of platelet, endothelial and smooth muscle cell fate. *RSC Adv.* **2014**, *4*, 212–221. [CrossRef]
22. Yang, Z.; Xiong, K.; Qi, P.; Yang, Y.; Tu, Q.; Wang, J.; Huang, N. Gallic Acid Tailoring Surface Functionalities of Plasma-Polymerized Allylamine-Coated 316L SS to Selectively Direct Vascular Endothelial and Smooth Muscle Cell Fate for Enhanced Endothelialization. *ACS Appl. Mater. Interfaces* **2014**, *6*, 2647–2656. [CrossRef] [PubMed]
23. Badhani, B.; Sharma, N.; Kakkar, R. Gallic acid: A versatile antioxidant with promising therapeutic and industrial applications. *RSC Adv.* **2015**, *5*, 27540–27557. [CrossRef]
24. Lim, K.S.; Park, J.-K.; Jeong, J.-O.; Bae, I.H.; Park, D.S.; Shim, J.W.; Kim, J.H.; Kim, H.K.; Kim, S.S.; Sim, O.S.; et al. Anti-Inflammatory Effect of Gallic Acid-Eluting Stent in a Porcine Coronary Restenosis Model. *Acta Cardiol. Sin.* **2018**, *34*, 224–232.
25. Diaz, M.N.; Frei, B.; Vita, J.A.; Keaney, J.F. Antioxidants and Atherosclerotic Heart Disease. *N. Engl. J. Med.* **1997**, *337*, 408–416. [CrossRef]
26. Zhang, H.; Luo, R.; Li, W.; Wang, J.; Maitz, M.F.; Wang, J.; Wan, G.; Chen, Y.; Sun, H.; Jiang, C.; et al. Epigallocatechin gallate (EGCG) induced chemical conversion coatings for corrosion protection of biomedical MgZnMn alloys. *Corros. Sci.* **2015**, *94*, 305–315. [CrossRef]
27. Zhang, B.; Yao, R.; Li, L.; Li, M.; Yang, L.; Liang, Z.; Yu, H.; Zhang, H.; Luo, R.; Wang, Y. Bionic Tea Stain–Like, All-Nanoparticle Coating for Biocompatible Corrosion Protection. *Adv. Mater. Interfaces* **2019**, *6*, 1900899. [CrossRef]
28. Zhang, B.; Yao, R.; Li, L.; Wang, Y.; Luo, R.; Yang, L.; Wang, Y. Green Tea Polyphenol Induced Mg2+-rich Multilayer Conversion Coating: Toward Enhanced Corrosion Resistance and Promoted in Situ Endothelialization of AZ31 for Potential Cardiovascular Applications. *ACS Appl. Mater. Interfaces* **2019**, *11*, 41165–41177. [CrossRef]
29. Chen, S.; Zhao, S.; Chen, M.; Zhang, X.; Zhang, J.; Li, X.; Zhang, H.; Shen, X.; Wang, J.; Huang, N. The anticorrosion mechanism of phenolic conversion coating applied on magnesium implants. *Appl. Surf. Sci.* **2019**, *463*, 953–967. [CrossRef]
30. Cipriano, A.F.; Sallee, A.; Tayoba, M.; Alcaraz, M.C.C.; Lin, A.; Guan, R.-G.; Zhao, Z.-Y.; Liu, H. Cytocompatibility and early inflammatory response of human endothelial cells in direct culture with Mg-Zn-Sr alloys. *Acta Biomater.* **2017**, *48*, 499–520. [CrossRef]
31. Lin, D.-J.; Hung, F.-Y.; Liu, H.-J.; Yeh, M.-L. Dynamic Corrosion and Material Characteristics of Mg-Zn-Zr Mini-Tubes: The Influence of Microstructures and Extrusion Parameters. *Adv. Eng. Mater.* **2017**, *19*, 1700159. [CrossRef]
32. Lin, D.-J.; Hung, F.-Y.; Yeh, M.-L.; Lui, T.-S. Microstructure-modified biodegradable magnesium alloy for promoting cytocompatibility and wound healing in vitro. *J. Mater. Sci. Mater. Electron.* **2015**, *26*, 1–10. [CrossRef] [PubMed]
33. Uan, J.-Y.; Yu, S.-H.; Lin, M.-C.; Chen, L.-F.; Lin, H.-I. Evolution of hydrogen from magnesium alloy scraps in citric acid-added seawater without catalyst. *Int. J. Hydrogen Energy* **2009**, *34*, 6137–6142. [CrossRef]
34. Seyfert, U.T.; Biehl, V.; Schenk, J. In vitro hemocompatibility testing of biomaterials according to the ISO 10993-4. *Biomol. Eng.* **2002**, *19*, 91–96. [CrossRef]

35. Garcia, E.J.; Oldoni, T.L.C.; De Alencar, S.M.; Reis, A.; Loguercio, A.D.; Grande, R.H.M. Antioxidant activity by DPPH assay of potential solutions to be applied on bleached teeth. *Braz. Dent. J.* **2012**, *23*, 22–27. [CrossRef] [PubMed]
36. Hirun, N.; Dokmaisrijan, S.; Tantishaiyakul, V. Experimental FTIR and theoretical studies of gallic acid–acetonitrile clusters. *Spectrochim. Acta Part A Mol. Biomol. Spectrosc.* **2012**, *86*, 93–100. [CrossRef]
37. Li, L.; Gao, J.; Wang, Y. Evaluation of cyto-toxicity and corrosion behavior of alkali-heat-treated magnesium in simulated body fluid. *Surf. Coat. Technol.* **2004**, *185*, 92–98. [CrossRef]
38. Hussein, A.S.; Ahmadun, F.-R.; Abdullah, N. In vitro degradation of poly (D, L-lactide-co-glycolide) nanoparticles loaded with linamarin. *IET Nanobiotechnol.* **2013**, *7*, 33–41. [CrossRef]
39. Htay, T.; Liu, M.W. Drug-eluting stent: a review and update. *Vasc. Heal. Risk Manag.* **2005**, *1*, 263–276. [CrossRef]
40. Ye, C.; Wang, J.; Zhao, A.; He, D.; Maitz, M.F.; Zhou, N.; Huang, N. Atorvastatin Eluting Coating for Magnesium-Based Stents: Control of Degradation and Endothelialization in a Microfluidic Assay and In Vivo. *Adv. Mater. Technol.* **2020**, *5*, 1900947. [CrossRef]

Publisher's Note: MDPI stays neutral with regard to jurisdictional claims in published maps and institutional affiliations.

 © 2020 by the authors. Licensee MDPI, Basel, Switzerland. This article is an open access article distributed under the terms and conditions of the Creative Commons Attribution (CC BY) license (http://creativecommons.org/licenses/by/4.0/).

Article

Antibacterial Films Based on Polylactide with the Addition of Quercetin and Poly(Ethylene Glycol)

Ewa Olewnik-Kruszkowska [1,*], Magdalena Gierszewska [1], Agnieszka Richert [2], Sylwia Grabska-Zielińska [1], Anna Rudawska [3] and Mohamed Bouaziz [4]

[1] Faculty of Chemistry, Chair of Physical Chemistry and Physicochemistry of Polymers, Nicolaus Copernicus University in Toruń, Gagarin 7 Street, 87-100 Toruń, Poland; mgd@umk.pl (M.G.); sylwia.gz@umk.pl (S.G.-Z.)
[2] Faculty of Biological and Veterinary Sciences, Chair of Genetics, Nicolaus Copernicus University in Toruń, Lwowska 1 Street, 87-100 Toruń, Poland; a.richert@umk.pl
[3] Faculty of Mechanical Engineering, Department of Production Engineering, Lublin University of Technology, 20-618 Lublin, Poland; a.rudawska@pollub.pl
[4] Electrochemistry and Environmental Laboratory, National Engineering School of Sfax, University of Sfax, BP1173, Sfax 3038, Tunisia; mohamed.bouaziz@fsg.rnu.tn
* Correspondence: olewnik@umk.pl; Tel.: +48-56-611-2210

Citation: Olewnik-Kruszkowska, E.; Gierszewska, M.; Richert, A.; Grabska-Zielińska, S.; Rudawska, A.; Bouaziz, M. Antibacterial Films Based on Polylactide with the Addition of Quercetin and Poly(Ethylene Glycol). *Materials* **2021**, *14*, 1643. https://doi.org/10.3390/ma14071643

Academic Editor: John T. Kiwi

Received: 25 February 2021
Accepted: 24 March 2021
Published: 27 March 2021

Publisher's Note: MDPI stays neutral with regard to jurisdictional claims in published maps and institutional affiliations.

Copyright: © 2021 by the authors. Licensee MDPI, Basel, Switzerland. This article is an open access article distributed under the terms and conditions of the Creative Commons Attribution (CC BY) license (https://creativecommons.org/licenses/by/4.0/).

Abstract: A series of new films with antibacterial properties has been obtained by means of solvent casting method. Biodegradable materials including polylactide (PLA), quercetin (Q) acting as an antibacterial compound and polyethylene glycol (PEG) acting as a plasticizer have been used in the process. The effect of quercetin as well as the amount of PEG on the structural, thermal, mechanical and antibacterial properties of the obtained materials has been determined. It was found that an addition of quercetin significantly influences thermal stability. It should be stressed that samples containing the studied flavonoid are characterized by a higher Young modulus and elongation at break than materials consisting only of PLA and PEG. Moreover, the introduction of 1% of quercetin grants antibacterial properties to the new materials. Recorded results showed that the amount of plasticizer did not influence the antibacterial properties; it does, however, cause changes in physicochemical properties of the obtained materials. These results prove that quercetin could be used as an antibacterial compound and simultaneously improve mechanical and thermal properties of polylactide-based films.

Keywords: polylactide; quercetin; antibacterial properties; food packaging

1. Introduction

In recent years, polymers have become one of the most common materials used for packaging, replacing materials such as glass, metal, paper and wood. It is estimated that about 40% of industrial packaging and about 50% of consumables packaging is made of polymers. Such a large interest in plastics as packaging materials is due to their many beneficial properties, such as low permeability and high mechanical strength [1–3].

Packaging materials, in particular those intended for food, have to meet a significant number of requirements. Most of all they have to ensure effective protection of the product against harmful factors (such as oxygen, UV radiation, moisture, bacteria and fungi) during food transport, storage and shelf life. For this reason materials with an additional function of so called active packaging are becoming more popular. The most popular include polymers modified with various substances characterized by such properties as absorption of moisture, oxygen or carbon dioxide. The group of active agents being introduced into the polymer matrix also includes substances with biocidal properties. The main benefit of using food packaging materials with antibacterial properties is the extension of products' shelf life and protection against bacteria and fungi, which can be hazardous to human health [2,4–6].

Currently, materials utilized in the production of packaging consist mainly of polyethylene, poly(ethylene terephthalate), polypropylene, poly(vinyl chloride) and polystyrene. Unfortunately, the use of polymer materials leads to the accumulation of huge amounts of waste that are harmful to the natural environment. This is due to their high durability and inherent resistance to biodegradation. Plastic waste accounts for about 30% of the weight of all the waste in the world. The growing problem of waste has recently resulted in an increased popularity and interest in biodegradable polymers. It seems to be very likely that in the immediate future they may completely replace traditional polymer materials. One of the polymers that will in all probability become an alternative to the currently used polymers of petrochemical origin is polylactide (PLA) [7,8]. It has to be stressed that polylactide is an aliphatic biodegradable polyester that can be obtained from renewable raw materials of plant origin [9]. The most popular materials used in an aim to manufacture lactic acid include corn, rice, barley, wheat or cassava [10].

In order to obtain antibacterial packaging materials based on polylactide, different compounds such as nanosilver, nisin, polyhexamethylene guanidine derivatives, cinnamon or tea tree essential oil [11–16] have been used. Among different active substances, flavonoids constitute a group of very interesting compounds characterized by varied antibacterial properties.

Molecular structure of flavonoids is characterized by two aromatic rings, which are connected by a three-carbon bridge. In the group of flavonoids we can distinguish: flavones, isoflavones, flavonols, flavanones, flavanols and anthocyanins. Flavonoids can, for example, be found in leaves, flowers, plant seeds, fruit (e.g., in citrus fruits, blueberries, berries, grapes, chokeberry), vegetables (e.g., in peppers, onions, tomatoes, broccoli) and in coffee and cocoa beans. It has to be noted that they possess many beneficial properties. They are characterized by anti-inflammatory, antiviral, antiatherosclerosis, antiallergic and anti-cancer properties. In addition, they perform a protective function, for instance deterring insects and inhibiting fungi, a common hazard to plants in which flavonoids can be found. The presence of these compounds also hinders the adverse effects of ultraviolet radiation [17–19].

One of the most popular flavonoids is quercetin—3,3′,4′,5,7-pentahydroxylflavone (Figure 1). It can be found in many types of fruit, vegetables, leaves, seeds and grains. Quercetin is also present in medicinal botanicals, including Ginkgo biloba and in varieties of honey from different plant sources. Recently scientists focused on its biological and antioxidative characteristics [20–23]. Moreover, it should also be mentioned that the flavonoid in question was used in the formulation of packaging films consisting mainly of ethylene-vinyl alcohol copolymer (EVOH) [24]. It was also scrutinized as an antioxidant and environmentally-friendly colored indicator of ageing time in a topas cyclo-olefin copolymer (ethylene-norbornene) [25]. Additional research devoted to the possible applications of quercetin has been described in the work of Kost et al. [26], where the fibers based on polylactide and β-cyclodextrin loaded with quercetin were used as dressing materials characterized with antibacterial properties. Other possible applications of the quercetin have been presented in the work of Hao et al. [27], where quercetin was encapsulated using chitosan-coated nanoliposomes.

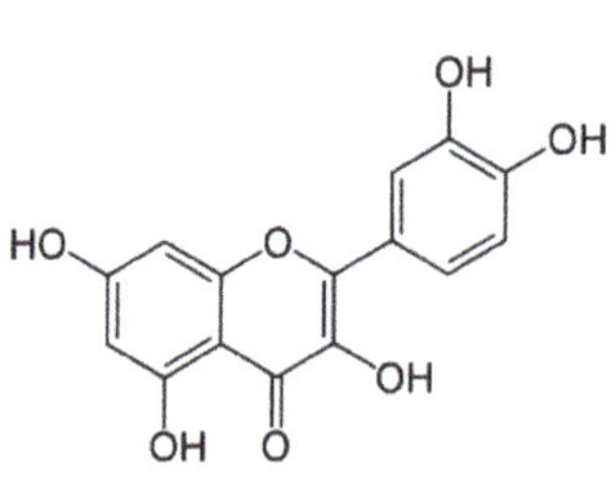

Figure 1. The structure of quercetin.

In the present work polylactide-based films with an addition of poly(ethylene glycol) acting as a plasticizer and quercetin used as a biocidal agent are suggested as new antibacterial packaging materials. Taking into account the properties of quercetin-infused PLA films, it has to be stressed that they present an enormous potential for application in the food industry.

2. Materials and Methods

2.1. Materials

Polylactide, type 2002D with melt flow rate 5–7 g/10 min (2.16 kg; 190 °C), with average molecular weight of 155,500 Da and content of monomeric units D and L equal to 3.5% and 96.5%, respectively was delivered in the form of pellets by Nature Works® (Minnetonka, MN, USA). Quercetin, an antibacterial compound, was purchased from Sigma-Aldrich (Steinheim, Germany). Poly(ethylene glycol) with an average molecular weight of 1500 (Sigma-Aldrich, Steinheim, Germany) was applied as the plasticizing agent. Acetone and chloroform were purchased from Avantor Performance Materials Poland S.A. (Gliwice, Poland).

In an aim to analyze antibacterial properties of the obtained materials, two bacterial reference strains were used in the study: Escherichia coli (ATCC 8739) and Staphylococcus aureus (ATCC 6538P) (Microbiologics®, St. Cloud, MN, USA). Moreover, an agar medium containing a composition [g/L]: tryptone peptone—15, phyton peptone—5, sodium chloride—5, agar—agar—15 was acquired from Oxoid (Hampshire, UK).

2.2. Formation of PLA-Based Films

In aim to obtain all of the PLA-based films, the solvent evaporation technique were applied. In the first stage, 1.5 g of pure and dry polylactide was dissolved in 35 mL of chloroform by vigorous mixing at room temperature. In the second stage, an appropriate amount of PEG (5% or 10% w/w of PLA, 0.075 g or 0.15 g of PEG, respectively) was added. Quercetin was dissolved in 15 mL of acetone and the mixture was then introduced into the solution containing PLA and PEG. The resulting solutions were cast onto clean glass plates, 145 mm in diameter and dried at ambient temperature for 48 h. Designations and compositions of the obtained materials are presented in Table 1.

Table 1. Compositions of the obtained materials (L-polylactide; P-PEG, Q-quercetin).

Sample	Quercetin Content [1] [wt.%]	PEG Content [1] [wt.%]
LP5	-	5
LP10	-	10
LP5Q1	1	5
LP10Q1	1	10
LP5Q2	2	5
LP10Q2	2	10

[1] Relative to PLA mass.

2.3. Fourier Transform Infrared Analysis Analysis

The Fourier transform infrared analysis of all studied materials was performed by means of a Nicolet iS10 (Thermo Fisher Scientific, Waltham, MA, USA). The spectra were recorded in the frequency range of 500–4000 cm^{-1} at a resolution of 4 cm^{-1} and scanned 64 times. The spectrum of quercetin in KBr disc form was obtained in the same conditions. All spectra were analyzed using the OMNIC 7.0 software (Thermo Fisher Scientific, Waltham, MA, USA).

2.4. Scanning Electron Microscopy (SEM)

Changes in the structure of the PLA-based films caused by an addition of PEG and quercetin were analyzed using a scanning electron microscope (Quanta 3D FEG, FEI

Company, Hillsboro, OR, USA). In an aim to obtain high quality images, the samples were covered with a thin layer of gold. Images of all samples were taken at 5000× magnification.

2.5. Atomic Force Microscopy (AFM)

The surface pictures of the PLA-based films were obtained by means of an AFM microscope with a scanning probe of the NanoScope MultiMode type (Veeco Metrology, Inc., Santa Barbara, CA, USA). Analyses were performed in the tapping mode, in air, at room temperature. Using a scan area of 5 × 5 μm and Nanoscope software (Veeco Metrology, Inc. Santa Barbara, CA, USA), the roughness parameters such as the root mean square (R_q) and arithmetical mean deviation of the assessed profile (R_a) were calculated.

2.6. Thermogravimetric Analysis

Thermogravimetric (TG) analyses of the PLA films with an addition of PEG and quercetin were performed on Simultaneous TGA-DTA Thermal Analysis type SDT 2960 (TA Instruments, London, UK). All measurements were carried out at a heating rate of 10 °C/min under air flow from room temperature to 600 °C.

2.7. Differential Scanning Calorimetry Method

Thermal analyses were carried out on DSC (Polymer Laboratories, Epsom, UK). Experiments were performed under nitrogen screening. The thermal response in the obtained materials was investigated at the temperatures ranging from 25 °C to 180 °C (1st heating cycle) and with a heating rate of 10 °C/min. The PL5 software version v5.40 (Polymer Laboratories, Epsom, UK) was applied in an aim to establish the detailed information extracted from the DSC data. The degree of crystallinity (X_m) was established by applying the following Equation (1) [28,29]:

$$X_m = \frac{\Delta H_m}{\Delta H^0 \cdot X_{PLA}} \cdot 100\% \tag{1}$$

where ΔH_m is the measured heat of fusion of sample, ΔH^0 is the heat of fusion of a 100% crystalline polylactide and ΔH^0 = 109 mJ/mg, X_{PLA} is the mass fraction of polylactide.

2.8. Mechanical Properties

The mechanical properties of the PLA-based films with and without an addition of quercetin were analyzed by means of the Instron 1193 machine (Instron Corp., Canton, OH, USA) test according to the PN-EN ISO 527-1, -3 standard [30,31]. The crosshead speed was 20 mm/min with an applied 100 N force. In the case of each type of the studied materials, at least five samples were analyzed. The obtained results allowed to establish the Young's modulus (E), elongation at break (ε) and tensile stress (σ_m).

2.9. Transparency

Transparency of the obtained materials was established based on the method presented in work of [32]. Absorbance of the polymeric films with and without quercetin at 600 nm (A_{600}) was measured by means of UV spectrophotometer (Ruili Analytical Instrument Company, Beijing, China). The analysed films were placed directly in a spectrophotometer test cell while an empty cell was used as the blank. The transparency (T) of the obtained materials was calculated according to the following Equation (2):

$$T = \frac{A_{600}}{d} \; [mm^{-1}], \tag{2}$$

where d is the film thickness [mm]. It should be noted that a higher transparency value is equivalent to lower transparency.

2.10. Colour Measurement

The changes in the color of the obtained films caused by the introduction of quercetin and PEG were studied by means of a MICRO-COLOR II LCM 6 (Dr Lange, Berlin, Germany) colorimeter. The CIE L*a*b* system was applied in aim to calculate the colour difference (ΔE) of materials. The following Equations (3) and (4) were used in order to establish the total values of ΔE and color intensity (C), respectively.

$$\Delta E = \sqrt{(L - L^*)^2 + (a - a^*)^2 + (b - b^*)^2}, \tag{3}$$

$$C = \sqrt{(a^*)^2 + (b^*)^2}, \tag{4}$$

where L is the component describing lightness, a represents the colors ranging from green ($-a$) to red ($+a$), while b represents a parameter change from blue ($-b$). The color of the control film (in this case a LP5 film), was expressed as L^* (lightness), a^* (redness/greenness) and b^* (yellowness/blueness) values [14]. In an attempt to obtain reliable data at least five measurements were performed for each of the samples, then average values were calculated. Moreover, based on the obtained results yellowness index (YI) was established using the Equation (5) described in the work of Pathare et al. [33]:

$$YI = \frac{142.86 \cdot b^*}{L^*}. \tag{5}$$

2.11. Analysis of Antibacterial Properties

Antibacterial properties were determined according to the ISO 20645:2006 standard: "Flat textile products. Determination of antibacterial activity. Diffusion method on an agar plate" [34]. Agar medium consisting of [g/L]: tryptone peptone—15, phyton peptone—5, sodium chloride—5, agar—agar—15 was poured onto each petri dish and allowed to gel. The medium was then infused with a bacterial culture at a concentration of 1.5×10^8 cfu/mL (0.5 McFarlanda). Centrally tested samples and control samples in the shape of a circle with a diameter of 25 ± 5 mm (four replicates) were placed on the dishes prepared in this way. Plates were incubated for 20 h at 37 ± 1 °C. After the end of the incubation time, the presence or absence of zones inhibiting the growth of microorganisms was determined. The width of the braking zone (H), i.e., the zone without bacteria near the edge of the sample, was calculated using the following Formula (6):

$$H = D - \frac{d}{2} \ [\text{mm}], \tag{6}$$

where:

H—braking zone width [mm],

D—total diameter of the working sample and width of the braking zone [mm],

d—diameter of the working sample [mm].

Stereoscopic microscope SZX 12 (Olympus, Tokyo, Japan) was used in an aim to establish the size of inhibition zones of bacterial growth as well as the extent of microorganism proliferation in contact of the studied films with agar. The appearance of agar plates after the studied materials had been removed was recorded by means of a SCAN® 1200 colony counter (Interscience, Saint-Nom-la-Bretèche, France). The scale shown in Table 2 was used to assess the potency of antibacterial properties.

Table 2. Antibacterial effect of antibacterial treatment (ISO 20645, 2006) [34].

Braking Zone [mm] the Average Value of Rise	Growth [a]	Description	Rating
>1	Lack	Inhibition zone above 1, no increase [b]	Good effect
0–1	Lack	Growth inhibition zone up to 1, no growth [b]	
0	Lack	No braking zone, no increase [c]	
0	Weak	Lack of braking zones, only some colonies limited growth almost completely stopped [d]	Limited effect
0	Average	No braking zone, height reduced to half compared to control [e]	
0	Strong	Lack of braking zones, the absence of a reduction in growth compared to the control, or only a slight reduction in growth	Insufficient effect

[a] Bacterial growth on the medium under the working sample. [b] The dynamometer range should only be partially taken into account in the calculations. An increase in the braking zone may be due to excess of active substance or unevenness of the substance in the article. [c] Lack of growth with a simultaneous lack of braking zone can be considered a good effect. A braking zone may not be possible due to limited diffusion. [d] "Almost as good as lack of growth"—an indication of limited efficiency. [e] Limited bacterial growth density means both the number of colonies and the diameter of the colonies.

3. Results and Discussion

3.1. FTIR Analysis

FTIR spectroscopy was used in an aim to establish the structure of the obtained materials with an addition of quercetin and different amounts of PEG. The FTIR spectrum of quercetin is shown in Figure 2, where the bands of characteristic groups can be observed. According to the literature [35,36] particular bands were assigned to the characteristic groups present in the quercetin structure. A broad band with the maximum at 3127 cm^{-1} belongs to the hydroxyl groups (phenolic -O-H stretching). The band at 1653 cm^{-1} indicates the presence of stretching vibrations of the -C=O group, while the bands at 1614 cm^{-1}, 1567 cm^{-1} and 1513 cm^{-1} belong to the C=C stretching bonds in the aromatic rings of quercetin [36,37]. In the spectrum of the neat quercetin an intense band at 1401 cm^{-1} is visible and can be ascribed to deformation vibrations of -OH groups. Additionally, in the discussed spectral bands, the maximum values of 1316 cm^{-1} and 1247 cm^{-1} correspond to the deformation vibrations of the -C-OH group. Other vibrations at 1167 cm^{-1} and 1092 cm^{-1} can be assigned to the anti-symmetrical and symmetrical stretching vibration of the -C-O-C group. The band observed at 997 cm^{-1} indicates the deformation vibrations of the -OH group, while the band at 933 cm^{-1} relates to the stretching vibrations of the -C-O. Other characteristic bands in the range between 884 cm^{-1} and 808 cm^{-1} correspond to the deformative vibrations of the -CH groups [35,37]. In Figures 3 and 4, the spectra of the materials consisting of PLA, PEG with or without addition of quercetin are depicted. The chemical structure of polylactide films is well known. It was also described in our previous publications [38]. Based on our study, it can be clearly seen that the bands at 3657 cm^{-1} and 3502 cm^{-1} correspond to -OH groups at the end of PLA chains. Absorption bands at 2996 cm^{-1} and 2947 cm^{-1} belong to the symmetrical and asymmetrical stretching vibrations of the -CH_3 group, while the band at 1763 cm^{-1} relates to the characteristic -C=O carbonyl group. Moreover, symmetrical stretching vibrations of -C-O-C were recorded at 1207 cm^{-1} and at 1127 cm^{-1}. The bands describing the stretching vibration of C-COO and the deformation vibration of CO can be seen at 873 cm^{-1} and 756 cm^{-1}.

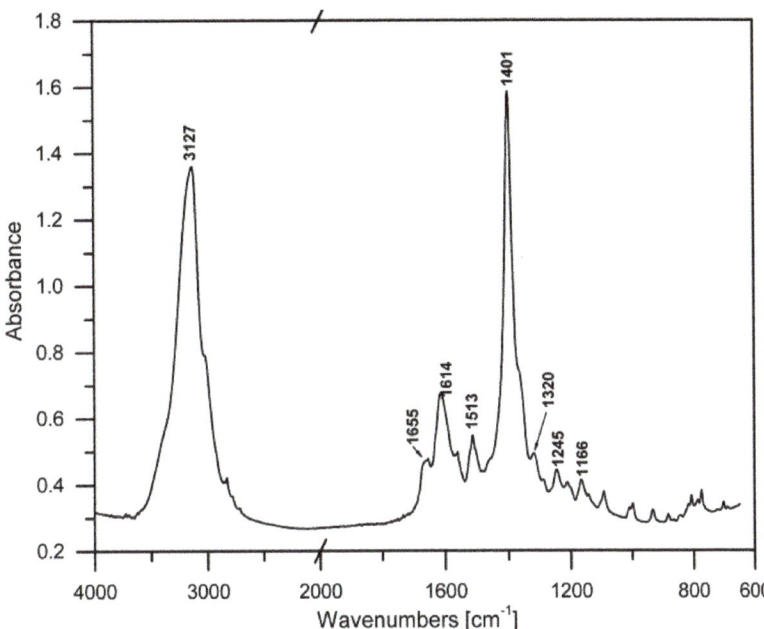

Figure 2. The FTIR spectrum of quercetin.

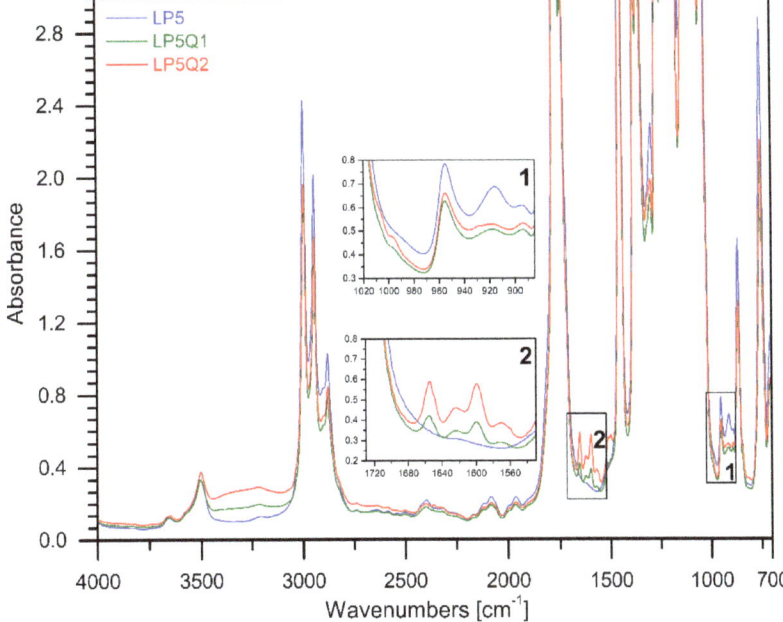

Figure 3. The FTIR spectra of materials consisting of PLA, PEG 5 wt.% with an addition of quercetin.

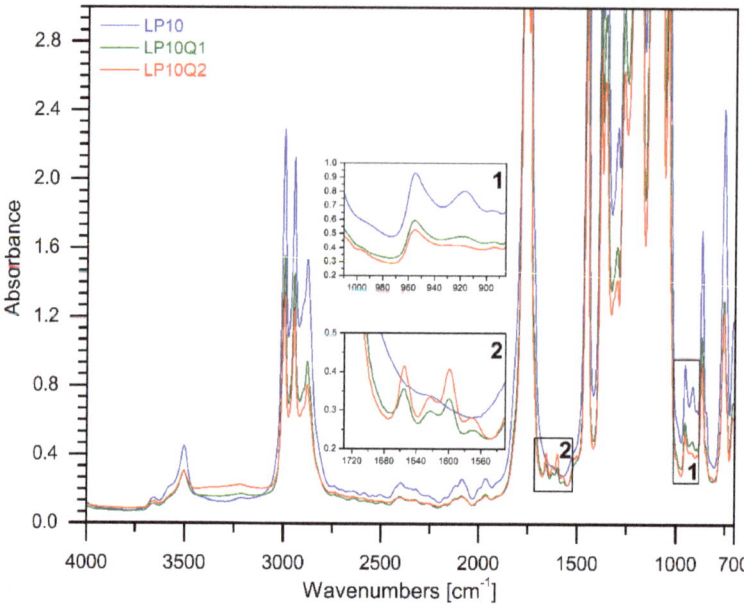

Figure 4. The FTIR spectra of materials consisting of PLA, PEG 10 wt.% with an addition of quercetin.

In the case of films based on PLA and PEG the typical vibrations caused by the -CH$_2$ group at 2878 cm^{-1} can be observed. Moreover, a peak at 3446 cm^{-1} for poly(ethylene glycol), which corresponds to the terminal hydroxyl group, has been recognized. Furthermore, it can be clearly noticed that the intensity of the band at 2878 cm^{-1} increases with the increase in the amount of PEG.

Introduction of quercetin into the PLA-PEG systems reveals a broad band at 3300 cm^{-1} which can be assigned to the -OH groups present in the structure of quercetin. Moreover, new bans at 1655 cm^{-1}, 1615 cm^{-1} and 1600 cm^{-1} corresponding respectively to stretching vibrations of the -C=O group and the C=C stretching bonds in the aromatic rings have been recognized. It has to be stressed that the intensities of the described new bands increased with the quantity of quercetin in the obtained films [37].

The spectra of the studied films after the addition of quercetin present certain changes in the intensity of bands at 915 cm^{-1} and 955 cm^{-1} belong to the amorphous and crystalline phases of PLA, respectively [39,40]. Increase in intensity of band at 915 cm^{-1} indicates that quercetin influences the crystallinity of the obtained materials.

3.2. SEM and AFM Analyses

Topography and surface morphology constitute factors which are taken into account when the potential application of materials is discussed. They are extremely important in the case of polymeric films used as food packaging as well as films with medical applications. In the present work we focused on the PLA-based materials with an addition of quercetin and poly(ethylene glycol). Obtained materials were characterized by antibacterial properties. In order to analyze their surface morphology atomic force microscopy and scanning electron microscopy were applied. The SEM images presented in Figure 5 reveal that samples of polylactide with an addition of 5 wt.% or 10 wt.% of poly(ethylene glycol), which plays the role of a plasticizer, were characterized by a smooth and flat surface, without any cracks or fissures. The same observation was made in the work of Jasim Ahmed at al. [41] where PLA-based films with an addition of 20 wt.% of PEG were studied. In the mentioned work, SEM photographs depicted good dispersion of PEG into the polylactide matrix, indicating that the obtained materials form homogeneous blends. Incorporation of quercetin into the PLA-PEG blends significantly influences the morphology of the obtained polymeric films It has to be taken into account, however, that the surfaces of films with a higher amount of PEG (10 wt.%) are characterized by larger convexities and deeper concavities of various shapes, evenly distributed throughout the surface in comparison with materials containing 5 wt.% of plasticizer.

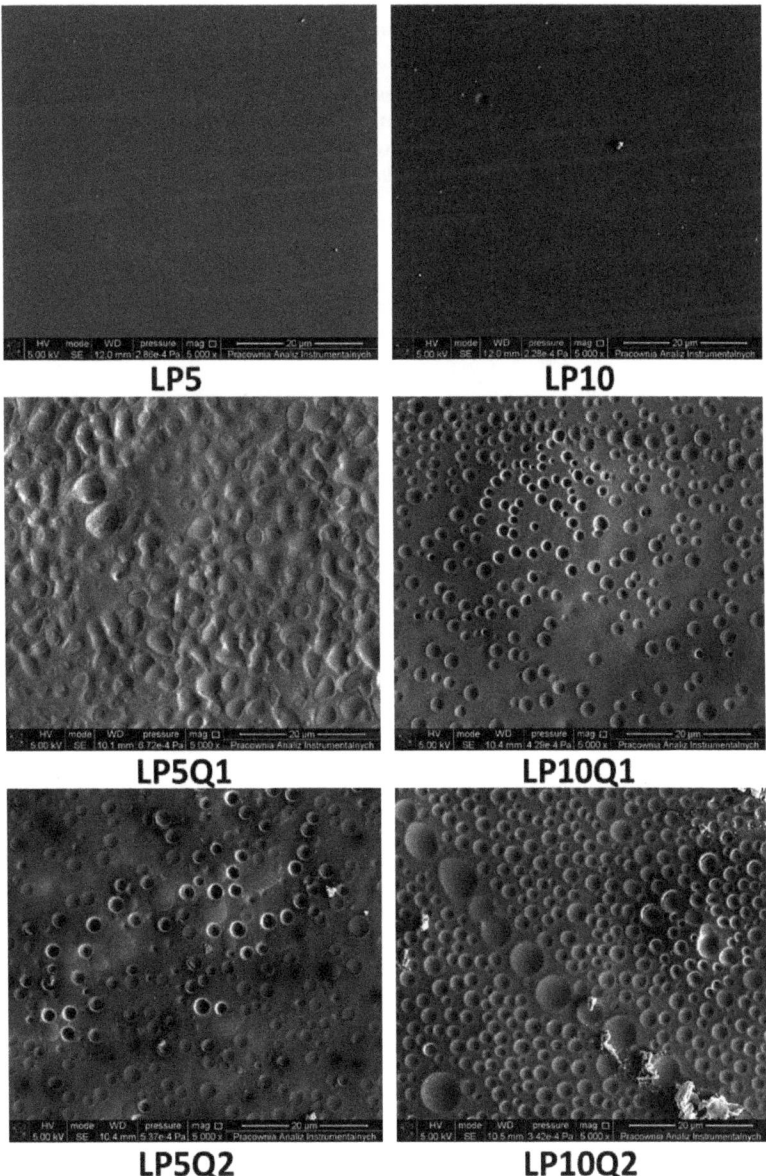

Figure 5. SEM images of obtained materials.

In an aim to establish the size and roughness of the studied surface, the AFM technique was applied. It is well known that roughness constitutes one of the factors which can significantly affect the antibacterial properties of the obtained materials. In Figure 6, three dimensional pictures of surfaces of PLA-based films are presented. It can be observed that the size of cavities formed on the surface of the obtained polymeric films containing quercetin, with an addition of 10 wt.% of PEG, are significantly enlarged. The diameter of the cavities in the case of LP5Q1 and LP5Q2 samples equaled 1.9 µm and 2.0 µm, respectively, while in the case of LP10Q1 and LP10Q2 materials the fissures measured 2.5 µm and 2.8 µm. Based on the presented images (Figure 6), the minimum depth of the formed cavities was also calculated. In the case of materials containing 5 wt.% of PEG (LP5Q1 and LP5Q2) the values of cavities' depth equaled 57 nm and 75 nm, while in the case of films with an addition of 10 wt.% of the plasticizer (LP10Q1 and LP10Q2) the depth of cavities increased to an extent of 60 nm and 80 nm, respectively. The results presented above confirm that the diameter of the formed fissures increases with the increase in the amount of PEG, while the depth of observed cavities increases with the increase in the amount of quercetin.

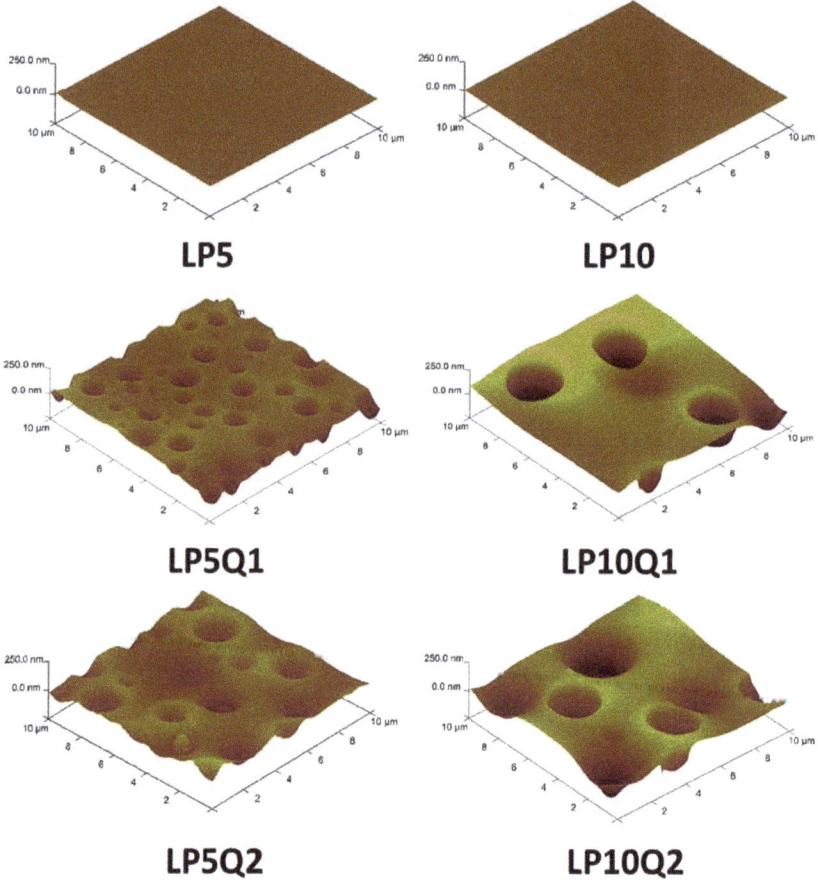

Figure 6. 3D AFM images of obtained materials.

Based on the obtained results the roughness parameters such as R_q (mean square deviation of surface roughness), R_a values (mean arithmetic deviation of the profile from the mean line) and R_{max} (maximum distance between the highest and lowest point of the recorded image) have been discussed (Table 3) [16].

Table 3. Roughness parameters of PLA and PLA-based films with and addition of quercetin and poly(ethylene glycol).

Sample	R_q [nm]	R_a [nm]	R_{max} [nm]
LP5	1.4 ± 0.1	1.0 ± 0.1	10.5 ± 0.5
LP10	3.8 ± 0.2	3.1 ± 0.2	23.1 ± 1.1
LP5Q1	32.2 ± 0.5	22.0 ± 0.4	232.0 ± 2.0
LP5Q2	34.1 ± 0.6	26.8 ± 0.5	237.0 ± 2.0
LP10Q1	69.5 ± 1.2	51.4 ± 1.0	344.0 ± 4.0
LP10Q2	71.9 ± 1.4	55.40 ± 1.0	364.0 ± 7.0

It needs to be stressed that the values of all of the analyzed roughness parameters significantly depend on the amount of PEG. Obtained results suggest a dependence between the increase of the amount of PEG and the roughness parameters discussed above which are characterized by significantly higher values. Taking into account the amount of quercetin used, values of R_a, R_q and R_{max} reveal that the same tendency as in the case of PEG. The introduction of 2 wt.% of quercetin, however, leads only to a slight increase in the values of roughness parameters in comparison with the change of values caused by the plasticizer.

3.3. Determination of Thermal Properties of Polylactide-Based Materials

In an aim to determine the effect of different amounts of plasticizer and quercetin on thermal stability of polylactide-based films, thermogravimetric analysis was conducted. Based on the obtained results, thermogravimetric (TG) curves for the samples LP5, LP5Q1, LP5Q2 are shown in Figure 7, while the TG curves relating to the samples LP10, LP10Q1 and LP10Q2 are depicted in Figure 8.

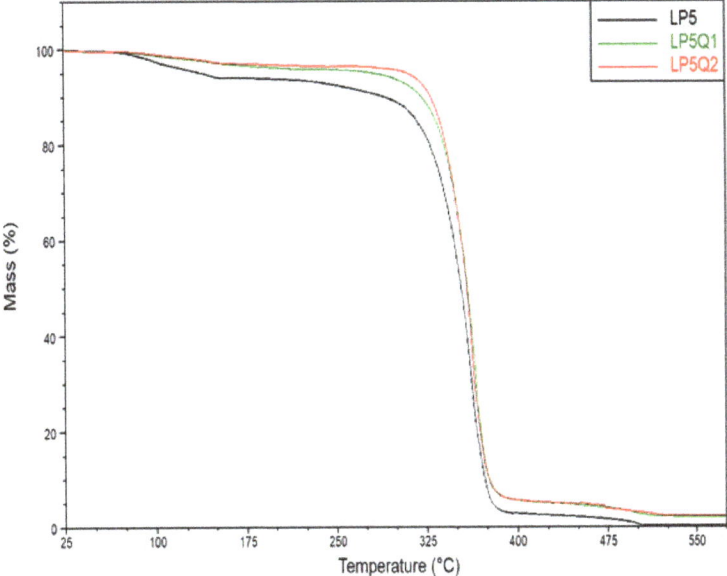

Figure 7. Thermogravimetric curves for polylactide-based materials with 5 wt.% of PEG.

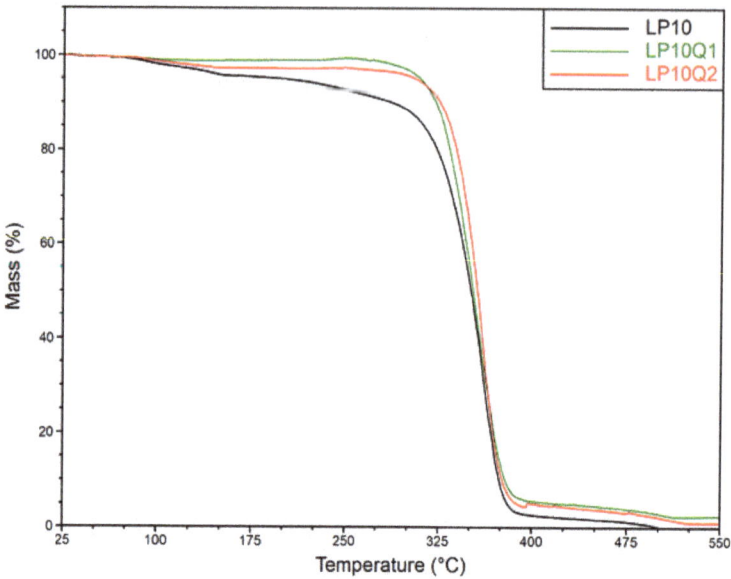

Figure 8. Thermogravimetric curves for polylactide-based materials with 10 wt.% of PEG.

It was previously discussed that the plasticizer as well as the biocidal agent can significantly affect the thermal stability of the polylactide-based films [16]. In the prevailing number of publications the temperatures at 5% mass loss are analyzed. In the work of Rhim et al. [42] and Tarach at al. [16], however, it was indicated that in the case of these types of materials, the mass loss of about 5% of the initial mass is connected with the evaporation of the solvent residue. As a result, the values of temperatures at 10%, 30% and 50% mass loss of all studied materials (signified as T10%, T30% and T50%, respectively) were selected and presented in Table 4. Taking into account the amount of the plasticizer used, it can be observed that with an increase in the amount of PEG the thermal stability of the PLA-PEG samples decreases. According to the work of Phaechamud [43], as well as Pielichowski [44], the observed reduction in thermal stability is the result of PEG decomposition, which occurs at a significantly lower temperature compared to polylactide.

Table 4. TG data for obtained materials with and without addition of quercetin.

Samples	Temperature (°C) at Mass Loss		
	10%	30%	50%
LP5	289.5 ± 0.4	340.1 ± 1.5	353.7 ± 1.5
LP10	287.2 ± 0.5	335.6 ± 1.3	350.9 ± 1.4
LP5Q1	322.6 ± 0.7	347.9 ± 1.0	358.5 ± 1.4
LP5Q2	329.8 ± 1.2	348.3 ± 1.0	358.3 ± 1.6
LP10Q1	320.0 ± 1.2	340.8 ± 1.3	352.4 ± 1.7
LP10Q2	325.5 ± 1.1	345.1 ± 1.3	356.1 ± 1.7

The incorporation of quercetin into the materials consisting of PLA and PEG, contributed to an increase in thermal stability of the obtained films. In thermograms presented in Figures 7 and 8 relating to the materials with the addition of 5 wt.% of PEG and 10 wt.% of plasticizer, respectively, a significant increase in thermal stability with the increase in an amount of biocidal agent can be observed. In the case of samples LP5Q1 and LP5Q2, the temperature for a 10% mass loss, compared to the LP5 sample, increased by 33.1 °C and 40.3 °C, respectively. The same tendency was observed in the case of samples based on PLA with 10 wt.% of PEG and containing quercetin. The 10% loss of mass in films containing

10 wt.% of PEG and modified with 1 and 2 wt.% of quercetin occurred at the temperature 32.8 °C and 38.3 °C higher than in the case of the PL10 material. The obtained results indicate that the most significant changes in thermal stability of the obtained materials can be observed after the addition of 1 wt.% of quercetin. The abovementioned results reveal that after incorporation of 2 wt.% of biocidal agent into the PL5 as well as the PL10 systems, the values of T10%, T30% and T50% also increase. The changes between 1 wt.% and 2 wt.% of quercetin, however, are not as significant as after addition of 1 wt.% of quercetin. It should be mentioned that an improvement of thermal stability of the obtained materials after the addition of quercetin was also observed in the work of Samper et al. [45]. In the mentioned work an addition of 0.25 wt.% of quercetin resulted in a remarkably higher stability of polypropylene (PP) in comparison with non-stabilized PP. According to the literature [37,45] this phenomenon can be attributed to the structure of the phenolic compound used in the procedure, especially to the number of hydroxyl groups. Moreover, it has been proved that in the case of quercetin there is an optimum amount which provides the best thermal stability. Further introduction of this substance beyond the optimum point does not provide additional benefits in terms of structural stabilization. Summarizing, based on the analysis of the above mentioned data (Table 4, Figures 7 and 8), it is reasonable to assume that the incorporation of quercetin significantly influences thermal stability of obtained materials. Obtained results are consistent with the data presented by the researchers mentioned above.

In order to determine changes in thermal parameters of the studied materials, caused by the addition of a plasticizer and quercetin, differential scanning calorimetry has been carried out (Figures 9 and 10).

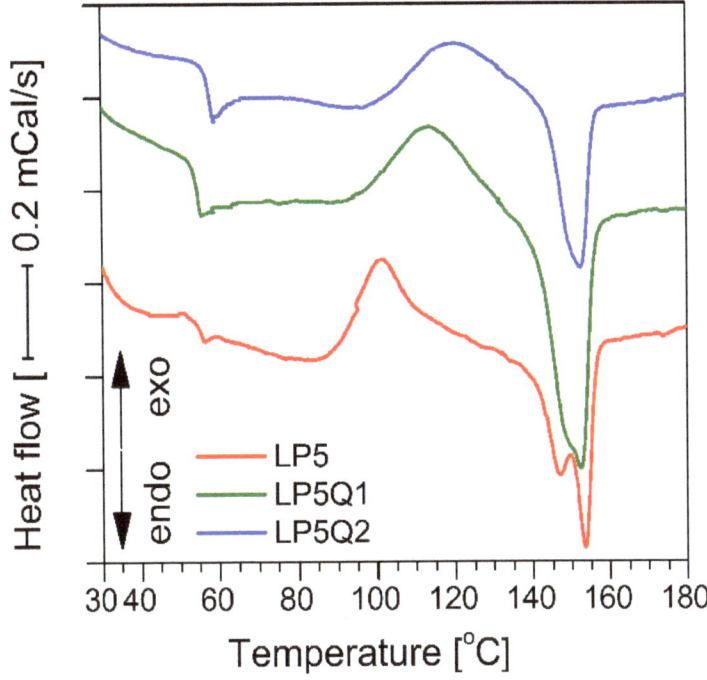

Figure 9. DSC thermograms of LP5 with and without an addition of quercetin.

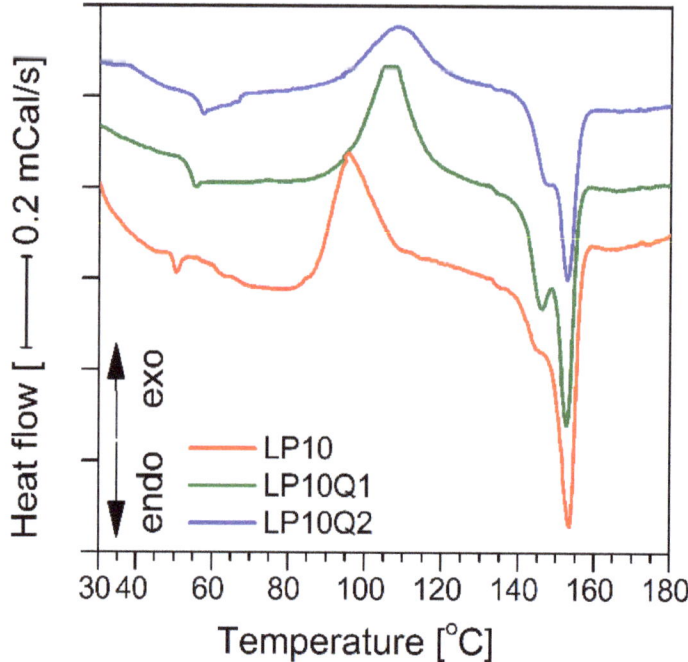

Figure 10. DSC thermograms of sample LP10 with and without an addition of quercetin.

The obtained parameters of the glass transition temperature (T_g), crystallization temperature (T_c), crystallization enthalpy (ΔH_c), melting point (T_m), melting enthalpy (ΔH_m) and the calculated values of the crystallization degree (χ) were presented in Table 5.

Table 5. DSC parameters for the all studied PLA-based materials.

Sample	T_g [°C]	T_c [°C]	ΔH_c [J/g]	T_m [°C]	ΔH_m [J/g]	χ_m [%]
LP5	54.6 ± 0.2	101.4 ± 0.5	18.4 ± 0.3	146.8/153.4 ± 0.2/0.4	24.1 ± 0.4	23.3
LP5Q1	54.5 ± 0.2	112.4 ± 0.7	16.1 ± 0.1	153.8 ± 0.2	22.0 ± 0.4	21.5
LP5Q2	58.6 ± 0.4	119.6 ± 0.8	15.4 ± 0.1	152.3 ± 0.2	21.0 ± 0.3	19.7
LP10	49.5 ± 0.1	95.5 ± 0.5	17.9 ± 0.2	153.5 ± 0.3	26.6 ± 0.5	27.1
LP10Q1	54.0 ± 0.2	106.5 ± 0.6	15.0 ± 0.2	146.2/152.6 ± 0.2/0.3	24.8 ± 0.4	25.6
LP10Q2	54.8 ± 0.2	108.9 ± 0.6	12.8 ± 0.1	146.7/152.8 ± 0.3/0.3	16.6 ± 0.2	17.3

Obtaining a homogeneous blend consisting of PLA and PEG depends to a large extent on the molecular mass of PEG. Information available in literature [46,47] indicates that blends of PLA with PEG1500 separate when the content of poly(ethylene glycol) exceeds 20 wt.%. For this reason it is reasonable to assume, that the obtained films present a homogeneous structure.

Based on the obtained results it can be seen that the increase in the amount of PEG causes a significant decrease in the glass transition temperature as well as crystallization temperature values. It was established by Piórkowska et al. [48] that the plasticizer molecules in the form of PEG significantly increase the free volume of the mixture, consequently allowing cooperative movements of macromolecules and, for this reason, decreasing the T_g value. Simultaneously, plasticization of PLA by means of PEG intensifies the crystallization of PLA and in this way lowers its T_c. Moreover, with the increase in PEG load, higher values of melting enthalpy and degree of crystallinity can be observed. The introduction of quercetin into the PLA-PEG systems results in an opposite effect. It has to be stressed that the presence of quercetin leads to a shift in glass transition temperature and crystallization temperature to higher values. Simultaneously, it can be observed that the values of both enthalpies, crystallization and melting, increase with the rise in the quercetin content. Furthermore, it should be noted, that quercetin has a profound impact on the degree of crystallinity. The direct correlation between the amount of quercetin and the degree of crystallinity of the obtained materials is apparent. This finding is consistent with the FTIR results, where the intensity of the band assigned to the crystalline phase decreases with the increase in quercetin content in the PLA-based films.

Moreover, the tests reveal that in the case of materials containing 10% wt of PEG, incorporation of quercetin causes the formation of different crystalline PLA forms. This phenomenon was widely described by Tabi et al. [49]. The formation of two crystalline forms is not observed in the case of polymeric films consisting of PLA, 5% PEG and quercetin, which is well worth noting.

3.4. Evaluation of Mechanical Properties

It is a commonly recognized fact that mechanical properties of packaging materials are extremely important [16]. In order to establish the impact different amounts of plasticizer and biocidal agent in form of quercetin have on the mechanical properties, the Young's modulus (E), tensile strength (σ_m) and elongation at break (ε) were determined. The values of Young's modulus and tensile strength, are shown in Figure 11a,b, respectively. In the case of a sample with a higher PEG content (LP10) a significant decrease of the Young's modulus and tensile strength in comparison with LP5 sample has been observed. These observations are consistent with results obtained by other researches [50,51] in relation to similar composition. During the study it has been established that the decrease in tensile strength, as well in Young's modulus values, can be attributed to poor stress transfer between the PLA and PEG phases [52]. An introduction of quercetin into polylactide-PEG systems significantly improves the properties indicated above. The PLA-PEG samples containing 1 as well as 2 wt.% of quercetin was characterized by a remarkable reinforcement of their structure. The highest values of mechanical properties were obtained in relation to the LP5Q2 sample. The value of Young's modulus for this material was about 2060 MPa while the maximum tensile strength reached about 32 MPa. The elongation at break (ε) values are presented in Figure 11c.

It can be clearly seen that an increase in the elasticity is correlated with the increase in the PEG content. It is well known that this phenomenon is related to reduction of intermolecular friction between the polymer molecules caused by the plasticizer molecules. The addition of quercetin allowed obtaining films with higher values of elasticity compared to the LP5 and LP10 samples. The highest value of the elongation at break was recorded for LP10 filled with 2.0 wt.% of quercetin. Moreover, it has to be stressed that in the case of the LP10Q2 sample the values of Young's modulus and elongation at breaking point are twice as high as for the LP10 material. Taking into account the obtained results the plasticizing effect of quercetin could potentially seem surprising. The same tendency, however, was reported in literature [53,54]. In the work of Luzi et al. [55] it was established that an improvement of the mechanical properties can be related to the inter-molecular interactions between hydrophilic groups of the polymer matrix and the polyhydroxyl groups of quercetin. In the work of K. Rubini et al. [53], where the influence of quercetin on the mechanical properties of gelatin-based films was evaluated, the values of the Young's modulus, the stress at break and the deformation at break increased significantly up to a point in which the flavonoid concentration of 1.5 wt.% was achieved. This limitation was caused by the non-homogeneous distribution of quercetin in gelatin. The susceptibility to deformation was also observed in the work of M. Latos-Brozio and A. Masek [54] where mechanical properties of PLA impregnated with quercetin and other substances of plant origin were discussed. Based on the obtained results and indicated literature, it is reasonable to assume that an incorporation of quercetin into the PLA matrix leads to an increase in mechanical properties of the obtained polymeric films.

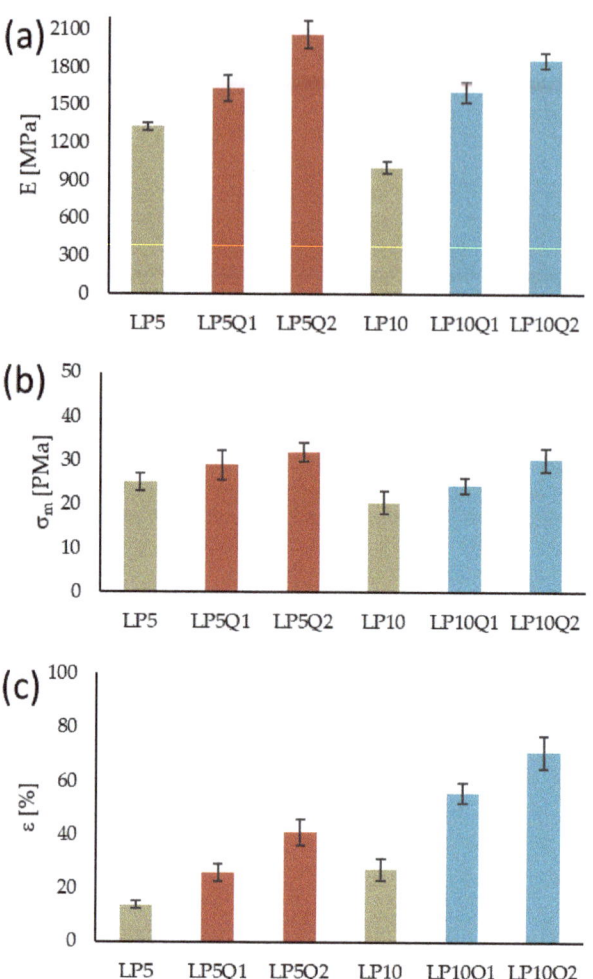

Figure 11. Mechanical properties of polylactide-based materials: (**a**) Young's modulus (E), (**b**) Tensile strength (σ_m), (**c**) Elongation at break (ε).

3.5. Thickness and Transparency of Studied Materials

The values of thickness and transparency were established based on the absorbance measurements and presented in Table 6. The obtained results indicate that an increase in the amount of PEG significantly increases the thickness of LP10 film in comparison with the LP5 sample. These findings are consistent with the results presented in literature. Anuar et al. [56] have established that an introduction of PEG into the polylactide matrix caused a diffusion of the plasticizer between the PLA chains. As a result, an increase of free volume between polylactide chains as well as an increase of thickness can be observed. Incorporation of quercetin into the PLA-PEG systems slightly decreases the thickness of the obtained materials, which is quite unexpected. In most cases, the introduction of different types of substances, such as essential oils, into the polymer matrix results in an increase of material thickness [16,57,58]. In the case of quercetin the opposite effect has been observed. Presumably, this can be connected to the interactions between –COOH groups of PLA and -OH of quercetin.

Table 6. Thickness and transparency values of studied materials.

Sample	Thickness [mm]	T [mm^{-1}]
LP5	0.071 ± 0.002	0.76 ± 0.1
LP10	0.086 ± 0.002	1.54 ± 0.1
LP5Q1	0.065 ± 0.003	3.34 ± 0.1
LP5Q2	0.066 ± 0.001	6.04 ± 0.1
LP10Q1	0.070 ± 0.004	3.92 ± 0.1
LP10Q2	0.080 ± 0.004	6.58 ± 0.2

Taking into account the results presented in Table 6, it was established that an addition of 10 wt.% of poly(ethylene glycol) significantly decreases the transparency in comparison with a polymeric film containing 5 wt.% of PEG. Observations indicated that the transparency of the LP5 films is twice as high as that of LP10. This is consistent with literature [58], where in the case of materials containing 20 wt.% of poly(ethylene glycol), the transparency value of the PLA-PEG system achieved a value of 2.63. Thus, it can be assumed the transparency of blends consisting of PLA and PEG significantly depends on the composition. An introduction of 1 wt.% as well as 2 wt.% of yellow quercetin significantly decreases transparency of the obtained materials. It should be stressed, however, that the decrease in transparency of films infused with the same amount of quercetin (LP5Q1, LP10Q1 as well as LP5Q2 and LP10Q2) has similar values regardless of the polymer compositions (LP5 and LP10). This phenomenon clearly indicates that in the case of studied materials transparency is mainly related to the amount of used quercetin rather than the applied PLA-PEG system. The obtained results suggest that polymeric films containing quercetin can be used as packaging materials, especially for products which are light sensitive. Summarizing, the application of materials consisting of PLA, PEG and quercetin will not allow for discoloration of products, loss of flavor or nutrients.

3.6. Differences in Colour of Studied Materials

In order to improve the product's appearance, which significantly influences the consumer's decision-making process in terms of purchase, the packaging materials are mixed with compounds which are able to change their color. In this paper the color of the polymeric films based on PLA with an addition of poly(ethylene glycol) as plasticizer and quercetin as antibacterial agent was analyzed. The values of measured parameters L, a, b and the calculated parameters such as total color difference (ΔE), chroma (C) and yellowness index (YI) are listed in Table 7.

Table 7. Color (L, a, b, ΔE, C, YI) parameters of obtained PLA-based materials.

Sample	L	a	b	ΔE	C	YI
LP5	91.7 ± 0.1	1.5 ± 0.1	−11.6 ± 0.2	—	11.7 ± 0.2	−18.1 ± 0.2
LP10	91.5 ± 0.1	1.5 ± 0.1	−12.5 ± 0.3	0.9 ± 0.2	12.6 ± 0.2	−19.5 ± 0.4
LP5Q1	89.5 ± 0.2	−5.0 ± 0.1	13.4 ± 0.2	25.9 ± 0.4	14.3 ± 0.3	21.4 ± 0.5
LP5Q2	88.2 ± 0.3	−8.9 ± 0.2	18.4 ± 0.5	31.9 ± 0.4	20.4 ± 0.7	29.8 ± 0.8
LP10Q1	89.2 ± 0.1	−6.6 ± 0.4	14.9 ± 0.3	27.8 ± 0.3	16.3 ± 0.7	23.9 ± 0.3
LP10Q2	87.9 ± 0.2	−8.6 ± 0.3	20.2 ± 0.4	33.6 ± 0.2	22.0 ± 0.5	32.8 ± 0.8

In the case of samples consisting of PLA and PEG, the color difference equals 0.9 and proves that an addition of a higher amount of plasticizer does not influence the color changes. According to literature [33] the values of ΔE lower than 3 indicate that only an experienced observer is able to notice the difference in color of the studied materials. The changes in color, however, were observed in the case of films comprising quercetin. The changes in color of the obtained materials were not intentional and were most likely caused by an addition of a yellow—quercetin which significantly influenced the appearance of the samples. It can be clearly seen that with an increasing amount of quercetin the PLA-based films seem to be darker. Moreover, the color shifts to yellow upon increasing the amount of quercetin. The most significant color changes are observed after introducing 2 wt.% of an antibacterial agent into the LP5 and PL10 systems. In the case of the LP5Q2 sample ΔE equaled 31.9, while for the LP10Q2 film ΔE equaled 33.6. The same tendency was observed in the case of topas cyclo-olefin copolymer (ethylene-norbornene) with an addition of quercetin described in the work of A. Masek et al. [25].

The second color parameter analyzed in this work was chroma (C) which is strongly connected with colorfulness. It has to be stressed that the higher the chroma values, the higher is the color intensity of the analyzed materials indicated by observers [33]. It is not surprising, that in the case of obtained materials with an addition of quercetin the chroma value significantly increased. This phenomenon is connected with the yellow color which is assigned to the conjugation in the B-ring cinnamoyl system and is typical of flavonoids [59]. Similar observations can be made in the case of the yellowness index (YI) values. Significant differences in b and, as a result, in C, ΔE and YI values were established between samples with and without quercetin. Summarizing, the addition of quercetin into PLA-PEG systems significantly influences the color parameters of the obtained materials.

3.7. Examination of Antibacterial Properties

It is a widely recognized fact that improving the antibacterial properties of biodegradable packaging materials prevents the development of food-damaging pathogens and significantly reduces the volume of plastic waste, by providing extended shelf life. In order to ensure antibacterial properties of packaging materials, different compounds are introduced into the polymer matrix. Depending on the type of active additive, they can be divided into chemoactive and bioactive substances. The most chemoactive ingredients include iron, titanium and zinc. The characteristic feature of the compounds mentioned above is reactivity with oxygen. It should be stressed, however, that all of them can affect the chemical composition of the product as well as the interior environment of the package. In addition, they can cause adverse health effects and hinder the recycling process. The indisputable disadvantages of chemoactive compounds lead to a more extensive application of bioactive additives derived from natural sources, such as polyphenols, essential oils and other natural extracts [15–17]. These substances play an important role in active packaging. As antioxidants, they allow to reduce production costs and eliminate the risk related to food safety. In the present work, the disk-diffusion method was applied in order to establish the capacity to combat microorganisms characterizing the obtained materials imbued with quercetin. In Table 8 the results of antibacterial properties are presented. In the case of the LP5 and LP10 films assigned as control samples, the inhibition of *S. Aureus* and *E. coli* growth was not observed.

Table 8. Results of antibacterial activity of the obtained PLA-based materials.

Sample	Diameter of Inhibition Zones of Bacteria Growth [mm]		Bacteria Growth in Direct Contact with Sample		Evaluation of Antibacterial Effect [1]	
	S. aureus	E. coli	S. aureus	E. coli	S. aureus	E. coli
LP5	0	0	Medium	Medium	Insufficient	Insufficient
LP10	0	0	Medium	Medium	Insufficient	Insufficient
LP5Q1	0	0	Lack	Weak	Good	Good
LP5Q2	1	0	Lack	Lack	Good	Good
LP10Q1	0	0	Lack	Weak	Good	Limited
LP10Q2	4	1	Lack	Lack	Good	Good

[1] in accordance with ISO 20645:2006 standard.

The same tendency was observed in literature [58,60]. Introduction of quercetin into the PLA-PEG systems significantly changed the antibacterial properties of the obtained materials (Figures S1 and S2). Moreover, it was established that quercetin creates a more significant obstacle for the microorganisms in the case of materials containing a higher amount of plasticizer (LP10Q2) (Figure 12). It is, therefore, reasonable to assume that the mentioned above beneficial change resulted from an increase of free volume between PLA chains caused by PEG. The effect of poly(ethylene glycol) on the antibacterial properties was also evaluated in our previous publication [16].

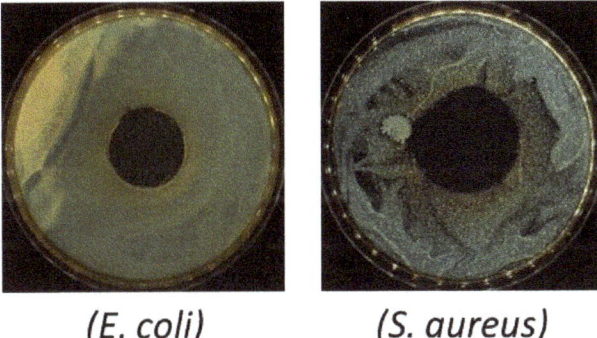

Figure 12. Photos of Bacteria (*E. coli* and *S. aureus*) growth in direct contact with LP10Q2 material.

In literature [61,62] quercetin is considered to be a biocidal agent. Ohemeng et al. [63] have established that quercetin is able to inhibit the *E. coli* DNA gyrase. Plaper et al. [64] showed that quercetin forms a bond with the GyrB subunit of *E. coli* DNA gyrase. Moreover, additional antibacterial mechanisms, triggered by quercetin, resulted in destabilizing of bacterial membrane functions were described by Mirzoeva et al. [65]. It should be stressed that the reported antibacterial activity of materials containing quercetin is consistent with the data published by Kost et al. [26]. Recent studies as well as the obtained results clearly indicate that quercetin has the potential to combat bacteria and for this reason the incorporation of quercetin contributed to an increase in antibacterial capacity of the obtained materials.

4. Conclusions

In summary, the influence of quercetin on the physico-chemical and antibacterial properties of PLA based materials was examined. Quercetin was introduced into two polymer systems consisting of polylactide and poly(ethylene glycol) (5 and 10 wt.%, respectively). The obtained materials containing quercetin were characterized by a significant improvement in thermal and mechanical properties. Simultaneously, it was also proved that the observed changes are a result of the inter-molecular interactions between hydrophilic groups of polylactide and poly(ethylene glycol) with the polyhydroxyl groups of quercetin. Moreover, it should be stressed that the introduction of quercetin allows obtaining antibacterial polymeric films. Furthermore, it can be clearly noticed that the yellow color of quercetin influences the change in the color and transparency of the studied materials. The films consisting of PLA, PEG and quercetin constitute promising bactericidal materials which can be applied as packaging materials.

Supplementary Materials: The following are available online at https://www.mdpi.com/article/10.3390/ma14071643/s1, Figure S1: Photos of Bacteria (*E. coli* and *S. aureus*) growth in direct contact with samples containing 5% wt. of PEG, Figure S2: Photos of Bacteria (*E. coli* and *S. aureus*) growth in direct contact with samples containing 10% wt. of PEG.

Author Contributions: Conceptualization, E.O.-K.; methodology, E.O.-K. and A.R. (Agnieszka Richert); software, E.O.-K., M.G., S.G.-Z., M.B. and A.R. (Anna Rudawska); validation, E.O.-K., M.G., A.R. (Agnieszka Richert); formal analysis, E.O.-K., M.G., M.B. and S.G.-Z.; investigation, E.O.-K., M.G., A.R. (Agnieszka Richert) and A.R. (Anna Rudawska); writing—original draft preparation E.O.-K.; writing—review and editing, E.O.-K., M.G.; visualization, E.O.-K., M.G; supervision, E.O.-K. All authors have read and agreed to the published version of the manuscript.

Funding: This research was funded by Nicolaus Copernicus University in Toruń.

Institutional Review Board Statement: Not applicable.

Informed Consent Statement: Not applicable.

Data Availability Statement: The data presented in this study are available on request from the corresponding author.

Conflicts of Interest: The authors declare no conflict of interest.

References

1. Rodríguez-Rojas, A.; Arango Ospina, A.; Rodríguez-Vélez, P.; Arana-Florez, R. What is the new about food packaging material? A bibliometric review during 1996–2016. *Trends Food Sci. Technol.* **2019**, *85*, 252–261. [CrossRef]
2. Barlow, C.Y.; Morgan, D.C. Polymer film packaging for food: An environmental assessment. *Resour. Conserv. Recycl.* **2013**, *78*, 74–80. [CrossRef]
3. Youssef, A.M.; El-Sayed, S.M. Bionanocomposites materials for food packaging applications: Concepts and future outlook. *Carbohydr. Polym.* **2018**, *193*, 19–27. [CrossRef]
4. Al-Tayyar, N.A.; Youssef, A.M.; Al-hindi, R. Antimicrobial food packaging based on sustainable bio-based materials for reducing foodborne pathogens: A review. *Food Chem.* **2020**, *310*, 125915. [CrossRef] [PubMed]
5. Cardoso, L.G.; Pereira Santos, J.C.; Camilloto, G.P.; Miranda, A.L.; Druzian, J.I.; Guimarães, A.G. Development of active films poly (butylene adipate co-terephthalate)—PBAT incorporated with oregano essential oil and application in fish fillet preservation. *Ind. Crop. Prod.* **2017**, *108*, 388–397. [CrossRef]
6. Sharma, R.; Jafari, S.M.; Sharma, S. Antimicrobial bio-nanocomposites and their potential applications in food packaging. *Food Control* **2020**, *112*, 107086. [CrossRef]
7. Armentano, I.; Bitinis, N.; Fortunati, E.; Mattioli, S.; Rescignano, N.; Verdejo, R.; Lopez-Manchado, M.A.; Kenny, J.M. Multifunctional nanostructured PLA materials for packaging and tissue engineering. *Prog. Polym. Sci.* **2013**, *38*, 1720–1747. [CrossRef]
8. Przybytek, A.; Sienkiewicz, M.; Kucińska-Lipka, J.; Janik, H. Preparation and characterization of biodegradable and compostable PLA/TPS/ESO compositions. *Ind. Crop. Prod.* **2018**, *122*, 375–383. [CrossRef]
9. Behera, K.; Chang, Y.-H.; Yadav, M.; Chiu, F.-C. Enhanced thermal stability, toughness, and electrical conductivity of carbon nanotube-reinforced biodegradable poly(lactic acid)/poly(ethylene oxide) blend-based nanocomposites. *Polymer* **2020**, *186*, 122002. [CrossRef]
10. Komesu, A.; de Oliveira, J.A.R.; da Silva Martins, L.H.; Maciel, M.R.W.; Filho, R.M. Lactic acid production to purification: A review. *BioResources* **2017**, *12*, 4364–4383. [CrossRef]
11. Bhargava, N.; Sharanagat, V.S.; Mor, R.S.; Kumar, K. Active and intelligent biodegradable packaging films using food and food waste-derived bioactive compounds: A review. *Trends Food Sci. Technol.* **2020**, *105*, 385–401. [CrossRef]
12. Gherasim, O.; Grumezescu, A.M.; Grumezescu, V.; Iordache, F.; Vasile, B.S.; Holban, A.M. Bioactive surfaces of polylactide and silver nanoparticles for the prevention of microbial contamination. *Materials* **2020**, *13*, 768. [CrossRef]
13. Holcapkova, P.; Hurajova, A.; Bazant, P.; Pummerova, M.; Sedlarik, V. Thermal stability of bacteriocin nisin in polylactide-based films. *Polym. Degrad. Stab.* **2018**, *158*, 31–39. [CrossRef]
14. Olewnik-Kruszkowska, E.; Gierszewska, M.; Jakubowska, E.; Tarach, I.; Sedlarik, V.; Pummerova, M. Antibacterial films based on PVA and PVA–chitosan modified with poly(hexamethylene guanidine). *Polymers* **2019**, *11*, 2093. [CrossRef] [PubMed]
15. Ahmed, J.; Mulla, M.; Arfat, Y.A. Application of high-pressure processing and polylactide/cinnamon oil packaging on chicken sample for inactivation and inhibition of *Listeria monocytogenes* and *Salmonella* Typhimurium, and post-processing film properties. *Food Control* **2017**, *78*, 160–168. [CrossRef]
16. Tarach, I.; Olewnik-Kruszkowska, E.; Richert, A.; Gierszewska, M.; Rudawska, A. Influence of tea tree essential oil and poly(ethylene glycol) on antibacterial and physicochemical properties of polylactide-based films. *Materials* **2020**, *13*, 4953. [CrossRef] [PubMed]
17. Biharee, A.; Sharma, A.; Kumar, A.; Jaitak, V. Antimicrobial flavonoids as a potential substitute for overcoming antimicrobial resistance. *Fitoterapia* **2020**, *146*, 104720. [CrossRef]
18. Kirschweng, B.; Tátraaljai, D.; Földes, E.; Pukánszky, B. Natural antioxidants as stabilizers for polymers. *Polym. Degrad. Stab.* **2017**, *145*, 25–40. [CrossRef]
19. Neri-Numa, I.A.; Arruda, H.S.; Geraldi, M.V.; Maróstica Júnior, M.R.; Pastore, G.M. Natural prebiotic carbohydrates, carotenoids and flavonoids as ingredients in food systems. *Curr. Opin. Food Sci.* **2020**, *33*, 98–107. [CrossRef]
20. Song, X.; Wang, Y.; Gao, L. Mechanism of antioxidant properties of quercetin and quercetin-DNA complex. *J. Mol. Model.* **2020**, *26*, 133. [CrossRef]
21. Masek, A.; Chrzescijanska, E. Effect of UV-A irradiation and temperature on the antioxidant activity of quercetin studied using ABTS, DPPH and electrochemistry methods. *Int. J. Electrochem. Sci.* **2015**, *10*, 5276–5290.
22. Parhi, B.; Bharatiya, D.; Swain, S.K. Application of quercetin flavonoid based hybrid nanocomposites: A review. *Saudi Pharm. J.* **2020**, *28*, 1719–1732. [CrossRef]

23. Tongdeesoontorn, W.; Mauer, L.J.; Wongruong, S.; Sriburi, P.; Rachtanapun, P. Physical and antioxidant properties of cassava starch–carboxymethyl cellulose incorporated with quercetin and TBHQ as active food packaging. *Polymers* **2020**, *12*, 366. [CrossRef]
24. López-de-Dicastillo, C.; Alonso, J.M.; Catalá, R.; Gavara, R.; Hernández-Muñoz, P. Improving the antioxidant protection of packaged food by incorporating natural flavonoids into ethylene–vinyl alcohol copolymer (EVOH) films. *J. Agric. Food Chem.* **2010**, *58*, 10958–10964. [CrossRef]
25. Masek, A.; Latos, M.; Piotrowska, M.; Zaborski, M. The potential of quercetin as an effective natural antioxidant and indicator for packaging materials. *Food Packag. Shelf Life* **2018**, *16*, 51–58. [CrossRef]
26. Kost, B.; Svyntkivska, M.; Brzeziński, M.; Makowski, T.; Piorkowska, E.; Rajkowska, K.; Kunicka-Styczyńska, A.; Biela, T. PLA/β-CD-based fibres loaded with quercetin as potential antibacterial dressing materials. *Colloids Surf. B Biointerfaces* **2020**, *190*, 110949. [CrossRef] [PubMed]
27. Hao, J.; Guo, B.; Yu, S.; Zhang, W.; Zhang, D.; Wang, J.; Wang, Y. Encapsulation of the flavonoid quercetin with chitosan-Coated nano-liposomes. *LWT-Food Sci. Technol.* **2017**, *85*, 37–44. [CrossRef]
28. Olewnik-Kruszkowska, E.; Nowaczyk, J.; Kadac, K. Effect of compatibilizig agent on the properties of polylactide and polylactide based composite during ozone exposure. *Polym. Test.* **2017**, *60*, 283–292. [CrossRef]
29. Chow, W.S.; Lok, S.K. Thermal properties of poly(lactic acid)/organo-montmorillonite nanocomposites. *J. Therm. Anal. Calorim.* **2009**, *95*, 627–632. [CrossRef]
30. ISO. *Plastics—Determination of tensile properties—Part 1: General principles*; 527-1:2020; International Organization for Standardization: Geneva, Switzerland, 2020.
31. ISO. *Plastics—Determination of tensile properties—Part 3: Test conditions for films and sheets*; 527-3:2019; International Organization for Standardization: Geneva, Switzerland, 2019.
32. Jakubowska, E.; Gierszewska, M.; Nowaczyk, J.; Olewnik-Kruszkowska, E. The role of a deep eutectic solvent in changes of physicochemical and antioxidative properties of chitosan-based films. *Carbohyd. Polym.* **2021**, *255*, 117527. [CrossRef]
33. Pathare, P.B.; Opara, U.L.; Al-Said, F.A.-J. Colour measurement and analysis in fresh and processed foods: A review. *Food Bioprocess Technol.* **2013**, *6*, 36–60. [CrossRef]
34. ISO. *Textile fabrics—Determination of antibacterial activity: Agar diffusion plate test*; 20645:2006; International Organization for Standardization: Geneva, Switzerland, 2006.
35. Basu, A.; Kundu, S.; Sana, S.; Halder, A.; Abdullah, M.F.; Datta, S.; Mukherjee, A. Edible nano-bio-composite film cargo device for food packaging applications. *Food Packag. Shelf Life* **2017**, *11*, 98–105. [CrossRef]
36. Yan, L.; Wang, R.; Wang, H.; Sheng, K.; Liu, C.; Qu, H.; Ma, A.; Zheng, L. Formulation and characterization of chitosan hydrochloride and carboxymethyl chitosan encapsulated quercetin nanoparticles for controlled applications in foods system and simulated gastrointestinal condition. *Food Hydrocoll.* **2018**, *84*, 450–457. [CrossRef]
37. Bai, R.; Zhang, X.; Yong, H.; Wang, X.; Liu, Y.; Liu, J. Development and characterization of antioxidant active packaging and intelligent Al3+-sensing films based on carboxymethyl chitosan and quercetin. *Int. J. Biol. Macromol.* **2019**, *126*, 1074–1084. [CrossRef]
38. Olewnik-Kruszkowska, E.; Koter, I.; Skopińska-Wiśniewska, J.; Richert, J. Degradation of polylactide composites under UV irradiation at 254nm. *J. Photochem. Photobiol. A* **2015**, *311*, 144–153. [CrossRef]
39. Moraczewski, K.; Pawłowska, A.; Stepczyńska, M.; Malinowski, R.; Kaczor, D.; Budner, B.; Gocman, K.; Rytlewski, P. Plant extracts as natural additives for environmentally friendly polylactide films. *Food Packag. Shelf Life* **2020**, *26*, 100593. [CrossRef]
40. Meaurio, E.; López-Rodríguez, N.; Sarasua, J.R. Infrared spectrum of poly(l-lactide): Application to crystallinity studies. *Macromolecules* **2006**, *39*, 9291–9301. [CrossRef]
41. Ahmed, J.; Hiremath, N.; Jacob, H. Antimicrobial, rheological, and thermal properties of plasticized polylactide films incorporated with essential oils to inhibit *Staphylococcus aureus* and *Campylobacter jejuni*. *J. Food Sci.* **2016**, *81*, E419–E429. [CrossRef]
42. Rhim, J.-W.; Mohanty, A.K.; Singh, S.P.; Ng, P.K.W. Effect of the processing methods on the performance of polylactide films: Thermocompression versus solvent casting. *J. Appl. Polym. Sci.* **2006**, *101*, 3736–3742. [CrossRef]
43. Phaechamud, T.; Chitrattha, S. Pore formation mechanism of porous poly(Dl-lactic acid) matrix membrane. *Mater. Sci. Eng. C* **2016**, *61*, 744–752. [CrossRef]
44. Pielichowski, K.; Flejtuch, K. Differential scanning calorimetry studies on poly(ethylene glycol) with different molecular weights for thermal energy storage materials. *Polym. Adv. Technol.* **2002**, *13*, 690–696. [CrossRef]
45. Samper, M.D.; Fages, E.; Fenollar, O.; Boronat, T.; Balart, R. The potential of flavonoids as natural antioxidants and UV light stabilizers for polypropylene. *J. Appl. Polym. Sci.* **2013**, *129*, 1707–1716. [CrossRef]
46. Younes, H.; Cohn, D. Phase separation in poly(ethylene glycol)/poly(lactic acid) blends. *Eur. Polym. J.* **1988**, *24*, 765–773. [CrossRef]
47. Baiardo, M.; Frisoni, G.; Scandola, M.; Rimelen, M.; Lips, D.; Ruffieux, K.; Wintermantel, E. Thermal and mechanical properties of plasticized poly(L-lactic acid). *J. Appl. Polym. Sci.* **2003**, *90*, 1731–1738. [CrossRef]
48. Piórkowska, E.; Kuliński, Z.; Gadzinowska, K. Plasticization of polylactide. *Polimery* **2009**, *54*, 83–90. [CrossRef]
49. Tabi, T.; Sajo, I.E.; Szabo, F.; Luyt, A.S.; Kovacs, J.G. Crystalline structure of annealed polylactic acid and its relation to processing. *Express Polym. Lett.* **2010**, *4*, 659–668. [CrossRef]

50. Mohapatra, A.K.; Mohanty, S.; Nayak, S.K. Properties and characterization of biodegradable poly(lactic acid) (PLA)/poly(ethylene glycol) (PEG) and PLA/PEG/organoclay: A study of crystallization kinetics, rheology, and compostability. *J. Thermoplast. Compos.* **2014**, *29*, 443–463. [CrossRef]
51. Sungsanit, K.; Kao, N.; Bhattacharya, S.N. Properties of linear poly(lactic acid)/polyethylene glycol blends. *Polym. Eng. Sci.* **2012**, *52*, 108–116. [CrossRef]
52. Holcapkova, P.; Hurajova, A.; Kucharczyk, P.; Bazant, P.; Plachy, T.; Miskolczi, N.; Sedlarik, V. Effect of polyethylene glycol plasticizer on long-term antibacterial activity and the release profile of bacteriocin nisin from polylactide blends. *Polym. Adv. Technol.* **2018**, *29*, 2253–2263. [CrossRef]
53. Rubini, K.; Boanini, E.; Menichetti, A.; Bonvicini, F.; Gentilomi, G.A.; Montalti, M.; Bigi, A. Quercetin loaded gelatin films with modulated release and tailored anti-oxidant, mechanical and swelling properties. *Food Hydrocolloid* **2020**, *109*, 106089. [CrossRef]
54. Latos-Brozio, M.; Masek, A. Impregnation of poly(lactic acid) with polyphenols of plant origin. *Fibres Text. East. Eur.* **2020**, *28*, 15–20. [CrossRef]
55. Luzi, F.; Pannucci, E.; Santi, L.; Kenny, J.M.; Torre, L.; Bernini, R.; Puglia, D. Gallic acid and quercetin as intelligent and active ingredients in poly(vinyl alcohol) films for food packaging. *Polymers* **2019**, *11*, 1999. [CrossRef] [PubMed]
56. Anuar, H.; Azlina, H.N.; Suzana AB, K.; Kaiser, M.R.; Bonnia, N.N.; Surip, S.N.; Abd Razak, S.B. Effect of PEG on impact strength of PLA hybrid biocomposite. In Proceedings of the 2012 IEEE Symposium on Business, Engineering and Industrial Applications, Bandung, Indonesia, 23–26 September 2012; pp. 473–476.
57. Qin, Y.; Li, W.; Liu, D.; Yuan, M.; Li, L. Development of active packaging film made from poly (lactic acid) incorporated essential oil. *Prog. Org. Coat.* **2017**, *103*, 76–82. [CrossRef]
58. Arfat, Y.A.; Ahmed, J.; Ejaz, M.; Mullah, M. Polylactide/graphene oxide nanosheets/clove essential oil composite films for potential food packaging applications. *Int. J. Biol. Macromol.* **2018**, *107*, 194–203. [CrossRef]
59. Cruz-Zúñiga, J.M.; Soto-Valdez, H.; Peralta, E.; Mendoza-Wilson, A.M.; Robles-Burgueño, M.R.; Auras, R.; Gámez-Meza, N. Development of an antioxidant biomaterial by promoting the deglycosylation of rutin to isoquercetin and quercetin. *Food Chem.* **2016**, *204*, 420–426. [CrossRef]
60. Ahmed, J.; Arfat, Y.A.; Bher, A.; Mulla, M.; Jacob, H.; Auras, R. Active chicken meat packaging based on polylactide films and bimetallic Ag–Cu nanoparticles and essential oil. *J. Food Sci.* **2018**, *83*, 1299–1310. [CrossRef]
61. Cushnie, T.P.T.; Lamb, A.J. Antimicrobial activity of flavonoids. *Int. J. Antimicrob. Agents* **2005**, *26*, 343–356. [CrossRef]
62. Rauha, J.-P.; Remes, S.; Heinonen, M.; Hopia, A.; Kähkönen, M.; Kujala, T.; Pihlaja, K.; Vuorela, H.; Vuorela, P. Antimicrobial effects of finnish plant extracts containing flavonoids and other phenolic compounds. *Int. J. Food Microbiol.* **2000**, *56*, 3–12. [CrossRef]
63. Ohemeng, K.A.; Podlogar, B.L.; Nguyen, V.N.; Bernstein, J.I.; Krause, H.M.; Hilliard, J.J.; Barrett, J.F. DNA gyrase inhibitory and antimicrobial activities of some diphenic acid monohydroxamides. *J. Med. Chem.* **1997**, *40*, 3292–3296. [CrossRef]
64. Plaper, A.; Golob, M.; Hafner, I.; Oblak, M.; Šolmajer, T.; Jerala, R. Characterization of quercetin binding site on DNA gyrase. *Biochem. Biophys. Res. Commun.* **2003**, *306*, 530–536. [CrossRef]
65. Mirzoeva, O.K.; Grishanin, R.N.; Calder, P.C. Antimicrobial action of propolis and some of its components: The effects on growth, membrane potential and motility of bacteria. *Microbiol. Res.* **1997**, *152*, 239–246. [CrossRef]

Article

Influence of Tea Tree Essential Oil and Poly(ethylene glycol) on Antibacterial and Physicochemical Properties of Polylactide-Based Films

Iwona Tarach [1], Ewa Olewnik-Kruszkowska [1,*], Agnieszka Richert [2], Magdalena Gierszewska [1] and Anna Rudawska [3]

1. Chair of Physical Chemistry and Physicochemistry of Polymers, Faculty of Chemistry, Nicolaus Copernicus University in Toruń, Gagarina 7 Street, 87-100 Toruń, Poland; tarach@doktorant.umk.pl (I.T.); mgd@umk.pl (M.G.)
2. Chair of Genetics, Faculty of Biological and Veterinary Sciences, Nicolaus Copernicus University in Toruń, Lwowska 1 Street, 87-100 Toruń, Poland; a.richert@umk.pl
3. Department of Production Engineering, Faculty of Mechanical Engineering, Lublin University of Technology, 20-618 Lublin, Poland; a.rudawska@pollub.pl
* Correspondence: olewnik@umk.pl; Tel.: +48-56-611-2210

Received: 5 October 2020; Accepted: 1 November 2020; Published: 4 November 2020

Abstract: The aim of the study was to establish the influence of poly(ethylene glycol) (PEG) on the properties of potential biodegradable packaging materials with antibacterial properties, based on polylactide (PLA) and tea tree essential oil (TTO). The obtained polymeric films consisted of PLA, a natural biocide, and tea tree essential oil (5–20 wt. %) was prepared with or without an addition of 5 wt. % PEG. The PLA-based materials have been tested, taking into account their morphology, and their thermal, mechanical and antibacterial properties against Staphylococcus aureus and Escherichia coli. It was established that the introduction of a plasticizer into the PLA–TTO systems leads to an increase in tensile strength, resistance to deformation, as well an increased thermal stability, in comparison to films modified using only TTO. The incorporation of 5 wt. % PEG in the PLA solution containing 5 wt. % TTO allowed us to obtain a material exhibiting a satisfactory antibacterial effect on both groups of representative bacteria. The presented results indicated a beneficial effect of PEG on the antibacterial and functional properties of materials with the addition of TTO.

Keywords: polylactide; tea tree essential oil; poly(ethylene glycol); packaging material; antibacterial films

1. Introduction

The proliferation of bacteria contributes to the spread of diseases, as well as food spoilage [1,2]. This group of microorganisms includes mainly strains such as *Bacillus cereus*, *Listeria monocytogenes*, *Salmonella typhimurium* and *Staphylococcus aureus* [3,4]. Due to the possibility of the undesirable contact of microorganisms with food, effective and safe methods of combating pathogenic bacteria are being sought [5]. For this reason, the food industry is still looking for new, innovative solutions to combat microorganisms' proliferation in packaged products, and to extend their shelf-life [6].

The manufacturing of packaging with antibacterial properties relies on the direct incorporation of biologically active substances into the polymer matrix [7]. In particular, the application of natural antibacterial agents for this purpose is noteworthy [8]. Essential oils (EO) are natural, microbiologically active substances with an extensive spectrum of activity [9]. Due to the acceptance of PLA and EO by the American Food and Drug Administration as "Generally Recognized As Safe GRAS", a significant development of studies of materials based on the above-mentioned compounds has taken place [10]. Ahmed et al. in their works described the results of detailed research devoted to films based on polylactide (PLA) with the addition of cinnamon oil [11,12]. The application of other essential oils, such as clove [4,13], garlic [9,12], as well as lemongrass, rosemary and bergamot essential oil [14], allowed the obtaining of PLA-based materials displaying antibacterial properties.

Tea tree oil (TTO) belongs to the group of essential oils exhibiting bactericidal properties. It is obtained during the distillation of *Melaleuca alternifolia* leaves [15]. TTO contains a mixture of terpenes (monoterpenes and sesquiterpenes) and their associated tertiary alcohols [16]. This essential oil is characterized by the capacity to inhibit the development of various microorganisms. As a result, it can be applied to effectively combat different types of respiratory and oral cavity infections, as well as skin diseases, and supports wound healing [5]. It should be stressed that TTO also exhibits anti-inflammatory, antiviral, antioxidant, antiseptic and antifungal properties [10]. The antibacterial potential of tea tree essential oil is mainly related to the presence of terpinen-4-ol and 1,8-cineole [8]. The other components of TTO synergistically intensify the antibacterial effect of individual compounds in the mixture, and ensure the low probability of bacterial resistance to the biocide in question [4]. The mechanism of combating microorganisms with tea tree essential oil is not yet fully understood. It is, however, known that TTO strongly interacts with the bacterial cell membrane and disrupts its vital functions [5]. The effectiveness of TTO has contributed to the extensive application of this substance in the cosmetics and pharmaceutical industries, as well as in medicine [16]. Moreover, publications devoted to the incorporation of TTO into the polymer matrix indicate the possibility of obtaining materials with satisfactory antibacterial properties [8,17,18]. It is noteworthy that PLA-based films with the addition of TTO have not been described in the literature so far.

Plasticizers are often used in order to increase the elasticity of polymer films [12]. Poly(ethylene glycol), poly(propylene glycol), lactic acid, and other polymers miscible with PLA count among the most common substances used to plasticize polylactide [9]. PEG is widely used, in particular due to its biodegradability. There are several works describing the characteristics of PLA-based materials with the addition of various essential oils [14], or PLA/PEG blends modified with these biocides [12,19]. Nevertheless, the influence of the plasticizer on the properties of a neat PLA film, as well as on PLA materials modified with tea tree essential oil, has not been studied yet.

The conducted research includes the development of new potential packaging films based on PLA and TTO, both with and without the addition of PEG as a plasticizer. The obtained films were characterized in terms of structure, as well as their antibacterial, mechanical and thermal properties. The novelty of the study was in determining the effect of the plasticizer on the functional properties, and the antibacterial effects against the *Staphylococcus aureus* and *Escherichia coli* strains, of PLA-based films modified with tea tree essential oil.

2. Materials and Methods

2.1. Materials

Polylactide (PLA) 2002D type (NatureWorks®, Minnetonka, MN, USA) with a melt flow rate of 5–7 g 10 min^{-1} (2.16 kg; 190 °C) and a density of 1.24 g cm^{-3} in the form of pellets was used to prepare polymer solutions. Tea tree essential oil (TTO) obtained from the leaves of Melaleuca alternifolia was purchased from Sigma-Aldrich Ltd., Poland. Poly(ethylene glycol) (PEG) with M_w = 1 500 g mol^{-1} (Sigma-Aldrich Ltd., Poznan, Poland) was used as a plasticizer. Chloroform (manufactured by Chempur, Piekary Śląskie, Poland) was used as a solvent.

2.2. Preparation of Films

The examined films were prepared using the solvent-casting method. For this purpose, polylactide pellets were dissolved in chloroform in an attempt to obtain a 3% (*w/v*) polymer solution. Subsequently 5 wt. %, 10 wt. % or 20 wt. % tea tree essential oil was added to the PLA solution (relative to the weight of polylactide used). In total, 5 wt. % PEG was introduced into the solutions to prepare plasticized PLA-based films. In order to obtain PLA-based materials, 50 mL of the prepared mixture was poured onto glass Petri dishes (14.5 cm in diameter) and left for 3 days to form a polymer film.

The film marked as L consists of neat polylactide. The PLA-based material containing 5 wt. % PEG was named as LP. The LTx symbol indicates samples modified only with tea tree essential oil (where x denotes the content of the biocide in relation to the mass of polylactide used to form a particular sample). Films with the addition of both PEG and 5 wt. %, 10 wt. % or 20 wt. % TTO were designated as LPT05, LPT10 and LPT20, respectively.

2.3. Methods of Analysis

2.3.1. Scanning Electron Microscopy

The morphology of the PLA-based films was studied by means of the Quanta 3D FEG scanning electron microscope (SEM, FEI Company, Hillsboro, OR, USA). Photographs of the topography of the samples were taken using the SE detector in 10,000-fold magnification. Before each of the analyses, the surfaces of the studied materials were sprayed with a layer of gold.

2.3.2. Atomic Force Microscopy

The roughness analyses of the obtained PLA materials, modified with TTO or TTO with a PEG addition, employed an atomic force microscope (AFM, NanoScope MultiMode, Veeco Metrology, Inc., Santa Barbara, CA, USA). Surface images of the studied films were obtained using an AFM microscope with an SPM scanning probe of the NanoScope MultiMode type (Veeco Metrology, Inc. Santa Barbara, CA, USA) in tapping mode. All analyses were conducted in air, at room temperature. Roughness parameters were calculated by applying the Nanoscope software for sample areas measuring 5 μm × 5 μm.

2.3.3. Thermogravimetry

Analyses of thermal stability of the PLA-based films were carried out by means of a Simultaneous TGA-DTA thermal analyzer of the SDT 2960 type (TA Instruments, London, UK) in the range of room temperature to 600 °C. All measurements were performed at a heating rate of 10 °C/min in an air stream.

2.3.4. Differential Scanning Calorimetry

The thermal properties (temperatures and enthalpy changes of individual phase transitions) of the studied materials were determined using a differential scanning calorimeter (Polymer Laboratories Ltd., Epsom, UK). Measurements were taken in the range of 30 to 180 °C, with a heating rate of 10 °C/min under nitrogen. On the basis of the obtained data, the degree of crystallinity (X_c) of the PLA-based materials was also calculated according to the following Formula (1):

$$X_c = \frac{\Delta H_m^{PLA} - \Delta H_c^{PLA}}{\Delta H^o \cdot X_{PLA}} \cdot 100\% \tag{1}$$

where ΔH_m^{PLA} is a change of melting enthalpy, ΔH_c^{PLA} is a change of enthalpy of cold crystallization, and X_{PLA} was marked as the mass fraction of polylactide in the tested sample. ΔH^o indicates the change of melting enthalpy of 100% crystalline PLA, the value of which, according to Sosnowski's work [20], was predetermined as 109 J/g.

2.3.5. Uniaxial Tensile Test

The examination of the mechanical properties of the obtained PLA-based films was performed using the Intron 1193 machine in accordance with the ISO 527-1:2020 [21] and ISO 527-3:2019 [22] standards. Paddle-shaped samples were stretched at a speed of 0.50 cm/min by means of the crosshead with an applied 50 N force. At least five measurements were made in the case of each of the studied films. Based on obtained data, the value percentage of the elongation at break and tensile strength were determined. The values of the Young's modulus were calculated from the rectilinear region of the registered curve during the stretching of samples.

2.3.6. Thickness and Transparency of Studied Materials

The thickness of the obtained materials was established as the average of five measurements performed using an Absolute Digimatic Indicator, Sylvac S229 Swiss (Yverdon, Switzerland), with a precision of 0.001 mm. In order to determine the transparency of the PLA-based films, transmittance at 600 nm (T_{600}) was measured by means of a UV spectrophotometer (Ruili Analytical Instrument Company, Beijing, China). The samples were cut into rectangles and placed in a spectrophotometer test cell. An empty cuvette was used as the blank. The transparency value (TV) of the PLA-based materials was calculated according to Equation (2) presented in the work of Arfat [19]:

$$TV = \frac{-\log T_{600}}{x} \tag{2}$$

where x is the film thickness (mm). Moreover, it should be stressed that a higher transparency value (TV) indicates the lower transparency of the studied materials.

2.3.7. Evaluation of Antibacterial Activity

In order to evaluate the antibacterial properties of the PLA-based films with an addition of TTO or TTO and a plasticizer, a disk-diffusion method was applied in accordance with the ISO 20645:2006 standard [23]. Agar plates were inoculated with representative strains of *Staphylococcus aureus* (ATCC 6538P) or *Escherichia coli* (ATCC 8739), and a bacteria concentration of 1.5×10^8 CFU/mL. Samples with a diameter of 25 ± 5 mm were placed onto an inoculated agar and incubated for 20 h at 37 ± 1 °C.

The size of the inhibition zones of bacterial growth, as well as the degree of microorganism proliferation in the direct contact between the studied films and agar, were examined using an Olympus SZX 12 stereoscopic microscope. Photographs of the examined materials were taken using an ARTRAY camera (ARTCAM 300MI model) in a 60-fold magnification mode. The appearances of the samples and agar plates after the investigated films had been removed was recorded by means of a SCAN® 1200 colony counter (Interscience, Saint-Nom-la-Bretèche, France).

3. Results and Discussion

3.1. Assessment of Film Morphology

The morphology of the PLA-based films was studied by means of scanning electron microscopy (Figure 1). The material consisting entirely of polylactide was characterized by a homogeneous and smooth surface. The PLA-based films with an addition of tea tree essential oil had cavities and holes which formed an irregular surface, similar to materials with an addition of oregano essential oil, as described in Javidi's work [24]. The materials described by Javidi et al., however, exhibited a less uniform structure than the PLA-TTO films. This can be attributed to the difference between the content and the extent of volatility displayed by TTO and the essential oil used in the mentioned work [24]. Moreover, PLA-based materials modified with tea tree essential oil had significantly fewer pores in their surface compared to the films based on gelatin with an addition of clove essential oil, as characterized by Ejaz [13]. The size as well as the number of pores on the surface of the obtained materials increased with the rise in the amount of TTO introduced into the PLA matrix (Figure 1). The factor which significantly affected the morphology of the films consisting of polylactide and tea tree essential oil was the partial volatilization of the biocide during the formation of the studied materials. Moreover, changes in the external structures of the PLA films modified only with TTO also resulted from the presence of the biocide droplets embedded between polymer chains. This mentioned effect was described in Qin's work [14] devoted to the formation of materials based on PLA and bergamot, lemongrass, rosemary and clove essential oils.

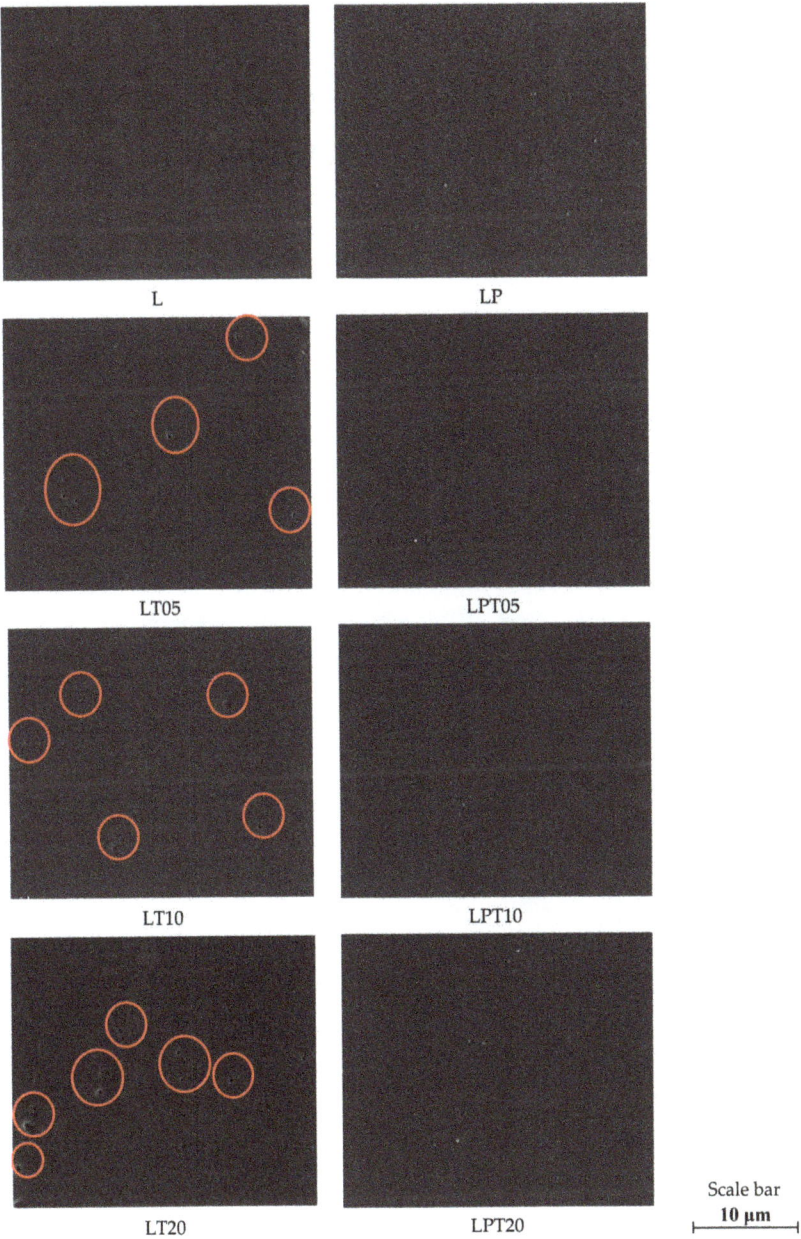

Figure 1. Structure of plasticized or unplasticized films based on polylactide (PLA) with the addition of different concentrations of TTO (magnification 10,000×).

On the other hand, the incorporation of the plasticizer into a neat PLA results in only a slight irregularity of the film surface (LP sample Figure 1). This is related to the phase separation of the PLA–PEG mixture during chloroform evaporation and the formation of small aggregates of plasticizer inside the polylactide matrix [25]. SEM photographs of the plasticized materials, based on polylactide with the addition of TTO, reveal convexities associated with the presence of PEG, which does not uniformly blend with PLA [26]. Moreover, in the cases of samples consisting of PLA, PEG and TTO pores were not formed while the materials were being dried. Based on the obtained results, it can be assumed that a reduction in the interactions between the PLA chains and PEG caused by the introduced molecules of TTO allows for preventing the flocculation and coalescence of essential oil during the formation of films. The magnitude of interactions between TTO and PLA plays a crucial role during the evaporation of the organic solvent [24]. For this reason, an insufficiently strong interaction between molecules of the essential oil and the polymer leads to the phase separation of this mixture. Therefore, it is justified to assume that the PLA and TTO mixtures without the addition of PEG during evaporation of chloroform may undergo the processes characteristic of two-phase dispersion systems, despite the solubility of the polymer and the essential oil in the mentioned solvent [8]. In order to confirm the occurrence of the flocculation, coalescence and creaming phenomena during the drying of the PLA-based films modified solely with TTO, SEM analyses of the cross-sections of all the tested materials were carried out (Figure 2).

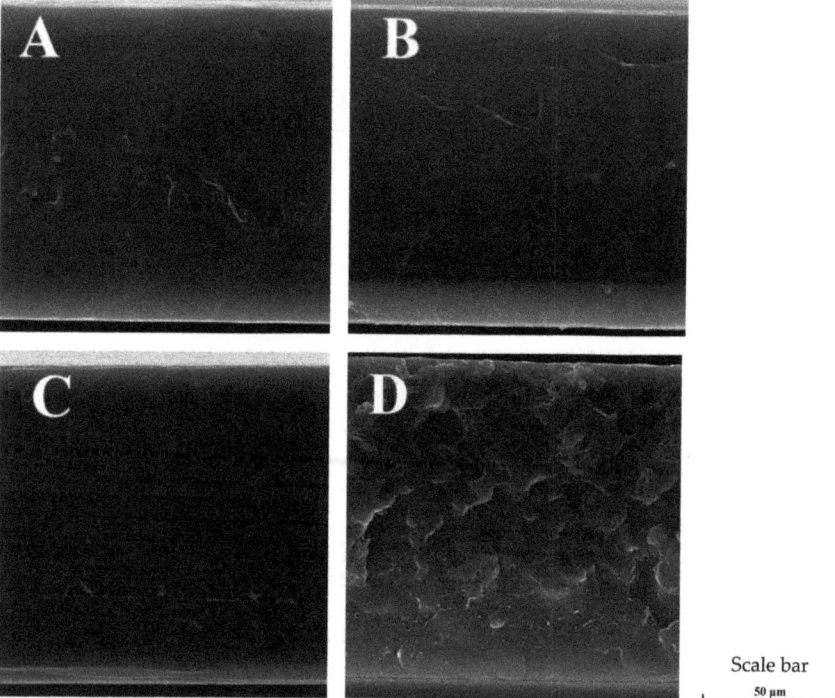

Figure 2. Cross-sections of films PLA modified with (**A**) 5%, (**B**) 10%, (**C**) 20% and (**D**) 50% TTO.

In the cases of films without the addition of PEG containing from 5% to 20% TTO, a compact structure of films was observed without the presence of pores in the materials. This leads to the conclusion that due to the high volatility of the incorporated biocidal substance, the migration of droplets of essential oil onto the surfaces of the materials did not result in the formation of a discontinuous structure. In order to verify that the creaming of the PLA–TTO mixtures containing higher amounts of biocidal substance during their drying had taken place, a PLA-based film with an addition of 50 wt. % tea tree essential oil was also imaged. Studies of that sample showed both the coalescence and the creaming of biocide droplets inside the PLA matrix during the evaporation of the solvent (Figure 2) [18,27]. These observations indicate the necessity of using essential oils in an amount less than 50%, in order to avoid the premature release of some volatile biocides from the polylactide film.

3.2. Examination of Surface Topography of PLA-Based Films

The capacity of materials to combat the proliferation of bacteria on their surface results from the presence of biologically active substances [28]. Moreover, in the case of packaging films, the effectiveness of the biocides introduced into the polymer matrix is affected by the structure of these materials—mainly by the actual surface of direct contact with the bacteria cells. In order to assess the roughness of the polymer, the film analyses carried out using AFM microscopy are extremely valuable [29]. Figure 3 depicts the values of the R_a (mean arithmetic deviation of the profile from the mean line), R_q (mean square deviation of surface roughness) and R_{max} (maximum distance between the highest and lowest points of the recorded image) parameters for the studied samples. The recorded AFM images indicate that due to the incorporation of tea tree essential oil into the PLA matrix, the surfaces of the films became more porous in comparison to the smooth surface of the control sample (pure PLA (Figure 3)). Taking into account the scales given in the presented results of microscopic analysis, the cavities captured on the cross-sections had sizes varying between about 0.10 µm and 0.25 µm (Figure 3). These pores were significantly smaller compared to the holes on the surfaces of films modified only with essential oil, observed in SEM photographs, whose sizes ranged between about 0.7 µm and 1.0 µm (Figure 1). The regularity of the surfaces of films with the addition of TTO slightly deteriorated as a result of the incorporation of the biocidal substance into the polymer matrix, which is indicated by the increase in the R_a roughness parameter from 1.16 µm to 1.74 µm for samples L and LT20, respectively (Figure 3).

Nevertheless, the R_q values presented in Figure 3 lead to the conclusion that the micro- and nano-holes are evenly spread on the film surface regardless of the amount of incorporated biocide. The values of the R_{max} parameter confirm that the cavities on the surface of the tested materials are of correspondingly larger size with the increase in the TTO content of the samples (Figure 3). This parameter, in the cases of the neat PLA film and the material with 20% TTO, equaled 14.30 nm and 41.10 nm, respectively. Similar observations of the increase in the roughness of films as a result of the addition of essential oil were made in works devoted to the modification of chitosan with clove essential oil [30] and Carum copticum essential oil [31]. Based on the works of Atarés [32], Javidi [24] and Reyes-Chaparro [30], however, it can be stated that the pores in the external layer of films modified with TTO appeared as a result of the occurrence of the phenomena of the coalescence and creaming of droplets of essential oil inside the polylactide matrix. PLA-based films with the addition of PEG and with or without TTO were also subjected to the AFM analysis.

The fact that the roughness of the film surface consisting of polylactide and poly(ethylene glycol) was greater than the irregularity of the neat polymer sample due to the presence of plasticizer aggregates results from the separation of the PLA and PEG phases [25] (Figure 3). The AFM analysis of the plasticized materials with the addition of tea tree essential oil showed that they had a rougher surface compared to the polylactide and poly(ethylene glycol) film. In addition, the films containing PEG and TTO had pores on the surface similar in size to the pores on the surface of PLA samples modified with TTO (Figure 3). The number of cavities observed in these materials, however, was significantly lower compared to the PLA-TTO systems (Figure 3). Based on the obtained results, it can be concluded that the decrease in the strength of interactions between polylactide chains, caused by the plasticizer applied in the process, impedes the flocculation and coalescence of TTO droplets in the polylactide solution.

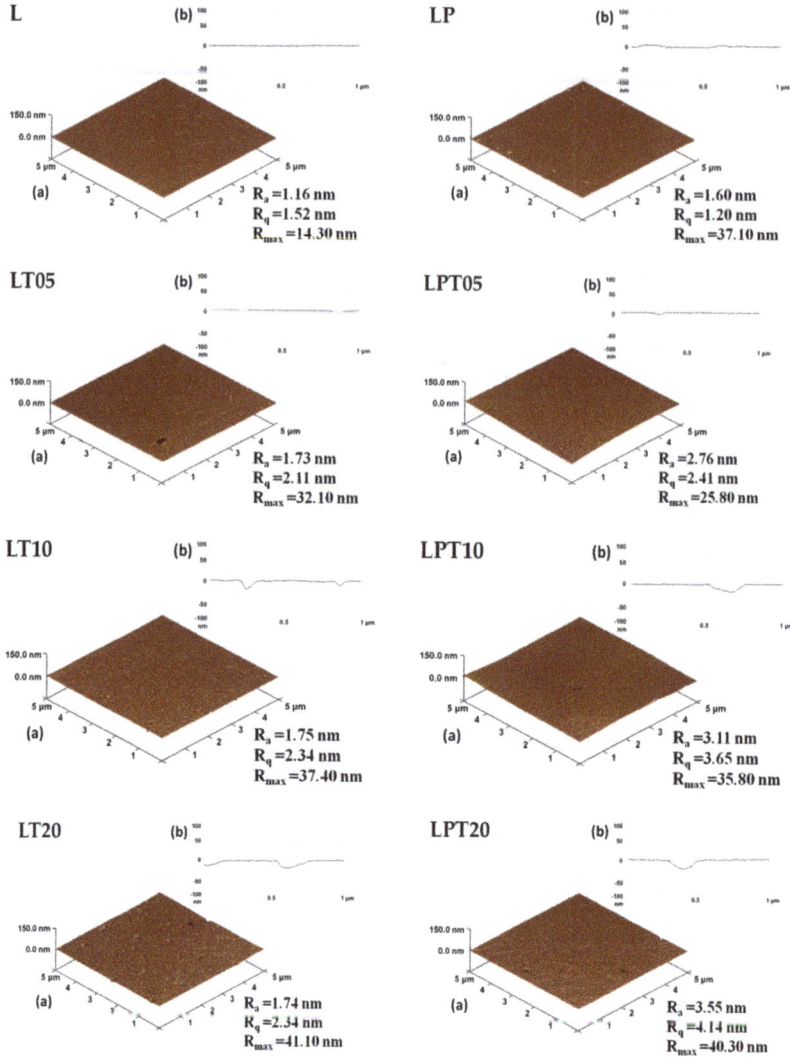

Figure 3. Roughness of PLA-based films ((**a**) 3D imaging and (**b**) cross-sections).

3.3. Determination of Thermal Properties of PLA-Based Films

The incorporation of both a biocidal substance as well as plasticizer can undoubtedly affect the thermal stability of polymer materials. For this reason, in order to establish the influence of introduced additives on the thermal stability of obtained films, all samples were subjected to thermogravimetric analysis. The temperatures at 5%, 10% and 50% mass loss of all investigated materials (signified as T5%, T10% and T50%, respectively) were listed in Table 1.

Table 1. Thermal properties of PLA-based materials.

Sample	$T_{5\%}$ (°C)	$T_{10\%}$ (°C)	$T_{50\%}$ (°C)	T_g^{PLA} (°C)	T_c^{PLA} (°C)	ΔH_c^{PLA} (J/g)	T_m^{PLA} (°C)	ΔH_m^{PLA} (J/g)	X_c^{PLA} (%)
L	273.2	308.7	342.9	56.5	124.4	−12.9	154.1	16.8	3.5
LT05	151.5	296.3	353.9	-	84.0	−11.6	153.3	23.7	11.6
LT10	140.9	307.3	355.2	-	75.8	−10.2	150.9	21.3	11.2
LT20	140.9	251.3	356.7	-	73.6	−9.3	151.3	18.8	10.5
LP	257.2	288.2	338.8	54.9	110.6	−23.0	149.0/154.8	23.6	0.6
LPT05	240.9	292.4	352.8	-	100.6	−23.2	145.0/153.1	24.9	1.6
LPT10	199.5	270.1	352.3	-	98.8	−22.3	144.3/152.1	23.7	1.4
LPT20	185.7	247.3	352.3	-	97.9	−21.9	140.5/150.4	23.0	1.2

$T_{5\%}, T_{10\%}, T_{50\%}$—temperatures at 5%, 10%, and 50% mass loss. $T_g^{PLA}, T_c^{PLA}, T_m^{PLA}$—glass transition, cold crystallization and melting temperatures. $\Delta H_c^{PLA}, \Delta H_m^{PLA}$—enthalpy of cold crystallization and melting processes. X_c^{PLA}—degree of crystallinity.

The comparison of the thermograms of materials consisting of PLA and PLA with an addition of PEG is shown in Figure 4a.

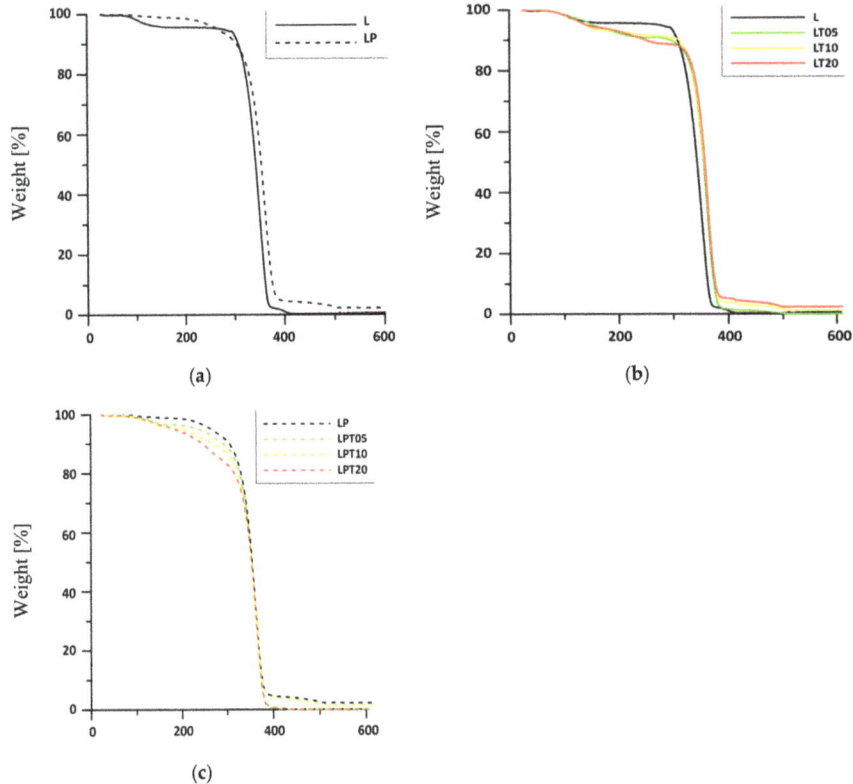

Figure 4. TG curves of (**a**) unplasticized and plasticized PLA film, (**b**) PLA films modified only with TTO and (**c**) PLA based materials with PEG and TTO addition.

The thermal degradation of these films took place in two stages [33]. In the range of the lower temperatures (100 °C–250 °C) for L and LP films, there was a comparable mass loss of the sample of about 4.5% of the initial mass. According to the work of Rhim et al. [34], this can be attributed to the evaporation of chloroform residue from the polymer matrix. Nevertheless, the mass of the L sample decreased more significantly in the mentioned temperature range compared to the PLA sample with the addition of the plasticizer. According to the work of Byun [35], the results could suggest a stronger interaction of polylactide chains with volatile substances (including chloroform) after the plasticization of PLA The second stage of the thermal degradation of the PLA film (L) involved the decomposition of polylactide chains. The incorporation of the plasticizer into PLA contributed to a decrease in thermal stability of this material compared to the sample consisting only of polylactide. This occurred as a result of PEG decomposition, which takes place at a much lower temperature than PLA [36]. A reduction in the value of the thermal degradation temperature and a gradual decrease in mass of the PLA sample with the addition of PEG at a temperature below 250 °C was also observed by Phaechamud [33].

In Figure 4b,c, thermograms for the PLA–TTO and PLA–PEG–TTO systems have been presented, respectively. The decomposition of samples modified only with biocide also occurred in two stages, as in the case of materials without the addition of a biocidal substance (Figure 4b). The loss of mass in films modified with TTO at 250 °C ranged from 8.2% in the LT5 film to 9.9% in the LT20 film. The described values of mass loss are higher in comparison to the control film (L), whose mass loss at the mentioned temperature was 5.2%. The registered percentage difference in the decrease of film mass resulted from the presence of essential oil in the PLA matrix.

The incorporation of the plasticizer into the polylactide matrix containing TTO led to a smaller and less violent decrease in the value of $T_{5\%}$ compared to films modified only with essential oil. The addition of PEG into the studied materials promoted the loosening of the PLA chains and resulted in the gradual release of volatile substances from the polymer matrix. A regular increase in mass loss can be observed on the thermograms of these films, depending on the increase in the amount of the biocidal substance in particular samples (Figure 4c).

In order to determine the influence of an individual additive on the temperatures and energy effects of the phase transitions of PLA in the obtained films, analyses were carried out using differential scanning calorimetry. The values of thermal properties obtained during studies are listed in Table 1. In the case of a neat PLA sample, the glass transition (T_g^{PLA}), cold crystallization (T_c^{PLA}) and melting (T_m^{PLA}) temperatures reached values equaling 56.5 °C, 124.4 °C and 154.1 °C, respectively (Table 1).

The introduction of tea tree essential oil into the polylactide matrix leads to a significant shift in T_g^{PLA} to lower values (Figure S1). Moreover, the values of the cold crystallization temperature of polylactide in the PLA–TTO films also decreased. Simultaneously, the crystallinity of materials increased with the rise in the content of biocidal substance (Table 1). The above-mentioned phenomena occur due to the presence of the essential oil within the polylactide matrix, as well as the interactions of the polymer with chloroform residues [35]. It should be noted that Qin [14] and Javidi [24] stated that the addition of an essential oil increases the mobility of polylactide chains and leads to their reorganization. On the other hand, according to Byun [35], the presence of chloroform molecules between PLA chains provides conditions preferential for the crystallization of the polymer to take place. Moreover, it should be stressed that the addition of only TTO improved the polymer chain's movements, facilitating the formation of the crystal structure. The addition of PEG with TTO in all likelihood influenced (increased) the distance between the macromolecules of PLA; for this reason, further increases in X_c were not observed.

Taking into account the samples consisting of PLA, TTO and PEG, it was observed that the incorporation of 5 wt. % PEG into the PLA film reduced the temperatures of all phase transitions (Table 1). It should be stressed that for all the samples containing a plasticizer in the form of PEG, a bimodal peak of melting was observed. This result can be associated with the formation of two crystal forms, as indicated in Tábi's work [37]. The phenomenon occurred most likely as a result of a localized effect of the plasticizer. In the case of the three-component materials, a significant decrease in the temperatures of phase transition (T_c^{PLA} and T_m^{PLA}) was observed compared to the LP results obtained in relation to the samples. Similar observations of a reduction in T_c^{PLA} and T_m^{PLA} were made by Arfat et al. [19] during studies on PLA/PEG films with an addition of clove essential oil. Moreover, the interactions between the chains of PLA, PEG and TTO ingredients allowed us to obtain materials characterized by a low degree of crystallinity compared to samples modified only by means of tea tree essential oil (Table 1).

3.4. Evaluation of Mechanical Properties

Due to the potential application of PLA-based materials in packaging, the analysis of the mechanical properties of films is extremely important [14]. The performed studies of the obtained materials resulted in determining such parameters as the Young's modulus (E), the tensile strength (σ_m), and the percentage relative elongation at break (ε_b). Values obtained in relation to the above-mentioned parameters have been presented in Figure 5.

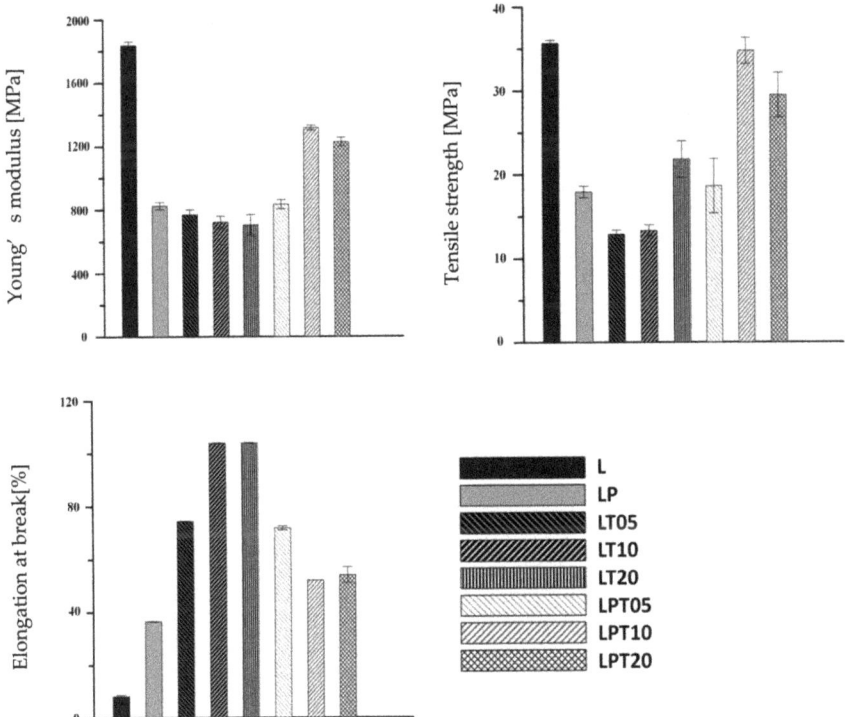

Figure 5. Comparison of mechanical properties of PLA based films.

The PLA sample was characterized by high rigidity and resistance to deformation. The value of the Young's modulus for this film was about 1835.3 MPa, while the maximum stress resistance reached 35.6 ± 0.4 MPa (Figure 5). Due to the brittleness of this material during static stretching, the sample broke at 8.3 ± 0.5% relative elongation. As indicated in other works [25,38], the incorporation of a plasticizer into PLA is intended to increase the flexibility of the obtained materials. According to this statement, an increase in the elasticity of the LP sample, and a simultaneous decrease in the E and σ_m compared to the pure PLA film, were observed.

With regard to the results described by Qin [14] and Javidi [24], the susceptibility to deformation increased with the incorporation of essential oil into the polylactide matrix. The film with an addition of 20 wt. % tea tree essential oil, however, had a higher tensile strength compared to other materials modified only with the biocide without the plasticizer. In addition, the samples containing a biocidal substance showed the highest ε_b out of all of the obtained films based on polylactide. The plasticizing effect of essential oil is responsible for the increase in the flexibility of the PLA-based materials, which was described in the literature [14,24]. The addition of PEG to a PLA solution containing TTO allowed for the obtaining of materials with higher values of Young's modulus and tensile strength compared to the samples modified with one of the additives (PEG or TTO) (Figure 5).

3.5. Thickness and Transparency of Studied PLA-Based Materials

Table 2 depicts the values of thickness as well as transparency, calculated based on the transmittance measurements. In relation to the obtained results, it was established that an addition of essential oil increases the thickness of PLA-based films with respect to the pure PLA film. The same tendency was observed after introducing PEG. The molecules of essential oil and PEG can diffuse between polymer chains and increase the free volume between them, as indicated in the work of Anuar et al. [39]. For this reason, the thickness of the obtained materials was higher compared to a neat polylactide-based film.

Table 2. Thickness and transparency values of PLA-based materials.

Sample	Thickness (mm)	TV (mm^{-1})
L	0.084 ± 0.002	0.75 ± 0.07
LT05	0.087 ± 0.002	0.96 ± 0.02
LT10	0.096 ± 0.004	1.12 ± 0.15
LT20	0.112 ± 0.007	1.33 ± 0.21
LP	0.094 ± 0.003	0.81 ± 0.06
LPT05	0.099 ± 0.004	1.22 ± 0.12
LPT10	0.108 ± 0.006	1.49 ± 0.06
LPT20	0.118 ± 0.005	1.66 ± 0.05

The average thickness of the films was used in order to calculate the transparency values according to Equation (2). Observations indicated that an addition of 5% of PEG does not significantly influence the transparency of the LP sample. In the case of samples containing 20% poly(ethylene glycol), described in the work of Arfat [19], the transparency value is higher (TV = 2.63) in comparison to the obtained LP film (TV = 0.81). This indicates that the transparency of the PLA–PEG materials depends on the composition of the mentioned blends. The other fact which must be taken into account is that the incorporation of tea tree oil into the polylactide, as well as the PLA–PEG system, slightly influences the transparency of the obtained materials. The obtained results suggest that droplets of tea tree oil were evenly distributed in the PLA matrix. For this reason, the extent of the light scattering effect of essential oil droplets in the polymer matrix was not significant [14,19]. Moreover, in the publications of Qin et al. [14], it was established that transparency depends primarily on the types of essential oils.

3.6. Examination of Antibacterial Properties

The capacity to combat microorganisms possessed by the obtained PLA films was assessed based on tests carried out by means of the disk-diffusion method. Table 3 summarizes the size of the bacteria-free zones around samples, and the intensity of bacteria proliferation on the surfaces of the samples. According to guidelines included in the ISO 20645:2006 standard, the antibacterial effects of individual materials have been defined as the potential to inhibit the growth of bacteria under conditions favorable for their proliferation (Figures S2 and S3).

Table 3. Results of antibacterial activity of the examined PLA-based films.

Sample	Diameter of Inhibition Zones of Bacteria Growth (mm)		Bacteria Growth in Direct Contact with Sample		Evaluation of Antibacterial Effect [1]	
	S. aureus	E. coli	S. aureus	E. coli	S. aureus	E. coli
L	0	0	medium	medium	insufficient	insufficient
LP	0	0	medium	medium	insufficient	insufficient
LT05	0	0	weak	lack	limited	good
LT10	0	0	weak	lack	limited	good
LT20	0	0	weak	weak	limited	limited
LPT05	0	0	lack	lack	good	good
LPT10	0	0	lack	lack	good	good
LPT20	1	0	lack	lack	good	good

[1] in accordance with ISO 20645:2006 standard.

Similarly to the method applied in Ahmed's [11] and Arfat's [19] research, in relation to a film consisting of neat PLA and PLA with an addition of PEG, used as control samples, the inhibition of the microorganisms' growth in contact with these materials was not observed. With the increase in the content of tea tree essential oil in PLA films, the antibacterial capacity of samples increased. Nevertheless, in the case of these materials, only a partial inhibition of Gram-positive bacteria was observed. Moreover, the incorporation of 20% TTO into the polymer matrix led to obtaining films with limited antibacterial effects against *S. aureus* and *E. coli*. The reduced capability to combat bacteria displayed by the PLA-based film with an addition of 20% tea tree essential oil, in comparison to other films based on polylactide modified with TTO, results from the premature release of a part of the biocide from the polylactide matrix while the film was being dried. This is due to the coalescence and creaming which occur in the mixture of the PLA solution and TTO [8,18]. The above-mentioned phenomena were observed in SEM photographs (Figure 2) and AFM images (Figure 3).

The incorporation of 5% of the plasticizer into the PLA films with an addition of TTO contributed to the rise in antibacterial potential of the obtained materials. This beneficial change resulted from the effect PEG on PLA, which depends on the reduction of the strength of interactions between polylactide molecules and the increase in PLA-biocide interactions [40]. These materials had enough of the essential oil to inhibit the *S. aureus* and *E. coli* proliferation on their surface. In the case of the film with an addition of PEG and 20% TTO, however, compared to all other materials containing a plasticizer and tea tree essential oil, the area of inhibition of *S. aureus* growth was observed (Table 2). According to the ISO 20645:2006 standard, the incorporation of 20% tea tree essential oil into the PLA/PEG film led to an excessive release of biocidal substances from the polylactide matrix. This also confirms previous conclusions indicating that the processes characteristic of two-phase dispersion systems occurred. In addition, it should be emphasized that regardless of the amount of added tea tree essential oil, all the developed materials containing TTO and the plasticizer were characterized by a positive antibacterial effect against both Gram-positive and Gram-negative bacteria. Based on the obtained results, it was determined that the incorporation of 5% tea tree essential oil and 5% PEG into the PLA matrix allows for obtaining of a film with satisfactory antibacterial properties. The performed analyses also indicate that the addition of PEG into the PLA solution contributed to the increase in the mobility of the polymer chains, which had a direct effect on the release of essential oil from the studied materials [40].

4. Conclusions

The studies of films modified only with tea tree essential oil as well as TTO and a plasticizer have shown the beneficial effect of poly(ethylene glycol) on the properties of the obtained materials.

The incorporation of PEG into the PLA matrix, with an addition of tea tree oil, limited the extent of the flocculation and coalescence of the biocidal substance during the drying of the polylactide-based films. In addition, the presence of PEG in PLA samples, modified by means of TTO, led to the increase in the space between the polylactide chains, which contributed to a gradual release of volatile substances. Moreover, it should be stressed that the incorporation of TTO and PEG into PLA contributed to an increase in its resistance to deformation and a rise in its tensile strength values, as well as a significant improvement in the flexibility of the obtained materials, compared to the neat polylactide sample. Furthermore, it can be clearly noted that the incorporation of a plasticizer in the form of poly(ethylene glycol) into PLA, with the addition of TTO, allows one to obtain a material with satisfactory antibacterial properties. In summarizing, this method of modification leads to the formation of bactericidal films based on polylactide with better functional properties compared to those displayed by unmodified PLA samples.

Supplementary Materials: The following are available online at http://www.mdpi.com/1996-1944/13/21/4953/s1. Figure S1: DSC thermograms of studied PLA-based materials. Figure S2: Photos of Bacteria (*E. coli* and *S. aureus*) growth in direct contact with samples consisted of PLA and TTO. Figure S3: Photos of Bacteria (*E. coli* and *S. aureus*) growth in direct contact with samples consisted of PLA, PEG and TTO.

Author Contributions: Conceptualization, E.O.-K. and I.T.; formal analysis, E.O.-K., A.R. (Agnieszka Richert), M.G.; investigation, E.O.-K., I.T.; methodology, E.O.-K., I.T., A.R. (Agnieszka Richert); supervision, E.O.-K., M.G. and A.R. (Anna Rudawska); visualization, E.O.-K., I.T., M.G. and A.R. (Anna Rudawska); writing—original draft, E.O.-K. and I.T.; writing—review and editing, E.O.-K. and M.G. All authors have read and agreed to the published version of the manuscript.

Funding: This research was funded by Nicolaus Copernicus University in Toruń.

Conflicts of Interest: The authors declare no conflict of interest.

References

1. Shi, C.; Zhao, X.; Yan, H.; Meng, R.; Zhang, Y.; Li, W.; Liu, Z.; Guo, N. Effect of tea tree oil on Staphylococcus aureus growth and enterotoxin production. *Food Control* **2016**, *62*, 257–263. [CrossRef]
2. Ahmed, J.; Mulla, M.Z.; Arfat, Y.A. Thermo-mechanical, structural characterization and antibacterial performance of solvent casted polylactide/cinnamon oil composite films. *Food Control* **2016**, *69*, 196–204. [CrossRef]
3. Lee, C.J.; Chen, L.W.; Chen, L.G.; Chang, T.L.; Huang, C.W.; Huang, M.C.; Wang, C.C. Correlations of the components of tea tree oil with its antibacterial effects and skin irritation. *J. Food Drug Anal.* **2013**, *21*, 169–176. [CrossRef]
4. Kwieciński, J.; Eick, S.; Wójcik, K. Effects of tea tree (Melaleuca alternifolia) oil on Staphylococcus aureus in biofilms and stationary growth phase. *Int. J. Antimicrob. Agents* **2009**, *33*, 343–347. [CrossRef] [PubMed]
5. Halcón, L.; Milkus, K. Staphylococcus aureus and wounds: A review of tea tree oil as a promising antimicrobial. *Am. J. Infect. Control* **2004**, *32*, 402–408. [CrossRef] [PubMed]
6. Ahmed, J.; Hiremath, N.; Jacob, H. Antimicrobial, Rheological, and Thermal Properties of Plasticized Polylactide Films Incorporated with Essential Oils to Inhibit Staphylococcus aureus and Campylobacter jejuni. *J. Food Sci.* **2016**, *81*, E419–E429. [CrossRef]
7. Mulla, M.; Ahmed, J.; Al-Attar, H.; Castro-Aguirre, E.; Arfat, Y.A.; Auras, R. Antimicrobial efficacy of clove essential oil infused into chemically modified LLDPE film for chicken meat packaging. *Food Control* **2017**, *73*, 663–671. [CrossRef]
8. Sánchez-González, L.; González-Martínez, C.; Chiralt, A.; Cháfer, M. Physical and antimicrobial properties of chitosan-tea tree essential oil composite films. *J. Food Eng.* **2010**, *98*, 443–452. [CrossRef]
9. Ahmed, J.; Hiremath, N.; Jacob, H. Efficacy of antimicrobial properties of polylactide/cinnamon oil film with and without high-pressure treatment against Listeria monocytogenes and Salmonella typhimurium inoculated in chicken sample. *Food Packag. Shelf Life* **2016**, *10*, 72–78. [CrossRef]
10. Wabner, D.; Geier, K.; Hauck, D. For a deeper understanding of tea tree oil: Fresh is best—Why we should only use fresh oil at any concentration. *Int. J. Aromather.* **2006**, *16*, 109–115. [CrossRef]
11. Ahmed, J.; Arfat, Y.A.; Bher, A.; Mulla, M.; Jacob, H.; Auras, R. Active Chicken Meat Packaging Based on Polylactide Films and Bimetallic Ag–Cu Nanoparticles and Essential Oil. *J. Food Sci.* **2018**, *83*, 1299–1310. [CrossRef]
12. Ahmed, J.; Hiremath, N.; Jacob, H. Antimicrobial efficacies of essential oils/nanoparticles incorporated polylactide films against L. monocytogenes and S. typhimurium on contaminated cheese. *Int. J. Food Prop.* **2017**, *20*, 53–67. [CrossRef]

13. Ejaz, M.; Arfat, Y.A.; Mulla, M.; Ahmed, J. Zinc oxide nanorods/clove essential oil incorporated Type B gelatin composite films and its applicability for shrimp packaging. *Food Packag. Shelf Life* **2018**, *15*, 113–121. [CrossRef]
14. Qin, Y.; Li, W.; Liu, D.; Yuan, M.; Li, L. Development of active packaging film made from poly (lactic acid) incorporated essential oil. *Prog. Org. Coat.* **2017**, *103*, 76–82. [CrossRef]
15. Hammer, K.A.; Carson, C.F.; Riley, T.V.; Nielsen, J.B. A review of the toxicity of Melaleuca alternifolia (tea tree) oil. *Food Chem. Toxicol.* **2006**, *44*, 616–625. [CrossRef]
16. Gallart-Mateu, D.; Largo-Arango, C.D.; Larkman, T.; Garrigues, S.; De la Guardia, M. Fast authentication of tea tree oil through spectroscopy. *Talanta* **2018**, *189*, 404–410. [CrossRef]
17. Nairetti, D.; Mironescu, M.; Tita, O. Antimicrobial activity of active biodegradable starch films on pathogenic microorganisms. *Ann. Rom. Soc. Cell Biol.* **2014**, *19*, 75–80.
18. Sánchez-González, L.; Vargas, M.; González-Martínez, C.; Chiralt, A.; Cháfer, M. Characterization of edible films based on hydroxypropylmethylcellulose and tea tree essential oil. *Food Hydrocoll.* **2009**, *23*, 2102–2109. [CrossRef]
19. Arfat, Y.A.; Ahmed, J.; Ejaz, M.; Mullah, M. Polylactide/graphene oxide nanosheets/clove essential oil composite films for potential food packaging applications. *Int. J. Biol. Macromol.* **2018**, *107*, 194–203. [CrossRef] [PubMed]
20. Sosnowski, S. Poly(L-lactide) microspheres with controlled crystallinity. *Polym. Guildf* **2001**, *42*, 637–643. [CrossRef]
21. ISO. *ISO 527-1:2020 Plastics—Determination of Tensile Properties—Part 1: General Principles*; International Organization for Standardization: Geneva, Switzerland, 2020.
22. ISO. *ISO 527-3:2019 Plastics—Determination of Tensile Properties—Part 3: Test Conditions for Films and Sheets*; International Organization for Standardization: Geneva, Switzerland, 2019.
23. ISO. *ISO 20645:2006 Textile Fabrics—Determination of Antibacterial Activity: Agar Diffusion Plate Test*; International Organization for Standardization: Geneva, Switzerland, 2006.
24. Javidi, Z.; Hosseini, S.F.; Rezaei, M. Development of flexible bactericidal films based on poly(lactic acid) and essential oil and its effectiveness to reduce microbial growth of refrigerated rainbow trout. *LWT Food Sci. Technol.* **2016**, *72*, 251–260. [CrossRef]
25. Saha, D.; Samal, S.K.; Biswal, M.; Mohanty, S.; Nayak, S.K. Preparation and characterization of poly(lactic acid)/poly(ethylene oxide) blend film: Effects of poly(ethylene oxide) and poly(ethylene glycol) on the properties. *Polym. Int.* **2019**, *68*, 164–172. [CrossRef]
26. Holcapkova, P.; Hurajova, A.; Kucharczyk, P.; Bazant, P.; Plachy, T.; Miskolczi, N.; Sedlarik, V. Effect of polyethylene glycol plasticizer on long-term antibacterial activity and the release profile of bacteriocin nisin from polylactide blends. *Polym. Adv. Technol.* **2018**, *29*, 2253–2263. [CrossRef]
27. Fernández-Pan, I.; Royo, M.; Ignacio Maté, J. Antimicrobial Activity of Whey Protein Isolate Edible Films with Essential Oils against Food Spoilers and Foodborne Pathogens. *J. Food Sci.* **2012**, *77*, M383–M390. [CrossRef]
28. Olewnik-Kruszkowska, E.; Gierszewska, M.; Jakubowska, E.; Tarach, I.; Sedlarik, V.; Pummerova, M. Antibacterial Films Based on PVA and PVA–Chitosan Modified with Poly(Hexamethylene Guanidine). *Polym. Basel* **2019**, *11*, 2093. [CrossRef]
29. Marinello, F.; La Storia, A.; Mauriello, G.; Passeri, D. Atomic Force microscopy techniques to investigate activated food packaging materials. *Trends Food Sci. Technol.* **2019**, *87*, 84–93. [CrossRef]
30. Reyes-Chaparro, P.; Gutierrez-Mendez, N.; Salas-Muñoz, E.; Ayala-Soto, J.G.; Chavez-Flores, D.; Hernández-Ochoa, L. Effect of the Addition of Essential Oils and Functional Extracts of Clove on Physicochemical Properties of Chitosan-Based Films. *Int. J. Polym. Sci.* **2015**, *2015*, 714254. [CrossRef]
31. Jahed, E.; Khaledabad, M.A.; Almasi, H.; Hasanzadeh, R. Physicochemical properties of Carum copticum essential oil loaded chitosan films containing organic nanoreinforcements. *Carbohydr. Polym.* **2017**, *164*, 325–338. [CrossRef]
32. Atarés, L.; Bonilla, J.; Chiralt, A. Characterization of sodium caseinate-based edible films incorporated with cinnamon or ginger essential oils. *J. Food Eng.* **2010**, *100*, 678–687. [CrossRef]
33. Phaechamud, T.; Chitrattha, S. Pore formation mechanism of porous poly(DL-lactic acid) matrix membrane. *Mater. Sci. Eng. C* **2016**, *61*, 744–752. [CrossRef]
34. Rhim, J.W.; Mohanty, A.K.; Singh, S.P.; Ng, P.K.W. Effect of the processing methods on the performance of polylactide films: Thermocompression versus solvent casting. *J. Appl. Polym. Sci.* **2006**, *101*, 3736–3742. [CrossRef]
35. Byun, Y.; Whiteside, S.; Thomas, R.; Dharman, M.; Hughes, J.; Kim, Y.T. The effect of solvent mixture on the properties of solvent cast polylactic acid (PLA) film. *J. Appl. Polym. Sci.* **2012**, *124*, 3577–3582. [CrossRef]

36. Pielichowski, K.; Flejtuch, K. Differential scanning calorimetry studies on poly(ethylene glycol) with different molecular weights for thermal energy storage materials. *Polym. Adv. Technol.* **2002**, *13*, 690–696. [CrossRef]
37. Tábi, T.; Sajó, I.E.; Szabó, F.; Luyt, A.S.; Kovács, J.G. Crystalline structure of annealed polylactic acid and its relation to processing. *Express Polym. Lett.* **2010**, *4*, 659–668. [CrossRef]
38. Baiardo, M.; Frisoni, G.; Scandola, M.; Rimelen, M.; Lips, D.; Ruffieux, K.; Wintermantel, E. Thermal and mechanical properties of plasticized poly(L-lactic acid). *J. Appl. Polym. Sci.* **2003**, *90*, 1731–1738. [CrossRef]
39. Anuar, H.; Azlina, H.N.; Suzana, A.B.K.; Kaiser, M.R.; Bonnia, N.N.; Surip, S.N.; Razak, S.B.A. Effect of PEG on impact strength of PLA hybrid biocomposite. In Proceedings of the ISBEIA 2012—IEEE Symposium on Business, Engineering and Industrial Applications (ISBEIA), Bandung, Indonesia, 23–26 September 2012; pp. 473–476.
40. Holcapkova, P.; Hurajova, A.; Bazant, P.; Pummerova, M.; Sedlarik, V. Thermal stability of bacteriocin nisin in polylactide-based films. *Polym. Degrad. Stab.* **2018**, *158*, 31–39. [CrossRef]

Publisher's Note: MDPI stays neutral with regard to jurisdictional claims in published maps and institutional affiliations.

© 2020 by the authors. Licensee MDPI, Basel, Switzerland. This article is an open access article distributed under the terms and conditions of the Creative Commons Attribution (CC BY) license (http://creativecommons.org/licenses/by/4.0/).

Article

Polylactide Films with the Addition of Olive Leaf Extract—Physico-Chemical Characterization

Sylwia Grabska-Zielińska [1,*], Magdalena Gierszewska [1,*], Ewa Olewnik-Kruszkowska [1] and Mohamed Bouaziz [2]

1. Department of Physical Chemistry and Physicochemistry of Polymers, Faculty of Chemistry, Nicolaus Copernicus University in Toruń, Gagarin 7 Street, 87-100 Toruń, Poland; olewnik@umk.pl
2. Electrochemistry and Environmental Laboratory, National Engineering School of Sfax, University of Sfax, BP1173, Sfax 3038, Tunisia; mohamed.bouaziz@fsg.rnu.tn
* Correspondence: sylwia.gz@umk.pl (S.G.-Z.); mgd@umk.pl (M.G.)

Abstract: The aim of this work was to obtain and characterize polylactide films (PLA) with the addition of poly(ethylene glycol) (PEG) as a plasticizer and chloroformic olive leaf extract (OLE). The composition of OLE was characterized by LC-MS/MS techniques. The films with the potential for using in the food packaging industry were prepared using a solvent evaporation method. The total content of the phenolic compounds and DPPH radical scavenging assay of all the obtained materials have been tested. Attenuated Total Reflectance-Fourier Transform Infrared Spectroscopy (FTIR-ATR) allows for determining the molecular structure, while Scanning Electron Microscopy (SEM) indicated differences in the films' surface morphology. Among other crucial properties, mechanical properties, thickness, degree of crystallinity, water vapor permeation rate (WVPR), and color change have also been evaluated. The results showed that OLE contains numerous active substances, including phenolic compounds, and PLA/PEG/OLE films are characterized by improved antioxidant properties. The OLE addition into PLA/PEG increases the material crystallinity, while the WVPR values remain almost unaffected. From these studies, significant insight was gained into the possibility of the application of chloroform as a solvent for both olive leaf extraction and for the preparation of OLE, PLA, and PEG-containing film-forming solutions. Finally, evaporation of the solvent from OLE can be omitted.

Keywords: polylactide films; olive leaves; extract; food packaging; antioxidants

Citation: Grabska-Zielińska, S.; Gierszewska, M.; Olewnik-Kruszkowska, E.; Bouaziz, M. Polylactide Films with the Addition of Olive Leaf Extract—Physico-Chemical Characterization. *Materials* **2021**, *14*, 7623. https://doi.org/10.3390/ma14247623

Academic Editor: Krzysztof Moraczewski

Received: 21 November 2021
Accepted: 9 December 2021
Published: 11 December 2021

Publisher's Note: MDPI stays neutral with regard to jurisdictional claims in published maps and institutional affiliations.

Copyright: © 2021 by the authors. Licensee MDPI, Basel, Switzerland. This article is an open access article distributed under the terms and conditions of the Creative Commons Attribution (CC BY) license (https://creativecommons.org/licenses/by/4.0/).

1. Introduction

Currently, the world is flooded with packaging that is not environmentally friendly. Therefore, in the process of producing packaging, especially packaging intended for food, more attention is being paid to packaging based on biodegradable polymers. Such materials include polylactide, which is easily decomposed in the presence of microorganisms or water, and at the same time, remains stable during exploitation. In addition to the biodegradability of packaging, materials characterized by antioxidative properties are also sought after, which are capable of extending the shelf life of products.

Active packaging includes systems that absorb or emit substances [1,2]. The absorbers often eliminate unwanted substances such as oxygen, ethylene, excess moisture, and certain smells and flavors. In contrast, the emitters exude substances with an antimicrobial or antioxidative activity. It should be emphasized that active packaging is a system that modifies the composition of food or interacts with it. Recent studies have allowed for the addition of natural extracts that can act as antioxidants, flavor and odor absorbents, and enzymatic and antimicrobial agents [3,4]. Fat oxidation is widely recognized as one of the most important mechanisms leading to the deterioration of foods containing triglycerides. Lipid oxidation leads to a shortening of the shelf life of food due to the highly undesirable, unfavorable changes in taste and/or odor, and to the deterioration in texture and nutritional quality [5]. For this reason, certain natural substances with antioxidative properties are incorporated into the package matrix.

The most popular groups of antimicrobial and antioxidative compounds incorporated into polymeric materials include essential oils, plant-derived substances, and other organic compounds. Essential oils that are incorporated into the polymer matrix because of their antioxidative or bactericidal properties comprise cinnamon, rosemary oil and clove [6–9], eugenol, [10–12], curcumin oil [13], and tea tree oil or tea seed oil [14,15], as well as oregano oil [16] or *Melaleuca alternifolia* essential oil [14]. The introduction of essential oil into the polymer matrix, in most cases, improves elongation at the break of the obtained materials, reduces the water vapor permeation rate, and results in antibacterial or antioxidative properties.

The other popular group of compounds characterized by antioxidative properties are flavonoids, with quercetin, catechin, and its derivatives being the most widely applied [17–20]. The flavonoids mentioned above are known as valuable antioxidative and antimicrobial substances. Moreover, they can be used as natural stabilizers and color indicators [21].

Other natural alternatives include different plant extracts. In the work of Ji-Hee Kim et al. [22], the antioxidative activity of extracts originating from the coffee residue in raw and cooked meat was studied. Different extracts were obtained using *Limnophila aromatica*, which is commonly used as a spice and a medicinal herb [23]. The results indicated that *L. aromatica* reduces oxidative stress and can be used in dietary applications. In the work of Badakhshan Mahdi-Pour [24], the antioxidative activity of methanol extracts obtained from different parts of *Lantana camara* was studied. The antioxidative effect of coconut shell extract was proven in the work of Tanwar et al. [25], while Martínez et al. [26] investigated the antioxidative and antimicrobial activity of rosemary, pomegranate, and olive extracts in fish patties. Currently, scientists have started to thoroughly study polymeric materials containing plant extracts [27–30]. Active packaging films consisting of polymer and a plant extract have started to be extremely attractive. This phenomenon is related to the natural origination of extracts, their antibacterial and antioxidative properties, as well as their compatibility with the polymer matrix and the availability of plants—the source of the extracts [31]. The preparation and antioxidative activities of gelatin films incorporated with Ginkgo biloba extract were described in the work of Hu et al. [32]. García et al. [33] analyzed films made of poly(ε-caprolactone) with the addition of almond skin extract.

Olive leaf extract is likely to be a promising antibacterial and antioxidative additive. It is well known that olive leaf extract contains polyphenols, namely oleuropein, hydroxytyrosol, tyrosol, verbacoside, rutin, apigenin-7-glucoside, and luteolin-7-glucoside [34]. In most cases, active compounds present in the olive leaves are extracted with different solvents. The type of used solvent constitutes a crucial factor on which the efficiency of the extraction depends. In the case of olive leaves the effectiveness of the following solvents and its mixtures has been studied: water [30,35], methanol-water (4:1 v/v) as well as (80:20, v/v) mixtures [36,37], ethanol [38] chloroform-methanol (50/50 v/v) mixture [34,39], acetone, ethanol and their aqueous forms (10–90%, v/v) [40], petroleum ether, dichloromethane, methanol [41]. Taking into account the antibacterial and antioxidative properties of olive leaf extracts, some researchers introduced them into the polymer matrix to obtain active packaging films. The effectiveness of antimicrobial packaging based on the carrageenan filled with olive leaf extract has been studied in the work of Thamiris Renata Martiny [30]. It was established that the olive leaf extracted in water allowed to obtain polymeric films highly effective against *E. coli*. In the work of Erdohan [39], olive leaf extract obtained in the mixture chloroform-methanol (50/50 v/v) was introduced into the polylactide-based film containing glycerol as the plasticizer. However, it should be stressed that polymeric films were formed from chloroform-ethanol mixtures. Introduction of aqua olive leaf extract into the carrageenan matrix significantly decreases elastic modulus of the obtained materials and simultaneously increases the value of elongation at break. Moreover, the obtained results show that the incorporation of the OLE into the polymeric film leads to an increase in the WVPR [30]. In the work of Erdohan et al. [39], polylactide with an addition of glycerol as plasticizer and different olive leaf extracts was analyzed. Based on the obtained data, it was established that the ethanol concentration on the solvent mixture significantly affected the WVPR as well as the mechanical properties. Incorporation of the olive leaf extract into the gelatin matrix improves antimicrobial and antioxidant activities of the films, however, it also increases the WVPR of the obtained films [42]. The same tendency was observed in the work of Silveira da Rosa et al. [43]. Moreover, a decrease in the Young's Modulus was also observed. It is well known that polylactide is not soluble in methanol or ethanol. The introduction of PLA and additives into the mixture of chloroform and methanol leads to the formation of an inhomogeneous structure. For this reason, the objective of our study was to develop PLA-based films with an addition of PEG as a plasticizer and olive leaf extract obtained in chloroform. The proposed method allows omitting the one step in the film-forming procedure, i.e., evaporation of solvent from OLE. The present study analyses the structural (FTIR and SEM), mechanical, and antioxidative properties of olive leaf extract incorporated PLA-based packaging materials. Moreover, the water vapor permeation rate and changes in colour were evaluated.

2. Materials and Methods

Acetone and chloroform were purchased from STANLAB (Lublin, Poland). Polylactide (type 2002D, average molecular weight = 79 kDa) was delivered in the form of pellets by Nature Works® (Minnetonka, MN, USA). Poly(ethylene glycol) (M_w = 1500 g·mol^{-1}) was purchased from Sigma-Aldrich (Steinheim, Germany).

2.1. Phenolic Analysis of Olive Leaf Extract

LC-MS/MS analysis was used to examine the olive leaf extract composition.

Olive leaves of the *O. europaea* tree (cultivar Chemlali) were collected from Sfax (Tunisia) in July 2019 and dried for 3 min in a JN-100 microwave dryer (Adasen, Jinan, China) (1200 W). The dry leaves were milled and stored at 4 °C. The extract was prepared from 2.5 wt.% olive leaf powder dispersion in chloroform, and was stirred for 1 h at 30 °C. The obtained solution was filtered using filter paper (Filtres RS, 8–11 mm, Madrid, Spain), and the residual solvent was evaporated at 40 °C under a vacuum. The obtained extract was freeze-dried and stored at 18 °C for the phenolic analysis.

The LC-UV-MS/MS analyses of the obtained extract were performed according to the method described in detail elsewhere [44].

2.2. Fabrication of Films

Thin films were obtained using the solvent evaporation method. First, the leaves from the olive tree (Tunisia) were harvested and dried. Then, they were ground in an electric pulverizer (IKA Werke GmbH and Co. KG, Staufen, Germany). An appropriate amount of olive leaf dust was poured with 50 cm^3 of chloroform, and the resulting mixture was mixed with a magnetic stirrer (Wigo, Pruszków, Poland) for 3 h and filtered. Then, 1, 3, and 5 wt.% extracts were prepared. Then, 1.5 g of polylactide was added to the olive leaf chloroformic extract and mixed 3 h with a magnetic stirrer. Then, 5 wt.% of poly(ethylene glycol), relative to polylactide, was added to the obtained solution and mixed with a magnetic stirrer to reach a homogenous mixture. Poly(ethylene glycol) was used as a plasticizer. The final solution was poured onto a glass petri dish with a known constant diameter and was evaporated at room temperature. The obtained films samples were dried 24 h at room temperature. Polylactide with a poly(ethylene glycol) film was used as a control sample.

2.3. The Total Content of Phenolic Compounds

The total content of phenolic compounds was evaluated using the Folin–Ciocâlteu method, with gallic acid as the standard, according to the Malik and Bradford procedure, with slight modifications [45]. The Folin–Ciocâlteu reagent was used as an oxidizing agent. The samples were immersed in a mixture of 0.5 mL of Folin–Ciocalteu reagent, 1 mL of

Na₂CO₃, and 8.5 mL of distilled water. The samples were incubated at 40 °C in the dark for 30 min. Next, the absorbance of the samples was measured spectrophotometrically at 725 nm using a UV-1800 spectrophotometer (UV-1800, Shimadzu, Reinach, Switzerland). The experiment was run in triplicate, and the results were expressed as a gallic acid equivalent using a five-point calibration curve as mg/mL.

2.4. DPPH Radical Scavenging Assay

The free radical scavenging activity of the PLA-based samples was measured in vitro using a DPPH reagent (2,2′-diphenyl-1-picrylhydrazyl, free radical, 95%; Alfa Aesar, Karlsruhe, Germany) [46]. First, 250 µM methanolic solution of DPPH was prepared. Next, small pieces (1 × 1 cm²) of films were prepared and placed in a 12-well plate. Then, 2 mL of a DPPH solution was added and the samples were kept for 60 min in the dark. As a control sample a DPPH solution left on the plate without contact with the films was used. After incubation, the absorbance was measured spectrophotometrically at 517 nm (UV-1800, Shimadzu, Reinach, Switzerland). All of the experiments were replicated five times. The antioxidant activity (RSA) was calculated using the following formula:

$$RSA = \frac{A_0 - A_{PB}}{A_0} \cdot 100\%, \quad (1)$$

where A_0 is the average absorbance of the DPPH solution without contact with the films, and A_{PB} is the average absorbance of the DPPH solution after contact with the film being tested.

2.5. FTIR-ATR Spectroscopy

Fourier transform infrared spectroscopy with attenuated total reflectance (FTIR-ATR) was used to evaluate the chemical structure of the obtained films. A Nicolet iS10 spectrophotometer (Nicolet iS10, ThermoFisherScientific, Waltham, MA, USA) with a germanium crystal was used for spectra recording. All of the spectra were recorded with the 64 scans and a resolution of 4 cm⁻¹, and were evaluated in the range of 400–4000 cm⁻¹.

2.6. Differential Scanning Calorimetry

Thermal analyses of all PLA-based films were performed using DSC equipment (DSC 204 F1 Phoenix, Netzsch, Germany) in the nitrogen atmosphere (gas flow 20 mL/min). Samples of ca. 3–5 mg were sealed in standard aluminum pans, then heated with a heating rate of 10 °C/min from 20 to 180 °C, cooled to 30 °C, equilibrated for 5 min at 30°, and heated again to 180 °C. The recorded data were analyzed with NETZSCH Proteus software (Version 6.1.0), and the degree of crystallinity (X_C) was calculated with the following equation [47]:

$$X_C = \frac{\Delta H_F}{\Delta H^o \cdot X_{PLA}} \cdot 100, \quad (2)$$

where ΔH_F is the heat of fusion of the analyzed sample representing the difference between PLA melting enthalpy (ΔH_m) and enthalpy of cold crystallization (ΔH_{cc}), ΔH^o is the heat of fusion of a fully (100%) crystalline PLA (ΔH^o = 109 J/g [48]), and X_{PLA} is the mass fraction of the PLA in the film.

2.7. Mechanical Properties and Thickness

The mechanical properties were analyzed at room temperature in the dry state using Zwick&Roell 0.5 testing machine (Zwick&Roell Group, Ulm, Germany). Mechanical properties were studied at a crosshead speed of 5 mm/min. The samples were cut using the same bone-shaper (50 mm length, 10 mm width).

The thickness of the samples was measured with an ultrameter type A-91 (Manufacture of Electronic Devices, Warsaw, Poland). For each kind of film, at least five samples were tested.

2.8. Water Vapor Permeation Rate (WVPR)

The water vapor permeation rate (WVPR) of PLA-based films was investigated. Three independent tests were performed for each film. First of all, the fresh desiccant was prepared ($CaCl_2$ dried at 110 °C for 24 h). Plastic containers of 24 mm in diameter containing a determined amount of desiccant were prepared, with the top covered tightly with the tested films. The containers with $CaCl_2$ but without covers were used as the control samples. The containers were placed in an oven at 30 °C with 75% relative humidity. After each 24 h to 7 days, the moisture penetration was determined based on the changes in the desiccant weight. The slopes of the steady-state (linear) portion of weight loss versus time curves were used to calculate the WVPR, according to the following equation:

$$\text{WVPR} = \frac{W}{A \cdot t} \left[g \cdot m^{-2} \cdot h^{-1} \right], \tag{3}$$

where W is the weight gained [g], A is the area of the film cover [m^2], and t is time [h].

2.9. Colour Change

The influence of OLE incorporation and the swelling process in the media of different pH on the color of the obtained films was studied using the MICRO-COLOR II LCM 6 (Dr. Lange, Berlin, Germany) colorimeter. The as-obtained films were immersed in 0.1 M HCL, 0.1 M NaOH, and distilled water for 7 days [46]. Then, films were taken out from the liquid and allowed to dry at room conditions (temperature and humidity). The color parameters of the dry as-prepared (before immersion) and the post-swelled (after immersion) films were measured. The CIE $L^*a^*b^*$ system was applied to calculate the color difference (ΔE) of the materials. To establish the total values of color difference (ΔE) and color intensity (C), the following formulas were used:

$$\Delta E = \sqrt{(L - L^*)^2 + (a - a^*)^2 + (b - b^*)^2}, \tag{4}$$

$$C = \sqrt{(a)^2 + (b)^2}, \tag{5}$$

where:
L—the component describing lightness,
a—represents the color ranging from green (−a) to red (+a),
b—represents the color ranging from blue (−b) to yellow (+b),
L^*, a^*, b^*—values characteristic for the control film (PLA/PEG), L^* (lightness), a^* (redness/greenness), and b^* (yellowness/blueness).

Yellowness index (YI) was calculated using the following equation [49]:

$$YI = \frac{142.86 \cdot b}{L}, \tag{6}$$

Each sample was analyzed five times and the mean values were calculated.

2.10. Morphology Observations—Scanning Electron Microscopy

Scanning Electron Microscopy was used to observe the surface and cross-section of the obtained films. Before each analysis, the surfaces of the studied films were sputtered with an Au thin layer. Photographs were taken using the Quanta 3D FEG scanning electron microscope (SEM, FEI Company, Hillsboro, OR, USA).

2.11. Statistical Analysis

The obtained data were statistically analysed using commercial software (GraphPad Prism 8.0.1.244, GraphPad Software, San Diego, CA, USA). The results, presented as a mean ± standard deviation (SD), were compared using one-way analysis of variance (one-way ANOVA). Multiple comparisons between the means were performed with the statistical significance set at $p \leq 0.05$. The results from the total content of phenolic compounds, the free radical scavenging activity, the mechanical properties, and thickness and water vapor permeation rate tests were subjected to statistical analysis.

3. Results and Discussion

3.1. The Total Content of Phenolic Compounds and DPPH Radical Scavenging Assay

There are several studies devoted to the extraction of olive waste and to the determination of the phenolic compound composition for the obtained extracts [40,50,51]. The analysis of these results indicate clearly that the presence of different phenolic compounds in olive leaf extract (OLE) depends on several factors [40,50–52]: olive tree cultivation, season, time of extraction, and solvent used. Thus, it is crucial to evaluate the composition of the particularly obtained OLE. Based on the LC-MS/MS analysis, the olive leaf extract composition was determined and is listed in Table 1.

Table 1. Phenolic compounds identified by LC-MS/MS using the negative ionization mode of Chemlali olive leaf cultivar extract, including LC-MS/MS parameters and MRM transitions.

N°	TR	M	Exp.[a] m/z for [M-H]$^-$	Chemical Formula	λ_{max} [nm]	Main Fragments via MS/MS	Proposed Compound
1	7.5	170.0215	169.0145	$C_7H_6O_5$	230; 270	125.0243; 124.0168; 97.0294; 73.0190	Gallic acid
2	11.8	154.063	153.0562	$C_8H_{10}O_3$	230; 278	123.0453; 109.0291	Hydroxytyrosol
3	18.0	180.0425	179.0351	$C_9H_8O_4$	295; 324	135.0450; 134.0370; 89.0339	Caffeic acid
4	19.5	640.2003	639.1933	$C_{29}H_{36}O_{16}$	230; 280; 324	621.1823; 179.0351;161.0247	β-hydroxyverbascoside III
5	20.38	610.1534	609.1471	$C_{27}H_{30}O_{16}$	357	463.0883; 301.0352; 300.0276; 178.9986; 151.0033	Quercetin 3-O-rutinoside (rutin)
6	21.1	624.2054	623.1986	$C_{29}H_{36}O_{15}$	234; 276	461.1652; 315.1071; 179.0349; 161.0242	Verbascoside
7	21.29	448.1006	447.0938	$C_{21}H_{20}O_{11}$	346	285.0393; 284.0317; 197.0611; 175.0397; 133.0288	Luteolin 7-O-glucoside
8	21.8	702.2371	701.2294	$C_{31}H_{42}O_{18}$	232; 282	539.1762; 437.1430; 377.1238; 345.0972; 307.0832; 275.0561; 223.0610; 179.0561; 149.0250; 139.0389; 113.0251	Oleuropein hexoside I
9	22.0	624.2054	623.1986	$C_{29}H_{36}O_{15}$	234; 278	461.1666; 315.1054; 179.0345; 161.0244	Isoverbascoside
10	23.0	432.1056	431.098	$C_{21}H_{20}O_{10}$	337	269.0453; 268.0376; 117.0348	Apigenin 7-O-glucoside
11	24.2	540.1843	539.1771	$C_{25}H_{32}O_{13}$	230; 280	403.1236; 377.1241; 371.0979; 327.0880; 307.0822; 275.0920; 223.0610; 179.0703; 165.0560; 149.0245; 139.0401	Oleuropein I
12	24.7	540.1843	539.1774	$C_{25}H_{32}O_{13}$	230; 280	403.1262; 377.1263; 371.1002; 327.0915; 307.0842; 275.0578; 223.0629; 179.0703; 165.0573; 149.0260; 139.0412	Oleuropein II

Table 1. Cont.

N°	TR	M	Exp. [a] m/z for [M-H]−	Chemical Formula	λ_{max} [nm]	Main Fragments via MS/MS	Proposed Compound
13	25.7	558.2309	557.224	$C_{26}H_{38}O_{13}$	280	513.2345; 345.1194; 327.1088; 227.1290; 185.1185; 183.0662; 121.0661	6′-O-[(2E)-2,6-dimethyl-8-hydroxy-2-octenoyloxy]-secologanoside
14	26.3	926.3056	925.2991	$C_{42}H_{54}O_{23}$	240; 284	893.2633; 763.2456; 745.2360; 693.2030; 539.1765; 521.1655; 377.1239; 307.0823	Jaspolyoside III

TR—time retention [min]; M—molar mass [g/mol]; λ_{max}—wavelength of maximum absorbance; m/z—the mass-to-charge ratio of the ion.

As can be seen, several phenolic compounds were detected in the extract. Oleuropein is the major phenolic compound in olive leaves. Its content varies from 17% to 23% depending upon the harvesting time of the leaves [53]. Among others substances present in chloroformic olive leaf extract, gallic acid, hydroxytyrosol, caffeic acid, rutin, verbascoside, luteolin 7-O-glucoside, oleuropein hexoside I, isoverbascoside, apigenin 7-O-glucoside, 6′-O-[(2E)-2,6-dimethyl-8-hydroxy-2-octenoyloxy]-secologanoside, and jaspolyoside III, were also found. All of the compounds given in Table 1 were previously reported in olive leaf extracts [54,55].

The total content of the phenolic compounds and the free radical scavenging activity of the obtained films with the olive leaf active substances extracted are shown in Table 2. It is well known that phenolic compounds can be found in all parts of the olive plant [56]. It is frequently reported that oleuropein is the most prominent and predominant phenolic compound in olive cultivars and can reach a concentration of 60–90 mg/g in dry leaves [57,58]. The quantitative analysis revealed the presence of phenolic compounds in each type of film, excluding the control film without olive leaf extract. As the concentration of olive leaf extract increases, the phenolic compound content increases, but no significant statistical difference ($p \leq 0.05$) was observed between PLA/PEG/1OLE and PLA/PEG/3OLE. The film with the highest content of OLE (PLA/PEG/5OLE) was also characterized by the highest phenolic compound concentration (0.637 ± 0.061 mg/mL). This trend is in line with expectations and with the results of other scientists. Da Rosa et al. [44] prepared biodegradable carrageenan films containing olive leaf extract by the evaporation method. They noticed a higher amount of phenolic compounds with the increasing amount of olive leaf extract. Albertos et al. [43] studied olive leaf-gelatin film properties and evaluated that the total soluble phenolic compounds content was higher for materials with a higher extract content.

Table 2. The total content of phenolic compounds (TP) and the free radical scavenging activity (RSA) of PLA films.

Specimen	TP [mg/mL]	RSA [%]
PLA/PEG	-	22.30 ± 2.06
PLA/PEG/1OLE	0.349 ± 0.071	56.26 ± 0.70 *
PLA/PEG/3OLE	0.498 ± 0.106	56.53 ± 1.35 *
PLA/PEG/5OLE	0.637 ± 0.061 [a]	56.90 ± 3.16 *[#]

[a]—Statistically significant difference vs. PLA/PEG/1OLE; *—Statistically significant difference vs. PLA/PEG (control); [#]—Statistically significant difference vs. PLA/PEG/1OLE.

Taking into consideration the free-radical scavenging (RSA) activity, it can be seen that polylactide film with the addition of olive leaf extract was characterized by a higher RSA than the non-modified polylactide film. These results prove that the obtained films with the addition of olive leaf extract exhibit improved antioxidant properties. However, it can also be observed that there were no statistically significant differences between the films with different contents of OLE (Table 2). Many scientists reported that as the active substance or extract concentration increases, the antioxidant activity also increases. Tymczewska et al. [59] reported the same trend—a non-modified film based on gelatin was characterized by lower antioxidant activity than the films modified by various contents of rapeseed meal extracts. Roy and Rhim [60] also noticed that the antioxidant activity increased with the increased concentration of the modifier (in this case, curcumin) in the material. A similar relationship was shown by Dou et al. [61] where gelatin/sodium alginate materials with the addition of various contents of tea tree polyphenols were tested. It can be assumed that free radical scavenging can occur (a) on the film surface and (b) in the solution. The (b) option is strongly dependent on the material swelling and the possible release of active substances. As PLA is not soluble in methanol, PLA-film swelling in this medium is very low and reduces the possibility of OLE components migration into the external solution during RSA testing. Due to this, it can be stated that free radical scavenging of PLA/PEG/OLE films is primarily an effect of the surface composition. It should be stressed out that there are minor differences in the total OLE content between PLA/PEG/OLE films prepared in this study. Moreover, as the solvent evaporation method was used, there can be slight differences in the composition of the film surface and the inner parts. All of the reasons mentioned above could be responsible for the minor differences in RSA observed for PLA/PEG/OLE films.

3.2. FTIR-ATR Spectroscopy

To evaluate the molecular structure of PLA/PEG and PLA/PEG/OLE films and to confirm the possible interactions between PLA, PEG, and OLE, FTIR-ATR analysis was performed (Figure 1). The molecular structure of the pristine PLA film was already analyzed with the FTIR technique and the results described elsewhere [62], confirming the presence of most characteristic functionals by vibrational bands at: 2999 cm^{-1} and 2945 cm^{-1} (ν_{as} and ν_s C-H in -CH$_3$), 1763 cm^{-1} (ν_s -C=O), 1451 cm^{-1} (bending asymmetrical C-H in -CH$_3$) 1207 and 1127 cm^{-1} (ν_s -C-O-C-), 873 cm^{-1} (ν -C-COO), and 756 cm^{-1} (deformation vibration CO). In the FTIR spectra of PLA/PEG (Figure 1a), almost the same absorption bands as in neat PLA were noticed. The resemblance of PLA and PLA/PEG spectra results from the structural similarity of PLA and PEG. However, as already proven by Yuniarto and coworkers [63], there are slight differences between intensity and bands position between PLA and PLA-PEG films. These confirm that the poly(ethylene glycol) (PEG) hydrophilic group interacts with the polar groups (e.g., C=O) of PLA, resulting in hydrogen bonding.

The spectrum of OLE (Figure 1b) revealed several bands characteristic for the presence of different active compounds listed in Table 1. Most of the OLE components possess characteristic functionals in their structure. Their occurrence was confirmed through the vibrational bands at 3292 and 1732 cm^{-1} (stretching nodes: N-H in amine, O-H in phenols, C=O in carboxylic acid and C-C in alkenes), 1155 cm^{-1} (stretching vibration: C OH in the protein or C-N in the amine), 2918 cm^{-1} (C-H stretching), 1614 cm^{-1} (indicate the fingerprint region of CO, C-O and O-H, C-O in amide I), and 1070 cm^{-1} (C-N stretching in amine). With a view of the differences in the composition of the OLE extracts connected to the olive leaf sources and processing, the recorded OLE spectra (Figure 1b) remained in agreement with others presented in the literature [64–66].

The comparison of the PLA/PEG, OLE, and PLA/PEG/5OLE spectra indicates the characteristic bands of both PLA/PEG and OLE in the final OLE extract-based polymeric films (Figure 1c); however, the most intense bands correspond to PLA/PEG. This can be due to the relatively low amount of OLE in the final film. In particular, the preparation of PLA/PEG films with the usage of OLE extract resulted in a new adsorption band at 1686 cm^{-1}. It can be assigned to the ν O-H of the OLE components; however, its position is visibly shifted in comparison to the spectra of pure OLE (i.e., 1614 cm^{-1}). Similarly, in the 2800–3100 cm^{-1} region, three bands were noticed: at 2994 cm^{-1} (corresponding to the asymmetric stretching of C-H in -CH$_3$ in PLA), 2918, and 2849 cm^{-1} (characteristic for OLE components). These observations allow for the suggestion of possible interactions between the chemicals present in the OLE and PLA/PEG matrix.

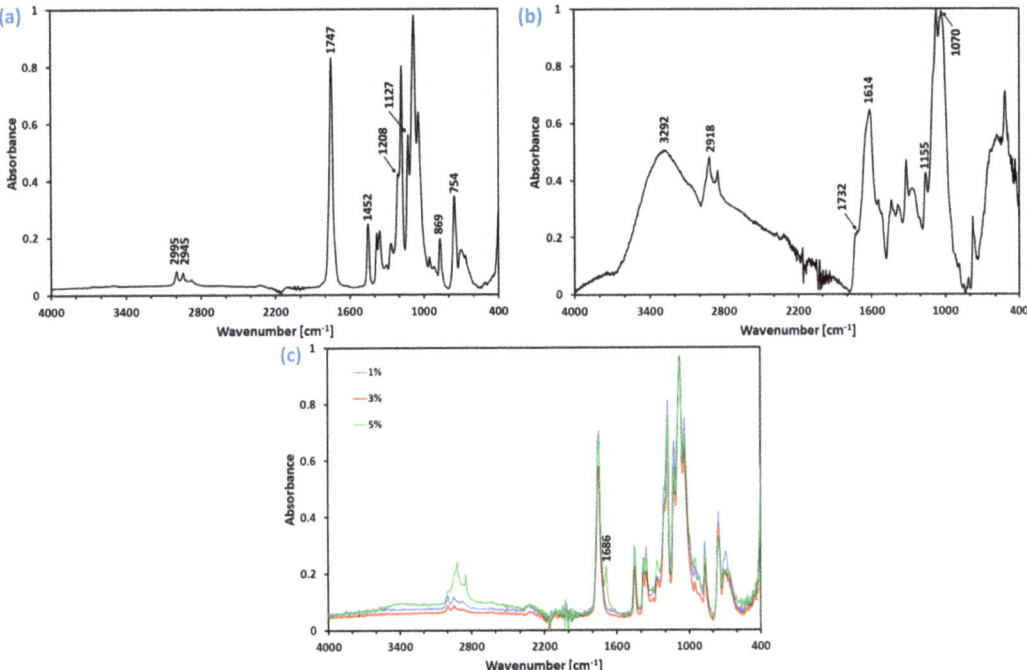

Figure 1. The FTIR-ATR spectra of the control samples (PLA/PEG (**a**) and OLE (**b**)) and polylactide films with the addition of olive leaf extract (**c**).

3.3. Mechanical Properties and Thickness

A tensile test coupled with thickness measurements was done to estimate the effect of OLE addition on the mechanical properties of PLA/PEG polymeric films. Stress–strain curves were used to evaluate Young's modulus (E_{mod}), elongation at break (Maximum deformation), and maximum force at break (F_{max}) (Figure 2).

It was proven several times in the literature [67–69], and also by us, that the addition of a PEG plasticizer into the PLA matrix causes an increase in the free volume inside the polymeric film, which in turn facilitates the movement of the polymeric chains. As already indicated in the literature [70], PEG affects the characteristics of PLA by disturbing the intermolecular forces. Thus, neat PLA/PEG films exhibit relatively high elongation at break and a maximum force at break (F_{max}) compared to pure PLA. As shown in Figure 2, an introduction of OL extract into the PLA/PEG matrix causes a reduction of maximum deformation, Young's modulus (E_{mod}), and F_{max} values. Moreover, the maximum deformation and E_{mod} depend on the amount of OLE, while F_{max} is almost independent of the OLE concentration. In this context, it can be stated that the incorporation of OLE slightly worsened the mechanical properties of PLA/PEG films.

In general, the literature indicates that the addition of natural essential oils causes an increase in the mechanical properties of pure PLA-based films [71]. Different observations have been made for PEG plasticized PLA materials. Tarach et al. [69] noticed an increase in maximum deformation after incorporating tea tree essential oil into PEG/PLA-based films. Similarly, Chieng et al. [72] registered maximum deformation values for poly(lactic acid) plasticized with poly(ethylene glycol) and epoxidized palm oil in comparison with neat poly(lactic acid) films. In turn, Vasile and coworkers [73] showed a decrease in E_{mod} after the addition of a low amount of rosemary extract into the PLA/PEG (80/20 w/w), while F_{max} was almost unaffected. This suggests that when OLE is added into an already plasticized PLA-matrix, both components (OLE and plasticizer) can act antagonistically on the polymer chain movement ability, and stay in agreement with the almost invisible changes in the FTIR spectra of PLA/PEG after OLE addition.

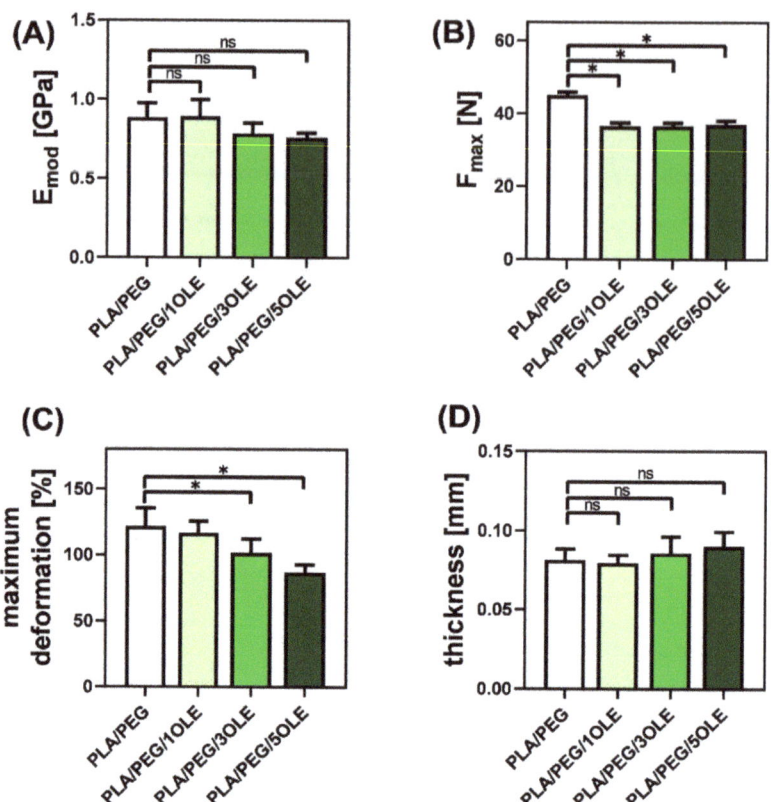

Figure 2. The mechanical properties: (**A**) Young modulus, (**B**) maximum force at break, (**C**) maximum deformation, and (**D**) thickness of the obtained films; *—statistically significant difference vs. control sample (PLA/PEG); ns—no statistically significant difference vs. control sample (PLA/PEG).

It is known that the mechanical properties of PLA-based materials are affected by different polymer properties, as well as by crystallinity [74,75]. In general, it is assumed that the greater the crystallinity, the harder the polymer, as crystallinity reduces the degree of freedom for the molecular chains to move. As we discussed earlier [69], the presence of chloroform within the PLA-based materials provides suitable conditions for the crystallization of the polymer. Moreover, based on the results presented in Table 3, it can be seen that the addition of OLE increases crystallinity, thus causing a reduction of maximum deformation of PLA/PEG/OLE materials. Based on the differences in crystallinity, which slightly changed after OLE incorporation, it can be assumed that OLE components act as a nucleus of crystallization, and finally reduce the maximum deformation. The current findings remain in agreement with the data reported by Malgorzata Latos-Brozio and Anna Masek [76], who proved that plant polyphenols can act as nucleating substances, increasing the crystallinity of PLA, and with the results of Javidi et al. [77], who stated that the essential oil facilitates the process of PLA crystallization.

Table 3. Water Vapor Permeation Rate (WVPR) and the degree of crystallinity (X_C) of PLA/PEG and OLE containing PLA/PEG films.

Specimen	WVPR [g·m^{-2}·h^{-1}]	X_C [%]
Control	42.75 ± 0.48	-
PLA/PEG	10.49 ± 0.81 [a]	0.72
PLA/PEG/1OLE	10.26 ± 0.65 [ab]	1.55
PLA/PEG/3OLE	9.07 ± 0.33 [ab]	1.76
PLA/PEG/5OLE	11.01 ± 1.61 [ab]	1.61

[a]—statistically significant difference vs. control sample; [b]—no statistically significant difference vs. PLA/PEG.

3.4. Water Vapor Permeation Rate (WVPR)

WVTR, according to the definition, is the steady rate at which H$_2$O vapor permeates through a barrier at a specific temperature and relative humidity (RH) in a given time. It is a crucial parameter in evaluating the food-package potential, as water affects the product's quality and shelf-life. Depending on the type of packaged food, WVPR values are expected to be relatively low or high, so there is no "good" value or range for WVPR.

Water barrier properties of PLA/PEG and OLE-containing PLA/PEG films were characterized by calculating the water vapor permeation rate (WVPR) values (Table 3). As can be seen, the application of PEG-plasticized polylactide film substantially reduces the water transport from the environment (of relative humidity RH = 75%) into the applied desiccant. However, the further addition of different amounts of OLE does not significantly affect the WVPR values.

As we already discussed [78], several factors affect the transport properties of polymeric films. The WVPR value depends on both the environmental factors (e.g., temperature, relative humidity, the difference in pressure, or concentration gradient across the film) and the material molecular and supramolecular characteristics (e.g., thickness and area, nature and M_w of the polymer, type, and content of additives).

The literature [79–81] indicates that, in general, neat PLA films are medium-to-low barrier materials to water vapor. The barrier properties of PLA-based films are highly influenced by their polymer chain orientation and packing, defined through crystallinity, crystal thickness, and morphology [82,83]. As indicated by Drieskens et al. [83], the crystallization of PLA has a positive effect on the barrier properties towards water and oxygen. The decrease of permeability in the case of a more crystalline polymer structure was also already described well by the Maxwell equation [84]. Generally, the reduction of gas and water vapor transport by crystallization can be explained as [83] (a) decrease in the amorphous phase that, opposite to the crystalline phase, is permeable to gases, and (b) increase in the tortuosity. In this context, according to the evaluated degree of crystallinity (Table 3), which increases with the addition and the content of OLE, a reduction of WVPR should be observed.

The reason for such unexpected changes in WVPR values, which do not vary with X_C, is probably connected with the nature of the additive (olive leaf extract). The primary phenolic compound found in this extract is oleuropein. For its extraction, simultaneously with other phenolic compounds, organic solvents of a higher polarity (e.g., methanol and ethanol) and water are proposed. It can be assumed that those components exhibit a higher affinity toward more polar solvents. Thus, the addition of OLE into the PLA/PEG matrix can cause an increase in water vapor diffusion. To summarize, it can be stated that an increase in the degree of crystallinity and the addition of OLE act antagonistically to the WVPR value. Finally, no substantial differences in WVPR were observed in the case of PLA/PEG and PLA/PEG/OLE films.

A similar effect of OLE addition on water transport properties was observed by Martiny et al. [30] for carrageenan/glycerol films. It should be added that researchers used a significant amount (62.5 wt.%) of OLE based on carrageenan mass. Contrary, García et al. [85] presented a reduction in water vapor permeation with the increasing amount of olive extract (OLE) in corn starch/glycerol films.

It is worth noting that in the above-discussed context, the OLE addition neither improved nor worsened the barrier properties of the PLA material toward the water.

3.5. Colour Change

In Table 4, CIEL*a*b* L, a, and b color parameters, as well as total color difference (ΔE) of PEG-PLA and OLE-doped PEG-PLA as-prepared (dry) and post-swelled films, are given. Moreover, the calculated color intensity and yellowness index (YI) are presented in Figure 3.

Table 4. Color (L, a, b, and ΔE) parameters of PLA-based films.

Colour Parameter	Conditions	PLA/PEG	PLA/PEG/1OLE	PLA/PEG/3OLE	PLA/PEG/5OLE
	Dry				
L	Dry	86.46 ± 0.05	66.64 ± 1.14	62.78 ± 0.90	50.28 ± 0.60
	HCl *	89.68 ± 0.12	72.60 ± 0.27	75.24 ± 1.14	58.70 ± 0.39
	Water *	89.42 ± 0.16	79.56 ± 0.08	67.52 ± 0.50	61.48 ± 0.84
	NaOH *	89.32 ± 0.07	71.34 ± 0.45	72.26 ± 0.79	70.02 ± 0.60
a	Dry	0.76 ± 0.05	−4.06 ± 0.08	−0.30 ± 0.13	5.34 ± 0.10
	HCl *	0.74 ± 0.10	−4.26 ± 0.08	−4.04 ± 0.08	−0.70 ± 0.17
	Water *	0.82 ± 0.04	−3.34 ± 0.05	−2.80 ± 0.14	−2.72 ± 0.19
	NaOH *	1.12 ± 0.04	−2.62 ± 0.04	−3.12 ± 0.15	−1.02 ± 0.04
b	Dry	−12.68 ± 0.10	29.08 ± 0.85	30.42 ± 0.16	28.64 ± 0.63
	HCl *	−13.36 ± 0.05	17.12 ± 0.29	15.34 ± 1.03	19.68 ± 0.18
	Water *	−13.28 ± 0.07	10.98 ± 0.07	20.84 ± 0.05	20.80 ± 0.15
	NaOH *	−13.88 ± 0.07	19.64 ± 0.54	22.78 ± 0.22	20.00 ± 1.07
ΔE	Dry	-	46.50 ± 0.32	49.20 ± 0.30	55.12 ± 0.23
	HCl *	-	35.30 ± 0.19	32.51 ± 0.52	45.32 ± 0.20
	Water *	-	26.52 ± 0.04	40.71 ± 0.25	44.22 ± 0.43
	NaOH *	-	38.23 ± 0.26	40.66 ± 0.19	39.06 ± 0.63

*—after immersing in HCl/Water/NaOH media.

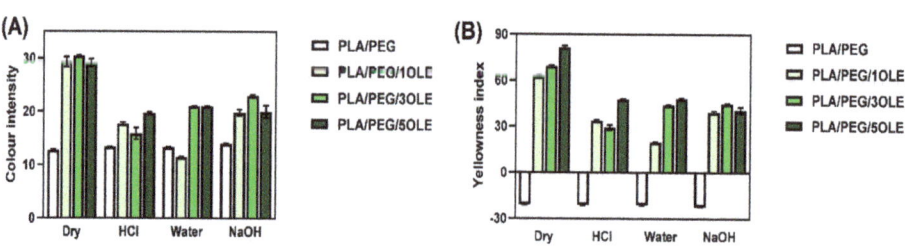

Figure 3. Color intensity (**A**) and Yellowness index (**B**) of PLA/PEG and PLA/PEG/OLE films.

All of the films were found to be visually homogeneous, slightly opaque, with a shade of blue or green. As shown (Table 4), the dry PLA/PEG film is characterized by a value close to 0 and $b = -12.86$ with a high $L = 86.46 \pm 0.05$ parameter. These data correlate with those presented by us earlier for PLA films plasticized with PEG [18], and indicate the white/light blue color of this film. The usage of olive leaf extracts in PLA film preparation resulted in polymeric materials (PLA/PEG/OLE) of almost an unaffected a value, but substantially contrasting L and b values. The higher the concentration of OLE, the lower L value that was measured, indicating reduced lightness of films, changing from 66.64 ± 1.14

to 50.28 ± 0.60, for PLA/PEG/1OLE to PLA/PEG/5OLE, respectively. The usage of OLE causes the b color parameter to change the sign from slightly negative (for PLA/PEG) to positive ($b \approx 30$ for PLA/PEG/OLE). All changes in the a, b, and L parameters confirm the visible color change ($\Delta E > 5$—observer notices two different colors) from light blue to dark green ("olive") (visualization in Table 4). The obtained results are consistent with the color of the olive leaf extracts (see Figure S3 in Supplementary Materials).

The human eye functions as a spectrometer analyzing the light reflected from the surface of a solid or passing through a liquid. If we consider white light as composed of a broad range of radiation wavelengths in the ultraviolet (UV), visible, and infrared (IR) portions of the spectrum, characterized by a wavelength, green color is observed when the substance absorbs in the red (780–620 nm) and the blue (495–455 nm), and green-blue when the substance absorbs orange-red (595–750 nm) [86,87].

Most of the colour organic substances belonging to pigments contain certain groups called chromophores [88]. Olive leaves' colour depends on the pigment composition and concentration, especially with reference to chlorophyll. As it was already given by Imen Tarchoune [89], the green colour of olive leaves and its extracts results from the presence of chlorophylls (absorbance 646–664 nm) and carotenoids (absorbance at 480 nm). As shown in Figure S3 (Supplementary Materials), the higher amount of olive leaf dust used, the more intense the green colour of extract observed. However, even if the colour intensity of PLA/PEG/OLE films was found higher in reflection to PLA/PEG, there is no visible correlation between the colour of extracts and color intensity of resulting PLA/PEG/OLE films (Figure 3A).

The changes in color can also be noticed through the differentiation in the yellowness index, which not only changed its sign when olive leaf extract was used for PLA films formation, but also substantially increased in comparison with the pristine PLA/PEG film (Figure 3B).

The changes in color of PLA-based films after contact with water, 0.1 M HCl, and 0.1 M NaOH solutions were also evaluated (Table 4). In the case of all of the tested films, contact with all applied media caused darkening (increase in L value). Similar observations were made by Michalska-Sionkowska [90] for fish skin collagen material modified with β-glucan. Simultaneously, a reduction of color intensity and yellowness were noticed. Among all CIEL*a*b* color parameters, a more remarkable change was found in the b value, which substantially decreased after film swelling. However, no correlation between solution pH and color change could be established.

Two possible phenomena can be responsible for the color changes after swelling. It can be assumed that some of the pigments diffuse from the polymeric matrix into an external solution during swelling. The diffusion strictly correlates with the degree of swelling value and is hindered when low swelling occurs. It should be mentioned that we did not observe extensive swelling in each of the chosen solutions. This observation stays in agreement with the non-ionic nature of PLA and the swelling experiments performed by others [91,92]. Moreover, in contrast with the observations (L value), the diffusion of pigments should result in a lightening of PLA/PEG/OLE films. Thus, we assume that the color changes are dominated mainly by the effect of the pH on the stability of the chlorophylls and carotenoids.

3.6. Morphology Observations—Scanning Electron Microscopy

The surface and cross-section of the obtained films based on polylactide with olive leaf extract and poly(ethylene glycol) as a plasticizer were observed by Scanning Electron Microscopy. To observe the cross-section, the films were frozen in liquid nitrogen and broken. In Figure 4, surface, and cross-section at 1000× (Figure 4A,B) and cross-section at 2500× (Figure 4C) magnification are presented. It can be seen that surface of polylactide film containing plasticizer (Figure 4A—PLA/PEG) is smooth and flat, without any damage, fissures, or cracks. Similar observation were made in our previous paper [18], for polylactide films with the addition of quercetin and poly(ethylene glycol). Jasim Ahmed et al. [93] also reported a smooth and flat surface of PEG-plasticized polylactide films. These observations suggest good mixing of the polylactide and poly(ethylene glycol) and homogeneity of its blend. Taking into account the cross-section of the PLA/PEG film (Figure 4B—PLA/PEG), cracks can be observed. This may be caused by sample preparation (breaking the film frozen in liquid nitrogen) and the cracking of the gold layer during visualization.

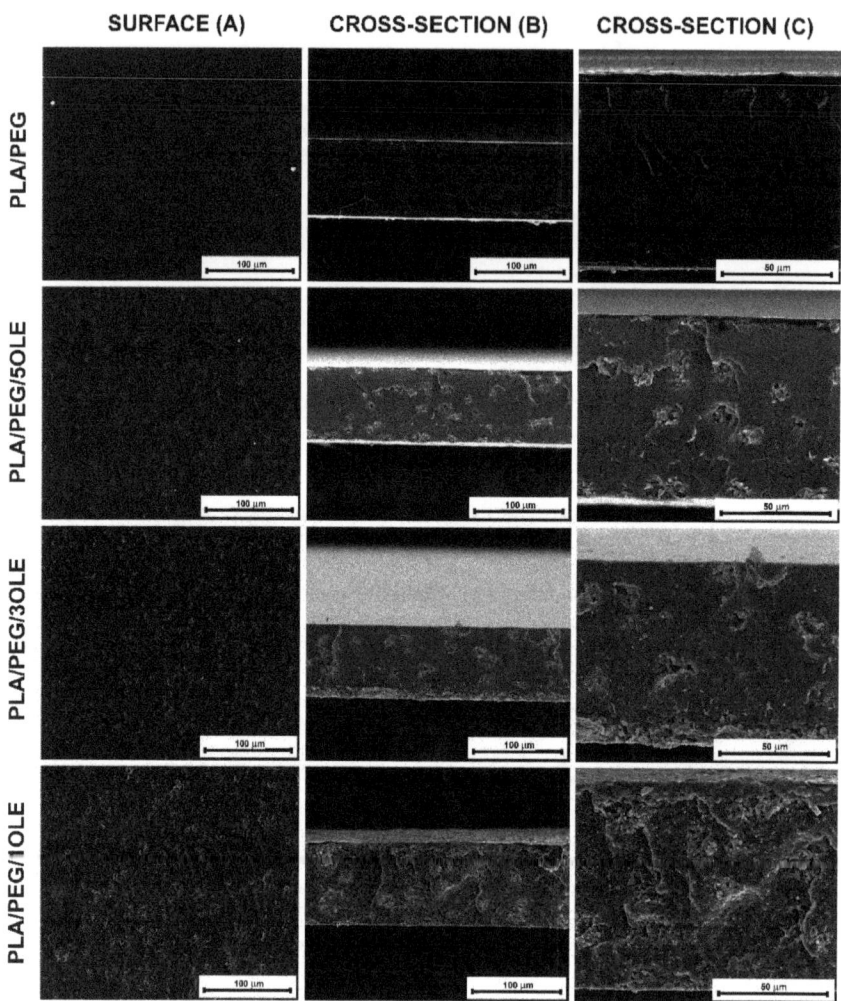

Figure 4. SEM images of PLA/PEG and PLA/PEG films with olive leaf extract addition: surface in 1000× magnification (**A**), and cross-section in 1000× (**B**) and 2500× (**C**) magnification.

When comparing films without the additive and with the addition of olive leaf extract, significant changes in morphology can be noticed. The surface of PLA/PEG/OLE is folded and inequalities can be observed. With increasing the extract concentration, a more uneven surface of the films was discovered. A similar conclusion was reached by Z. Javidi et al. [77] for polylactide films with the addition of 0.5 and 1.5 % (w/w) of oregano oil (OOil). It was stated that the addition of a higher amount of OOil into the PLA caused the presence of discontinuities in the resultant films, suggesting that the interaction between oregano oil and PLA was not sufficiently strong to prevent phase separation between both components during the solvent evaporation. The presence of a discontinuous phase in the PLA films with the addition of different essential oils (bergamot, lemongrass, rosemary, and clove) was also observed by Qin et al. [71]. This discontinuous structure would increase the roughness of the films.

Additionally, small holes and cavities were found in the films with OLE addition, as in our previous paper [69] describing the influence of tea tree essential oil on polylactide-based films. However, these surface irregularities were evenly distributed—which might suggest even and reasonably well-mixed ingredients. The incorporation of OLE reduced the compactness of the film. SEM observations stayed in agreement with the FTIR analysis, which did not confirm the strong interactions between the OLE and PLA/PEG film.

When it comes to the thickness of the films—as it can be seen in SEM images (Figure 4B) and in Figure 2D (mechanical properties), where the results for thickness measurements were given—no statistically significant differences in thickness were observed, depending on the concentration of the extract.

4. Conclusions

Polymer films with poly(ethylene glycol) as a plasticizer and olive leaf extract were successfully obtained by the solvent evaporation method. It was proven that the addition of OLE positively influences the antioxidant properties of PLA/PEG-based materials. This is due to the composition of the olive leaf extract, which mainly consists of phenolic substances, including oleuropein.

The incorporation of OLE affected the final material crystallinity, which in turn influenced the water vapor permeation rate and mechanical properties. Due to the nature of OLE components, no differences in WVPR between PLA/PEG and PLA/PEG/OLE were observed. The color of the polymeric films was strongly dependent on the presence and content of the extract used and changed during contact with different solutions.

The application of chloroform as a solvent for olive leaf extraction allows for, without additional solvent-evaporation steps, preparing OLE, PLA, and PEG containing film-forming solutions. Thus, the main benefit of the proposed method relies on using the same solvent for both extraction and film formation.

To more strictly characterize the obtained films and to prove their high potential as an active packaging material, additional antimicrobial and storage experiments have already been performed. Their results will be the object of a separate manuscript.

Supplementary Materials: The following are available online at https://www.mdpi.com/article/10.3390/ma14247623/s1, Figure S1. Base peak chromatogram of Chemlali olive cultivar leaf extract, Figure S2. DSC thermograms of PLA/PEG and PLA/PEG/OLE films, Figure S3. Real photo of fresh olive leaf extracts in chloroform, Table S1. DSC thermal data of PLA/PEG and PLA/PEG/OLE films.

Author Contributions: Conceptualization, S.G.-Z. and E.O.-K.; methodology, S.G.-Z. and M.G.; software, S.G.-Z., M.G. and E.O.-K.; validation, S.G.-Z., M.G. and E.O.-K.; formal analysis, S.G.-Z., M.G. and E.O.-K.; investigation, S.G.-Z., M.G. and M.B.; resources, S.G.-Z., M.G., E.O.-K. and M.B.; data curation, S.G.-Z. and M.G.; writing—original draft preparation, S.G.-Z., M.G., E.O.-K. and M.B.; writing—review and editing, S.G.-Z., M.G., E.O.-K. and M.B.; visualization, S.G.-Z. and M.G.; supervision, E.O.-K.; project administration, S.G.-Z. and M.G.; funding acquisition, S.G.-Z. and M.G. All authors have read and agreed to the published version of the manuscript.

Funding: This research was funded by the Dean of Faculty of Chemistry (Nicolaus Copernicus University in Toruń) grant no. 492/2020 and by the "Excellence Initiative—Research University—Debuts" programme, 1st edition (Nicolaus Copernicus University in Toruń).

Institutional Review Board Statement: Not applicable.

Informed Consent Statement: Not applicable.

Data Availability Statement: The data presented in this study are available on request from the corresponding author.

Acknowledgments: The authors would like to thank Beata Kaczmarek-Szczepańska for help in preparing the DPPH radical scavenging assay and Erragued Rim for preparing phenolic analysis of olive leaf extract.

Conflicts of Interest: The authors declare no conflict of interest.

References

1. Prasad, P.; Kochhar, A. Active Packaging in Food Industry: A Review. *IOSR J. Environ. Sci. Toxicol. Food Technol.* **2014**, *8*, 1–7. [CrossRef]
2. Bastarrachea, L.J.; Wong, D.E.; Roman, M.J.; Lin, Z.; Goddard, J.M. Active Packaging Coatings. *Coatings* **2015**, *5*, 771–791. [CrossRef]
3. Dobrucka, R.; Cierpiszewski, R. Active and Intelligent Packaging Food-Research and Development-a Review. *Pol. J. Food Nutr. Sci.* **2014**, *64*, 7–15. [CrossRef]
4. Peighambardoust, S.H.; Fasihnia, S.H.; Peighambardoust, S.J.; Pateiro, M.; Domínguez, R.; Lorenzo, J.M. Active Polypropylene-Based Films Incorporating Combined Antioxidants and Antimicrobials: Preparation and Characterization. *Foods* **2021**, *10*, 722. [CrossRef]
5. Ayala, A.; Muñoz, M.F.; Argüelles, S. Lipid Peroxidation: Production, Metabolism, and Signaling Mechanisms of Malondialdehyde and 4-Hydroxy-2-Nonenal. *Oxidative Med. Cell. Longev.* **2014**, *2014*, 360438. [CrossRef]
6. Chen, C.; Xu, Z.; Ma, Y.; Liu, J.; Zhang, Q.; Tang, Z.; Fu, K.; Yang, F.; Xie, J. Properties, Vapour-Phase Antimicrobial and Antioxidant Activities of Active Poly(Vinyl Alcohol) Packaging Films Incorporated with Clove Oil. *Food Control* **2018**, *88*, 105–112. [CrossRef]
7. Stepczyńska, M. Influence of Active Compounds on the Degradation of Polylactide Biocomposites. *Polim. Polym.* **2019**, *64*, 410–416. [CrossRef]
8. Fiore, A.; Park, S.; Volpe, S.; Torrieri, E.; Masi, P. Active Packaging Based on PLA and Chitosan-Caseinate Enriched Rosemary Essential Oil Coating for Fresh Minced Chicken Breast Application. *Food Packag. Shelf Life* **2021**, *29*, 100708. [CrossRef]
9. do Nascimento, L.D.; de Moraes, A.A.B.; da Costa, K.S.; Galúcio, J.M.P.; Taube, P.S.; Costa, C.M.L.; Cruz, J.N.; de Andrade, E.H.A.; de Faria, L.J.G. Bioactive Natural Compounds and Antioxidant Activity of Essential Oils from Spice Plants: New Findings and Potential Applications. *Biomolecules* **2020**, *10*, 988. [CrossRef]
10. Gülçin, I.; Elmastaş, M.; Aboul-Enein, H.Y. Antioxidant Activity of Clove Oil—A Powerful Antioxidant Source. *Arab. J. Chem.* **2012**, *5*, 489–499. [CrossRef]
11. Bezerra, D.P.; Militão, G.C.G.; de Morais, M.C.; de Sousa, D.P. The Dual Antioxidant/Prooxidant Effect of Eugenol and Its Action in Cancer Development and Treatment. *Nutrients* **2017**, *9*, 1367. [CrossRef]
12. Gülçin, I. Antioxidant Activity of Eugenol: A Structure-Activity Relationship Study. *J. Med. Food* **2011**, *14*, 975–985. [CrossRef] [PubMed]
13. Fernández-Marín, R.; Fernandes, S.C.M.; Sánchez, M.Á.A.; Labidi, J. Halochromic and Antioxidant Capacity of Smart Films of Chitosan/Chitin Nanocrystals with Curcuma Oil and Anthocyanins. *Food Hydrocoll.* **2022**, *123*. [CrossRef]
14. Cazón, P.; Antoniewska, A.; Rutkowska, J.; Vázquez, M. Evaluation of Easy-Removing Antioxidant Films of Chitosan with Melaleuca Alternifolia Essential Oil. *Int. J. Biol. Macromol.* **2021**, *186*, 365–376. [CrossRef] [PubMed]
15. Liu, G.; Zhu, W.; Zhang, J.; Song, D.; Zhuang, L.; Ma, Q.; Yang, X.; Liu, X.; Zhang, J.; Zhang, H.; et al. Antioxidant Capacity of Phenolic Compounds Separated from Tea Seed Oil in Vitro and in Vivo. *Food Chem.* **2022**, *371*, 131122. [CrossRef]
16. Wu, M.; Zhou, Z.; Yang, J.; Zhang, M.; Cai, F.; Lu, P. ZnO Nanoparticles Stabilized Oregano Essential Oil Pickering Emulsion for Functional Cellulose Nanofibrils Packaging Films with Antimicrobial and Antioxidant Activity. *Int. J. Biol. Macromol.* **2021**, *190*, 433–440. [CrossRef]
17. Pawłowska, A.; Stepczyńska, M. Natural Biocidal Compounds of Plant Origin as Biodegradable Materials Modifiers. *J. Polym. Environ.* **2021**, 1–26. [CrossRef] [PubMed]
18. Olewnik-Kruszkowska, E.; Gierszewska, M.; Richert, A.; Grabska-Zielińska, S.; Rudawska, A.; Bouaziz, M. Antibacterial Films Based on Polylactide with the Addition of Quercetin and Poly(Ethylene Glycol). *Materials* **2021**, *14*, 1643. [CrossRef]
19. Masek, A.; Olejnik, O. Aging Resistance of Biocomposites Crosslinked with Silica and Quercetin. *Int. J. Mol. Sci.* **2021**, *22*, 10894. [CrossRef]
20. Latos-Brozio, M.; Masek, A. Natural Polymeric Compound Based on High Thermal Stability Catechin from Green Tea. *Biomolecules* **2020**, *10*, 1191. [CrossRef]
21. Masek, A. Flavonoids as Natural Stabilizers and Color Indicators of Ageing for Polymeric Materials. *Polymers* **2015**, *7*, 1125–1144. [CrossRef]
22. Kim, J.H.; Ahn, D.U.; Eun, J.B.; Moon, S.H. Antioxidant Effect of Extracts from the Coffee Residue in Raw and Cooked Meat. *Antioxidants* **2016**, *5*, 21. [CrossRef] [PubMed]
23. Do, Q.D.; Angkawijaya, A.E.; Tran-Nguyen, P.L.; Huynh, L.H.; Soetaredjo, F.E.; Ismadji, S.; Ju, Y.H. Effect of Extraction Solvent on Total Phenol Content, Total Flavonoid Content, and Antioxidant Activity of Limnophila Aromatica. *J. Food Drug Anal.* **2014**, *22*, 296–302. [CrossRef]
24. Mahdi-Pour, B.; Jothy, S.L.; Latha, L.Y.; Chen, Y.; Sasidharan, S. Antioxidant Activity of Methanol Extracts of Different Parts of Lantana Camara. *Asian Pac. J. Trop. Biomed.* **2012**, *2*, 960–965. [CrossRef]
25. Tanwar, R.; Gupta, V.; Kumar, P.; Kumar, A.; Singh, S.; Gaikwad, K.K. Development and Characterization of PVA-Starch Incorporated with Coconut Shell Extract and Sepiolite Clay as an Antioxidant Film for Active Food Packaging Applications. *Int. J. Biol. Macromol.* **2021**, *185*, 451–461. [CrossRef]

26. Martínez, L.; Castillo, J.; Ros, G.; Nieto, G. Antioxidant and Antimicrobial Activity of Rosemary, Pomegranate and Olive Extracts in Fish Patties. *Antioxidants* **2019**, *8*, 86. [CrossRef] [PubMed]
27. Rambabu, K.; Bharath, G.; Banat, F.; Show, P.L.; Cocoletzi, H.H. Mango Leaf Extract Incorporated Chitosan Antioxidant Film for Active Food Packaging. *Int. J. Biol. Macromol.* **2019**, *126*, 1234–1243. [CrossRef]
28. Kanatt, S.R. Development of Active/Intelligent Food Packaging Film Containing Amaranthus Leaf Extract for Shelf Life Extension of Chicken/Fish during Chilled Storage. *Food Packag. Shelf Life* **2020**, *24*, 100506. [CrossRef]
29. Cejudo Bastante, C.; Casas Cardoso, L.; Fernández Ponce, M.T.; Mantell Serrano, C.; Martínez de la Ossa-Fernández, E.J. Characterization of Olive Leaf Extract Polyphenols Loaded by Supercritical Solvent Impregnation into PET/PP Food Packaging Films. *J. Supercrit. Fluids* **2018**, *140*, 196–206. [CrossRef]
30. Martiny, T.R.; Raghavan, V.; de Moraes, C.C.; da Rosa, G.S.; Dotto, G.L. Bio-Based Active Packaging: Carrageenan Film with Olive Leaf Extract for Lamb Meat Preservation. *Foods* **2020**, *9*, 1759. [CrossRef]
31. Gonelimali, F.D.; Lin, J.; Miao, W.; Xuan, J.; Charles, F.; Chen, M.; Hatab, S.R. Antimicrobial Properties and Mechanism of Action of Some Plant Extracts against Food Pathogens and Spoilage Microorganisms. *Front. Microbiol.* **2018**, *9*, 1639. [CrossRef]
32. Hu, X.; Yuan, L.; Han, L.; Li, S.; Song, L. Characterization of Antioxidant and Antibacterial Gelatin Films Incorporated with: Ginkgo Biloba Extract. *RSC Adv.* **2019**, *9*, 27449–27454. [CrossRef]
33. García, A.V.; Serrano, N.J.; Sanahuja, A.B.; Garrigós, M.C. Novel Antioxidant Packaging Films Based on Poly(ε-Caprolactone) and Almond Skin Extract: Development and Effect on the Oxidative Stability of Fried Almonds. *Antioxidants* **2020**, *9*, 629. [CrossRef] [PubMed]
34. Erdohan, Z.Ö.; Turhan, K.N. Olive Leaf Extract and Usage for Development of Antimicrobial Food Packaging. In *Science against Microbial Pathogens: Communicating Current Research and Technological Advances*; Mendez-Vilas, A., Ed.; Formatex Research Center: Bodojaz, Spain, 2011; pp. 1094–1101.
35. Amaro-Blanco, G.; Delgado-Adámez, J.; Martín, M.J.; Ramírez, R. Active Packaging Using an Olive Leaf Extract and High Pressure Processing for the Preservation of Sliced Dry-Cured Shoulders from Iberian Pigs. *Innov. Food Sci. Emerg. Technol.* **2018**, *45*, 1–9. [CrossRef]
36. Bouaziz, M.; Sayadi, S. Isolation and Evaluation of Antioxidants from Leaves of a Tunisian Cultivar Olive Tree. *Eur. J. Lipid Sci. Technol.* **2005**, *107*, 497–504. [CrossRef]
37. Testa, B.; Lombardi, S.J.; Macciola, E.; Succi, M.; Tremonte, P.; Iorizzo, M. Efficacy of Olive Leaf Extract (*Olea europaea* L. cv Gentile di Larino) in Marinated Anchovies (*Engraulis encrasicolus*, L.) Process. *Heliyon* **2019**, *5*, e01727. [CrossRef] [PubMed]
38. Ahmed, A.M.; Rabii, N.S.; Garbaj, A.M.; Abolghait, S.K. Antibacterial Effect of Olive (*Olea europaea* L.) Leaves Extract in Raw Peeled Undeveined Shrimp (*Penaeus Semisulcatus*). *Int. J. Vet. Sci. Med.* **2014**, *2*, 53–56. [CrossRef]
39. Erdohan, Z.Ö.; Çam, B.; Turhan, K.N. Characterization of Antimicrobial Polylactic Acid Based Films. *J. Food Eng.* **2013**, *119*, 308–315. [CrossRef]
40. Altiok, E.; Bayçin, D.; Bayraktar, O.; Ülkü, S. Isolation of Polyphenols from the Extracts of Olive Leaves (*Olea europaea* L.) by Adsorption on Silk Fibroin. *Sep. Purif. Technol.* **2008**, *62*, 342–348. [CrossRef]
41. Kiritsakis, K.; Kontominas, M.G.; Kontogiorgis, C.; Hadjipavlou-Litina, D.; Moustakas, A.; Kiritsakis, A. Composition and Antioxidant Activity of Olive Leaf Extracts from Greek Olive Cultivars. *J. Am. Oil Chem. Soc.* **2010**, *87*, 369–376. [CrossRef]
42. Albertos, I.; Avena-Bustillos, R.J.; Martín-Diana, A.B.; Du, W.X.; Rico, D.; McHugh, T.H. Antimicrobial Olive Leaf Gelatin Films for Enhancing the Quality of Cold-Smoked Salmon. *Food Packag. Shelf Life* **2017**, *13*, 49–55. [CrossRef]
43. da Rosa, G.S.; Vanga, S.K.; Gariepy, Y.; Raghavan, V. Development of Biodegradable Films with Improved Antioxidant Properties Based on the Addition of Carrageenan Containing Olive Leaf Extract for Food Packaging Applications. *J. Polym. Environ.* **2020**, *28*, 123–130. [CrossRef]
44. Frikha, N.; Bouguerra, S.; Kit, G.; Abdelhedi, R.; Bouaziz, M. Smectite Clay KSF as Effective Catalyst for Oxidation of M-Tyrosol with H_2O_2 to Hydroxytyrosol. *React. Kinet. Mech. Catal.* **2019**, *127*, 505–521. [CrossRef]
45. Malik, N.S.A.; Bradford, J.M. Changes in Oleuropein Levels during Differentiation and Development of Floral Buds in "Arbequina" Olives. *Sci. Hortic.* **2006**, *110*, 274–278. [CrossRef]
46. Kaczmarek, B. Improving Sodium Alginate Films Properties by Phenolic Acid Addition. *Materials* **2020**, *13*, 2895. [CrossRef]
47. Chow, W.S.; Lok, S.K. Thermal Properties of Poly(Lactic Acid)/Organo-Montmorillonite Nanocomposites. *J. Therm. Anal. Calorim.* **2009**, *95*, 627–632. [CrossRef]
48. Sosnowski, S. Poly(L-Lactide) Microspheres with Controlled Crystallinity. *Polymer* **2001**, *42*, 637–643. [CrossRef]
49. Pathare, P.B.; Opara, U.L.; Al-Said, F.A.J. Colour Measurement and Analysis in Fresh and Processed Foods: A Review. *Food Bioprocess Technol.* **2013**, *6*, 36–60. [CrossRef]
50. Altinyay, Ç.; Levent Altun, M. HPLC Analysis of Oleuropein in *Olea europaea* L. *J. Fac. Pharm. Ank. Univ.* **2006**, *35*, 1–11. [CrossRef]
51. Japón-Luján, R.; Luque-Rodríguez, J.M.; Luque De Castro, M.D. Multivariate Optimisation of the Microwave-Assisted Extraction of Oleuropein and Related Biophenols from Olive Leaves. *Anal. Bioanal. Chem.* **2006**, *385*, 753–759. [CrossRef]
52. Malik, N.S.A.; Bradford, J.M. Recovery and Stability of Oleuropein and Other Phenolic Compounds during Extraction and Processing of Olive (*Olea europaea* L.) Leaves. *J. Food Agric. Environ.* **2008**, *6*, 8–13.
53. Afaneh, I.; Yateem, H.; Al-Rimawi, F. Effect of Olive Leaves Drying on the Content of Oleuropein. *Am. J. Anal. Chem.* **2015**, *6*, 246–252. [CrossRef]

54. Medina, E.; Romero, C.; García, P.; Brenes, M. Characterization of Bioactive Compounds in Commercial Olive Leaf Extracts, and Olive Leaves and Their Infusions. *Food Funct.* **2019**, *10*, 4716–4724. [CrossRef] [PubMed]
55. Markhali, F.S.; Teixeira, J.A.; Rocha, C.M.R. Olive Tree Leaves-A Source of Valuable Active Compounds. *Processes* **2020**, *8*, 1177. [CrossRef]
56. Omar, S.H. Oleuropein in Olive and Its Pharmacological Effects. *Sci. Pharm.* **2010**, *78*, 133–154. [CrossRef]
57. le Tutour, B.; Guedon, D. Antioxidative Activities of *Olea europaea* Related Phenolic Compounds. *Phytochemistry* **1992**, *31*, 1173–1178. [CrossRef]
58. Soler-Rivas, C.; Espín, J.C.; Wichers, H.J. Oleuropein and Related Compounds. *J. Sci. Food Agric.* **2000**, *80*, 1013–1023. [CrossRef]
59. Tymczewska, A.; Furtado, B.U.; Nowaczyk, J.; Hrynkiewicz, K.; Szydłowska-Czerniak, A. Development and Characterization of Active Gelatin Films Loaded with Rapeseed Meal Extracts. *Materials* **2021**, *14*, 2869. [CrossRef] [PubMed]
60. Roy, S.; Rhim, J.W. Preparation of Antimicrobial and Antioxidant Gelatin/Curcumin Composite Films for Active Food Packaging Application. *Colloids Surf. B Biointerfaces* **2020**, *188*, 110761. [CrossRef] [PubMed]
61. Dou, L.; Li, B.; Zhang, K.; Chu, X.; Hou, H. Physical Properties and Antioxidant Activity of Gelatin-Sodium Alginate Edible Films with Tea Polyphenols. *Int. J. Biol. Macromol.* **2018**, *118*, 1377–1383. [CrossRef]
62. Olewnik-Kruszkowska, E.; Koter, I.; Skopińska-Wiśniewska, J.; Richert, J. Degradation of Polylactide Composites under UV Irradiation at 254 Nm. *J. Photochem. Photobiol. A Chem.* **2015**, *311*, 144–153. [CrossRef]
63. Yuniarto, K.; Purwanto, Y.A.; Purwanto, S.; Welt, B.A.; Purwadaria, H.K.; Sunarti, T.C. Infrared and Raman Studies on Polylactide Acid and Polyethylene Glycol-400 Blend. *AIP Conf. Proc.* **2016**, *1725*, 020101. [CrossRef]
64. Ikhmal, W.M.K.W.M.; Yasmin, M.Y.N.; Fazira, M.F.M.; Rafizah, W.A.W.; Wan Nik, W.B.; Sabri, M.G.M. Anticorrosion Coating Using Olea Sp. Leaves Extract. *IOP Conf. Ser. Mater. Sci. Eng.* **2018**, *344*, 012028. [CrossRef]
65. Nasir, G.A.; Mohammed, A.K.; Samir, H.F. Biosynthesis and Characterization of Silver Nanoparticles Using Olive Leaves Extract and Sorbitol. *Iraqi J. Biotechnol.* **2016**, *15*, 22–32.
66. Khalil, M.M.H.; Ismail, E.H.; El-Magdoub, F. Biosynthesis of Au Nanoparticles Using Olive Leaf Extract. *Arab. J. Chem.* **2012**, *5*, 431–437. [CrossRef]
67. Mohapatra, A.K.; Mohanty, S.; Nayak, S.K. Effect of PEG on PLA/PEG Blend and Its Nanocomposites: A Study of Thermo-Mechanical and Morphological Characterization. *Polym. Compos.* **2014**, *35*, 283–293. [CrossRef]
68. Bijarimi, M.; Ahmad, S.; Rasid, R.; Khushairi, M.A.; Zakir, M. Poly(Lactic Acid)/Poly(Ethylene Glycol) Blends: Mechanical, Thermal and Morphological Properties. *AIP Conf. Proc.* **2016**, *1727*, 020002. [CrossRef]
69. Tarach, I.; Olewnik-Kruszkowska, E.; Richert, A.; Gierszewska, M.; Rudawska, A. Influence of Tea Tree Essential Oil and Poly(Ethylene Glycol) on Antibacterial and Physicochemical Properties of Polylactide-Based Films. *Materials* **2020**, *13*, 4953. [CrossRef]
70. Sobiesiak, M.; Podkościelna, B.; Sevastyanova, O. Thermal Degradation Behavior of Lignin-Modified Porous Styrene-Divinylbenzene and Styrene-Bisphenol A Glycerolate Diacrylate Copolymer Microspheres. *J. Anal. Appl. Pyrolysis* **2017**, *123*, 364–375. [CrossRef]
71. Qin, Y.; Li, W.; Liu, D.; Yuan, M.; Li, L. Development of Active Packaging Film Made from Poly (Lactic Acid) Incorporated Essential Oil. *Prog. Org. Coat.* **2017**, *103*, 76–82. [CrossRef]
72. Chieng, B.W.; Ibrahim, N.A.; Then, Y.Y.; Loo, Y.Y. Mechanical, Thermal, and Morphology Properties of Poly(Lactic Acid) Plasticized with Poly(Ethylene Glycol) and Epoxidized Palm Oil Hybrid Plasticizer. *Polym. Eng. Sci.* **2016**, *56*, 1169–1174. [CrossRef]
73. Vasile, C.; Stoleru, E.; Darie-Niţa, R.N.; Dumitriu, R.P.; Pamfil, D.; Tarţau, L. Biocompatible Materials Based on Plasticized Poly(Lactic Acid), Chitosan and Rosemary Ethanolic Extract I. Effect of Chitosan on the Properties of Plasticized Poly(Lactic Acid) Materials. *Polymers* **2019**, *11*, 941. [CrossRef]
74. Casalini, T.; Rossi, F.; Castrovinci, A.; Perale, G. A Perspective on Polylactic Acid-Based Polymers Use for Nanoparticles Synthesis and Applications. *Front. Bioeng. Biotechnol.* **2019**, *7*, 259. [CrossRef]
75. Farah, S.; Anderson, D.G.; Langer, R. Physical and Mechanical Properties of PLA, and Their Functions in Widespread Applications—A Comprehensive Review. *Adv. Drug Deliv. Rev.* **2016**, *107*, 367–392. [CrossRef] [PubMed]
76. Latos-Brozio, M.; Masek, A. The Application of (+)-Catechin and Polydatin as Functional Additives for Biodegradable Polyesters. *Int. J. Mol. Sci.* **2020**, *21*, 414. [CrossRef]
77. Javidi, Z.; Hosseini, S.F.; Rezaei, M. Development of Flexible Bactericidal Films Based on Poly(Lactic Acid) and Essential Oil and Its Effectiveness to Reduce Microbial Growth of Refrigerated Rainbow Trout. *LWT-Food Sci. Technol.* **2016**, *72*, 251–260. [CrossRef]
78. Jakubowska, E.; Gierszewska, M.; Nowaczyk, J.; Olewnik-Kruszkowska, E. The Role of a Deep Eutectic Solvent in Changes of Physicochemical and Antioxidative Properties of Chitosan-Based Films. *Carbohydr. Polym.* **2021**, *255*, 117527. [CrossRef]
79. Rojas-Lema, S.; Quiles-Carrillo, L.; Garcia-Garcia, D.; Melendez-Rodriguez, B.; Balart, R.; Torres-Giner, S. Tailoring the Properties of Thermo-Compressed Polylactide Films for Food Packaging Applications by Individual and Combined Additions of Lactic Acid Oligomer and Halloysite Nanotubes. *Molecules* **2020**, *25*, 1976. [CrossRef]
80. Mohsen, A.H.; Ali, N.A. Mechanical, Color and Barrier, Properties of Biodegradable Nanocomposites Polylactic Acid/Nanoclay. *J. Bioremediation Biodegrad.* **2018**, *9*. [CrossRef]
81. Halász, K.; Hosakun, Y.; Csóka, L. Reducing Water Vapor Permeability of Poly(Lactic Acid) Film and Bottle through Layer-by-Layer Deposition of Green-Processed Cellulose Nanocrystals and Chitosan. *Int. J. Polym. Sci.* **2015**, *2015*. [CrossRef]

82. Maharana, T.; Mohanty, B.; Negi, Y.S. Melt-Solid Polycondensation of Lactic Acid and Its Biodegradability. *Prog. Polym. Sci.* **2009**, *34*, 99–124. [CrossRef]
83. Drieskens, M.; Peeters, R.; Mullens, J.; Franco, D.; Iemstra, P.J.; Hristova-Bogaerds, D.G. Structure versus Properties Relationship of Poly(Lactic Acid). I. Effect of Crystallinity on Barrier Properties. *J. Polym. Sci. Part B Polym. Phys.* **2009**, *47*, 2247–2258. [CrossRef]
84. Guo, Y.; Yang, K.; Zuo, X.; Xue, Y.; Marmorat, C.; Liu, Y.; Chang, C.C.; Rafailovich, M.H. Effects of Clay Platelets and Natural Nanotubes on Mechanical Properties and Gas Permeability of Poly (Lactic Acid) Nanocomposites. *Polymer* **2016**, *83*, 246–259. [CrossRef]
85. García, A.V.; Álvarez-Pérez, O.B.; Rojas, R.; Aguilar, C.N.; Garrigós, M.C. Impact of Olive Extract Addition on Corn Starch-Based Active Edible Films Properties for Food Packaging Applications. *Foods* **2020**, *9*, 1339. [CrossRef]
86. Schanda, J. *Colorimetry-Understanding the CIE System*; John Wiley & Sons, Inc.: Hoboken, NJ, USA, 2007; ISBN 978-0-470-04904-4.
87. Lindon, J.C.; Tranter, G.E.; Holmes, J.L. *Encyclopedia of Spectroscopy and Spectrometry*; Elsevier: Amsterdam, The Netherlands, 2000; ISBN 0-12-226680-3.
88. Sklar, A.L. Theory of Color of Organic Compounds. *J. Chem. Phys.* **1937**, *5*, 669–681. [CrossRef]
89. Tarchoune, I.; Sgherri, C.; Eddouzi, J.; Zinnai, A.; Quartacci, M.F.; Zarrouk, M. Olive Leaf Addition Increases Olive Oil Nutraceutical Properties. *Molecules* **2019**, *24*, 545. [CrossRef] [PubMed]
90. Michalska-Sionkowska, M.; Warżyńska, O.; Kaczmarek-Szczepańska, B.; Łukowicz, K.; Osyczka, A.M.; Walczak, M. Preparation and Characterization of Fish Skin Collagen Material Modified with β-Glucan as Potential Wound Dressing. *Materials* **2021**, *14*, 1322. [CrossRef]
91. Bond Tee, Y.; Talib, R.A.; Abdan, K.; Chin, L.; Kadir Basha, R.; Faezah, K.; Yunos, M. Effect of Aminosilane Concentrations on the Properties of Poly(Lactic Acid)/Kenaf-Derived Cellulose Composites. *Polym. Polym. Compos.* **2017**, *25*, 63–76.
92. Sharp, J.S.; Forrest, J.A.; Jones, R.A.L. Swelling of Poly(DL-Lactide) and Polylactide-Co-Glycolide in Humid Environments. *Macromolecules* **2001**, *34*, 8752–8760. [CrossRef]
93. Ahmed, J.; Hiremath, N.; Jacob, H. Antimicrobial, Rheological, and Thermal Properties of Plasticized Polylactide Films Incorporated with Essential Oils to Inhibit Staphylococcus Aureus and Campylobacter Jejuni. *J. Food Sci.* **2016**, *81*, E419–E429. [CrossRef] [PubMed]

Article

Preparation and Characterization of Novel Microgels Containing Nano-SiO$_2$ and Copolymeric Hydrogel Based on Poly (Acrylamide) and Poly (Acrylic Acid): Morphological, Structural and Swelling Studies

Tannaz Soltanolzakerin Sorkhabi [1], Mehrab Fallahi Samberan [1,*], Krzysztof Adam Ostrowski [2], Tomasz M. Majka [3], Marcin Piechaczek [2] and Paulina Zajdel [2]

[1] Department of Chemical Engineering, Ahar Branch, Islamic Azad University, Ahar P.O. Box 5451116714, Iran; tannazsoltanzakeri@gmail.com
[2] Faculty of Civil Engineering, Cracow University of Technology, 24 Warszawska Str., 31-155 Cracow, Poland; krzysztof.ostrowski.1@pk.edu.pl (K.A.O.); marcin.jaroslaw.piechaczek@gmail.com (M.P.); paulina.zajdel1@pk.edu.pl (P.Z.)
[3] Department of Chemistry and Technology of Polymers, Faculty of Chemical Engineering and Technology, Cracow University of Technology, Warszawska 24, 31-155 Cracow, Poland; tomasz.majka@pk.edu.pl
* Correspondence: m-fallahi@iau-ahar.ac.ir

Abstract: In this paper, novel microgels containing nano-SiO$_2$ were prepared by in situ copolymerization using nano-SiO$_2$ particles as a reinforcing agent, nanosilica functional monomer (silane-modified nano-SiO$_2$) as a structure and morphology director, acrylamide (AAm) as a monomer, acrylic acid (AAc) as a comonomer, potassium persulfate (KPS) as a polymerization initiator, and N,N'-methylene bis (acrylamide) (MBA) as a crosslinker. In addition, a conventional copolymeric hydrogel based on poly (acrylamide/acrylic acid) was synthesized by solution polymerization. The microgel samples, hydrogel and nanoparticles were characterized by transmission electron microscopy (TEM), field emission scanning electron microscopy (FESEM), Fourier transform infrared (FTIR) spectroscopy, thermogravimetric analysis (TGA) and differential scanning calorimetry (DSC). A FESEM micrograph of copolymeric hydrogel showed the high porosity and 3D interconnected microstructure. Furthermore, FESEM results demonstrated that when nano-SiO$_2$ particles were used in the AAm/AAc copolymerization process, the microstructure and morphology of product changed from porous hydrogel to a nanocomposite microgel with cauliflower-like morphology. According to FESEM images, the copolymerization of AAm and AAc monomers with a nanosilica functional monomer or polymerizable nanosilica particle as a seed led to a microgel with core–shell structure and morphology. These results demonstrated that the polymerizable vinyl group on nano-SiO$_2$ particles have controlled the copolymerization and the product morphology. FTIR analysis showed that the copolymeric chains of polyacrylamide (PAAm) and poly (acrylic acid) (PAAc) were chemically bonded to the surfaces of the nano-SiO$_2$ particles and silane-modified nano-SiO$_2$. The particulate character of microgel samples and the existence of long distance among aggregations of particles led to rapid swelling and increasing of porosity and therefore increasing of degree of swelling.

Keywords: copolymer; hydrogel; microgel; nanocomposite; nanosilica

Citation: Sorkhabi, T.S.; Samberan, M.F.; Ostrowski, K.A.; Majka, T.M.; Piechaczek, M.; Zajdel, P. Preparation and Characterization of Novel Microgels Containing Nano-SiO$_2$ and Copolymeric Hydrogel Based on Poly (Acrylamide) and Poly (Acrylic Acid): Morphological, Structural and Swelling Studies. *Materials* **2022**, *15*, 4782. https://doi.org/10.3390/ma15144782

Academic Editor: Krzysztof Moraczewski

Received: 26 May 2022
Accepted: 4 July 2022
Published: 8 July 2022

Publisher's Note: MDPI stays neutral with regard to jurisdictional claims in published maps and institutional affiliations.

Copyright: © 2022 by the authors. Licensee MDPI, Basel, Switzerland. This article is an open access article distributed under the terms and conditions of the Creative Commons Attribution (CC BY) license (https://creativecommons.org/licenses/by/4.0/).

1. Introduction

Polymeric hydrogels have a particular characteristic that make them unique materials. Hydrogels are a three-dimensional network of loosely crosslinked polymers that can absorb and retain several times even thousands of times weight of aqueous fluids due to the presence of large amounts of hydrophilic groups such as carboxyl (–COOH), sulfonic acid (–SO$_3$H), amine (–NH$_2$) and hydroxyl (–OH) [1–3]. The presence of these functional groups in the structure of hydrogels contribute to the hydrophilic character and water absorbency of hydrogels [4]. Meanwhile, microgels are hydrogel particles with their size ranging from nanometers to micrometers [5]. These particles can form colloids. Hydrogels are classified into three groups based on their composition. This group includes multi-polymer interpenetrating polymeric hydrogel, copolymeric hydrogels and homopolymeric hydrogels. Hydrogels are also divided into four groups based on the network electrical charges including nonionic, ionic, amphoteric electrolyte, and zwitterion [6,7]. Hydrogels as environmental-sensitive materials can show volume transition in response to the surrounding environment (e.g., light, electric and magnetic field, temperature, pH, enzymes) [8–10]. These kind of revolutionary materials are widely used in environmental and separation systems (heavy metal ions remover [11]), physiological hygiene products (baby diapers and sanitary towels [12]), tissue engineering (material for scaffolds [13]), contact lenses [14], sensors and actuators [15], supercapacitors [16], slow release fertilizer [17], pharmaceuticals (controlled drug delivery [18]), agriculture and forestry (water conservation retention [19]), cosmetic industry [20] and civil engineering (sealing rod, cement [21]). During the past two decades, organic/inorganic nanocomposites have been developed to help improve the properties of conventional hydrogel and overcome their weaknesses [22]. These novel types of strong materials with excellent properties such as thermal stability and high swelling ratio are generally organic macromolecule composites with inorganic materials such as nanoparticles with nanoscale size and high surface area [23].

AAc acid and AAm are among the main monomers used in the preparation and production of commercial hydrogels. These monomers are common materials for preparing absorbent materials in industry due to their desired swelling properties. AAms as a small molecular weight are the most commonly used hydrogels [24]. Aam-based hydrogels show significant volume transition in response to external (physical and chemical) stimuli [3]. Propenoic acid or AAc monomer is crosslinked in the single or multi-component polymerization system to produce hydrogels with high water absorbing capacity. AAc has the ability of connection to the vinyl group due to its carboxylic acid group. The presence of this ionizable carboxylic acid group aids in increasing the ionic strength and sensitivity to the pH of the prepared hydrogel samples. Meanwhile, AAc monomers are used with a combination of some other polymers such as polyacrylamide to synthesize different forms of hydrogels [3,25]. In recent years, nanocomposites, which consist of polymers and nanomaterials, gained considerable attention due to synergistic effects among their components. The presence of nanomaterials in nanocomposite could result in superior properties such as high mechanical strength, good barrier properties, improved thermal stability, and so on [26]. Among the nanocomposites, polymer/nano-SiO_2 nanocomposites have attracted much attention due to the large surface area and smooth surface of nano-SiO_2 particles. The presence of silanol and siloxane groups on the nano-SiO_2 surface lead to the improvement of particles' hydrophilicity, which promote the compatibility with polymer chains. In addition, nano-SiO_2 can function as structure and morphology directors in nanocomposite synthesis. The results of many studies have indicated that the performance of nanocomposite material could be considerably improved by combination or copolymerization with a functional monomer containing nano-SiO_2 [27]. The nanocomposite materials containing nano-SiO_2 such as nylon 6 [28], polyaniline [29], styrene butadiene rubber [30], polyimide [31], polyethylene terephthalate [32], may show more satisfactory thermal stability, toughness, and strength.

Nanocomposite systems can be synthesized by various synthesis routes, thanks to the ability to combine different ways to introduce each phase. The organic component can be introduced as (1) a precursor, which can be a monomer or an oligomer, (2) a preformed linear polymer in solution, emulsion, or molten states, or (3) a polymer network, chemically or physically cross-linked. The mineral part can be introduced as (1) a precursor for example tetraethylorthosilicate (TEOS) or (2) preformed nanoparticles. Organic or inorganic polymerization generally becomes necessary if at least one of the starting materials is a precursor. This leads to three general techniques for the preparation of polymer/SiO_2

nanocomposites according to the starting materials and processing methods: blending, sol-gel processes, and in situ polymerization [27].

In the following article, we mainly focus on the preparation of microgels containing nano-SiO$_2$ with in situ copolymerisation of acrylic acid (AAc) and acrylamide (AAm) as monomers and methylene-bis-acrylamide (MBA) as a crosslinking agent in the presence of nano-SiO$_2$ particles. The effect of vinyl functionalization of nano-SiO$_2$ with vinyltri-ethoxysilane on some properties of the prepared microgel samples, such as morphology, thermal stability, glass transition and swelling degree, was also investigated. The grafting mechanism, including the proposed structure of the synthesized samples and a schematic representation of the method for the synthesis of microgel samples and copolymeric hydrogel are shown in Schemes 1–3. In addition, in this work, synthesis, characterization and swelling studies of copolymeric hydrogel-based on poly (acrylamide) and poly (acrylic acid) have been carried out. The novelty of this research is simple preparation and investigation of surface modification of nano-SiO$_2$ on microgel particles' morphology and their swelling properties.

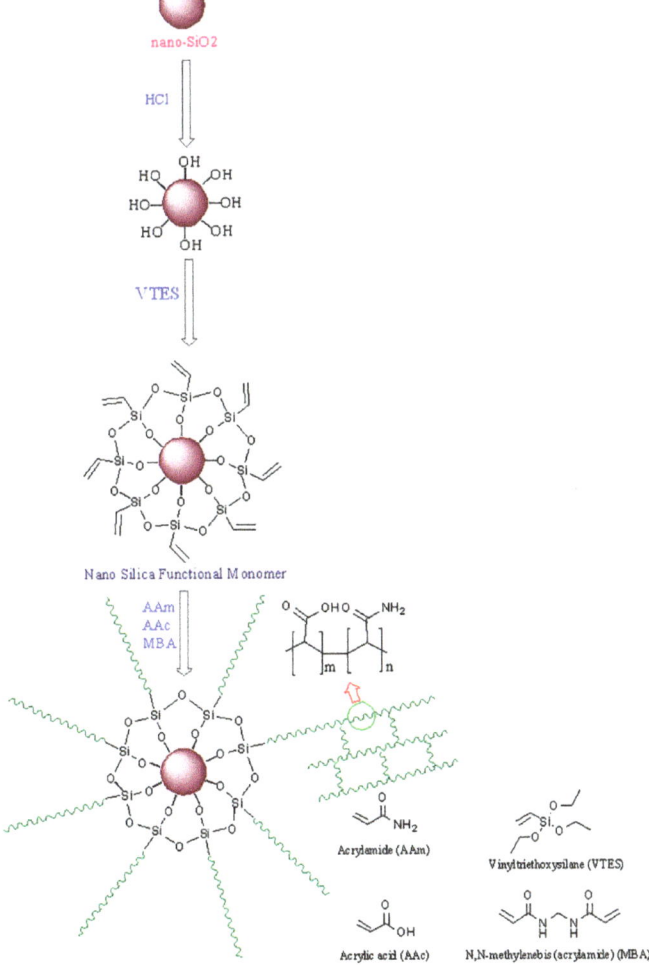

Scheme 1. Synthesis of core-shell microgel.

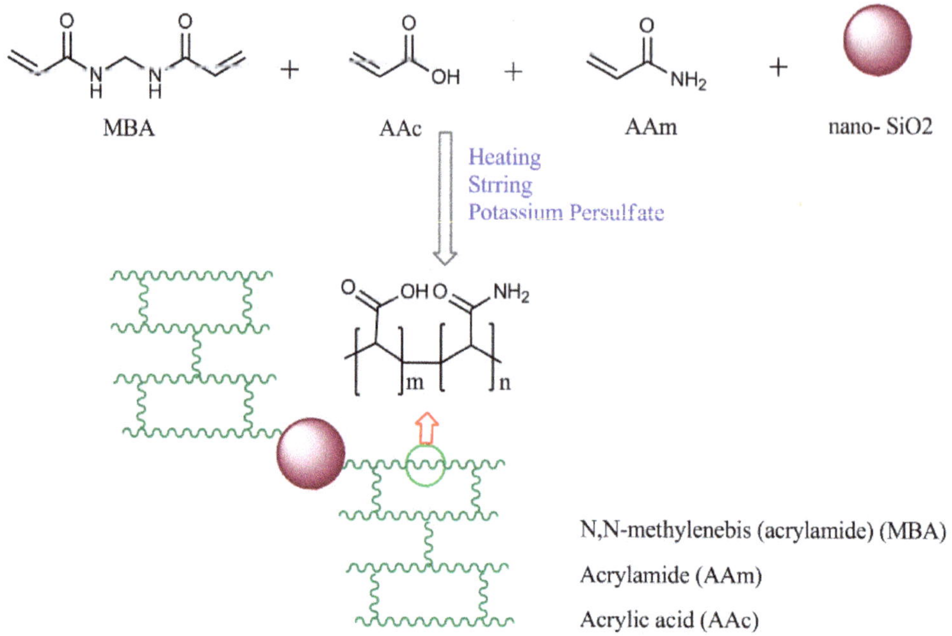

Scheme 2. Synthesis of nanocomposite microgel.

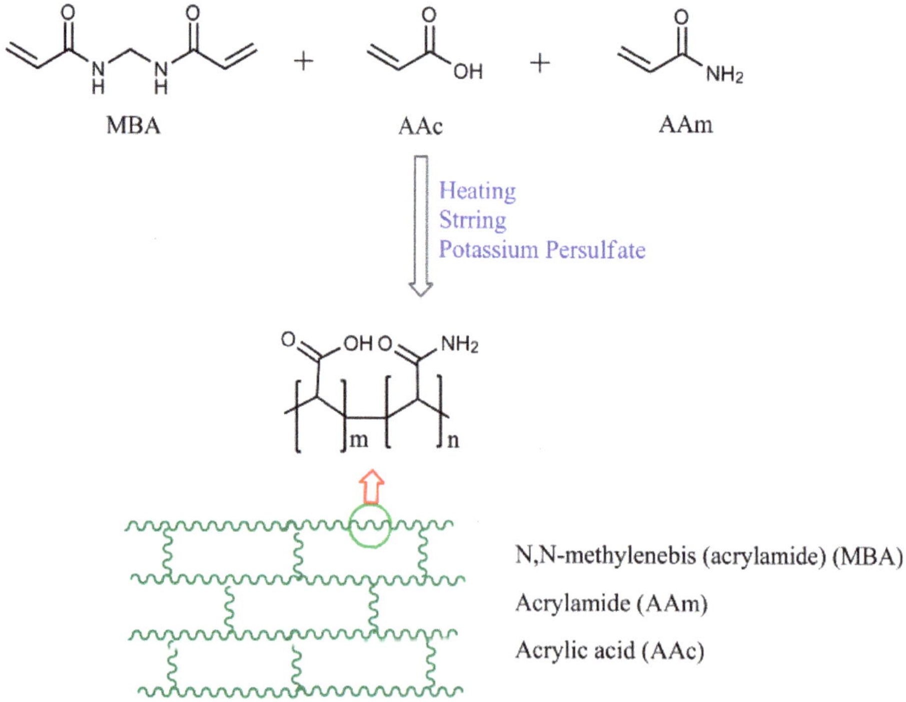

Scheme 3. Synthesis of copolymeric hydrogel.

2. Experimental Section

2.1. Materials

Acrylamide (C_3H_5NO, AAm, Merck, Darmstadt, Germany, purity: ≥99%), acrylic acid ($C_3H_4O_2$, AA, Merck, Darmstadt, Germany, purity: ≥99%) and N,N'-methylenebis(acrylamide) ($C_7H_{10}N_2O_2$, MBA, Sigma-Aldrich, St. Louis, MO, USA, purity: ≥99%), potassium persulfate ($K_2S_2O_8$, KPS, Panreac Química, Barcelona, Spain, purity: ≥98%), sodium dodecyl sulfate (SDS, Merck, Germany, purity: ≥90%), nanosilica particles (SiO_2, Us Nano, Houston, TX, USA, purity: ≥99%), hydrochloric acid (HCl, Merck, Darmstadt, Germany, purity = 37%), vinyltriethoxysilane ($NaC_{12}H_{25}SO_4$, VTES, Energy Chemical, Shanghai, China, purity: ≥98%), ammonia (NH_4OH, Merck, Germany, purity = 25%), ethanol (C_2H_6O, Merck, Germany, purity = 99.99%) and distilled water were used in current research.

2.2. Synthesis of Nanosilica Functional Monomer

According to Scheme 1, a 1 g sample of nano-SiO_2 particles was dispersed in 50 mL of hydrochloric acid, HCl aqueous solution (5%, v/v) in a 250 mL beaker. The solution was left stirred for 1 h at room temperature. The activated nanoparticles were obtained by centrifugation of solution, washing with distilled water and vacuum drying at room temperature. Then activated nano-SiO_2 (1 g) was added into the stirred solution of ethanol (60 mL), distilled water (10 mL) and ammonia (1.0 mL) in the round-bottom flask. The mixture was sonicated (24 kHz) for 30 min to better disperse of the nanoparticles. Then, 1.0 mL of vinyltriethoxysilane was added into the stirred solution. The reaction was started at 30 ± 2 °C and stopped after 12 h. The product was nanosilica functional monomer (silane-modified nano-SiO_2) which was separated by centrifugation and washed with distilled water.

2.3. Synthesis of Core—Shell Microgels

The above synthesized nanosilica functional monomer was used as a seed (core) for the synthesis of the AAm/AAc/nanosilica core-shell microgel (Scheme 1). In addition, the nano-SiO_2 particles were used in the preparation of AAm/AAc/nano-SiO_2 nanocomposite microgel (Scheme 2). For the synthesis of the AAm/AAc/nanosilica core-shell microgel, firstly, nanosilica functional monomer (500 mg) was dispersed in 100 mL of distilled water by ultrasonic waves (24 kHz) for 30 min. Then, AAm (3.5 g, 49.24 mM), AAc (1.5 g, 20.81 mM), MBA (0.54 g, 3.50 mM ≅ 5% of total monomers) and KPS (198.8 mg, 0.736 mM ≅ 1% of total monomers) as radical polymerization initiator were added to the prepared nanosilica functional monomer aqueous solution and the reaction started at 65 ± 2 °C under stirring conditions and the inert gas, nitrogen (N_2) atmosphere. After 4 h the reaction was stopped.

2.4. Synthesis of Nanocomposite Microgels

An in situ free radical polymerization technique was used to synthesize AAm/AAc/nano-SiO_2 nanocomposite microgel. For the synthesis of the nanocomposite microgel sample, 500 mg of nano-SiO_2 particles (without surface activation and modification) was poured into 100 mL of distilled water and sonicated to better aid the dispersion and then SDS surfactant (25 mg) was added to help the stabilization of the dispersed nano-SiO_2 particles. After preparation of the nano-SiO_2 colloidal solution, 3.5 g AAm (49.24 mM) and 1.5 mg AAc (20.81 mM) as comonomer, 0.54 g MBA (3.50 mM ≅ 5% of total monomers) as crosslinking agent, 198.8 mg KPS (0.736 mM ≅ 1% of total monomers) as polymerization initiator were also added to it. The obtained mixture was stirred at 65 ± 2 °C under the inert gas, nitrogen (N_2) atmosphere. The reaction was complete after 4 h.

2.5. Prepration of Polyacrylamide and Copolymeric Hydrogel

The copolymerization of AAm/AAc in the presence of MBA crosslinker and the absence of nano-SiO_2 and the nanosilica functional monomer was carried out by using the method and values mentioned in the previous section. To study the water uptake, all synthesized products dried after ethanol washing. Then, they were crushed and sieved to get uniform particle sizes. In addition, the same process was applied to prepare homopolymer hydrogel based on polyacrylamide (PAAm).

2.6. Dynamic Swelling Studies

The pulverized hydrogel, nanocomposite and core-shell microgels in the uniform particle sizes were washed with distilled water and ethanol to remove unreacted starting materials, free oligomers, free polymer and copolymer chains and ungrafted nano-SiO$_2$ particles. The dynamic swelling experiment was performed by measuring the water weight of the dried samples. A small amount of superabsorbent samples was taken (0.1 g) and placed in the three beakers. Then, 500 mL of aqueous buffer solution was poured into the beakers. After 5 min the swollen samples were separated by using a filter and then the wet weight was measured carefully. The weight gain as a function of time was taken as the swelling measurement. According to the following Equation (1), the swelling ratio was expressed as the percent weight ratio of the water held in the hydrogel to the dry sample at any instant during swelling.

$$\text{Swelling ratio (\%)} = \frac{W_t - W_d}{W_d} \times 100 \qquad (1)$$

where W_t and W_d are the weight of the swollen sample at time t and the weight of the dry sample at time 0, respectively.

2.7. Methods

After coating the samples with gold film (thickness $\cong$ 10 nanometers) to obtain a good quality image because the polymers prepared here are not conductive, the morphology of the freeze-dried samples of homopolymer hydrogel, copolymeric hydrogel, nanocomposite and core-shell microgels were examined by a field emission scanning electron microscope (FESEM, Hitachi model S-4160,Daypetronic Company, Tokyo, Japan) at magnifications of 100,000× g and 70,000× g. Nano-SiO$_2$ particles were viewed using a Zeiss Leo 906 (Carl Zeiss Inc., Jena, Germany) transmission electron microscope (TEM) at a magnification of 100,000× g. The FTIR spectra of synthesized samples were recorded on Tensor 27 FTIR spectrometers (Bruker Optik GmbH, Ettlingen, Germany) using KBr discs and under strictly constant conditions in the region of 400–4000 cm^{-1}. About 5.3 mg of the dry samples were taken and thermogravimetric analysis (TGA)/derivative thermogravimetry (DTG) testing was conducted using an STA409C131F thermogravimetric analyzer (TGA, NETZSCH Company, Germany) in a temperature range of 30 °C–620 °C under the inert gas. In addition, the thermal behavior of samples was determined using a differential scanning calorimeter (DSC-200F3, NETZSCH Company, Germany). Around 4 mg of each sample encapsulated in an aluminum pan. Then the pan was heated from 25 to 200 °C at a heating rate of 10 °C/min under nitrogen purge to measure their glass transition temperature (T$_g$) according to ASTM 3418-15 standard.

3. Results and Discussion

3.1. Morphological Studies

One of the substantial features of the microgels and hydrogels is their morphologies, which determine their applications and crucial properties such as porosity, water permeation and swelling capacity. TEM results of the nano-SiO$_2$ particles are shown in Figure 1. According to TEM image shown in Figure 1, the mean size of the particles was 23 ± 4.6 nm and no particle aggregation was observed.

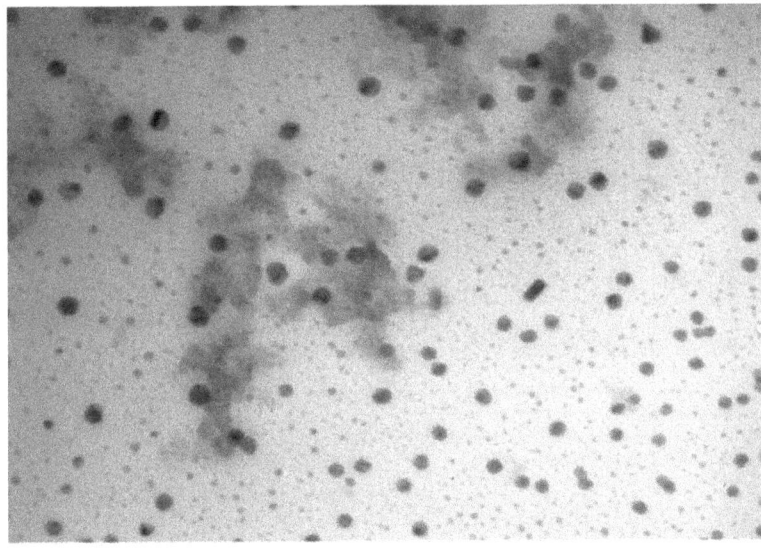
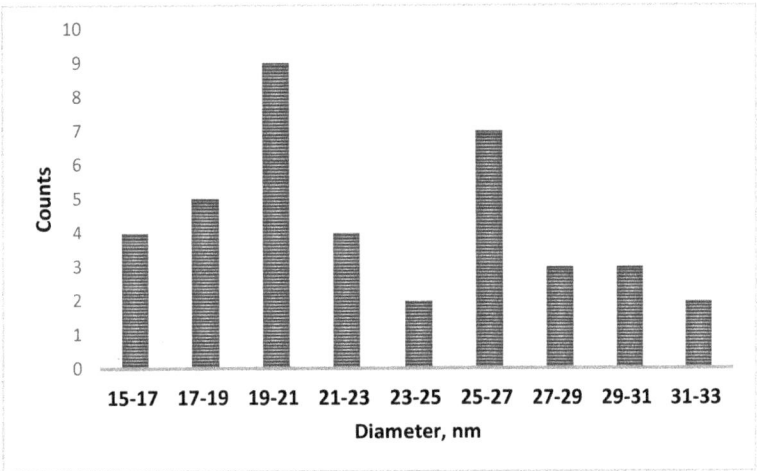

Figure 1. TEM image and bar graphs of diameter ranges of nano-SiO$_2$.

As shown in Figure 2, the copolymerization of AAm and AAc in the presence of MBA crosslinking agent and the absence of nano-SiO$_2$ particles resulted in conventional hydrogel formation. The FESEM micrographs of freeze-dried copolymeric hydrogel showed the high porosity and 3D interconnected microstructures like other reported polymeric hydrogel structures. The porosity formation and interconnectivity of the microchannels in the hydrogel structures could be assigned to the crosslinking polymerization in the presence of solvent that dissolves the monomers, but causes precipitation of the formed polymer. Either bulk or solution polymerization can synthesize the hydrogel material. However, bulk polymerization yields a glassy and optically transparent gel with no porosity. In contrast, the solution polymerization produces a hydrogel with porous structures.

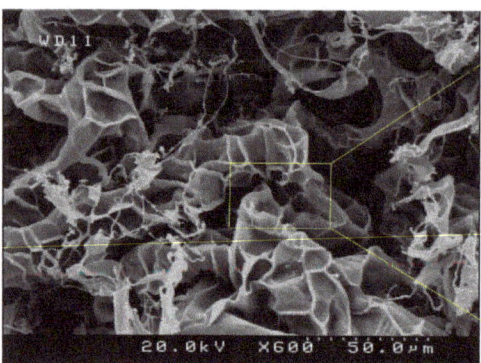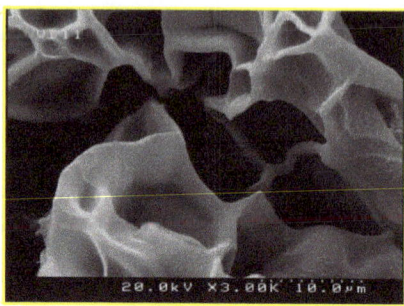

Figure 2. FESEM micrographs of freeze-dried copolymeric hydrogel.

As described in the experimental section, when nano-SiO_2 particles were added to the AAm/AAc copolymerization system, the microstructure and morphology of products were obviously changed from porous hydrogel to nanocomposite microgel with cauliflower-like morphology as shown in Figure 3. FESEM micrographs of the nanocomposite microgel indicated the homogenous dispersion and uniform distribution of the nano-SiO_2 particles in the AAm/AAc copolymer matrix and most of the nanoparticles are individual and some of them are observable at the surface of the nanocomposite microgel sample. Although the results showed that the AAm/AAc copolymer chains were grafted on the nano-SiO_2 particles and covalent bonding formed between them. However, the synthesis of AAm/AAc copolymer containing nano-SiO_2 particles did not result in the formation of complete core-shell morphology. The formation of this special micro-nano structures and the resulting morphology demonstrated that the nano-SiO_2 particles have also had a co-crosslinking role and have helped three-dimensional structure formation. It is worth mentioning that elimination of MBA crosslinking agent from this copolymerization system, the resulting product did not exhibit any hydrogel properties and a copolymeric solution with very low gel content achieved.

Figure 3. FESEM micrographs of freeze-dried nanocomposite microgel.

The FESEM image of Figure 4 reveals that the copolymerization of AAm and AAc monomores with nanosilica functional monomer or polymerizable nanosilica particle as seed led to core-shell structure. These results demonstrated that a polymerizable vinyl group on nano-SiO$_2$ particles not only have worked as a co-crosslinking agent but also as seed have controlled the copolymerization and the product morphology. Core-shell morphology development and shell growth can be summarized as follows. The polymerization of hydrophilic AAm and AAc monomers started from the surface of the vinyl modified nano-SiO$_2$ to form an oligomer chain containing shell. During the copolymerization and growth of shell, the propagating copolymer of AAm and AAc anchored on the surface of the nano-SiO$_2$ and led to the appearance of core–shell structure. In addition, the aggregation of particles to form core–shell clusters can be attributed to this fact that with growth of a shell layer on the seeds. The adjacent core-shell particles are connected to each other by growing and living chains of AAm/AAc copolymer. Furthermore, as shown in Figure 5, the synthesized homopolymer hydrogel of PAAm did not have any porosity in the FESEM micrographs.

Figure 4. FESEM micrographs of freeze-dried core-shell microgel.

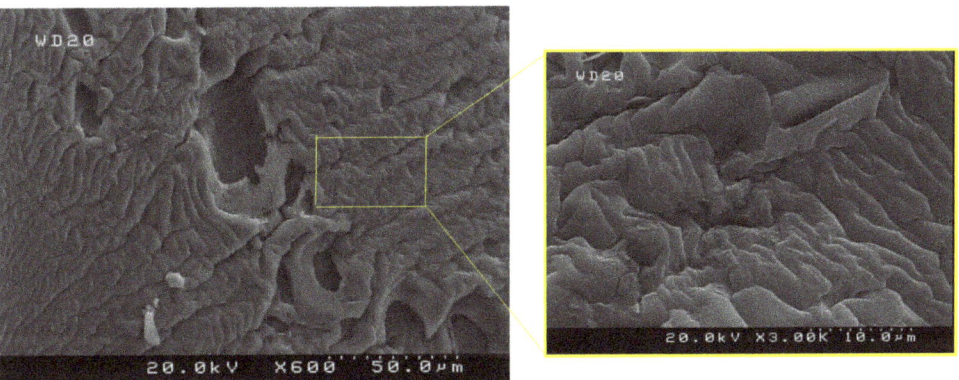

Figure 5. FESEM micrographs of freeze-dried homopolymer hydrogel of PAAm.

3.2. FTIR Analysis

The characterization of chemical structure of synthesized samples was carried out by FTIR analysis. The spectra of nano-SiO$_2$ particles, poly (acrylic acid) (PAAc), PAAm, copolymeric hydrogel of AAm and AAc, nanosilica functional monomer, nanocomposite microgel with cauliflower-like morphology and core-shell microgel were compared and the corresponding results are depicted in Figure 6. The structure of pure PAA was confirmed by absorptions at $\nu = 3302$ cm^{-1} (–OH hydroxyl groups) and $\nu = 1658$ cm^{-1} (–C=O carbonyl groups). The peaks at about 1159 cm^{-1} are attributed to –CO in –COOH of PAA. The characteristic absorption bands of pure PAAm were observed at 3445 cm^{-1} and 1540 cm^{-1} for the N–H and 1639 cm^{-1} for the C=O carbonyl group in the structure of the amide group. In the spectrum of the AAm/AAc copolymer hydrogel, the peak observed at 3422 cm^{-1} corresponds to N-H and O-H stretching. The absorbance at 2921 cm^{-1} is assigned to –C-H stretching of the acrylate group. The peak at 1542 cm^{-1} and at 1661 cm^{-1} are assigned to C=O stretching of the acrylamid groups and acrylate groups, respectively. These absorbance bands in the AAm/AAc copolymer indicated the successful synthesis of copolymeric hydrogel based on PAAm and PAAc. In addition, in Figure 6, the weak bands at 3400 and 1637 cm^{-1} are due to the O–H group on the surface of nano-SiO$_2$, and the strong peak observed at 1100 cm^{-1} in the nano-SiO$_2$ spectrum is due to the Si-O-Si bonds. Additionally, the bands at 471 cm^{-1} and 814 cm^{-1} in nano-SiO$_2$ spectrum represent Si-O bending vibration and stretching vibration, respectively. The strong peak at 3300 to 3400 cm^{-1} and the new peak at 3025 cm^{-1} corresponded to large numbers of Si-O-H group and =CH stretching vibrations in nano-SiO$_2$ functional monomer, respectively, which was appeared after the surface modification of nano- SiO$_2$ particles. Furthermore, the absorption peak at 654 cm^{-1} is due to the stretching vibrations of Si–C in the structure of microgel with cauliflower-like morphology and core-shell microgel. This suggests that the copolymeric chains of PAAc and PAAm was chemically bonded to the surface of the nano-SiO$_2$ particles.

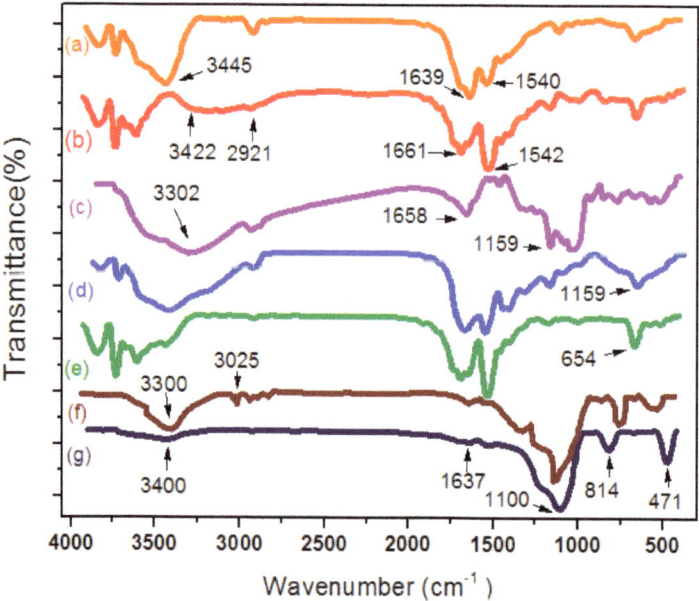

Figure 6. FTIR spectra for: PAAm (**a**), copolymeric hydrogel (**b**), PAAc (**c**), nanocomposite microgel (**d**), core-shell microgel (**e**), nano- SiO$_2$ functional monomer (**f**), nano- SiO$_2$ particles (**g**).

3.3. Thermal Stability Analysis

Thermal stability of synthesized copolymeric hydrogel and microgel samples were investigated by TGA/DTG at 30–620 °C with N_2 in the inert atmosphere. Representative TG thermograms of samples along with derivative thermograms (DTG) curve are shown in Figure 7. It is clearly seen from Figure 6 that the weight of the samples continuously decreases as the temperature increases. According to these TGA profiles, three stages of weight loss were observed for both samples of microgels (nanocomposite and core-shell) and two stages of weight loss was seen for copolymeric hydrogel. For both copolymeric hydrogel and microgels, minor weight loss was observed at temperatures less than 240 °C. These weight losses can be attributed to the anhydride formation and the evaporation of volatile solvent or entrapped water in the structure of samples [33]. The second stage of thermal degradation as the main weight loss was observed at 340 °C for the copolymeric hydrogel, 374 °C for the microgel with cauliflower-like morphology and 378 °C for the core-shell microgel. This was assigned to the thermal decomposition of the functional groups in the three copolymeric samples (amide groups in AAm and carboxyl groups in AAc). The final decomposition stage of thermal degradation of the samples at high temperatures can be attributed to the degradation of the C–C bonds in the side and main chain of these three copolymeric samples and destroying of their structures [34].

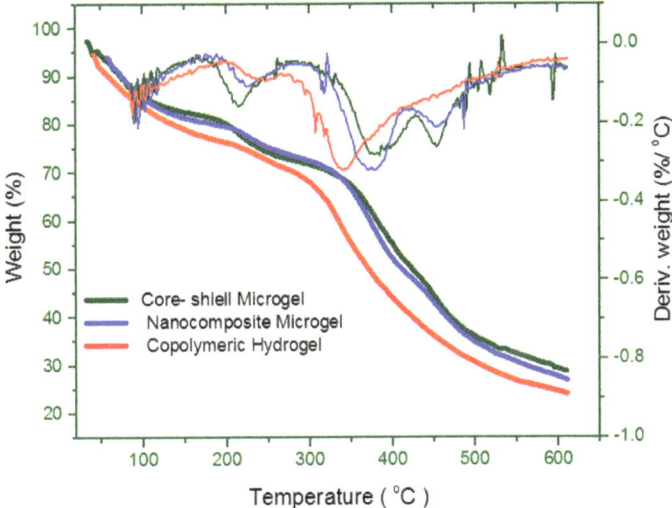

Figure 7. TGA curves of synthesized samples.

The results showed that the addition of nanosilica particles increased the thermal stability of the microgel samples. At 600 °C, the residual weight percent of copolymeric hydrogels (without the addition of nanosilica particles) was 29.3%. The residual weight percent of the nanocomposite microgel with cauliflower-like morphology was 33.3%. The residual weight percent of core-shell microgel was 31.6%. Briefly, these results indicated that the presence of nanosilica particles resulted in the improvement of the thermal stability of microgel samples compared to pure copolymer.

The differential scanning calorimetry (DSC) analysis of the copolymeric hydrogel and microgel samples are shown in Figure 8. The endothermic peaks in the DSC curves of the three samples correspond to their glass transition temperature (T_g) and volatilization of bound water. For pure copolymeric hydrogel, the glass transition temperature is observed around 88 °C. The T_g of nanocomposite microgel with cauliflower-like morphology and core-shell microgel are about 91 and 96, respectively. It is clear that all samples as a random copolymer exhibit only one T_g. The occurrence of a single peak (T_g) can be attributed to the miscibility between PAAm and PAAc and the formation intermolecular H-bonding between these polymers. Furthermore, the addition of the nano-SiO$_2$ particles shifts the T$_g$ of samples slightly to high temperatures resulting in more stable samples. Because there are attractive forces and covalent bonds between the nanosilica and copolymer and the graft and adsorption of copolymer chains on the nanosilica surface decreases their mobility. In addition, the endothermic processes of the samples from 240 °C to 262 °C can be attributed to the volatilization of bound water in the hydrophilic structure of samples.

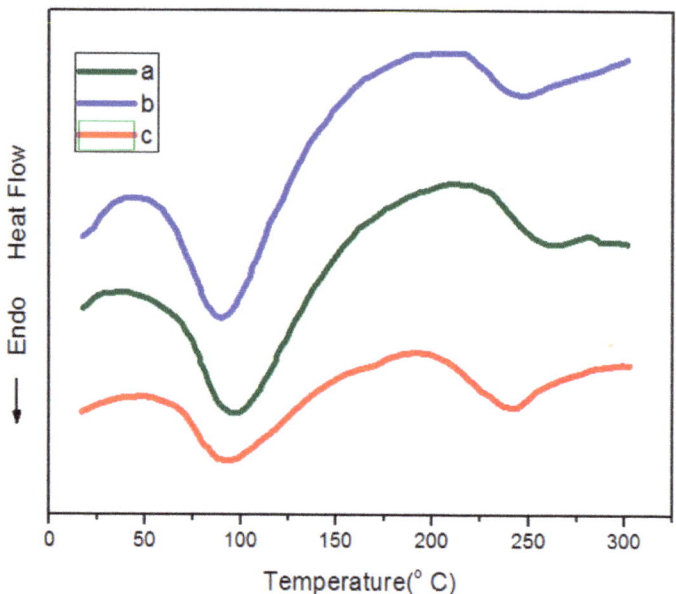

Figure 8. DSC thermograms of the core-shell microgel (**a**), the nanocomposite microgel (**b**) and the copolymeric hydrogel (**c**).

3.4. Swelling Studies

In order to study the swelling behavior, samples were allowed to swell to equilibrium in buffer solution of pH 7.4 at a room temperature and the swelling kinetics of these samples were investigated. The dynamic swelling behavior of the samples are shown in Figure 9. As can be seen from their swelling behavior in this figure, the swelling of samples increases with time until a certain point when it becomes constant. These constant values are taken as the equilibrium swelling and are 333% for the copolymeric hydrogel, 405% for the microgel with cauliflower-like morphology and 430% for the core-shell microgel. The ability of water absorbency of the samples prepared in this study arises from the two hydrophilic functional groups, -COOH of the AAc and -CONH$_2$ of the AAm units attached to the copolymeric backbone of these samples. While the resistance of these samples with many hydrophilic functional groups to dissolution arises from the presence of the MBA crosslinker and three-dimensional network structure. It is well known that PAAm is nonionic and insensitive to pH of the medium but PAAc is a pH- sensitive polymer. All our samples formed by the homogenous copolymer of AAc and AAm with MBA as a crosslinker where the carboxylic acidic groups of AA which bound to the copolymer chains made the samples pH sensitive. Thus, at a pH lower than the dissociation constant, pK$_a$ of PAAc (about 4.3) the degree of ionization of the carboxylic acid group is small and most of them are in –COOH form which can form a hydrogen bond with -COONH$_2$ side groups of AAm units leading to shrinking of hydrogel or microgel samples. In contrast, at neutral or basic pH (or pH greater than 4.3) such as a pH value of 7.4 in this study, AAc units in backbone of prepared copolymeric samples are negatively charged due to the ionization and the deprotonation of COOH groups. Thus, all AAc/AAm copolymeric samples resulted in expansion of networks and swell to a great degree at this pH condition (pH = 7.4) because of the electrostatic repulsion among the carboxylate anions (–COO⁻). Meanwhile, as shown in Figure 2, the copolymeric hydrogels show a porous network structure in character that makes it easier for water to diffuse in or out of the gel matrix. The porosity that itself has a great influence on swelling behavior can be generated by electrostatic repulsive forces among the similarly charged carboxyl groups along copolymeric segments during the copolymerization process. Whereas as shown in Figure 5 there are no porosity in the FESEM micrographs of homopolymer hydrogel of PAAm and it has a relatively dense structure. In addition, from the swelling plots in Figure 8, it is clear that the water absorbency of the microgel samples is much more than the copolymeric hydrogel.

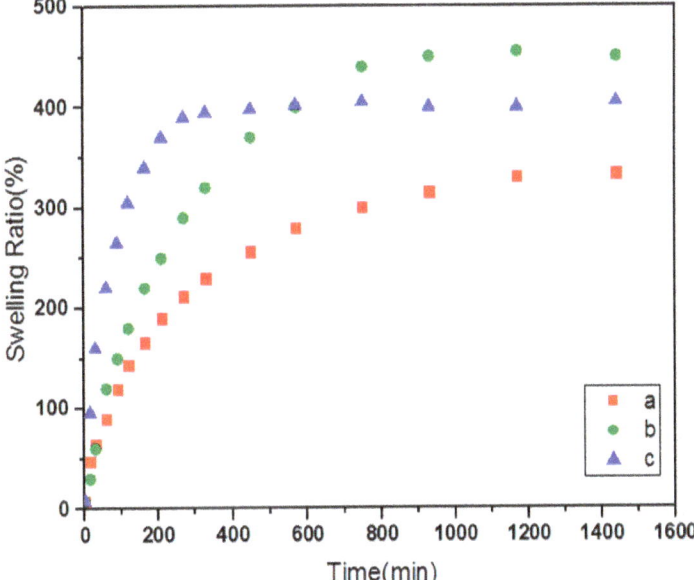

Figure 9. Swelling behavior of the copolymeric hydrogel (**a**), the core-shell microgel (**b**), and the nanocomposite microgel (**c**).

According to the morphology study and the FESEM results in Figures 2–4, particulate character of microgels and the existence of long distance among aggregations of particles lead to rapid swelling and the increase of porosity and therefore increasing the degree of swelling. The differences in the kinetics of swelling for the core-shell microgel and the nanocomposite microgel can be attributed to this fact that according to Schemes 1 and 2 nanosilica functional monomer as seed have not been played as crosslinking agent between chains and particles but nano-SiO_2 particles in the nanocomposite microgel have had the crosslinking role. As a result, the swelling rate of the core-shell microgel was similar to the copolymeric hydrogel due to a similar degree of crosslinking and chemical similarity of the shell and copolymeric sample. The high swelling rate and low swelling ratio of nanocomposite microgel in comparison with the core-shell microgel could be because of the smaller particles and high degree of crosslinking, respectively.

4. Conclusions

Synthesis, characterization, and morphological, structural and swelling studies of novel microgels containing nano-SiO_2 and copolymeric hydrogel based on PAAm and PAAc was successfully conducted. The FESEM micrograph of freeze-dried copolymeric hydrogel show the high porosity and 3D interconnected microstructures. The interconnectivity of the microchannels in the hydrogel structures could be assigned to the MBA crosslinking of poly (AAm-co-AAc) chains. It was found that introducing only a small amount of nanosilica into the copolymerization system can change the morphology of products from porous hydrogel (hydrogel particle). The presence of silanol and siloxane groups on the nano-SiO_2 surface led to better control of morphology of the microgel and formation of complete core-shell due to hydrophilicity, compatibility and because of polymerizable vinyl group of VTES on nanosilica particles. TGA revealed that the presence of nanosilica particles in nanocomposite microgel with cauliflower-like morphology and core-shell microgel resulted in the improvement of the thermal stability compared to copolymeric hydrogel. The glass transition temperature (T_g) for pure copolymeric hydrogel, nanocomposite microgel and core-shell microgel were observed by DSC around 88, 91 and 96 °C, respectively. In addition, water absorbency of the microgel samples was much more than copolymeric hydrogel due to their particulate character. High swelling rate and low swelling ratio of nanocomposite microgel in comparison with the core-shell microgel could be because of the smaller particles and high degree of crosslinking, respectively.

Author Contributions: T.S.S.—conceptualization, writing the original and revised draft; modelling, visualization, project administration; M.F.S.—supervision, designing the experiments, processing the experimental data, interpreting the results, writing the original draft, editing, review, project administration; K.A.O.—review, writing the original draft, editing, funding; T.M.M.—processing the experimental data, interpreting the results, writing the original draft, M.P.—writing original draft, visualization; P.Z.—writing revised draft, visualization, editing. All authors have read and agreed to the published version of the manuscript.

Funding: This research was funded by the Cracow University of Technology.

Institutional Review Board Statement: Not applicable.

Informed Consent Statement: Not applicable.

Data Availability Statement: The data presented in this article are available within the article.

Conflicts of Interest: The authors declare no conflict of interest.

References

1. Erceg, T.; Dapčević-Hadnađev, T.; Hadnađev, M.; Ristić, I. Swelling kinetics and rheological behaviour of microwave synthesized poly(acrylamide-co-acrylic acid) hydrogels. *Colloid Polym. Sci.* **2020**, *299*, 11–23. [CrossRef]
2. Sorkhabi, T.S.; Samberan, M.F.; Ostrowski, K.A.; Majka, T.M. Novel Synthesis, Characterization and Amoxicillin Release Study of pH-Sensitive Nanosilica/Poly(acrylic acid) Macroporous Hydrogel with High Swelling. *Materials* **2022**, *15*, 469. [CrossRef] [PubMed]
3. Sennakesavan, G.; Mootakhdemim, M.; Dkhar, L.; Seyfoddin, A.; Fatihhi, S. Acrylic acid/acrylamide based hydrogels and its properties-A review. *Polym. Degrad. Stab.* **2020**, *180*, 109308. [CrossRef]
4. Sun, N.; Ji, R.; Zhang, F.; Song, X.; Xie, A.; Liu, J.; Zhang, M.; Niu, L.; Zhang, S. Structural evolution in poly(acrylic-co-acrylamide) pH-responsive hydrogels by low-field NMR. *Mater. Today Commun.* **2019**, *22*, 100748. [CrossRef]
5. De Lima, C.S.A.; Balogh, T.S.; Varca, J.P.R.O.; Varca, G.H.C.; Lugão, A.B.; Camacho-Cruz, L.A.; Bucio, E.; Kadlubowski, S.S. An Updated Review of Macro, Micro, and Nanostructured Hydrogels for Biomedical and Pharmaceutical Applications. *Pharmaceutics* **2020**, *12*, 970. [CrossRef]
6. Bustamante-Torres, M.; Romero-Fierro, D.; Arcentales-Vera, B.; Palomino, K.; Magaña, H.; Bucio, E. Hydrogels Classification According to the Physical or Chemical Interactions and as Stimuli-Sensitive Materials. *Gels* **2021**, *7*, 182. [CrossRef]
7. Ahmed, E.M. Hydrogel: Preparation, characterization, and applications: A review. *J. Adv. Res.* **2015**, *6*, 105–121. [CrossRef]
8. Bashir, S.; Hina, M.; Iqbal, J.; Rajpar, A.H.; Mujtaba, M.A.; Alghamdi, N.A.; Wageh, S.; Ramesh, K.; Ramesh, S. Fundamental Concepts of Hydrogels: Synthesis, Properties, and Their Applications. *Polymers* **2020**, *12*, 2702. [CrossRef]
9. Qiu, Y.; Park, K. Environment-sensitive hydrogels for drug delivery. *Adv. Drug Deliv. Rev.* **2001**, *53*, 321–339. [CrossRef]
10. Khan, S.; Ullah, A.; Ullah, K.; Rehman, N.-U. Insight into hydrogels. *Des. Monomers Polym.* **2016**, *19*, 456–478. [CrossRef]
11. Perumal, S.; Atchudan, R.; Edison, T.; Babu, R.; Karpagavinayagam, P.; Vedhi, C. A Short Review on Recent Advances of Hydrogel-Based Adsorbents for Heavy Metal Ions. *Metals* **2021**, *11*, 864. [CrossRef]
12. Haque, M.O.; Mondal, M.I.H. Cellulose-Based Hydrogel for Personal Hygiene Applications. In *Cellulose-Based Superabsorbent Hydrogels*; Polymers and Polymeric Composites: A Reference Series; Mondal, M., Ed.; Springer: Cham, Switzerland, 2018.

13. Tomić, S.; Nikodinović-Runić, J.; Vukomanović, M.; Babić, M.M.; Vuković, J.S. Novel Hydrogel Scaffolds Based on Alginate, Gelatin, 2-Hydroxyethyl Methacrylate, and Hydroxyapatite. *Polymers* **2021**, *13*, 932. [CrossRef]
14. Tran, N.-P.-D.; Yang, M.-C. Synthesis and Characterization of Silicone Contact Lenses Based on TRIS-DMA-NVP-HEMA Hydrogels. *Polymers* **2019**, *11*, 944. [CrossRef]
15. Ehrenhofer, A.; Binder, S.; Gerlach, G.; Wallmersperger, T. Multisensitive Swelling of Hydrogels for Sensor and Actuator Design. *Adv. Eng. Mater.* **2020**, *22*, 2000004. [CrossRef]
16. Wang, M.; Chen, Q.; Li, H.; Ma, M.; Zhang, N. Stretchable and Shelf-Stable All-Polymer Supercapacitors Based on Sealed Conductive Hydrogels. *ACS Appl. Energy Mater.* **2020**, *3*. [CrossRef]
17. Shen, Y.; Wang, H.; Liu, Z.; Li, W.; Liu, Y.; Li, J.; Wei, H.; Han, H. Fabrication of a water-retaining, slow-release fertilizer based on nanocomposite double-network hydrogels via ion-crosslinking and free radical polymerization. *J. Ind. Eng. Chem.* **2020**, *93*, 375–382. [CrossRef]
18. Li, J.; Mooney, D.J. Designing hydrogels for controlled drug delivery. *Nat. Rev. Mater.* **2016**, *1*, 16071. [CrossRef]
19. Cheng, D.; Liu, Y.; Yang, G.; Zhang, A. Water- and Fertilizer-Integrated Hydrogel Derived from the Polymerization of Acrylic Acid and Urea as a Slow-Release N Fertilizer and Water Retention in Agriculture. *J. Agric. Food Chem.* **2018**, *66*, 5762–5769. [CrossRef]
20. Mitura, S.; Sionkowska, A.; Jaiswal, A.K. Biopolymers for hydrogels in cosmetics: Review. *J. Mater. Sci. Mater. Med.* **2020**, *31*, 1–14. [CrossRef]
21. Krafcik, M.J.; Macke, N.D.; Erk, K.A. Improved Concrete Materials with Hydrogel-Based Internal Curing Agents. *Gels* **2017**, *3*, 46. [CrossRef]
22. Ma, X.; Zhang, B.; Cong, Q.; He, X.; Gao, M.; Li, G. Organic/inorganic nanocomposites of ZnO/CuO/chitosan with improved properties. *Mater. Chem. Phys.* **2016**, *178*, 88–97. [CrossRef]
23. Rafieian, S.; Mirzadeh, H.; Mahdavi, H.; Masoumi, M.E. A review on nanocomposite hydrogels and their biomedical applications. *Sci. Eng. Compos. Mater.* **2019**, *26*, 154–174. [CrossRef]
24. Meshram, I.; Kanade, V.; Nandanwar, N.; Ingle, P. Super-Absorbent Polymer: A Review on the Characteristics and Application. *Int. J. Adv. Res. Chem. Sci.* **2020**, *7*, 8–21. [CrossRef]
25. Lv, Q.; Shen, Y.; Qiu, Y.; Wu, M.; Wang, L. Poly(acrylic acid)/poly(acrylamide) hydrogel adsorbent for removing methylene blue. *J. Appl. Polym. Sci.* **2020**, *137*, 49322. [CrossRef]
26. Sen, M. *Nanocomposite Materials. Nanotechnology and the Environment*; IntechOpen: London, UK, 2020. [CrossRef]
27. Zou, H.; Wu, S.; Shen, J. Polymer/Silica Nanocomposites: Preparation, Characterization, Properties, and Applications. *Chem. Rev.* **2008**, *108*, 3893–3957. [CrossRef]
28. Li, Y.; Yu, J.; Guo, Z.-X. The influence of interphase on nylon-6/nano-SiO2 composite materials obtained from in situ polymerization. *Polym. Int.* **2003**, *52*, 981–986. [CrossRef]
29. Xia, H.; Wang, Q. Preparation of conductive polyaniline/nanosilica particle composites through ultrasonic irradiation. *J. Appl. Polym. Sci.* **2003**, *87*, 1811–1817. [CrossRef]
30. Rueda, L.I.; Anton, C.C. Effect of the textural characteristics of the new silicas on the dynamic properties of styrene-butadiene rubber (SBR) vulcanizates. *Polym. Compos.* **1988**, *9*, 204–208. [CrossRef]
31. Wang, X.; Zhao, X.; Wang, M.; Shen, Z. The effects of atomic oxygen on polyimide resin matrix composite containing nano-silicon dioxide. *Nucl. Instruments Methods Phys. Res. Sect. B Beam Interact. Mater. Atoms* **2006**, *243*, 320–324. [CrossRef]
32. Zheng, J.; Cui, P.; Tian, X.; Zheng, K. Pyrolysis studies of polyethylene terephthalate/silica nanocomposites. *J. Appl. Polym. Sci.* **2006**, *104*, 9–14. [CrossRef]
33. Ali, A.E.-H.; Shawky, H.; El Rehim, H.A.; Hegazy, E. Synthesis and characterization of PVP/AAc copolymer hydrogel and its applications in the removal of heavy metals from aqueous solution. *Eur. Polym. J.* **2003**, *39*, 2337–2344. [CrossRef]
34. Giraldo, L.J.; Giraldo, M.A.; Llanos, S.; Maya, G.; Zabala, R.D.; Nassar, N.N.; Franco, C.A.; Alvarado, V.; Cortés, F.B. The effects of SiO$_2$ nanoparticles on the thermal stability and rheological behavior of hydrolyzed polyacrylamide based polymeric solutions. *J. Pet. Sci. Eng.* **2017**, *159*, 841–852. [CrossRef]

Article

Development of Eco-Sustainable PBAT-Based Blown Films and Performance Analysis for Food Packaging Applications

Arianna Pietrosanto *[ ], Paola Scarfato [ ], Luciano Di Maio [ ] and Loredana Incarnato

Department of Industrial Engineering, University of Salerno, Via Giovanni Paolo II, 132, 84084 Fisciano (SA), Italy; pscarfato@unisa.it (P.S.); ldimaio@unisa.it (L.D.M.); lincarnato@unisa.it (L.I.)
* Correspondence: arpietrosanto@unisa.it

Received: 9 November 2020; Accepted: 25 November 2020; Published: 27 November 2020

Abstract: In this work, eco-sustainable blown films with improved performance, suitable for flexible packaging applications requiring high ductility, were developed and characterized. Films were made by blending two bioplastics with complementary properties—the ductile and flexible poly(butylene-adipate-*co*-terephthalate) (PBAT) and the rigid and brittle poly(lactic acid) (PLA)—at a 60/40 mass ratio. With the aim of improving the blends' performance, the effects of two types of PLA, differing for viscosity and stereoregularity, and the addition of a commercial polymer chain extender (Joncryl®), were analyzed. The use of the PLA with a viscosity ratio closer to PBAT and lower stereoregularity led to a finer morphology and better interfacial adhesion between the phases, and the addition of the chain extender further reduced the size of the dispersed phase domains, with beneficial effects on the mechanical response of the produced films. The best system composition, made by the blend of PBAT, amorphous PLA, and the compatibilizer, proved to have improved mechanical properties, with a good balance between stiffness and ductility and also good transparency and sealability, which are desirable features for flexible packaging applications.

Keywords: biodegradable polymers; PBAT/PLA; blown films; food packaging; toughness

1. Introduction

Currently, the packaging sector is among the major consumers of plastic materials and, in this field, conventional non-biodegradable polymers are widely employed for their desirable properties. However, they become a major source of waste after use due to their poor biodegradability. Therefore, the use of biodegradable polymers for packaging applications represent an effective strategy to decrease the quantity of plastics waste sent to landfill and facilitate bio-waste collection and organic recycling, therefore reducing the plastics disposal problems [1–3].

Poly(butylene adipate-*co*-terephthalate) (PBAT) is a biodegradable random copolymer, consisting of aromatic and aliphatic chains. Among biodegradable polymers, it stands out for its very high ductility and flexibility, which make PBAT particularly interesting for packaging applications such as plastic bags and wraps. However, its poor stiffness, low transparency, and low seal strength until now have limited its use [4]. In this context, the melt blending of PBAT with another bioplastic could represent an effective and economic way to improve its properties without compromising its biodegradability. Poly(lactic acid) (PLA), is a bio-based and biodegradable polyester with good processability and interesting properties in the packaging field that could also be customized by varying the relative content of the D and L isomers [5,6]. It has high transparency and complementary mechanical properties to PBAT, exhibiting high stiffness but also high brittleness [7,8]. Therefore, the melt blending of these two polymers, by varying the mass ratio of PLA and PBAT in the blend, can be a useful strategy to modulate and tailor the performance of the final product from a rigid and brittle material (100% PLA) to a ductile and flexible one (100% PBAT) [9]. Several authors have explored this possibility, mainly with the aim to reduce the brittleness of blends having PLA as a matrix phase, obtaining interesting advantages in terms of the mechanical performances of such blends. Commercial examples of PLA- and PBAT-based blends are already available on the market under the trade name of Ecovio; however, their high price limits the diffusion of these materials on a large-scale [10]. Furthermore, since the exact composition of these blends is confidential, the study of the factors affecting PLA/PBAT blends properties can be of great importance in order to further enhance their performance, therefore improving their diffusion in the large-scale market. In this context, since PLA and PBAT are not thermodynamically miscible, although they have very close solubility parameters, researchers have also evidenced the necessity to control the morphology and the interfacial adhesion between the phases in order to gain optimized properties of the final product [11].

Among the different strategies, the incorporation of multifunctional chain extenders containing epoxy groups proved to be an effective way to improve the compatibility between these polymers [12–16]. Joncryl® is a commercial food-grade multifunctional epoxy chain extender, specifically designed for biodegradable polymers and PET. Its compatibilization and chain extension mechanisms are obtained by the formation of in situ block copolymers, through the reaction of the epoxy groups with the terminal groups of polyesters, specifically by epoxy ring-opening and subsequent hydrogen abstraction from hydroxyl and carboxylic acid groups [17,18]. The average number of epoxy groups per chains (functionality), which usually ranges between 4 and 9, influence the final branching degree of the formed copolymer and therefore, the processability of the final system, as previously reported [18].

Several studies, performed on PBAT/PLA blends with PLA as a matrix phase, demonstrated that the incorporation of this chain extender, at concentrations ranging from 0.25 to 1 wt %, leading to the formation of extended and branched chains and, at the same time, to the formation of a PLA–Joncryl–PBAT copolymer, which is placed at the interface between the two phases, thus enhancing the interfacial adhesion [19,20] and the final performances of the resulting blends [20–23]. However, few works deal with the effect of Joncryl® in blends with PBAT as a matrix phase and the results reported in the literature have not shown a relevant improvement of the mechanical properties of the resulting systems [24–27].

Moreover, other strategies could also be adopted in order to improve the performance of immiscible blends. In fact, it is widely known that, among several factors, the stereoregularity of a component polymer and the relative viscosity of the two phases can considerably affect the compatibility and the properties of blend systems, as demonstrated for several conventional and biodegradable blend systems [28–30]. As regards PLA and PBAT blends, the effect of PLA tacticity on the blend performance has not been considered until now, and the effect of the viscosity ratio was evaluated only by Lu et al. [31]. They observed that in blends made by dispersed PBAT in the PLA matrix (PBAT/PLA 30/70 *w/w*), an increase in the dispersed phase/matrix viscosity ratio led to an increase in size of PBAT domains and to a decrease in the interfacial tension between the two polymers. However, also in this case, there are no studies on the effect of the viscosity ratio for blends of PLA and PBAT in which PBAT is the matrix phase.

In this work, we focused our attention on PBAT/PLA blends with a high content of PBAT: these systems could be of great importance for applications where high ductility is required, such as packaging applications at low storage temperature, as demonstrated in our previous work [9]. In particular, here, we intend to investigate the effects of both PLA stereoregularity and PLA/PBAT relative viscosity on the morphology and properties of blend systems, in which PBAT is the matrix phase. To this aim, two commercial types of PLA, with different viscosities and stereoregularity, have been used as blend constituents. Moreover, the effect of the addition of Joncryl® chain extender on the developed morphologies and properties of the produced films was also considered.

2. Materials and Methods

2.1. Materials

PLA 4032D (semicrystalline, D-isomer content = 1.5 wt %, Mw ~241,700 g/mol, specific gravity = 1.24 g/cm^3, T_m = 155–170 °C), named as PLA1, and PLA 4060D (amorphous, D-isomer content = 12 wt %, Mw ~190,000 g/mol, specific gravity = 1.24 g/cm^3), named as PLA2, were supplied by NatureWorks LLC (Minnetonka, MN USA). Ecoworld PBAT 009 (density = 1.26 g/cm^3, T_m = 110–120 °C), composed of 29 wt % of adipic acid, 26 wt % of terephthalic acid, and 45 wt % of 1,4-butanediol, was manufactured by Jin Hui Zhaolong (Lüliang, China). A multifunctional epoxy chain extender named as Joncryl ADR-4368C (referred to as Joncryl in the following), with Mw = 6800 g/mol, epoxy equivalent weight = 285 g/mol, and functionality >4, was supplied by BASF (Ludwigshafen, Germany). All the materials comply with USA FDA and EU regulations for food contact.

2.2. Preparation of the Films

PLA1, PLA2, and PBAT pellets were dried under vacuum at 70 °C for 16 h prior to processing. The polymers and the chain extender were mixed at the compositions reported in Table 1. Each mixture was melt blended in a Collin ZK25 co-rotating twin-screw extruder (COLLIN Lab & Pilot Solutions GmbH, Maitenbeth, Germany, D = 25 mm, L/D = 42) at a constant speed of 100 rpm (mass flow equal to 51–53 g/min) and with a temperature profile ranging from 140 to 180 °C from the hopper to the die. Then, the neat polymers and the two blends were dried under vacuum at 70 °C for 16 h before processing. In order to ensure a homogeneous distribution of the material in the extruder head, the blown films were prepared using two extruders GIMAC (Caserta, Italy, D = 12 mm, L/D = 24) of a multilayer co-extrusion blown film plant. The processing temperature at different zones was set from 190 to 135 °C, the screw speed was 25 rpm (mass flow equal to 17–18 g/min), and the take-up speed was 3 m/min. Films were produced with a blow-up ratio (BUR) and a take-up ratio (TUR) equal to 1.7 and 20, respectively, and an average thickness of 23 ± 0.8 µm.

Table 1. Blends compositions.

Sample	PBAT Content (phr)	PLA 4032 Content (phr)	PLA 4060 Content (phr)	Joncryl Content (phr)
PBAT	100	-	-	-
PLA1	-	100	-	-
PLA2	-	-	100	-
PBAT/PLA1	60	40	-	-
PBAT/PLA1 + J	60	40	-	1
PBAT/PLA2	60	-	40	-
PBAT/PLA2 + J	60	-	40	1

2.3. Films Characterization

The rheological properties in oscillatory mode of the extruded pellets of the neat materials and the blends were measured using an ARES rotational rheometer (Rheometrics, Inc., Piscataway, NJ, USA). Samples were dried under vacuum at 70 °C for 16 h prior to testing. Tests were performed with a parallel-plate geometry (d = 25 mm) with a gap of 1mm at 180 °C under a nitrogen atmosphere. A strain sweep test was initially conducted to guarantee the linear viscoelastic regime for each formulation. Thus, all the frequency sweep tests were performed with a strain equal to 5% and with a frequency ranging from 0.1 to 100 rad/s.

The morphology of the blends was analyzed using a field emission scanning electron microscope (FESEM) (LEO 1525 model, Carl Zeiss SMT AG, Oberkochen, Germany). First, film samples were cryo-fractured and then, coated with a thin gold layer (Agar Auto Sputter Coater mod. 108 A, Stansted, UK) at 30 mA for 160 s to improve their conductivity. After, their cross sections parallel to the transversal direction (TD) were scanned by FESEM.

Thermal analysis was carried out using a Differential Scanning Calorimeter (DSC mod. 822, Mettler Toledo, Columbus, OH, USA) under a nitrogen gas flow (100 mL/min). Three scans were performed; samples were heated from −70 to 200 °C with a speed of 10 °C/min and held at 200 °C for 5 min. After, they were cooled at −70 at 10 °C/min and heated again to 200 °C at 10 °C/min. The crystallinity degree of PLA, Xc, was calculated as follows:

$$X_c = (\Delta H_m - \Delta H_{cc})/(\Delta Hm^0 \times \varphi_i) \times 100 \qquad (1)$$

where ΔH_m and ΔH_{cc} (J/g) are the heat of melting and heat of cold crystallization, respectively, ΔHm_0 is equal to 93.6 J/g [9] for PLA, and φ_i is the relative weight fraction of PLA in the blend.

Tensile testing of the blown films was performed on a SANS dynamometer equipped with a 100 N load cell. The rectangular shape specimens (width = 12.7 mm and length = 30 mm) were extended at a crosshead speed set according to ASTM D822 standard [32]. The mechanical properties were evaluated in the machine direction (MD). All the data are the average of at least ten measurements.

The transparency of the films was evaluated according to ASTM D1746-03 [33]. Films were cut into rectangular shapes and placed on the internal side of a spectrophotometer cell. Then, the transmittance was measured using a UV–vis spectrophotometer (Lambda 800, PerkinElmer, Waltham, MA, USA) at 560 nm. Three replicates of each film were tested. The percent transparency (TR) was calculated as follows:

$$TR = T_r/T_0 \times 100 \qquad (2)$$

where T_r is the transmittance with the specimen in the beam and T_0 is the transmittance with no specimen in the beam.

Hot-tack strength was evaluated using a heat-sealing machine (mod. HSG-C by Brugger, Munich, Germany) equipped with a Hot Tack Device. A pair of film ribbons with 15 mm width were fitted between two heated bars and hot-pressed together at 80 N for 0.5 s, at 85 °C, according to ASTM F1921 [34]. The hot-tack data were measured just after the films were heat-sealed and, while still hot, they were pulled apart, recording the weight required to separate the two sealed surfaces. The result is an average of three specimens.

3. Results and Discussion

3.1. Rheological Analysis

PBAT, PLA1, and PLA2 resins were submitted for rheological measurements in order to investigate the differences in their flow behavior. The complex viscosity (η^*) and storage modulus (G') curves, obtained at 180 °C, are compared in Figure 1a,b. The graphs of Figure 1 show that all the neat polymers exhibit at low frequency a Newtonian plateau and a shear thinning behavior at higher ω values. Both types of PLA have complex viscosity markedly higher than that of PBAT in the whole analyzed frequency range. Moreover, PLA1 has higher viscosity and storage modulus than PLA2, as expected. In fact, PLA1 has lower D-lactide content (higher stereoregularity), which increases the secondary forces in the polymer melt, and higher molecular weight [35,36]. Computing the viscosity ratios (i.e., viscosity of dispersed phase / viscosity of matrix phase) of the two pairs of polymer melts at different frequencies, we obtained $\eta^*_{PLA1}/\eta^*_{PBAT}$ from 5.4 (ω = 0.1 rad/s) to 4.8 (ω = 100 rad/s) and $\eta^*_{PLA2}/\eta^*_{PBAT}$ from 1.7 (ω = 0.1 rad/s) to 2.0 (ω = 100 rad/s). Differences in the viscosity values can affect the morphology of the resulting blend system [29–31]. If the viscosity ratio of the melts is much higher than one, a coarse morphology of the blend should be expected, while if it is close to one, it is possible to achieve a finer morphology. Therefore, in this case, a better mixing can be supposed for the blends with PLA2 than those with PLA1.

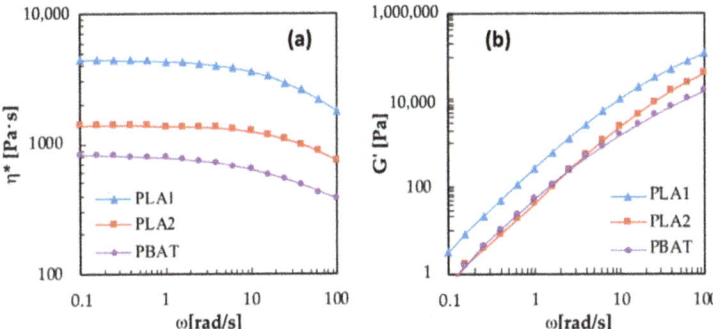

Figure 1. (a) Complex viscosity and (b) Storage Modulus of the neat polymers.

The complex viscosity and storage modulus of the blends are reported in Figure 2a,b. Both the uncompatibilized blends showed, at 0.1 rad/s, a complex viscosity between those of the respective neat materials. Particularly, the PBAT/PLA2 blend had a lower complex viscosity with respect to PBAT/PLA1 in all the analyzed frequency range, according to the different viscosities of the neat PLA resins. Moreover, the first system also exhibited a more accentuated shear thinning behavior compared to the second one and to the respective neat materials, which could be attributed to the occurrence of reactions (e.g., transesterification) between PLA2 and PBAT, which resulted in a wider molecular weight distribution [37,38].

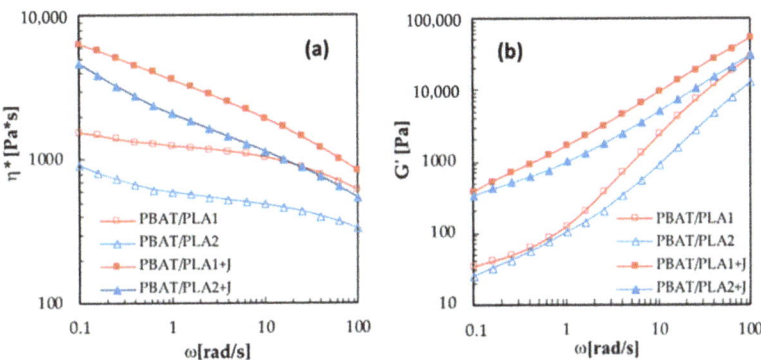

Figure 2. (a) Complex viscosity and (b) storage modulus of the blends.

For both the blends, the addition of the chain extender resulted in an increase in the complex viscosity values at all frequencies and in a strong enhancement of the shear thinning behavior, as predictable from the literature [19–21]. It is widely known that zero-shear viscosity is related to the molecular weight of the polymer and the enhanced shear thinning behavior is attributable to the increase in chain branching and molecular weight distribution. Therefore, our rheological analysis suggests that in the used processing conditions, the addition of the multifunctional epoxy compatibilizer led to the formation of the PLA–Joncryl–PBAT copolymer, having an increased length and a more branched structure compared to the PBAT and PLA in the uncompatibilized blends, as found also by others in reactive extrusion experiments of PLA and PBAT with an epoxy compatibilizer [13,18,20].

As reported in Figure 2b, the storage modulus of PBAT/PLA1 blend was also higher with respect to PBAT/PLA2, according to the storage modulus of the respective neat materials. However, the PBAT/PLA1 blend exhibited a shoulder of G' at low frequency values, which is due to the additional elastic response generated by the surface tension of the dispersed domains in the continuous matrix [9], which suggests higher surface tension between PBAT and PLA1 with respect to PLA2.

The addition of the compatibilizer led to an increase in the storage moduli for both the systems, which was more relevant for low frequency values, as a result of the longer relaxation times of the compatibilized blends, which are owed by the formation of longer and more branched chains and to a more entangled structure [19].

Moreover, the incorporation of Joncryl led to a reduction in the shoulder at low frequency observed for the blend containing PLA1, therefore suggesting that the presence of the chain extender led to a reduction in the surface tension between PLA1 and PBAT [15].

In Figure 3, the storage modulus is plotted versus the dissipative one for all the blended systems.

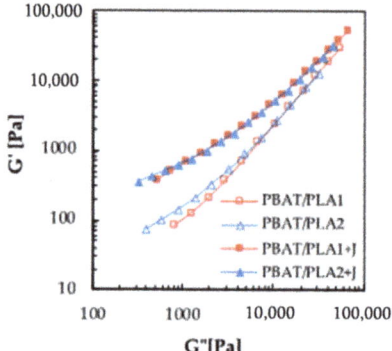

Figure 3. Plot of G' versus G'' for the blended systems.

It has been demonstrated that the plot of G' versus G'' can be used as a criterion of compatibility of a blended system, since it gives composition-independent correlations for compatible blends and composition-dependent correlations for incompatible blends [39]. According to this affirmation, as reported in Figure 3, both the compatibilized systems showed the same correlation between G' and G'', while the G'–G'' correlations were not coincident for the un-compatibilized blends, which is a further confirmation of the compatibilization effect of Joncryl. Moreover, the PBAT/PLA2 system exhibited a G'–G'' correlation closer to the compatibilized blend with respect to the PBAT/PLA1 one, suggesting, together with the complex viscosity and storage modulus curves, a higher compatibility between PLA2 and PBAT compared to PLA1.

3.2. Morphology

The effects of both the type of PLA and the incorporation of Joncryl compatibilizer on the PBAT/PLA blend morphology were investigated by means of FESEM analysis. The images taken on cryo-fractured film sections are reported in Figure 4. Both the uncompatibilized blends (Figure 4a,b) showed the typical two-phase morphology of immiscible systems, with PLA domains dispersed in the PBAT matrix with quite uniform distribution and average size. However, in the PBAT/PLA1 blend, the droplets had bigger dimensions (>2 μm) than in the PBAT/PLA2 one (<1 μm) and were pulled out by cryo-fracturing, leaving empty cavities in the PBAT matrix. These findings are coherent with the rheological results. In fact, since PLA2 has a complex viscosity closer to PBAT compared to PLA1, it was possible to have, in the same process conditions, a better mixing for the blend containing PLA2 that was traduced in a finer and more homogeneous morphology, with fewer and smaller voids and a more ambiguous interface compared to the PBAT/PLA1 blend, indicative of a better interfacial adhesion. Two factors can contribute to these findings. One is the formation of PLA–PBAT mixed chains (copolymers) that can be generated during the melt blending of PBAT and PLA: the reaction is influenced by the melt viscosity ratio of the neat polymers and by their molecular mobility in the melt state, as demonstrated by other authors [37]. The other is the lower stereoregularity of PLA2 compared to PLA1. Because of this, PLA2 has a more flexible polymer chain; therefore, it can be hypothesized that the higher segmental mobility may enhance the polar group accessibility, consequently increasing the possibility of hydrogen bonding to PBAT. Additional analyses are necessary to clarify this point, which will be investigated in further work.

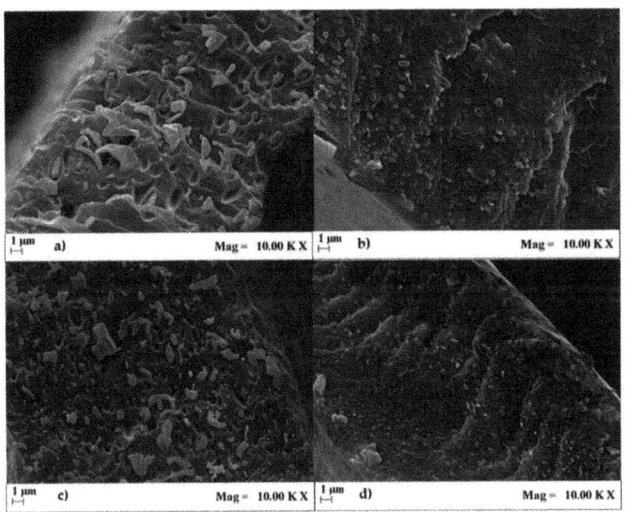

Figure 4. SEM picture of the fracture surfaces of (**a**) PBAT/PLA1, (**b**) PBAT/PLA2, (**c**) PBAT/PLA1 + J, and (**d**) PBAT/PLA2 + J.

Both the compatibilized blends (Figure 4c,d) kept the two-phase morphology; however, the addition of Joncryl led to a relevant reduction in the dispersed phase size without changing its shape, compared to the respective uncompatibilized blend. This indicated a decrease in the interfacial tension and a higher compatibility between the two constituents due to the in situ formation of the PLA–Joncryl–PBAT copolymer based on the combination of PLA and PBAT chains, which is placed at the interface between the two phases, as demonstrated by others on blends of PLA and PBAT of different compositions [16,18,24]. Moreover, the effect of the chain extender seemed more evident for blends containing PLA1 than for those containing PLA2, likely owing to the weaker interactions between PLA1 and PBAT that could be enhanced by the addition of the compatibilizer.

3.3. Thermal Properties

The thermal properties of a polymeric system considerably influence the performance of the final product. Therefore, the thermal properties of the films were also investigated. The thermograms and the main thermal parameters of the films related to the first heating scan are reported in Figure 5 and Table 2, respectively.

From the thermograms, it is evident that PLA1 is a semi-crystalline polymer, while PLA2 is completely amorphous; differences in their thermal behavior owe to their different content of D-isomer (see Materials section). They both exhibited a glass transition temperature around 60 °C and PLA1 also showed a cold crystallization and a melting peak at 97 and 170 °C, respectively. A small exothermic peak just before the melting point can be observed in the thermogram of PLA1, which is related to the transition from the α' crystal form to the more stable α ones [40]. Neat PBAT exhibited a lower glass transition temperature compared to PLA, and two melting points—the first one related to the melting of the butylene adipate fraction (around 49 °C) and the second one to the co-crystallization of butylene adipate units into butyl terephthalate crystals (around 111 °C) [41].

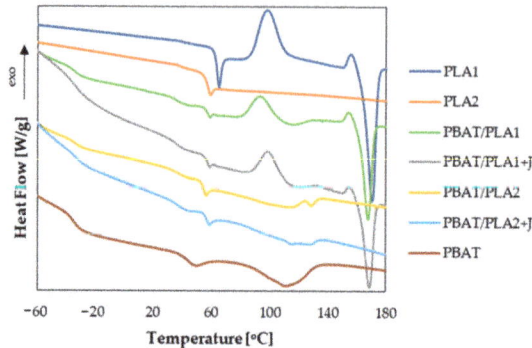

Figure 5. Thermograms of the film related to the first heating scan.

Table 2. The main thermal properties of the films related to the first heating scan: Glass Transition Temperature (T_g), Cold Crystallization Temperature (T_{cc}), Cold Crystallization Enthalpy (ΔH_{cc}), Melting Temperature (T_m), Melting Enthalpy (ΔH_m), and Crystallinity degree (X_c).

Sample	T_g^{PBAT} (°C)	T_g^{PLA} (°C)	T_{cc} (°C)	ΔH_{cc} (J/g)	T_{m1}^{PBAT} (°C)	T_{m2}^{PBAT} (°C)	ΔH_m^{PBAT} (J/g)	T_m^{PLA} (°C)	ΔH_m^{PLA} (J/g)	X_c^{PLA} (%)
PBAT	−35.4	-	-	-	48.8	110.6	17.2	-	-	-
PLA1	-	63.4	97.5	28.5	-	-	-	170.1	30.1	1.7
PLA2	-	56.3	-	-	-	-	-	-	-	-
PBAT/PLA1	−34.1	57.1	92.6	8.1	41.1	115.4	2.3	167.1	13.8	15.2
PBAT/PLA1 + J	−33.9	57.2	98.1	8.9	41.0	116.5	1.7	168.2	14.1	13.9
PBAT/PLA2	−33.3	54.6	-	-	42.3	111.8–124.1	7.2	-	-	-
PBAT/PLA2 + J	−34.2	56.3	-	-	42.5	114.4–125.3	5.3	-	-	-

All the blends showed both the glass transitions of the neat polymers, representative of their immiscibility [9,42], and all the main thermal transitions of the respective components. As regards the un-compatibilized blends, the blending with PBAT led to a reduction in the cold crystallization temperature of PLA1 and an increase in its crystallinity degree, since PBAT increases the crystallization rate of PLA, as previously reported [25]. However, the melting peak of PBAT in the PBAT/PLA1 film was partially hidden by the cold crystallization of PLA; therefore, the calculation of its crystallinity degree was not possible.

The blends containing PLA2 exhibited a double PBAT endothermic peak, which can be attributable to the formation of two separate crystalline phases resembling those of the homopolymers: polybutylene adipate and poly butyl terephthalate [41]. Since this isodimorphic behavior of PBAT in PBAT/PLA2 blends is different from the neat PBAT and in the blends containing PLA1, which exhibited one melting peak at 111–115 °C, it can be supposed that PLA2 has influenced the crystallization behavior of PBAT, as a confirmation of the highest interaction between PBAT and PLA2.

The presence of Joncryl led to an increase in the cold crystallization temperature of PLA1 for the PBAT/PLA1 + J film and to a slight decrease in the melting enthalpies of PBAT for both the compatibilized blends, which is attributable to the increase in the molecular weight and branching density of the polymers that hindered the crystallization process [22,25]. Moreover, the presence of the compatibilizer also increased the glass transition of PLA2, a sign of the reduced mobility of the amorphous chain segments, which have a more rigid structure [43].

3.4. Mechanical Properties

The results obtained from the tensile tests are reported in Figure 6. It is evident that PLA and PBAT have complementary mechanical properties: the first is rigid and brittle, while the latter is flexible and tough [9]. Moreover, comparing the two types of PLA, it can be observed that PLA1 has a higher elastic modulus and yield stress than PLA2 too, which is related to the amorphous nature of PLA2 due to its high D-isomer content that led to a worsening of the materials' rigidity [44].

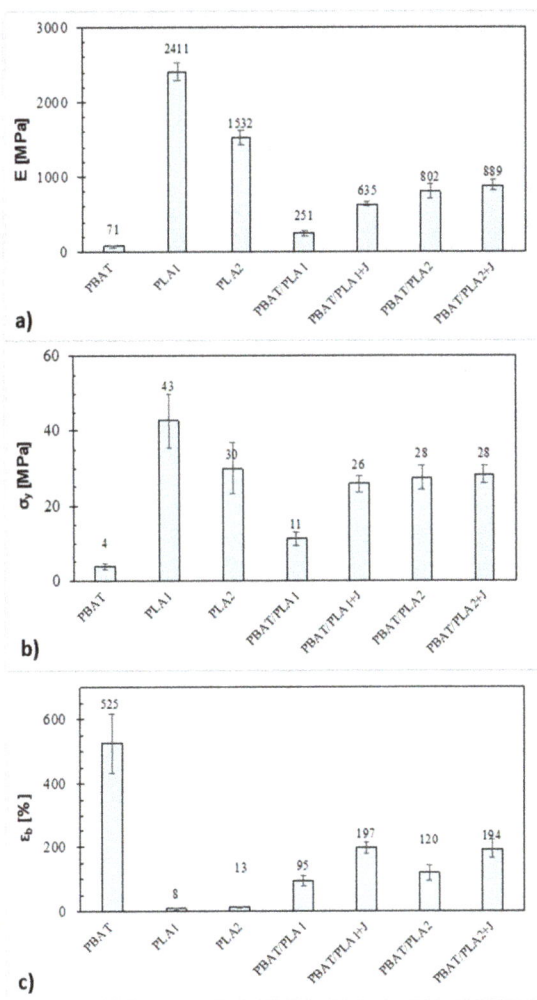

Figure 6. (a) Elastic modulus, (b) yield stress, and (c) elongation at break of the films.

All the blends showed mechanical properties between those of the corresponding neat materials. From the comparison of two uncompatibilized systems, it is evident that the PBAT/PLA2 blend, although based on the less rigid PLA, had a considerably higher elastic modulus (800 MPa) and yield stress (28 MPa) than the PBAT/PLA1 one, which exhibited an elastic modulus and yield stress of 251 and 11MPa, respectively. Since both the elastic modulus and the yield stress are greatly affected by the nature of the interface of a multi-phase system and the blend morphology [45], this result is a further confirmation of a better compatibility and higher strength of the interactions between PBAT and PLA2 with respect to PLA1, as observed in the rheological and morphological analysis. These findings point out that the viscosity ratio and the nature of the interface between the blend constituents had a more important role than the individual resin properties in determining the blend strength and stiffness. Nevertheless, as regards the ductility of the films, the use of different PLA did not considerably affect the elongation at break.

With the addition of Joncryl, the elastic modulus and yield stress of the system containing PLA1 were more than tripled and doubled, respectively, since the addition of the chain extender enhanced the strength of the interactions between PLA1 and PBAT and therefore, considerably improved the interface of the resulting system, while, for the system containing PLA2, the effect of the compatibilizer was not so effective. In particular, the elastic modulus slightly increased, and the yield stress did not change, because the PBAT/PLA2 blend already showed a good phase adhesion, also without the addition of the compatibilizer, as revealed by the morphological analysis. However, although greatly improved, the stiffness and the strength of the PBAT/PLA1 + J blend (E = 635 MPa and σ_y = 26 MPa) were still lower than those of PBAT/PLA2 + J (E = 889 MPa and σ_y = 28 MPa).

Moreover, the addition of Joncryl also led to an improvement in the ductility of the systems, whose value was doubled for both the blends, owing to the reduction in the dispersed phase domains, as found in the SEM analysis.

Up to now, similar benefits due to Joncryl addition were reported only for PLA/PBAT with PLA as a matrix phase [20–22,25,26], whereas investigations on these blends with PLA as a dispersed phase showed a worsening of the mechanical performance. Only Nunes et al. [24] reported that the addition of 0.5 wt % Joncryl led to an improvement of the ductility of the systems, but with a negative impact on the materials' rigidity, since the concentration used was not enough to increase the interfacial adhesion between the phases. Therefore, from the comparison of the present results with the literature data, it turns out that the relative viscosity between the phases, the content of the compatibilizer, and the blending process conditions play a key role in determining the effectiveness of the chain extension reaction, the morphology, and thus, the final performance of the resulting system.

On the whole, among all the blends investigated in this study, the system that allowed the achievement of the best mechanical properties was PBAT/PLA2 + J, which exhibited a good compromise between stiffness and ductility.

3.5. Optical Properties

Transparency is an important physical property of packaging films, and transparent film materials are highly desirable for a number of packaging applications [46]. The transparency values are displayed in Table 3. Both the types of PLA films exhibited high transmittance; in particular, PLA2 was slightly more transparent than PLA1 due to its amorphous nature. On the other hand, light transmission of the PBAT film was almost prevented, in accordance with the literature data [47]. The blends showed transparency values between those of the neat materials. PBAT/PLA1 films had lower transparency than PBAT/PLA2 and the difference in the transparency values of the blends was markedly higher than the neat PLAs. This is attributable to the fact that, as reported in the thermal analysis, blending semi-crystalline PLA with PBAT, led to an increase in its crystallinity degree, therefore resulting in a lower transparency of the PBAT/PLA1 films with respect to PBAT/PLA2.

Moreover, the addition of the compatibilizer slightly increased the transparency of both the systems due to the reduction in the crystallinity degree of the polymers.

Table 3. Transparency of the films.

Sample	Transparency (%)
PBAT	5.8 ± 0.7
PLA1	89.7 ± 0.2
PLA2	91.0 ± 0.8
PBAT/PLA1	8.2 ± 0.1
PBAT/PLA1 + J	12.4 ± 0.1
PBAT/PLA2	40.7 ± 0.7
PBAT/PLA2 + J	43.0 ± 0.9

The transparency values of the PBAT/PLA2 films, with and without the compatibilizer, were comparable with those reported for commonly used non-biodegradable plastics such as polyethylene (PE) [48]; therefore, even if they are lower than those of PLA, they can be considered as acceptable for packaging application requiring see-through properties.

3.6. Hot-Tack Measurements

Sealability is one of the key performance requirements for flexible packaging that allows packages to be made at high packaging speeds and keeps the product secure [49].

As reported in Table 4, amorphous PLA exhibited the highest seal strength, while the semi-crystalline one proved to be not sealable. In fact, it is widely known that the sealability of a material is also linked to its crystallinity degree and amorphous polymers have higher chain mobility on the surface of the film, leading to higher diffusion and thus, higher adhesion strength [50].

Table 4. Hot-tack measurements of the films.

Sample	Hot-Tack Strength (g/15 mm)
PBAT	125 ± 5
PLA1	-
PLA2	650 ± 10
PBAT/PLA1	-
PBAT/PLA1 + J	-
PBAT/PLA2	610 ± 10
PBAT/PLA2 + J	600 ± 15

Consequently, since the seal strength of PBAT was markedly lower than PLA2, blends containing PLA1 had a seal strength lower than 100 g/15 mm, while PBAT/PLA2 blends proved to have a seal strength close to PLA2, which was not considerably influenced by the incorporation of the compatibilizer. Few works have reported on the sealability of bioplastics, particularly, as regards PBAT/PLA blends, only Tabasi et al. [51] investigated the hot-tack behavior of PBAT/PLA blends with PLA as a matrix phase, obtaining higher values of the hot-tack strength due to the higher content of PLA in the blend. However, the hot-tack strength of PBAT/PLA2 blends was comparable with those reported for PE films [52]; therefore, PBAT/PLA2 films proved to also have adequate seal properties as flexible packaging materials.

4. Conclusions

In this work, eco-sustainable PBAT/PLA blown packaging films, having high toughness and being suitable for direct food contact, were successfully produced.

In order to optimize film performances, PBAT/PLA blends in a 60/40 mass ratio were produced starting from two commercial types of PLA with different viscosities and stereoregularity, with and without the use of the multifunctional epoxy chain extender Joncryl, so as to investigate the effects of the different viscosity ratio and stereoregularity of the polymer melts and the chain extension reaction on the blend constituent compatibility and on the final film properties.

Rheological and morphological investigations have shown that the PBAT/PLA2 blend, whose PLA had a lower stereoregularity and a viscosity closer to PBAT than the PLA1 blend, has a finer dispersion and distribution of the dispersed PLA phase and a stronger interfacial adhesion between phases. Consequently, the PBAT/PLA2 system shows better mechanical performance than the corresponding blend with PLA1, especially in terms of stiffness, even if based on the less rigid PLA2.

Moreover, the use of an amorphous PLA, such as PLA2, allowed the obtaining of better transparency and hot-tack strength, which are other key properties of flexible packaging films.

For both the investigated blend systems, but to a greater extent for the PBAT/PLA1 blend, the addition of Joncryl promoted interactions between the two constituents. This resulted in a finer morphology and better mechanical response of the compatibilized systems, in terms of elastic modulus, yield stress, and elongation at break. However, only minor changes were measured in terms of transparency and hot-tack strength.

On the whole, the best performing system was the PBAT/PLA2 + J film, which exhibited the best compromise in terms of stiffness and ductility, with mechanical performances (E = 889 MPa, σ_y = 28 MPa, ϵ_b = 194%) and transparency and hot-tack strength (43% and 600 g/15mm, respectively), comparable with those reported for commonly used non-biodegradable plastics.

Author Contributions: Conceptualization, L.I.; data curation, A.P. and L.D.M.; formal analysis, L.D.M. and P.S.; funding acquisition, L.I.; investigation, A.P.; supervision, P.S. and L.I.; writing—original draft, A.P.; writing—review and editing, P.S. and L.I. All authors have read and agreed to the published version of the manuscript.

Funding: This research received no external funding.

Acknowledgments: Thanks to BASF for providing Joncryl ADR-4368C.

Conflicts of Interest: The authors declare no conflict of interest.

References

1. PlasticsEurope. Plastic-the Facts 2019: An Analysis of European Plastics Production, Demand and Waste Data. 2019. Available online: https://www.plasticseurope.org/en/resources/publications/1804-plastics-facts-2019 (accessed on 2 September 2020).
2. Balart, R.; Montanes, N.; Dominici, F.; Boronat, T.; Torres-Giner, S. Environmentally Friendly Polymers and Polymer Composites. *Materials* **2020**, *13*, 4892. [CrossRef] [PubMed]
3. Apicella, A.; Scarfato, P.; Di Maio, L.; Incarnato, L. Sustainable Active PET Films by Functionalization with Antimicrobial Bio-Coatings. *Front. Mater.* **2019**, *6*, 243. [CrossRef]
4. Ferreira, F.V.; Cividanes, L.S.; Gouveia, R.F.; Lona, L.M.F. An overview on properties and applications of poly(butylene adipate-co-terephthalate)–PBAT based composites. *Polym. Eng. Sci.* **2019**, *59*, 7–15. [CrossRef]
5. Ncube, L.K.; Ude, A.U.; Ogunmuyiwa, E.N.; Zulkifli, R.; Beas, I.N. Environmental Impact of Food Packaging Materials: A Review of Contemporary Development from Conventional Plastics to Polylactic Acid Based Materials. *Materials* **2020**, *13*, 4994. [CrossRef]
6. Scarfato, P.; Di Maio, L.; Milana, M.R.; Giamberardini, S.; Denaro, M.; Incarnato, L. Performance properties, lactic acid specific migration and swelling by simulant of biodegradable poly(lactic acid)/nanoclay multilayer films for food packaging. *Food Addit. Contam. A* **2017**, *34*, 1730–1742. [CrossRef]
7. He, H.; Wang, G.; Chen, M.; Xiong, C.; Li, Y.; Tong, Y. Effect of Different Compatibilizers on the Properties of Poly (Lactic Acid)/Poly (Butylene Adipate-Co-Terephthalate) Blends Prepared under Intense Shear Flow Field. *Materials* **2020**, *13*, 2094. [CrossRef]
8. Aliotta, L.; Vannozzi, A.; Panariello, L.; Gigante, V.; Coltelli, M.B.; Lazzeri, A. Sustainable micro and nano additives for controlling the migration of a biobased plasticizer from PLA-based flexible films. *Polymers* **2020**, *12*, 1366. [CrossRef]
9. Pietrosanto, A.; Scarfato, P.; Di Maio, L.; Nobile, M.R.; Incarnato, L. Evaluation of the suitability of poly(lactide)/poly(butylene-adipate-co-terephthalate) blown films for chilled and frozen food packaging applications. *Polymers* **2020**, *12*, 804. [CrossRef]
10. Hamad, K.; Kaseem, M.; Ayyoob, M.; Joo, J.; Deri, F. Polylactic acid blends: The future of green, light and tough. *Prog. Polym. Sci.* **2018**, *85*, 83–127. [CrossRef]

11. Su, S.; Duhme, M.; Kopitzky, R. Uncompatibilized PBAT/PLA Blends: Manufacturability, Miscibility and Properties. *Materials* **2020**, *13*, 4897. [CrossRef]
12. Al-Itry, R.; Lamnawar, K.; Maazouz, A. Improvement of thermal stability, rheological and mechanical properties of PLA, PBAT and their blends by reactive extrusion with functionalized epoxy. *Polym. Degrad. Stab.* **2012**, *97*, 1898–1914. [CrossRef]
13. Corre, Y.M.; Duchet, J.; Reignier, J.; Maazouz, A. Melt strengthening of poly (lactic acid) through reactive extrusion with epoxy-functionalized chains. *Rheol. Acta* **2011**, *50*, 613–629. [CrossRef]
14. Li, X.; Yan, X.; Yang, J.; Pan, H.; Gao, G.; Zhang, H.; Dong, L. Improvement of compatibility and mechanical properties of the poly(lactic acid)/poly(butylene adipate-co-terephthalate) blends and films by reactive extrusion with chain extender. *Polym. Eng. Sci.* **2018**, *58*, 1868–1878. [CrossRef]
15. Zhang, N.; Zeng, C.; Wang, L.; Ren, J. Preparation and Properties of Biodegradable Poly(lactic acid)/Poly(butylene adipate-co-terephthalate) Blend with Epoxy-Functional Styrene Acrylic Copolymer as Reactive Agent. *J. Polym. Environ.* **2013**, *21*, 286–292. [CrossRef]
16. Freitas, A.L.P.D.L.; Tonini Filho, L.R.; Calvão, P.S.; de Souza, A.M.C. Effect of montmorillonite and chain extender on rheological, morphological and biodegradation behavior of PLA/PBAT blends. *Polym. Test.* **2017**, *62*, 189–195. [CrossRef]
17. Torres-Giner, S.; Montanes, N.; Boronat, T.; Quiles-Carrillo, L.; Balart, R. Melt grafting of sepiolite nanoclay onto poly(3-hydroxybutyrate-co-4-hydroxybutyrate) by reactive extrusion with multi-functional epoxy-based styrene-acrylic oligomer. *Eur. Polym. J.* **2016**, *84*, 693–707. [CrossRef]
18. Quiles-Carrillo, L.; Montanes, N.; Lagaron, J.M.; Balart, R.; Torres-Giner, S. In Situ Compatibilization of Biopolymer Ternary Blends by Reactive Extrusion with Low-Functionality Epoxy-Based Styrene–Acrylic Oligomer. *J. Polym. Environ.* **2018**, *27*, 10. [CrossRef]
19. Al-Itry, R.; Lamnawar, K.; Maazouz, A. Reactive extrusion of PLA, PBAT with a multi-functional epoxide: Physico-chemical and rheological properties. *Eur. Polym. J.* **2014**, *58*, 90–102. [CrossRef]
20. Al-Itry, R.; Lamnawar, K.; Maazouz, A. Rheological, morphological, and interfacial properties of compatibilized PLA/PBAT blends. *Rheol. Acta* **2014**, *53*, 501–517. [CrossRef]
21. Wang, X.; Peng, S.; Chen, H.; Yu, X.; Zhao, X. Mechanical properties, rheological behaviors, and phase morphologies of high-toughness PLA/PBAT blends by in-situ reactive compatibilization. *Compos. Part B Eng.* **2019**, *173*, 107028. [CrossRef]
22. Al-Itry, R.; Lamnawar, K.; Maazouz, A. Biopolymer blends based on poly (lactic acid): Shear and elongation rheology/structure/blowing process relationships. *Polymers* **2015**, *7*, 939–962. [CrossRef]
23. Wang, Y.; Fu, C.; Luo, Y.; Ruan, C.; Zhang, Y.; Fu, Y. Melt synthesis and characterization of Poly(L-lactic acid) chain linked by multifunctional epoxy compound. *J. Wuhan Univ. Technol. Mater. Sci. Ed.* **2010**, *25*, 774–779. [CrossRef]
24. Nunes, E.; de Souza, A.G.; dos Rosa, S.D. Effect of the Joncryl® ADR Compatibilizing Agent in Blends of Poly(butylene adipate-co-terephthalate)/Poly(lactic acid). *Macromol. Symp.* **2019**, *383*, 1800035. [CrossRef]
25. Li, X.; Ai, X.; Pan, H.; Yang, J.; Gao, G.; Zhang, H.; Yang, H.; Dong, L. The morphological, mechanical, rheological, and thermal properties of PLA/PBAT blown films with chain extender. *Polym. Adv. Technol.* **2018**, *29*, 1706–1717. [CrossRef]
26. Arruda, L.C.; Magaton, M.; Bretas, R.E.S.; Ueki, M.M. Influence of chain extender on mechanical, thermal and morphological properties of blown films of PLA/PBAT blends. *Polym. Test.* **2015**, *43*, 27–37. [CrossRef]
27. Schneider, J.; Manjure, S.; Narayan, R. Reactive modification and compatibilization of poly(lactide) and poly(butylene adipate-co-terephthalate) blends with epoxy functionalized-poly(lactide) for blown film applications. *J. Appl. Polym. Sci.* **2016**, *133*, 43310. [CrossRef]
28. Yoshie, N.; Azuma, Y.; Sakurai, M.; Inoue, Y. Crystallization and compatibility of poly(vinyl alcohol)/poly(3-hydroxybutyrate) blends: Influence of blend composition and tacticity of poly(vinyl alcohol). *J. Appl. Polym. Sci.* **1995**, *56*, 17–24. [CrossRef]
29. Mbarek, S.; Jaziri, M.; Chalamet, Y.; Carrot, C. Effect of the viscosity ratio on the morphology and properties of pet/hdpe blends with and without compatibilization. *J. Appl. Polym. Sci.* **2010**, *117*, 1683–1694. [CrossRef]
30. Ostafinska, A.; Fortelný, I.; Hodan, J.; Krejčíková, S.; Nevoralová, M.; Kredatusová, J.; Kruliš, Z.; Kotek, J.; Šlouf, M. Strong synergistic effects in PLA/PCL blends: Impact of PLA matrix viscosity. *J. Mech. Behav. Biomed. Mater.* **2017**, *69*, 229–241. [CrossRef]

31. Lu, X.; Zhao, J.; Yang, X.; Xiao, P. Morphology and properties of biodegradable poly (lactic acid)/poly (butylene adipate-co-terephthalate) blends with different viscosity ratio. *Polym. Test.* **2017**, *60*, 58–67. [CrossRef]
32. ASTM D882-18. *Standard Test Method for Tensile Properties of Thin Plastic Sheeting*; ASTM International: West Conshohocken, PA, USA, 2018.
33. ASTM D1746-03. *Standard Test Method for Transparency of Plastic Sheeting*; ASTM International: West Conshohocken, PA, USA, 2003.
34. ASTM F1921/F1921M-12. *Standard Test Methods for Hot Seal Strength (Hot Tack) of Thermoplastic Polymers and Blends Comprising the Sealing Surfaces of Flexible Webs*; ASTM International: West Conshohocken, PA, USA, 2018.
35. Puchalski, M.; Kwolek, S.; Szparaga, G.; Chrzanowski, M.; Krucinska, I. Investigation of the influence of PLA molecular structure on the crystalline forms (α'' and α) and Mechanical Properties of Wet Spinning Fibres. *Polymers* **2017**, *9*, 18. [CrossRef] [PubMed]
36. Di Maio, L.; Garofalo, E.; Scarfato, P.; Incarnato, L. Effect of polymer/organoclay composition on morphology and rheological properties of polylactide nanocomposites. *Polym. Compos.* **2015**, *36*, 1135–1144. [CrossRef]
37. Coltelli, M.B.; Toncelli, C.; Ciardelli, F.; Bronco, S. Compatible blends of biorelated polyesters through catalytic transesterification in the melt. *Polym. Degrad. Stab.* **2011**, *96*, 982–990. [CrossRef]
38. Signori, F.; Coltelli, M.-B.; Bronco, S. Thermal degradation of poly(lactic acid) (PLA) and poly(butylene adipate-co-terephthalate) (PBAT) and their blends upon melt processing. *Polym. Degrad. Stab.* **2009**, *94*, 74–82. [CrossRef]
39. Dae Han, C.; Chuang, H.-K. Criteria for Rheological Compatibility of Polymer Blends. *J. Appl. Polym. Sci.* **1985**, *30*, 4431–4454. [CrossRef]
40. Androsch, R.; Schick, C.; Di Lorenzo, M.L. Melting of conformationally disordered crystals (α' phase) of poly(L-lactic acid). *Macromol. Chem. Phys.* **2014**, *215*, 1134–1139. [CrossRef]
41. Cranston, E.; Kawada, J.; Raymond, S.; Morin, F.G.; Marchessault, R.H. Cocrystallization model for synthetic biodegradable poly(butylene adipate-co-butylene terephthalate). *Biomacromolecules* **2003**, *4*, 995–999. [CrossRef]
42. Gigante, V.; Cinelli, P.; Righetti, M.C.; Sandroni, M.; Tognotti, L.; Seggiani, M.; Lazzeri, A. Evaluation of mussel shells powder as reinforcement for pla-based biocomposites. *Int. J. Mol. Sci.* **2020**, *21*, 5364. [CrossRef]
43. Sohel, M.A.; Mondal, A.; Sengupta, A. Effect of Physical Aging on Glass Transition and Enthalpy Relaxation in PLA Polymer Filament. *Mater. Sci. Biophys.* **2018**, 191–194.
44. Farah, S.; Anderson, D.G.; Langer, R. Physical and mechanical properties of PLA, and their functions in widespread applications—A comprehensive review. *Adv. Drug Deliv. Rev.* **2016**, *107*, 367–392. [CrossRef]
45. Bartczak, Z.; Galeski, A. Mechanical Properties of Polymer Blends. In *Polymer Blends Handbook*; Utracki, L.A., Wilkie, C.A., Eds.; Springer: Dordrecht, The Netherlands, 2014; pp. 1203–1297. ISBN 978-94-007-6064-6.
46. Lee, J.W.; Son, S.M.; Hong, S.I. Characterization of protein-coated polypropylene films as a novel composite structure for active food packaging application. *J. Food Eng.* **2008**, *86*, 484–493. [CrossRef]
47. Wang, L.F.; Rhim, J.W.; Hong, S.I. Preparation of poly(lactide)/poly(butylene adipate-co-terephthalate) blend films using a solvent casting method and their food packaging application. *LWT Food Sci. Technol.* **2016**, *68*, 454–461. [CrossRef]
48. Moreno-Vásquez, M.J.; Plascencia-Jatomea, M.; Ocaño-Higuera, V.M.; Castillo-Yáñez, F.J.; Rodríguez-Félix, F.; Rosas-Burgos, E.C.; Graciano-Verdugo, A.Z. Engineering and antibacterial properties of low-density polyethylene films with incorporated epigallocatechin gallate. *J. Plast. Film Sheeting* **2017**, *33*, 413–437. [CrossRef]
49. Butler, T.I.; Morris, B.A. PE-Based Multilayer Film Structures. In *Plastic Films in Food Packaging. Plastic Design Library*; Ebnesajjad, S., Ed.; William Andrew Publishing: Waltham, MA, USA, 2013; pp. 21–52.
50. Tabasi, R.Y.; Najarzadeh, Z.; Ajji, A. Development of high performance sealable films based on biodegradable/compostable blends. *Ind. Crops Prod.* **2015**, *72*, 206–213. [CrossRef]

51. Tabasi, R.Y.; Ajji, A. Tailoring heat-seal properties of biodegradable polymers through melt blending. *Int. Polym. Process.* **2017**, *32*, 606–613. [CrossRef]
52. Shih, H.H.; Wong, C.M.; Wang, Y.C.; Huang, C.J.; Wu, C.C. Hot Tack of Metallocene Catalyzed Polyethylene and Low-Density Polyethylene Blend. *J. Appl. Polym. Sci.* **1999**, *73*, 1769–1773. [CrossRef]

Publisher's Note: MDPI stays neutral with regard to jurisdictional claims in published maps and institutional affiliations.

 © 2020 by the authors. Licensee MDPI, Basel, Switzerland. This article is an open access article distributed under the terms and conditions of the Creative Commons Attribution (CC BY) license (http://creativecommons.org/licenses/by/4.0/).

Article

Grafted Lactic Acid Oligomers on Lignocellulosic Filler towards Biocomposites

Anna Czajka [1,*], Radosław Bulski [1], Anna Iuliano [2], Andrzej Plichta [2], Kamila Mizera [3] and Joanna Ryszkowska [1]

[1] Faculty of Materials Science and Engineering, Warsaw University of Technology, Wołoska 141, 02-507 Warsaw, Poland; radoslaw.bulski.stud@pw.edu.pl (R.B.); joanna.ryszkowska@pw.edu.pl (J.R.)
[2] Faculty of Chemistry, Warsaw University of Technology, Noakowskiego 3, 00-664 Warsaw, Poland; anna.iuliano@pw.edu.pl (A.I.); andrzej.plichta@pw.edu.pl (A.P.)
[3] Central Institute for Labour Protection—National Research Institute, Czerniakowska 16, 00-701 Warsaw, Poland; kamiz@ciop.pl
* Correspondence: anna.czajka2.dokt@pw.edu.pl

Abstract: Lactic acid oligomers (OLAs) were in situ synthesized from lactic acid (LAc) and grafted onto chokeberry pomace (CP) particleboards by direct condensation. Biocomposites of poly (lactic acid) (PLA) and modified/unmodified CP particles containing different size fractions were obtained using a mini-extruder. To confirm the results of the grafting process, the FTIR spectra of filler particles were obtained. Performing ^{1}HNMR spectroscopy allowed us to determine the chemical structure of synthesized OLAs. The thermal degradation of modified CP and biocomposites were studied using TGA, and the thermal characteristics of biocomposites were investigated using DSC. In order to analyse the adhesion between filler particles and PLA in biocomposites, SEM images of brittle fracture surfaces were registered. The mechanical properties of biocomposites were studied using a tensile testing machine. FTIR and ^{1}HNMR analysis confirmed the successful grafting process of OLAs. The modified filler particles exhibited a better connection with hydrophobic PLA matrix alongside improved mechanical properties than the biocomposites with unmodified filler particles. Moreover, a DSC analysis of the biocomposites with modified CP showed a reduction in glass temperature on average by 9 °C compared to neat PLA. It confirms the plasticizing effect of grafted and ungrafted OLAs. The results are promising, and can contribute to increasing the use of agri-food lignocellulosic residue in manufacturing biodegradable packaging.

Keywords: lignocellulosic material; chemical modification; poly(lactic acid) composites; in situ polymerization; grafting biocomposites; biodegradable composites

1. Introduction

These days, environmental issues are becoming key when designing plastic materials. Following the EU Strategy for Plastics in the Circular Economy [1], the goal is to achieve the sustainable management of the product at every stage of its life (extraction, manufacture, use, and disposal). After use, the material should be reused in accordance with the circular economy where nothing is wasted.

The reuse of agri-food lignocellulosic residue fits with this idea. Lignocellulosic residue from the agri-food industry is used as an animal feed to produce biogas or extracts. An interesting form of managing this waste is its use as a natural filler in polymer composites. Such a strategy was proposed by R. Turco et al. [2], where they developed a composite based on poly (lactic acid) (PLA) reinforced by epoxidized oil and presscake waste fibers from oil extraction of the *Cynara cardunculus* plant. Rocha et al. compatibilized natural lignocellulosic residues such as sugarcane bagasse, maçaranduba, and pinus through starch also as a PLA fillers [3]. Mysiukiewicz and Barczewski also used linseed cake to make green composites on the PLA matrix [4]. Many authors have reported using lignocellulosic fibers (not only of waste origin) as fillers for plastics. The most interesting and advantageous in terms of ecological aspects is its use as a filler for biodegradable plastic, such as PLA [5], polyalkanolates (PHA) [6,7]. According to Mohanty et al., a biocomposite created with a biodegradable polymer matrix with a biodegradable filler should allow the obtainment of a biodegradable composite [8]. Using lignocellulosic fillers is also common for fossil plastic, such as polypropylene (PP) [9] and polyethylene (PE) [10,11].

The most promising polymer to replace petroleum-based plastics is PLA, which is a biodegradable aliphatic semi-crystalline polyester obtained in the industry by the ring-opening polymerization (ROP) of lactide (LA) [12]. LA is formed by the dimerization-cyclisation of lactic acid oligomers, which are produced by the condensation of lactic acid (LAc). The latter is usually obtained by the fermentation of agricultural raw materials (saccharides). Due to the non-petroleum origin of LAc, LA, and PLA, and polymer's ability to biodegradation [13], it is considered as a "double green" polymer. PLA and LA copolymers (e.g., block copolymers [14]) may be used for biomedical or pharmaceutical [15–18] applications. However, the main market of these polymers is packaging [19,20].

The main purpose of adding lignocellulosic filler in particleboards is to cut down on material cost. However, reinforcement by lignocellulosic fibers such as jute [21,22] or flax [23] may improve the mechanical properties compared to the unfilled polymer matrix. The most basic ingredients of lignocellulose filler are cellulose, hemicellulose, lignin, waxes, pectin, and water-soluble ingredients. The content of these components may vary depending on the growing conditions and the methods determining their content [24].

Cellulose is considered to be the main component of the fiber backbone. It is a polysaccharide with a semi-crystalline structure of D-glucopyranose units. It provides strength, rigidity, and structural stability of the fibers. Hemicellulose is an amorphous, highly branched polysaccharide, associated with cellulose possibly by hydrogen bonds. It is mainly made of hexoses and pentoses of varied chemical structures. The highly polar and hydrophilic nature of cellulose and hemicellulose is due to the presence of a large amount of the hydroxyl group. Lignin, as well as hemicellulose, is an amorphous polymer composed of phenylpropane units [24–26]. Thanks to the high content of aromatic rings, it shows a higher hydrophobic character. Additionally, its location between cellulose and hemicellulose chains protect against environmental conditions such as temperature and humidity [27].

Unfortunately, lignocellulosic fillers have a number of disadvantages due to their chemical structure. Due to the presence of numerous OH groups, these fillers are characterized by high water absorption. The highly hydrophilic surface has poor adhesion to the hydrophobic polymer matrix in the polymer composite. This causes the deterioration of the mechanical properties, changes in the dimensions of the composite, and water absorption. One of the methods of improving these properties is the chemical modification of lignocellulosic fillers. Its main purpose is to activate the OH group or introduce new groups. As a result, the surface of the filler becomes more hydrophobic, which increases adhesion to the polymer matrix and even meshes with it [24].

Researchers use various chemical treatments. For example, the alkaline processing of roselle and sugar palm reinforced thermoplastic polyurethane showed that fiber surface modification improved hybrid composites' mechanical, physical, and thermal properties [28]. A.K. Bledzki et al. [29] proved that the acetylation of flax fiber can reduce water absorption by 42%. The esterification of cellulose with citric acid by X. Cui et al. [30] showed enhanced flexural modulus and stress.

In 2012 alone, 469,200 tons of waste was generated in Poland during the processing and preservation of fruit and vegetables [31]. According to *Statistics Poland* data, more than 50,000 tons of chokeberry were produced in Poland in 2018 [32]. A crucial part of becoming more eco-friendly is reducing the carbon footprint associated with transporting wastes. The reuse of agri-food waste is also significant in protecting the environment. In order to reduce the transport and management costs for locally produced waste, chokeberry pomace (CP) was chosen as a filler.

In this paper, biocomposites based on PLA and modified lignocellulosic filler in the form of CP, which are residue from the food industry, were prepared. The purpose of the modification is to increase its hydrophobicity and the compatibility of the filler with the polymer matrix. The modification consisted of the reaction of free hydroxyl groups present in lignin, cellulose, and hemicellulose with carboxyl groups (esterification reaction) of LAc and its oligomers (OLAs) by the direct condensation of LAc on CP fibers. Additionally, free OLAs formed in condensation polymerization (ungrafted to the lignocellulosic backbone) could act as a PLA plasticizer. Furthermore, the influence of the particle size of lignocellulosic fillers on thermal and mechanical properties was investigated. The main reason was to check whether the change in the proportion of modified filler fraction after modification and grinding affects the properties. It is worth emphasizing that the modification process was made without using organic solvents, which increases the eco-friendliness of the process. This type of modification (by esterification), known as *grafting*, has been the subject of J. Ambrosio-Martín et al.'swork [33] who modified bacterial cellulose nanowhiskers (bacterial CNW). R. Patwa et al. [34] applied a similar procedure for bacterial cellulose modification. Using OLAs as a plasticizer was reported by scientists [35–37]. N. Burgos et al. showed that OLAs with a molar mass around 1000 g/mol can be used as a biodegradable plasticizer for PLA, replacing conventional plasticizers [36]. The use of a PLA plasticizer seems necessary due to the high glass transition temperature (T_g = 50–60 °C), which causes brittleness and stiffness at room temperature [38].

2. Materials and Methods

2.1. Materials

Chokeberry pomace (CP) was kindly supplied by the company AGROPOL Sp. z o.o., Góra Kalwaria, Poland. DL-LAc (racemic mixture) was supplied as an 85 wt% aqueous solution by Sigma Aldrich (St. Louis, MO, USA). $SnCl_2 \cdot 2H_2O$ by Sigma Aldrich (St. Louis, MO, USA) was used as a catalyst. DMSO-d_6 (0.03% TMS, min. 99.8% deuterization degree) was used as a solvent for ^{1}HNMR analysis. Chloroform stabilized with amylene, supplied by Pol-Aura (Olsztyn, Poland) was used for Soxhlet extraction. Poly (lactic acid) (PLA) (Ingeo^{TM} 2003D; M_n = 108 kg/mol; 4 wt% D-isomer) from NatureWorks LLC (Minnetonka, MN, USA) was used as polymeric matrix. All chemicals were used as supplied without any purification.

2.2. Preparation of Chokeberry Pomace Filler

Pre-dried CP (40 °C, 10 days) was milled in sieve mill MUKF-10 (Młynpol P.P.H., Wyszków, Poland) using a sieve with a mesh size of 200 μm. Milled CP was subjected to modification reaction. Before introducing unmodified CP filler into the polymer matrix, it was fractionated using a sieve machine Haver EML 200 Premium Remote (Haver & Boecker OHG, Oelde, Germany) using a sieve with mesh size: 250, 125, and 63 μm. The fraction proportion in the milled CP filler is shown in Table 1. In prepared biocomposites in further analysis, this fraction proportions was **called "Mix"**.

Table 1. Fraction proportions of milled CP filler (Mix).

Type of Filler	Fraction, μm	Fraction Content, %	Name of Fraction Proportion
CP	250–125	32.7	Mix
	125–63	49.1	
	<63	18.2	

For thermal and structural comparison to modified CP filler, 1.5 g of CP sample was purified using 220 mL of chloroform in Soxhlet extraction for 10 h and dried in 60 °C for 8 h. This step provided a fat-free sample, the same as after purification of modified CP filler (CP-pure).

2.3. Grafting of Chokeberry Pomace Using LAc and OLAs

The modification was carried out in a 1:1 ratio (*w/w*) (CP filler: DL-LAc) and 90 ppm of catalyst (for the sum of the reactants). The water content of the CP filler was carried out using a moisture analyzer Axis ATS147-2 from Axis sp. z o. o. (Gdańsk, Poland) (program details: 105 °C; ending parameters: at least 3 measurements with a constant weight every 5 s). The amount of used chemicals is shown in Table 2. All chemicals were added to the two-liter glass reactor. After 2 h of mixing (with constant stirring) under reflux, the solvent was distilled both in atmospheric and under reduced pressure. Then, the condensation process was carried out for 8 h at 160 °C within the pressure range 0.11 to 0.02 mbar. The scheme of the reaction process is shown in Figure 1.

Table 2. Modification details.

CP Filler, g	Water Content in CP Filler, %	DL-Lac [1], g	$SnCl_2 \cdot 2H_2O$ [1], mg	Total Water, g
180	4.0	180	39.4	630

[1] With respect to the 100% reactants.

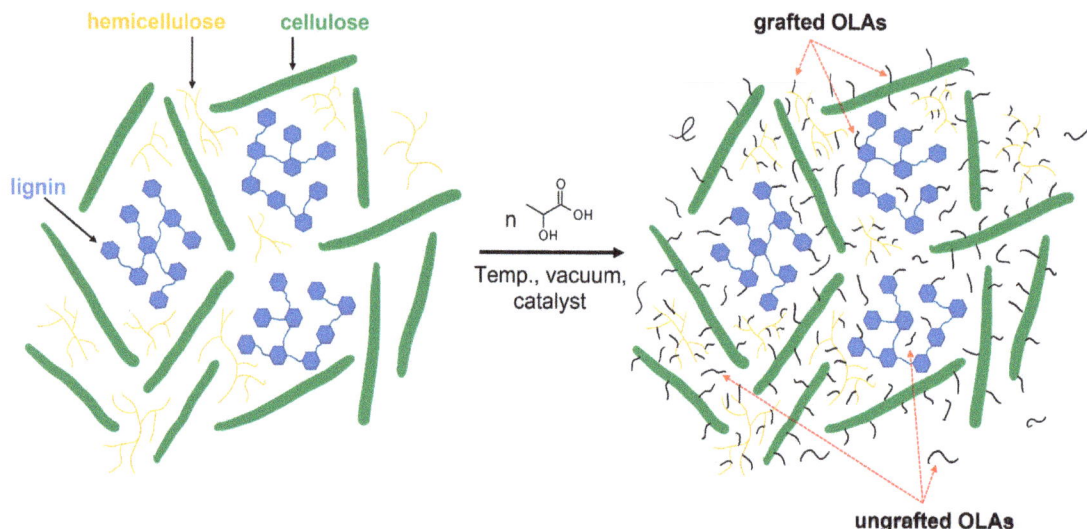

Figure 1. Scheme of chemical modification.

After the modification, the modified filler agglomerated into hard bulks that required shredding. Due to this, bulks of the modified CP filler (CP-g-OLA) were pre-crushed using a hammer and crushed into the powder using a mortal grinder Pulverisette 2 from Fritsch GmbH (Idar-Oberstein, Germany). Grinded CP-g-OLA was fractioned in the same way as unmodified CP filler (Section 2.2). The fraction proportions in the milled CP-g-OLA filler are shown in Table 3. In prepared biocomposites in further analysis, this fraction proportions was called **"Mix 1"**.

Table 3. Fraction proportions of grinded CP filler after the modification (Mix 1).

Type of CP Filler	Fraction, μm	Fraction Content, %	Type of Mix
CP-g-OLA	250–125	43.5	Mix 1
	125–63	33.4	
	<63	23.1	

To determine the ungrafted OLAs content in the CP-g-OLA and purify the CP-g-OLA for further analysis (CP-g-OLA-pure), the sample was purified in the same way as CP-pure (Section 2.2).

2.4. Preparation of Biocomposites

Biocomposites were made using a mini-extruder Haake MiniLab II (Thermo Fisher Scientific, Waltham, MA, USA) equipped with recycle channel and co-rotating conical system of two screws. The mix of filler and PLA granules (5.5 g) were added to the mini-extruder during 3 min of loading time, and it was then mixed for 20 min at a screw speed of 25 rpm in cycle mode in 170 °C. Then, a cylindrical profile formed using Ø 1 mm die was collected for further analysis.

Types of prepared composites are shown in Table 4.

Table 4. Manufactured biocomposites composition.

Sample Name	Type of Composite	Filler Quantity, wt%	PLA Quantity, wt%	Ungrafted OLAs Quantity, wt%	Fraction of the Filler, μm
PLA	Neat PLA	0	100	0	-
<63 CP/PLA	CP/PLA	30	70	0	<63
63–125 CP/PLA	CP/PLA	30	70	0	63–125
Mix CP/PLA	CP/PLA	30	70	0	Mix
Mix1 CP/PLA	CP/PLA	30	70	0	Mix 1
<63 CP-g-OLA/PLA	CP-g-OLA/PLA	30	67.09	2.91 [2]	<63
63–125 CP-g-OLA/PLA	CP-g-OLA/PLA	30	67.09	2.91 [2]	63–125
Mix CP-g-OLA/PLA	CP-g-OLA/PLA	30	67.09	2.91 [2]	Mix
Mix1 CP-g-OLA/PLA	CP-g-OLA/PLA	30	67.09	2.91 [2]	Mix 1

[2] Quantity of ungrafted OLAs, LAc, and LA where the main component is OLAs (for more details look at Table 5).

It is worth noticing that CP-g-OLA was not purified before the formation of the biocomposites due to the use of ungrafted OLAs containing minor quantities of LAc and LA. The amount of ungrafted OLAs in the biocomposite composition is shown in Table 4, determined in Soxhlet extraction. Also, the CP filler was not purified before the formation of the biocomposites.

Table 5. ^{1}HNMR results.

Sample	$\overline{DP}_{OLA}$	M_n, g/mol	[COOH]/[OH]	X_{LA}	X_{LAc}	X_{OLA}
CP-g-OLA	1.9	281	0.62	0.11	0.09	0.80

2.5. Characterisation of Fillers and Biocomposites

The surface morphology of the CP fillers and brittle fracture surface of biocomposites was determined using Hitachi TM3000 SEM (Hitachi Group, Tokyo, Japan). The applied accelerating voltage was 15 kV. All samples were coated using a Polaron SC7640 sputter coater (Quorum Technologies Ltd., Laughton, UK) for 80 s at 10 mA and 1.5 kV with gold and palladium before SEM imaging.

The chemical structure of the CP and CP-g-OLA-pure fillers was studied with the Fourier transform infrared spectroscopy (FTIR) using a Nicolet 6700 spectrometer (Thermo Electron Corporation, Waltham, MA, USA). Spectral data were collected as a sum of 64 scans in the 4000–400 cm^{-1} range with manual baseline correction and CO_2 correction. The results were analyzed using the OMNIC 8.2.0 software by ThermoFisher Scientific Inc.

The chemical structure of OLAs was performed by proton nuclear magnetic resonance spectroscopy (^{1}HNMR) using the equipment: Varian NMR System 500 (Varian, Inc., Palo Alto, CA, USA). The applied frequency was 500 MHz, measured at room temperature. Before measurement, around 80 mg of CP-g-OLA was diluted in 1.5 mL of DMSO-d$_6$ and filtered.

Thermogravimetric analysis (TGA) of the CP-pure, CP-g-OLA, CP-g-OLA-pure, and biocomposites was performed using TGA Q500 (TA Instruments, New Castle, PA, USA). The sample weight was around 10 mg. The sample was heated in ramp procedure (10 °C/min) from room temperature to 650 °C in a nitrogen atmosphere. Data analysis was performed using the Universal Analysis 2000 software, version 4.7 A, by TA Instruments. The measurements were performed using 3 samples.

The scanning of differential calorimetry (DSC) of the biocomposites was performed using DSC Q1000 (TA Instruments, New Castle, PA, USA). A sample weight of 6 ± 0.2 mg was sealed in an aluminum pan. A sample was heated from room temperature to 190 °C with a heating speed of 10 °C/min (first heating cycle); cooling to -80 °C at 5 °C/min (first cooling cycle) and heating to 190 °C at 10 °C/min (second heating cycle) in an inert atmosphere. For further characterization, a second heating cycle was used. The Universal Analysis 2000 software version 4.7 A by TA Instruments was used to determine thermal parameters. The measurements were performed using 3 samples.

The water absorption of modified and unmodified CP fillers was defined as the weight change using TGA. To dry out the fillers, a sample of around 10 mg was heated to 60 °C and held for 180 min. After 24 h, the procedure was repeated.

The mechanical properties of the biocomposites were carried out using Instron 5566 (Instron Norwood, MA, USA) (load cell of 1 kN) tensile testing machine. Cylindrical profiles obtained as a result of extruding were cut into 80 mm long samples and weighed to determine the tex factor. The speed of the tensile strength test was 20 mm/min. The measuring base was 20 mm. The cross-section was determined as a linear mass (tex). To change the unit to MPa, the result in N/tex was multiplied by 900. The measurements were performed using at least 15 samples.

3. Results

3.1. Analysis of the Chokeberry Pomace Fillers

3.1.1. Chemical Structure of the Fillers

FTIR measurements characterized both CP-pure and CP-g-OLA-pure (after elimination of ungrafted residue) fillers to confirm the chemical reaction between the LAc and hydroxyl groups of lignin, cellulose, and hemicellulose. FTIR spectra are shown in Figure 2. Analyzing the CP-g-OLA-pure filler spectrum shows a peak at around 3200 cm^{-1} assigned to the OH stretching vibration decreased compared to ungrafted CP filler. It is a result of the esterification of the hydroxyl group by LAc, which results in a reduction in OH group concentration. At the same time, the peak around 1735 cm^{-1} assigned to the C=O stretching vibration increased, proving the formation of an ester bond. Y. Luan et al. obtained similar results during grafting cellulose acetate by ROP process [39], the same as A. Goffin et al., who grafted CNW by the ROP process [40]. Additionally, the FTIR spectrum of CP-g-OLA-pure shows new peaks around 1200 and 1090 cm^{-1} assigned to the C-O stretching vibration of OLAs (PLA) backbone [41]. This evidence of chemical structure supports the success of the esterification of hydroxyl groups with LAc and/or OLAs, and incorporating OLAs chains into the macromolecules forming lignocellulosic fillers' backbones.

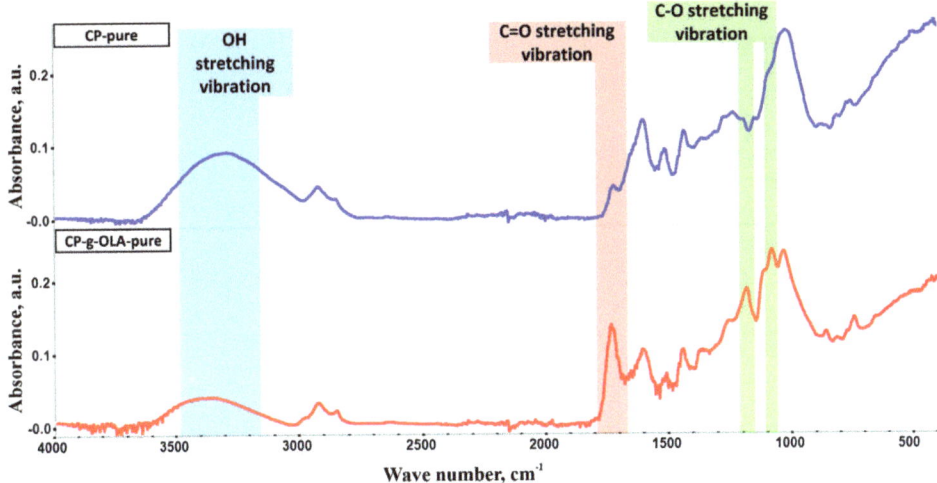

Figure 2. FTIR spectra of CP-pure and CP-g-OLA-pure.

3.1.2. Scanning Electron Microscopy of the Fillers

Figure 3a shows the CP filler after milling. There are significant particle size dispersion and shape differences. It is probably related to the heterogeneous composition of the chokeberry pomace, where you can find stems, fruits, leaves, seeds, etc. Figure 3b presents the CP filler after the modification process. The lower quantity of fibrous particles is likely due to the subsequent crushing of the filler after modification.

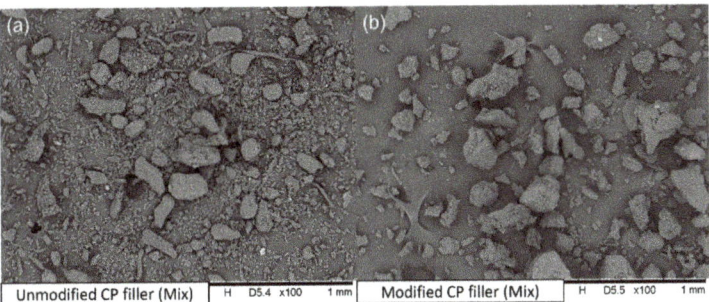

Figure 3. SEM pictures of (**a**) unmodified CP filler; (**b**) modified CP filler.

3.1.3. Proton Nuclear Magnetic Resonance Analysis

The determination of the degree of polymerization of OLAs ($\overline{DP}_{OLA}$) (grafted and ungrafted), carboxyl group to hydroxyl group ratio ([COOH]/[OH]), the mole fraction of LA (χ_{LA}), LAc (χ_{LAc}), and OLAs (χ_{OLA}) (grafted and ungrafted) was performed using ^{1}HNMR analysis (Table 5, Figure 4). $\overline{DP}_{OLA}$ value was determined from the equation:

$$\overline{DP}_{OLA} = \sum integral\ CH / integral\ CH(OH\ end) \qquad (1)$$

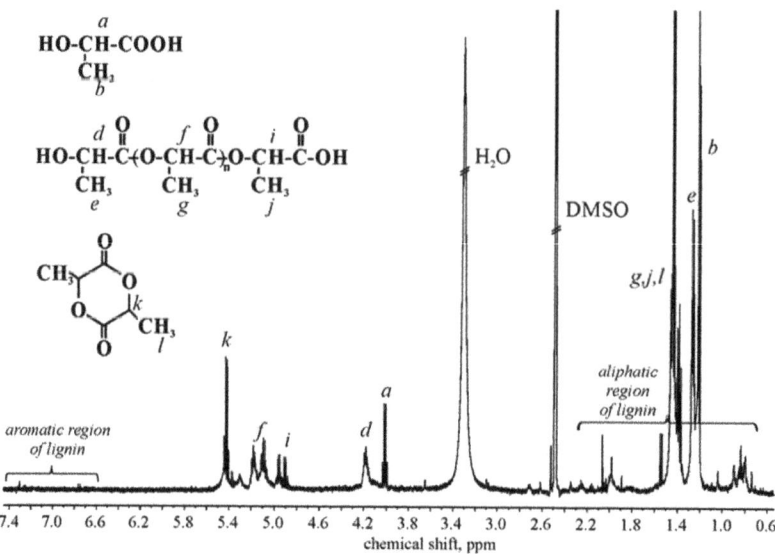

Figure 4. ^{1}HNMR spectrum of CP-g-OLA.

According to J. Espartero et al., signals from OLAs, LA, and LAc have been assigned in Figure 4 [42]. A [COOH]/[OH] value deviating from 1 suggests that in the spectrum, some protons of OLAs grafted to the lignocellulosic backbone are also seen. The signals of protons present in the backbone of hemicellulose and cellulose are not visible in the spectra due to the insolubility of the backbone in a deuterated solvent, so we are probably dealing with a dispersion that is not recordable under these ^{1}H NMR conditions. W. Zhao et al. received signals in the ^{1}HNMR spectra in the DMSO-d_6 of hemicellulose units [43], grafted cellulose (at 80 °C) [44], and lignin [45]. However, it is hard to assign protons to detected signals. In this case, probably aromatic and aliphatic lignin units are shown in shift range 6–8 ppm and 0.5–2.75 ppm, respectively [45]. The presence and content of LA ($\chi_{LA} = 0.11$) is related to used catalyst and process conditions. In the obtained composition, some unreacted LAc is also detected.

3.1.4. Thermal Degradation Analysis of the Fillers

TGA analysis was used in order to characterize the fillers (Figure 5a,b).

Based on the TG and DTG curves obtained in TGA analysis, the temperature of 2 and 5% mass loss ($T_{2\%}$ and $T_{5\%}$, respectively), maximum degradation temperature (T_{max}), maximum degradation rate in the third stage of degradation, (V_{max3}), and the amount of residue after combustion at 650 °C (R_{650}) were determined for all fillers (Table 6).

An analysis of CP-pure filler (Figure 5a) shows multiple peak with three stages of degradation: T_{max3}, T_{max4}, and T_{max5}, which are assigned to hemicellulose, cellulose/lignin, and lignin degradation, respectively [46,47]. The shoulder with T_{max2} is probably related to tannin degradation [48]. J. Lisperguer et al. obtained a degradation peak for tannin obtained from *Acacia dealbata* at 258 °C [48]. The chemical composition of this specific CP (with no purification) was determined in the previous report by researchers from the Warsaw University of Technology [49]. Table 7 shows the content of cellulose, hemicellulose, lignin, and raw fat in CP filler.

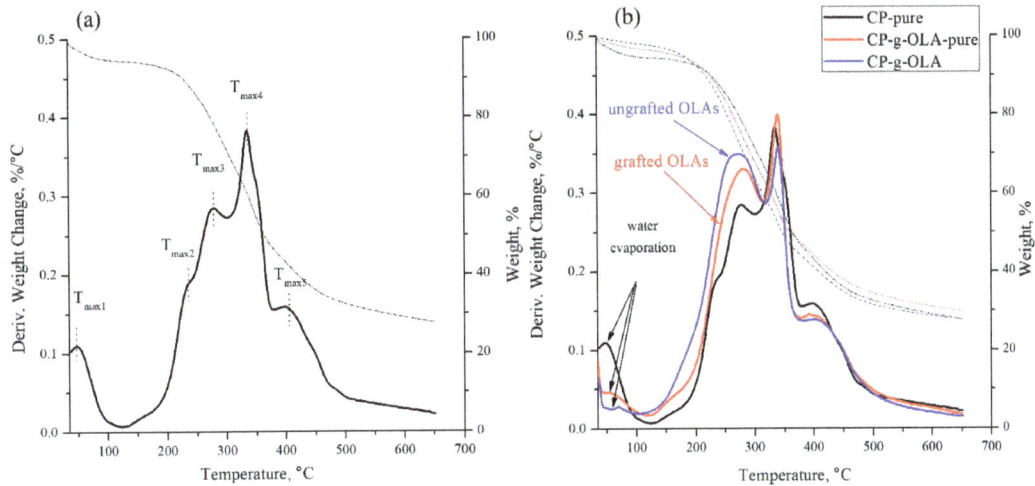

Figure 5. TG and DTG curves of (**a**) CP-pure filler; (**b**) overlaid TG and DTG curves of CP-pure, CP-g-OLA, CP-g-OLA-pure fillers.

Table 6. TGA results of the fillers.

Sample	The Characteristic Temperature of Thermal Decomposition, °C							V_{max3}, %/°C	R_{650}, %
	$T_{2\%}$	$T_{5\%}$	T_{max1}	T_{max2}	T_{max3}	T_{max4}	T_{max5}		
CP-pure	42 ± 4	75 ± 8	52 ± 2	237 ± 3	278 ± 0	335 ± 1	398 ± 1	0.29 ± 0.01	30 ± 1
CP-g-OLA	106 ± 4	169 ± 10	76 ± 7	-	262 ± 11	341 ± 1	404 ± 3	0.40 ± 0.05	25 ± 3
CP-g-OLA-pure	71 ± 6	172 ± 3	67 ± 10	-	283 ± 1	338 ± 1	394 ± 2	0.34 ± 0.01	30 ± 1

Table 7. Chemical composition of CP filler [49].

Filler	Raw Fat, %	Cellulose, %	Hemicellulose, %	Lignin, %
CP	7.3	20.6	21.7	58.0

The high content of lignin may positively impact the modification process because of the shielding for hemicellulose and cellulose against humidity and high temperature, as we mentioned at the beginning of the following discussion [27]. The hydroxyl groups which may react with the LAc and OLAs are related to hemicellulose, lignin, and amorphous cellulose. The crystalline cellulose structure is too closely packed (because of hydrogen bonds), which prevents reacting with those hydroxyl groups. Therefore, a high amount of amorphous units (hemicellulose, lignin) content may help to increase the efficiency of the esterification [29,50].

$T_{2\%}$ and $T_{5\%}$ of CP-pure filler are low (42 ± 4 °C and 71 ± 6 °C, respectively), and it is due to the water evaporation in the weight loss step of maximum temperature T_{max1}. After the modification process, $T_{2\%}$ value increases significantly, and it can be related to the reduced water absorption by modified, more hydrophobic fillers. The difference in 'DTG's curve shape of CP-g-OLA and CP-g-OLA-pure, shown in Figure 5b, in the temperature range around 150–320 °C, is contributed to the grafted and ungrafted OLAs degradation peak superimposed on the peak from tannin and hemicellulose degradation. It is up to a peak with a maximum temperature of approximately 280 °C, and a maximum degradation rate in average range 0.34–0.40%/°C.

Y. Guo et al. also reported an additional T_{max} peak in temperature around 230 °C, corresponding to grafted PLLA chains into the cellulose backbone [51]. N. Burgos et al.

synthesized OLAs as a plasticization system for PLA, and also obtained a similar T_{max} degradation of OLAs [36].

In the next step, in the range of 320–400 °C, cellulose and lignin are degraded. The temperature at which the maximum decomposition rate is achieved in this stage is approximately—340 °C.

After the degradation of the CP filler before and after modification at 650 °C, a significant amount of the sample mass remains-approximately 30%. These are probably the tarry residues of lignin pyrolytic degradation.

3.1.5. Water Absorption Analysis

To determine whether the filler has changed its character to be more hydrophobic after the modification, the water absorption was determined. Table 8 shows the water absorption of CP, CP-g-OLA, and CP-g-OLA-pure fillers after 24 h of exposure in equal conditions.

Table 8. Water absorption results.

Sample	Water Content, %
CP	5.5
CP-g-OLA-pure	3.6
CP-g-OLA	1.8

The results show increasing hydrophobicity of the CP filler after the carried out modification both in purified and impurified (with ungrafted OLAs) fillers for 34.5 and 67.3%, respectively. It confirmed that OLAs have a hydrophobic effect on the surface of lignocellulosic fillers by blocking the hydroxyl groups, which was confirmed by FTIR analysis. Additionally, ungrafted OLAs also have a significant impact on improving the hydrophobicity of the lignocellulosic 'filler's surface. The ungrafted OLAs are probably physically bound to the surface of the modified filler, and they are shielding hydroxyl groups, which reduces water absorption.

3.2. Analysis of Biocomposites

3.2.1. Preparation of Biocomposites

Eight samples of biocomposites comprising 67.09 or 70% of PLA matrix and 30% of various filler fractions or mixtures were prepared via melt mixing in a laboratory extruder at 170 °C for 20 min. A cylindrical profile formed using Ø 1 mm die was collected for further analysis. Detailed information can be found in the Section 2.4.

3.2.2. Scanning Electron Microscopy of the Biocomposites

In order to determine the adhesion between filler particles and the PLA matrix, SEM images of brittle fracture surfaces of biocomposites were obtained (Figure 6).

During the analysis of brittle fracture surfaces of all samples containing particles of unmodified CP filler, it was found that the particles exhibited poor bonding with the PLA matrix. Red-dotted lines in the example SEM images of <63 CP/PLA and Mix CP/PLA samples (Figure 6a,b) indicate areas where CP filler particles can be spotted. Particles are visibly separated from the polymer matrix, and there are no visible indications of good wettability on the particles' surface by PLA. Moreover, some particles were extracted from the matrix while preparing brittle fractures and left voids (yellow-dotted lines in SEM images in Figure 6a,b). However, surface analysis of brittle fractures of all samples containing CP-g-OLA filler showed that particles exhibited better compatibility and adhesion with the PLA matrix. In the example SEM images of 63–125 CP-g-OLA/PLA and Mix CP-g-OLA/PLA samples (Figure 6c,d), black-dotted lines mark particles, and green arrows point out areas where particles show optimal wettability by PLA matrix. A similar observation was made by Cui et al. during a study involving a modification of cellulose with citric acid [30]. In addition, the lack of <63 μm particles (which are present in the fraction proportions

"Mix"; Table 1; Figure 3) on the surface of Mix CP-g-OLA/PLA suggests that these particles were well incorporated into the PLA matrix, proving the effects of modification on better adhesion of CP filler particles with the hydrophobic polymer.

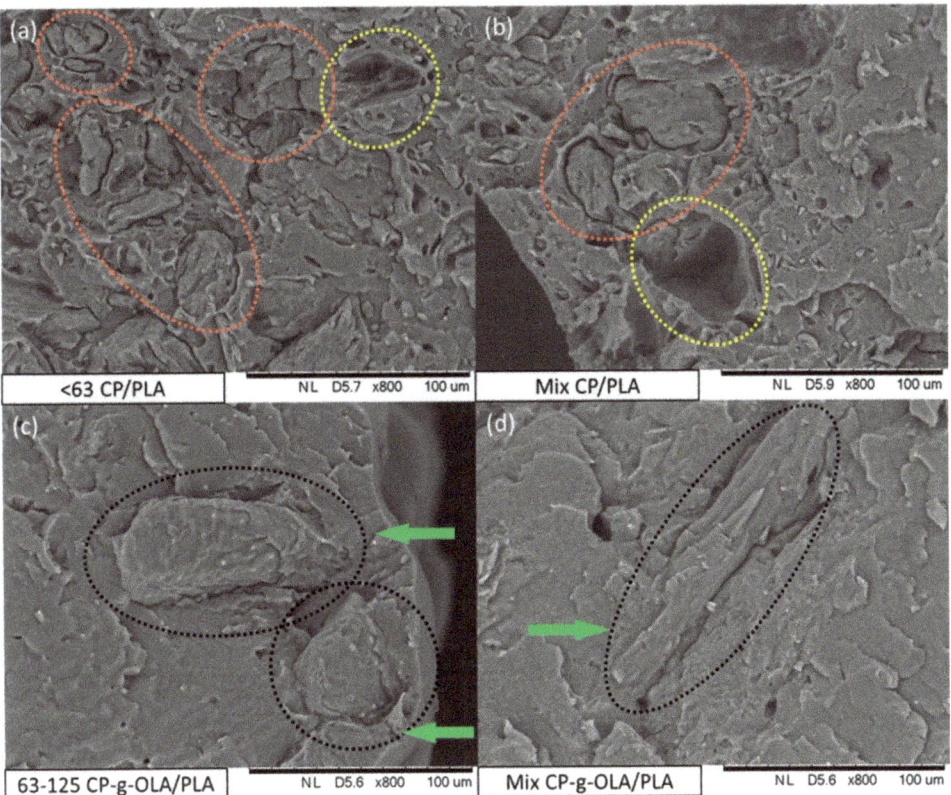

Figure 6. SEM images of (**a**) <63 CP/PLA; (**b**) Mix CP/PLA; (**c**) 63–125 CP-g-OLA/PLA; (**d**) Mix CP-g-OLA/PLA biocomposites.

3.2.3. Thermal Degradation Analysis of the Composites

Based on the TG and DTG curves obtained in the TGA analysis, the maximum degradation rate (T_{max}), the degradation rate (V_{max}), and a temperature of 2%, a 5% material loss ($T_{2\%}$ and $T_{5\%}$) was determined for all biocomposites (Table 9). Representative TG and DTG curves are shown in Figure 7.

TGA analysis showed lower $T_{2\%}$, $T_{5\%}$, T_{max}, and V_{max} for all biocomposites compared to the unfilled PLA matrix. After adding the lignocellulosic filler, a negative impact on thermal stability is also widely reported [38,52]. Composites are expected to have lower thermal stability due to the presence of lignocellulose filler with lower thermal stability than PLA [52]. The modification and particle size of the lignocellulosic filler do not significantly affect the thermal properties of the analysed biocomposites. Some exceptions are $T_{2\%}$ and $T_{5\%}$, which are around 20–30 °C lower than composites with an unmodified filler. Similar results were obtained by N. Burgos et al., and are contributed to the low molar mass of OLAs, which makes OLAs more volatile [36]. The wide shoulder in the range of 200–300 °C and the lower V_{max} of biocomposites compared to neat PLA is also related to the content of lignocellulosic filler and OLAs.

Table 9. TGA results of PLA and biocomposites.

Sample	$T_{2\%}$, °C	$T_{5\%}$, °C	T_{max}/V_{max}, °C/%/°C
PLA	305 ± 1	320 ± 1	360 ± 1/2.78 ± 0.05
<63 CP/PLA	184 ± 1	251 ± 2	350 ± 3/2.15 ± 0.13
63–125 CP/PLA	201 ± 1	259 ± 6	347 ± 12/1.93 ± 0.19
Mix CP/PLA	194 ± 4	254 ± 5	321 ± 5/1.66 ± 0.32
Mix1 CP/PLA	198 ± 2	254 ± 12	350 ± 6/2.01 ± 0.11
<63 CP-g-OLA/PLA	173 ± 1	231 ± 1	346 ± 2/2.08 ± 0.02
63–125 CP-g-OLA/PLA	178 ± 0	234 ± 3	343 ± 4/1.79 ± 0.20
Mix CP-g-OLA/PLA	174 ± 2	231 ± 1	346 ± 2/1.96 ± 0.04
Mix1 CP-g-OLA/PLA	181 ± 4	230 ± 1	343 ± 1/1.69 ± 0.34

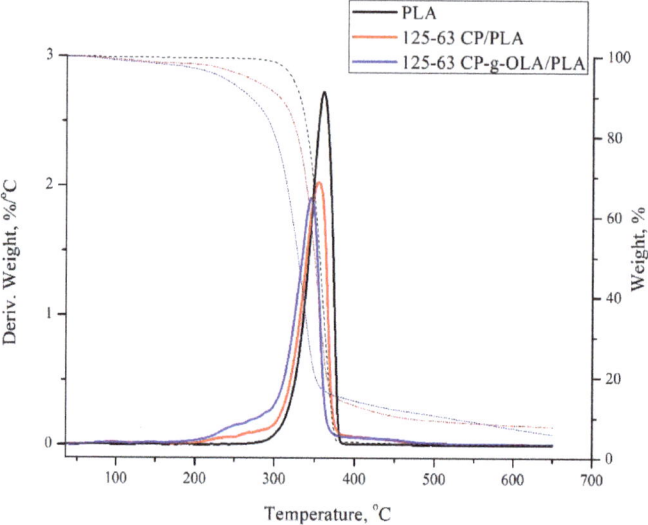

Figure 7. TG and DTG curves of PLA and biocomposites.

3.2.4. Differential Scanning Calorimetry Analysis

Glass transition temperature (T_g), cold crystallization temperature (T_{cc}), and melting temperature (T_m) were determined for all biocomposites using DSC analysis. The crystallinity of the samples (X_c) was calculated according to the following Equation (2) [2,53,54]:

$$X_c, \% = \frac{\Delta H_m - \Delta H_{cc}}{\Delta H_m^0 \cdot \omega_{PLA}} \cdot 100 \qquad (2)$$

where ΔH_{cc} is a cold crystallization enthalpy; ΔH_m is a melting enthalpy; ΔH_m^0 is a melting enthalpy of fully crystalline PLA, which is 93.6 J/g [55], and ω_{PLA} is a weight fraction of PLA in the biocomposites (Table 4) or in neat PLA. Table 10 and Figure 8 show the results of DSC analysis.

The introduction of lignocellulosic filler into the polymer matrix should increase the T_g due to the limitation of the mobility of PLA chains by natural fibers embedded in the matrix [54]. In this case, the T_g of neat PLA was found at 60 ± 0 °C; it decreased 2–3 °C after introducing unmodified CP filler. This behavior can be caused by the presence of raw fat in the filler structure (Table 7), which can act as plasticizer of the amorphous phase of PLA. An important finding from this analysis is the significant reduction of T_g after the introduction of the modified CP filler into the PLA matrix (average reduction of 8 to 9 °C). R. Avolio et al. have shown that 10% of OLAs may cause the drop of T_g by a similar value [35]. In our paper, the total amount of ungrafted OLAs is 2.91% (Table 4). It can be deduced that grafted OLAs also have a plasticizing effect on the PLA matrix, and moreover, the effectiveness of plasticization with OLAs is correlated with their chain length, which in this case is rather low. A. Goffin et al. also obtained a plasticizing effect for pure CNW-g-PLA [40]. To summarize, the decreasing of the T_g is connected with the presence of ungrafted and grafted OLAs. It confirms the plasticizing effect of the chain of OLAs in this composition. Additionally, DSC curves showed only one T_g in the analyzed temperature range, which shows no macroscopic phase separation and good compatibility between the PLA and OLAs in analyzed biocomposites [35].

Table 10. Results of DSC analysis.

Sample	T_g, °C	T_{cc}, °C	T_{m1}/T_{m2}, °C	X_c, %
PLA	60 ± 0	115 ± 0	149 ± 0/-	1.8 ± 0.3
<63 CP/PLA	58 ± 0	121 ± 1	149 ± 0/154 ± 1	3.4 ± 0.5
63–125 CP/PLA	58 ± 0	121 ± 2	149 ± 1/154 ± 1	4.0 ± 1.2
Mix CP/PLA	57 ± 0	118 ± 1	148 ± 1/154 ± 1	3.3 ± 1.1
Mix1 CP/PLA	58 ± 0	120 ± 1	149 ± 0/153 ± 1	4.4 ± 1.1
<63 CP-g-OLA/PLA	52 ± 1	117 ± 4	145 ± 2/151 ± 1	4.2 ± 0.9
63–125 CP-g-OLA/PLA	51 ± 2	118 ± 6	145 ± 3/151 ± 1	2.5 ± 0.3
Mix CP-g-OLA/PLA	51 ± 3	118 ± 7	145 ± 3/151 ± 1	5.1 ± 0.3
Mix1 CP-g-OLA/PLA	52 ± 1	121 ± 4	146 ± 1/152 ± 1	4.4 ± 0.4

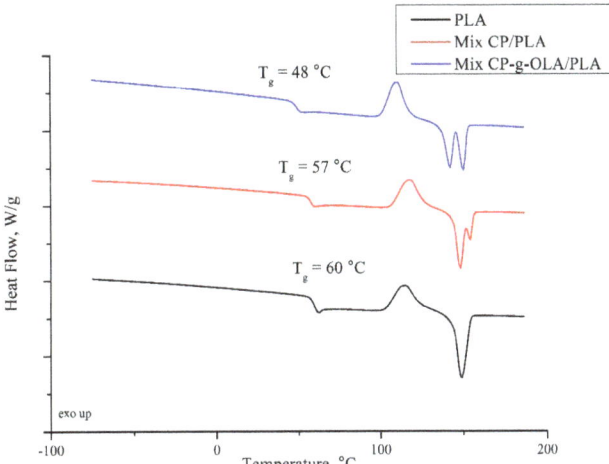

Figure 8. DSC curves of PLA, Mix CP-g-OLA/PLA, and Mix CP/PLA biocomposites.

The addition of unmodified CP filler increase T_{cc} which can correspond to the limitation of PLA's chains mobility. Introduction OLAs to the PLA matrix generally caused a slight decrease in T_{m1}. This phenomenon is also related to the plasticizing effect of OLAs and increasing PLA chain mobility [35]. Except for the unfilled PLA matrix, samples were characterized by a double melting peak (T_{m1} and T_{m2}). This effect is widely known in the literature as α' phase melting (T_{m1}), and its recrystallisation to α and re-melting at higher temperature (T_{m2}) [56]. Lower T_{m2} is characterized in all grafted CP fillers, which could contribute to higher nucleation of crystallization by CP-g-OLA because of the higher mobility of PLA chain. It is clearly shown that the introduction of the filler affects the process of nucleation, thus increasing the X_c polymer matrix [11]. According to the literature, OLAs also have an increasing impact on the degree of crystallinity [35].

3.2.5. Mechanical Properties

Table 11 shows Young's modulus and tensile strength of unmodified and modified biocomposites.

Table 11. Mechanical properties of biocomposites.

Sample	Tensile Strength, MPa
PLA	47.6 ± 2.7
<63 CP/PLA	27.18 ± 0.99
63–125 CP/PLA	26.19 ± 1.08
Mix CP/PLA	26.37 ± 1.08
Mix1 CP/PLA	25.92 ± 0.90
<63 CP-g-OLA/PLA	30.06 ± 0.99
63–125 CP-g-OLA/PLA	27.63 ± 1.08
Mix CP-g-OLA/PLA	27.81 ± 1.17
Mix1 CP-g-OLA/PLA	27.63 ± 1.62

All of the biocomposites are characterized by a lower tensile strength than neat PLA. It is quite an obvious observation, given the degree of filling (30%) of the composite and the poor adhesion of the unmodified filler to the matrix. A. Dufresne et al. also observed this effect after adding lignocellulosic flour to the PHBV [6]. An upward trend is clearly visible in the case of the tensile strength of composites with modified fillers compared to unmodified fillers. The increase is slight, and often on the verge of standard deviation in most cases. It is important to notice that generally, in this case, plasticizers and OLAs reduce tensile strength [35,36,57]. In this composition, good adhesion and wettability (which is proven by SEM analysis, Figure 6c,d) may compensate for the reduction in tensile strength caused by the addition of the plasticizer. The highest increase in tensile strength of the biocomposites with a modified filler (comparing with biocomposites with unmodified filler) is shown in the biocomposites with a filler in the size fraction <63 μm and is around 10%. This result can be justified as a greater development of the surface area, which results in a higher number of hydroxyl groups on the surface, able to react with OLAs and LAc [58]. Figure 9 presents this result.

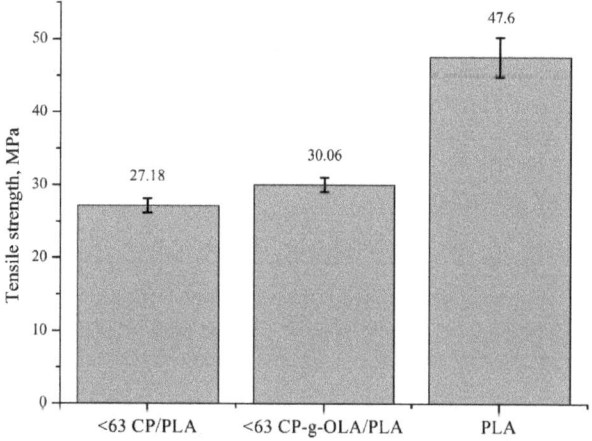

Figure 9. Tensile strength of neat PLA, <63 CP/PLA and <63 CP-g-OLA/PLA.

4. Conclusions

In this paper, CP, which is a food-based residue, was modified using in situ formed OLAs to increase the adhesion between the filler particles and PLA matrix. The esterification reaction results in the mixture of ungrafted and grafted OLAs to CP particles. FTIR spectra confirmed the success of the grafting process. SEM images confirm better adhesion between PLA matrix and modified filler than with unmodified CP filler. DSC results showed the plasticization effect of ungrafted and grafted OLAs, with decreasing T_g on average by 9 °C. The water absorption of a modified CP filler is lower than an unmodified filler by 67.3%, and confirms participation in blocking hydroxyl groups also by ungrafted OLAs. Thanks to a better connection of modified CP filler to PLA matrix, the mechanical properties of biocomposites were less deteriorated than in the case of an unmodified natural filler. Mechanical and thermal tests have shown that the use of a mixture of fractions of a different size (Mix 1) than the original (Mix) does not have a significant effect on the properties of biocomposites. However, it seems that the smallest particles (less than 63 μm) have the best impact on the properties of obtained biocomposites. The research results suggest that the produced biocomposites with a natural plasticizer will have potential applications in the production of food packaging.

Author Contributions: Conceptualization, A.C. and A.P.; formal analysis, A.C., R.B., A.I. and K.M.; funding acquisition, J.R.; investigation, A.C., R.B., A.I., A.P. and K.M.; methodology, A.C., A.I. and A.P.; project administration, J.R.; resources, A.P. and J.R.; supervision, A.P. and J.R.; validation, A.C. and J.R.; visualization, A.C., R.B., A.I. and A.P.; writing—original draft, A.C.; writing—review and editing, A.P. and J.R. All authors have read and agreed to the published version of the manuscript.

Funding: This research was funded by Materials Technologies project granted by Warsaw University of Technology under the program Excellence Initiative: Research University (ID-UB). Project number: 504044961090.

Institutional Review Board Statement: Not applicable.

Informed Consent Statement: Not applicable.

Data Availability Statement: The data presented in this study are available on request from the corresponding author.

Conflicts of Interest: The authors declare no conflict of interest.

References

1. COM/2020/98 Final. Communication from the Commission to the European Parliament, the Council, the European Economic and Social Committee and the Committee of the Regions. A new Circular Economy Action Plan for a Cleaner and More Competitive Europe. 11 March 2020. Available online: https://www.eumonitor.eu/9353000/1/j9vvik7m1c3gyxp/vl6vh7khf4n9 (accessed on 1 November 2021).
2. Turco, R.; Zannini, D.; Mallardo, S.; Dal Poggetto, G.; Tesser, R.; Santagata, G.; Malinconico, M.; Di Serio, M. Biocomposites based on Poly(lactic acid), *Cynara cardunculus* seed oil and fibrous presscake: A novel eco-friendly approach to hasten PLA biodegradation in common soil. *Polym. Degrad. Stab.* **2021**, *188*, 109576. [CrossRef]
3. Rocha, D.B.; de Souza, A.G.; Szostak, M.; Rosa, D.d.S. Polylactic acid/Lignocellulosic residue composites compatibilized through a starch coating. *Polym. Compos.* **2020**, *41*, 3250–3259. [CrossRef]
4. Mysiukiewicz, O.; Barczewski, M. Utilization of linseed cake as a postagricultural functional filler for poly(lactic acid) green composites. *J. Appl. Polym. Sci.* **2019**, *136*, 47152. [CrossRef]
5. Oksman, K.; Skrifvars, M.; Selin, J.-F. Natural fibres as reinforcement in polylactic acid (PLA) composites. *Compos. Sci. Technol.* **2003**, *63*, 1317–1324. [CrossRef]
6. Dufresne, A.; Le Dupeyre, D.; Paillet, M. Lignocellulosic Flour-Reinforced Poly(hydroxybutyrate-co-valerate) Composites. *J. Appl. Polym. Sci.* **2002**, *87*, 1302–1315. [CrossRef]
7. Bourban, C.; Karamuk, E.; De Fondaumière, M.J.; Ruffieux, K.; Mayer, J.; Wintermantel, E. Processing and characterization of a new biodegradable composite made of a PHB/V matrix and regenerated cellulosic fibers. *J. Environ. Polym. Degrad.* **1997**, *5*, 159–166. [CrossRef]
8. Mohanty, A.K.; Misra, M.; Hinrichsen, G. Biofibres, biodegradable polymers and biocomposites: An overview. *Macromol. Mater. Eng.* **2000**, *276–277*, 1–24. [CrossRef]

9. Joseph, P.V.; Joseph, K.; Thomas, S.; Pillai, C.K.S.; Prasad, V.S.; Groeninckx, G.; Sarkissova, M. The thermal and crystallisation studies of short sisal fibre reinforced polypropylene composites. *Compos. Part A Appl. Sci. Manuf.* **2003**, *34*, 253–266. [CrossRef]
10. Salasinska, K.; Ryszkowska, J. The effect of filler chemical constitution and morphological properties on the mechanical properties of natural fiber composites. *Compos. Interfaces* **2014**, *22*, 39–50. [CrossRef]
11. Salasinska, K.; Polka, M.; Gloc, M.; Ryszkowska, J. Natural fiber composites: The effect of the kind and content of filler on the dimensional and fire stability of polyolefin-based composites. *Polimery* **2016**, *61*, 255–265. [CrossRef]
12. Madhavan Nampoothiri, K.; Nair, N.R.; John, R.P. An overview of the recent developments in polylactide (PLA) research. *Bioresour. Technol.* **2010**, *101*, 8493–8501. [CrossRef] [PubMed]
13. Kliem, S.; Kreutzbruck, M.; Bonten, C. Review on the biological degradation of polymers in various environments. *Materials* **2020**, *13*, 4586. [CrossRef]
14. Florjańczyk, Z.; Jóźwiak, A.; Kundys, A.; Plichta, A.; Dębowski, M.; Rokicki, G.; Parzuchowski, P.; Lisowska, P.; Zychewicz, A. Segmental copolymers of condensation polyesters and polylactide. *Polym. Degrad. Stab.* **2012**, *97*, 1852–1860. [CrossRef]
15. Im, S.-H.; Im, D.-H.; Park, S.-J.; Chung, J.-J.; Jung, Y.; Kim, S.-H. Stereocomplex polylactide for drug delivery and biomedical applications: A review. *Molecules* **2021**, *26*, 2846. [CrossRef] [PubMed]
16. Plichta, A.; Kowalczyk, S.; Kamiński, K.; Wasyłeczko, M.; Więckowski, S.; Olędzka, E.; Nałęcz-Jawecki, G.; Zgadzaj, A.; Sobczak, M. ATRP of Methacrylic Derivative of Camptothecin Initiated with PLA toward Three-Arm Star Block Copolymer Conjugates with Favorable Drug Release. *Macromolecules* **2017**, *50*, 6439–6450. [CrossRef]
17. Domańska, I.M.; Oledzka, E.; Sobczak, M. Sterilization process of polyester based anticancer-drug delivery systems. *Int. J. Pharm.* **2020**, *587*, 119663. [CrossRef] [PubMed]
18. Perinelli, D.R.; Cespi, M.; Bonacucina, G.; Palmieri, G.F. PEGylated polylactide (PLA) and poly (lactic-co-glycolic acid) (PLGA) copolymers for the design of drug delivery systems. *J. Pharm. Investig.* **2019**, *49*, 443–458. [CrossRef]
19. Hong, L.G.; Yuhana, N.Y.; Zawawi, E.Z.E. Review of bioplastics as food packaging materials. *AIMS Mater. Sci.* **2021**, *8*, 166–184. [CrossRef]
20. de Oliveira, W.Q.; de Azeredo, H.M.C.; Neri-Numa, I.A.; Pastore, G.M. Food packaging wastes amid the COVID-19 pandemic: Trends and challenges. *Trends Food Sci. Technol.* **2021**, *116*, 1195–1199. [CrossRef]
21. Kumar, A.; Srivastava, A. Preparation and Mechanical Properties of Jute Fiber Reinforced Epoxy Composites. *Ind. Eng. Manag.* **2017**, *6*, 4–7. [CrossRef]
22. Hong, C.-K.; Kim, N.; Kang, S.-L.; Nah, C.; Lee, Y.-S.; Cho, B.-H.; Ahn, J.-H. Mechanical properties of maleic anhydride treated jute fibre/polypropylene composites. *Plast. Rubber Compos.* **2008**, *37*, 325–330. [CrossRef]
23. Li, X.; Panigrahi, S.; Tabil, L.G. A study on flax fiber-reinforced polyethylene biocomposites. *Appl. Eng. Agric.* **2009**, *25*, 525–531. [CrossRef]
24. Li, X.; Tabil, L.G.; Panigrahi, S. Chemical treatments of natural fiber for use in natural fiber-reinforced composites: A review. *J. Polym. Environ.* **2007**, *15*, 25–33. [CrossRef]
25. Kabir, M.M.; Wang, H.; Lau, K.T.; Cardona, F. Chemical treatments on plant-based natural fibre reinforced polymer composites: An overview. *Compos. Part B Eng.* **2012**, *43*, 2883–2892. [CrossRef]
26. Gírio, F.M.; Fonseca, C.; Carvalheiro, F.; Duarte, L.C.; Marques, S.; Bogel-Łukasik, R. Hemicelluloses for fuel ethanol: A review. *Bioresour. Technol.* **2010**, *101*, 4775–4800. [CrossRef]
27. Al-Maharma, A.Y.; Al-Huniti, N. Critical review of the parameters affecting the effectiveness of moisture absorption treatments used for natural composites. *J. Compos. Sci.* **2019**, *3*, 27. [CrossRef]
28. Radzi, A.M.; Sapuan, S.M.; Jawaid, M.; Mansor, M.R. Effect of Alkaline Treatment on Mechanical, Physical and Thermal Properties of Roselle/Sugar Palm Fiber Reinforced Thermoplastic Polyurethane Hybrid Composites. *Fibers Polym.* **2019**, *20*, 847–855. [CrossRef]
29. Bledzki, A.K.; Mamun, A.A.; Lucka-Gabor, M.; Gutowski, V.S. The effects of acetylation on properties of flax fibre and its polypropylene composites. *Express Polym. Lett.* **2008**, *2*, 413–422. [CrossRef]
30. Cui, X.; Ozaki, A.; Asoh, T.-A.; Uyama, H. Cellulose modified by citric acid reinforced Poly(lactic acid) resin as fillers. *Polym. Degrad. Stab.* **2020**, *175*, 109118. [CrossRef]
31. Kasztelan, A.; Kierepka, M. Oddziaływanie Przemysłu Spożywczego Na Środowisko W Polsce. *Rocz. Nauk. Stow. Ekon. Rol. Agrobiz.* **2014**, *16*, 109–116. (In Polish)
32. Statistics Poland, Agriculture Department. *Production of Agricultural and Horticultural Crops in 2018*; Statistics Poland: Warsaw, Poland, 2019.
33. Ambrosio-Martín, J.; Fabra, M.J.; Lopez-Rubio, A.; Lagaron, J.M. Melt polycondensation to improve the dispersion of bacterial cellulose into polylactide via melt compounding: Enhanced barrier and mechanical properties. *Cellulose* **2015**, *22*, 1201–1226. [CrossRef]
34. Patwa, R.; Saha, N.; Sáha, P.; Katiyar, V. Biocomposites of poly(lactic acid) and lactic acid oligomer-grafted bacterial cellulose: It's preparation and characterization. *J. Appl. Polym. Sci.* **2019**, *136*, 47903. [CrossRef]
35. Avolio, R.; Castaldo, R.; Gentile, G.; Ambrogi, V.; Fiori, S.; Avella, M.; Cocca, M.; Errico, M.E. Plasticization of poly(lactic acid) through blending with oligomers of lactic acid: Effect of the physical aging on properties. *Eur. Polym. J.* **2015**, *66*, 533–542. [CrossRef]

36. Burgos, N.; Tolaguera, D.; Fiori, S.; Jiménez, A. Synthesis and Characterization of Lactic Acid Oligomers: Evaluation of Performance as Poly(Lactic Acid) Plasticizers. *J. Polym. Environ.* **2014**, *22*, 227–235. [CrossRef]
37. Martin, O.; Avérous, L. Poly(lactic acid): Plasticization and properties of biodegradable multiphase systems. *Polymer* **2001**, *42*, 6209–6219. [CrossRef]
38. Masirek, R.; Kulinski, Z.; Chionna, D.; Piorkowska, E.; Pracella, M. Composites of poly(L-lactide) with hemp fibers: Morphology and thermal and mechanical properties. *J. Appl. Polym. Sci.* **2007**, *105*, 255–268. [CrossRef]
39. Luan, Y.; Wu, J.; Zhan, M.; Zhang, J.; Zhang, J.; He, J. "One pot" homogeneous synthesis of thermoplastic cellulose acetate-graft-poly(l-lactide) copolymers from unmodified cellulose. *Cellulose* **2013**, *20*, 327–337. [CrossRef]
40. Goffin, A.-L.; Raquez, J.-M.; Duquesne, E.; Siqueira, G.; Habibi, Y.; Dufresne, A.; Dubois, P. From interfacial ring-opening polymerization to melt processing of cellulose nanowhisker-filled polylactide-based nanocomposites. *Biomacromolecules* **2011**, *12*, 2456–2465. [CrossRef]
41. Chieng, B.W.; Ibrahim, N.A.; Yunus, W.M.Z.W.; Hussein, M.Z. Poly(lactic acid)/poly(ethylene glycol) polymer nanocomposites: Effects of graphene nanoplatelets. *Polymers* **2014**, *6*, 93–104. [CrossRef]
42. Espartero, J.L.; Rashkov, I.; Li, S.M.; Manolova, N.; Vert, M. NMR analysis of low molecular weight poly(lactic acid)s. *Macromolecules* **1996**, *29*, 3535–3539. [CrossRef]
43. Zhao, W.; Glavas, L.; Odelius, K.; Edlund, U.; Albertsson, A.-C. A robust pathway to electrically conductive hemicellulose hydrogels with high and controllable swelling behavior. *Polymer* **2014**, *55*, 2967–2976. [CrossRef]
44. Onwukamike, K.N.; Tassaing, T.; Grelier, S.; Grau, E.; Cramail, H.; Meier, M.A.R. Detailed Understanding of the DBU/CO_2 Switchable Solvent System for Cellulose Solubilization and Derivatization. *ACS Sustain. Chem. Eng.* **2018**, *6*, 1496–1503. [CrossRef]
45. Mainka, H.; Täger, O.; Körner, E.; Hilfert, L.; Busse, S.; Edelmann, F.T.; Herrmann, A.S. Lignin—An alternative precursor for sustainable and cost-effective automotive carbon fiber. *J. Mater. Res. Technol.* **2015**, *4*, 283–296. [CrossRef]
46. Xu, F.; Yu, J.; Tesso, T.; Dowell, F.; Wang, D. Qualitative and quantitative analysis of lignocellulosic biomass using infrared techniques: A mini-review. *Appl. Energy* **2013**, *104*, 801–809. [CrossRef]
47. Cozzani, V.; Lucchesi, A.; Stoppato, G.; Maschio, G. A new method to determine the composition of biomass by thermogravimetric analysis. *Can. J. Chem. Eng.* **1997**, *75*, 127–133. [CrossRef]
48. Lisperguer, J.; Saravia, Y.; Vergara, E. Structure and thermal behavior of tannins from Acacia dealbata bark and their reactivity toward formaldehyde. *J. Chil. Chem. Soc.* **2016**, *61*, 3188–3190. [CrossRef]
49. Leszczyńska, M.; Malewska, E.; Ryszkowska, J.; Kurańska, M.; Gloc, M.; Leszczyński, M.K.; Prociak, A. Vegetable fillers and rapeseed oil-based polyol as natural raw materials for the production of rigid polyurethane foams. *Materials* **2021**, *14*, 1772. [CrossRef]
50. Tserki, V.; Zafeiropoulos, N.E.; Simon, F.; Panayiotou, C. A study of the effect of acetylation and propionylation surface treatments on natural fibres. *Compos. Part A Appl. Sci. Manuf.* **2005**, *36*, 1110–1118. [CrossRef]
51. Guo, Y.; Liu, Q.; Chen, H.; Wang, X.; Shen, Z.; Shu, X.; Sun, R. Direct grafting modification of pulp in ionic liquids and self-assembly behavior of the graft copolymers. *Cellulose* **2013**, *20*, 873–884. [CrossRef]
52. Espinach, F.X.; Boufi, S.; Delgado-Aguilar, M.; Julián, F.; Mutjé, P.; Méndez, J.A. Composites from poly(lactic acid) and bleached chemical fibres: Thermal properties. *Compos. Part B Eng.* **2018**, *134*, 169–176. [CrossRef]
53. Jia, S.; Yu, D.; Zhu, Y.; Wang, Z.; Chen, L.; Fu, L. Morphology, crystallization and thermal behaviors of PLA-based composites: Wonderful effects of hybrid GO/PEG via dynamic impregnating. *Polymers* **2017**, *9*, 528. [CrossRef] [PubMed]
54. Avérous, L.; Le Digabel, F. Properties of biocomposites based on lignocellulosic fillers. *Carbohydr. Polym.* **2006**, *66*, 480–493. [CrossRef]
55. Barczewski, M.; Mysiukiewicz, O.; Hejna, A.; Biskup, R.; Szulc, J.; Michałowski, S.; Piasecki, A.; Kloziński, A. The effect of surface treatment with isocyanate and aromatic carbodiimide of thermally expanded vermiculite used as a functional filler for polylactide-based composites. *Polymers* **2021**, *13*, 890. [CrossRef] [PubMed]
56. Backes, E.H.; Pires, L.d.N.; Costa, L.C.; Passador, F.R.; Pessan, L.A. Analysis of the degradation during melt processing of pla/biosilicate® composites. *J. Compos. Sci.* **2019**, *3*, 52. [CrossRef]
57. Ge, H.; Yang, F.; Hao, Y.; Wu, G.; Zhang, H.; Dong, L. Thermal, mechanical, and rheological properties of plasticized poly(L-lactic acid). *J. Appl. Polym. Sci.* **2013**, *127*, 2832–2839. [CrossRef]
58. Amin, F.R.; Khalid, H.; Zhang, H.; Rahman, S.; Zhang, R.; Liu, G.; Chen, C. Pretreatment methods of lignocellulosic biomass for anaerobic digestion. *AMB Express* **2017**, *7*, 72. [CrossRef]

Article
Assessment of the Decomposition of Oxo- and Biodegradable Packaging Using FTIR Spectroscopy

Florentyna Markowicz and Agata Szymańska-Pulikowska *

Institute of Environmental Engineering, Wrocław University of Environmental and Life Sciences, pl. Grunwaldzki 24, 50-363 Wrocław, Poland; florentyna.markowicz@upwr.edu.pl
* Correspondence: agata.szymanska-pulikowska@upwr.edu.pl

Abstract: The strength and resistance of plastics at the end of their service life can hinder their degradation. The solution to this problem may be materials made of biodegradable and oxo-biodegradable plastics. The aim of this research was to determine the degree and nature of changes in the composition and structure of composted biodegradable and oxo-biodegradable bags. The research involved shopping bags and waste bags available on the Polish market. The composting of the samples was conducted in an industrial composting plant. As a result of the research, only some of the composted samples decomposed. After composting, all samples were analysed using FTIR (Fourier Transformation Infrared) spectroscopy. Carbonyl index and hierarchical cluster analysis method was used to detect similarities between the spectra of the new samples. The analysis of the obtained results showed that FTIR spectroscopy is a method that can be used to confirm the degradation and detect similarities in the structure of the analysed materials. The analysis of spectra obtained with the use of FTIR spectroscopy indicated the presence of compounds that may be a potential source of compost contamination. Plastics with certificates confirming their biodegradability and compostability should be completely biodegradable, i.e., each element used in their production should be biodegradable and safe for the environment.

Keywords: biodegradable and oxo-biodegradable packaging; polymers; MSW composting plant; FTIR spectroscopy

1. Introduction

Plastic materials are widely used in many industries due to their strength, environmental resistance and flexibility. The growing demand for plastics poses the problem of the increasing amount of waste from plastics at the end of their service life [1]. The characteristics of plastics that used to be an advantage, at the end of their service life are the source of decomposition problems [2]. The solution to this problem may be materials made of biodegradable and oxo-biodegradable plastics [3]. Biodegradable materials should decompose under the influence of macro- and microorganisms. Oxo-biodegradable plastics are also biodegradable, but initiating biodegradation requires "abiotic degradation", e.g., through the use of heat energy or UV radiation [4,5]. Oxo-biodegradable plastics contain additives that are responsible for the initiation of decomposition (prooxidants). The most common additives on the market are d2w® (biodegradable plastic technology) or TDPA® (Totally Degradable Plastic Additives). The rate of biodegradation is influenced by many factors, e.g., chemical character of the polymer, environmental conditions, microbial population activity [6,7]. According to the manufacturers' assumptions [8–10], biodegradable plastics must fulfil designated functions, that is, be (for example) durable and flexible, and at the same time, at the end of their service life, they should be biodegradable in the environment, i.e., in litter, landfill, compost or soil.

Biodegradable and oxo-biodegradable materials are used to produce waste disposal bags and food/shopping bags. Such products are often used to collect bio-waste because of the information provided by producers about "compostability", "environmental degradability" of their materials. These products are indeed certified in accordance with European standards, such as EN 13432, which stipulates that bags should be 90% biodegradable within six months. Consumers who use them believe they are helping to protect the environment by using waste disposal bags. Together with bio-waste, the bags are sent to a waste treatment plant (e.g., a composting plant). However, packaging should only be allowed for biological treatment if it actually decomposes under composting conditions [11]. Scientific research on the decomposition of plastics is very often performed in conditions significantly different from real ones [12,13]. It happens that in laboratory conditions there are changes in the structure of the material, but its total decomposition is not proven [14,15], or may concern only a part of the polymers, which are its components [16]. In the event of partial decomposition of the packaging, contamination may remain in the environment, e.g., in the form of microplastics. In addition, even small changes in the structure of plastics (such as discoloration or porosity), may indicate that micro-particles and, with them, contaminants in the form of e.g., toxic elements are released into the environment. Therefore, it is necessary to develop the technical standards specifying how to conduct biodegradation tests in the environment, taking into account real conditions (depending on whether the degradation takes place in water, soil or landfill). The results of the tests carried out in this way should make it possible to determine the time and degree of decomposition, depending on environmental conditions [17].

One method that is suitable for the analysis of samples consisting of different materials is Fourier transform infrared spectroscopy (FTIR) [18–22]. This method has been used, among others, for the identification of polymers in marine waste and even those found in animal organisms. In the case of waste materials, FTIR spectroscopy has been used to determine compost maturity, characterise humic substances [23], present in compost and anaerobic decomposing waste, and identify unknown materials present in waste dumped from a landfill site [24]. The conducted research has shown that with the FTIR method it is possible to obtain a lot of information on samples of complex composition [25]. Due to the possibility of conducting research in a relatively simple and non-destructive way, FTIR spectroscopy can also be used to identify plastics and track changes in their composition [26]. Even on the discoloured surface of plastics, it is possible to observe changes confirming the presence of microorganisms and traces of biodegradation [27]. The degree of degradation of plastics can be assessed on the basis of a decrease in the intensity of the bands indicating the presence of C–H bonds, or the appearance of new bands indicating the presence of oxygen connections, such as C=O, C–O, O–H, O–C=O and C=C [28–30].

The aim of the research was to determine the degree and nature of changes in the composition and structure of composted (in real conditions) biodegradable and oxobiodegradable shopping bags and waste bags. The research was carried out using FTIR spectroscopy, which is often used to analyse the structure and degree of decomposition of various materials. The research involved shopping bags and waste bags available on the market, used by consumers to collect the biodegradable fraction of municipal waste, collected selectively for composting processes. Depending on the degree of decomposition of these products, substances included in the plastics, as well as microplastics resulting from the degradation and defragmentation of the film, may be released into the compost to the environment. On the other hand, insufficiently decomposed plastics (larger fragments) will contaminate the compost, preventing its sale and use.

2. Materials and Methods

Shopping bags and waste bags generally available in Poland were selected for the research. According to the information provided on the packaging by the producers, the samples were conducted from biodegradable or oxo biodegradable materials (foil). Some of the samples were partially or completely coloured (in green, brown, orange or black), while the rest were white.

Twelve samples were selected for the research. Table 1 presents information about the analysed samples.

Table 1. The most important information about the samples selected for research.

Sample Number	Type of Polymer	Additional Manufacturer Information
1	Biodegradable, compostable	There is no need to remove from the stream of bio-waste at the composting industry
2	Biodegradable, compostable	It is degraded in composting conditions
3	Biodegradable, compostable	Made on the basis of corn and potato starch
4	Oxo-biodegradable, d2w® additive	It is decomposed under the influence of oxygen, UV and heat, use within 18 months
5	Oxo-biodegradable, TDPA additive	It is subject to accelerated decomposition
6	Biodegradable, LDPE and sugar cane	Sugar cane content above 85%
7	Oxo-biodegradable, d2w® additive	The bag is 100% biodegradable
8	Oxo-biodegradable, HDPE	The bag is oxo-biodegradable by 100%
9	Oxo-biodegradable, d2w® additive	It has the Oxo-biodegradable Plastics Association mark
10	Oxo-biodegradable, HDPE	The bag is oxo-biodegradable by 100%
11	Biodegradable, compostable	Bags for biodegradable waste
12	Oxo-biodegradable, HDPE	Bags for organic waste

From selected packages, samples were prepared in the form of sheets with an area of approximately 500 cm^2. Each sheet was sandwiched between two layers of glass fibre mesh with a mesh size of ca 1 × 1 mm (the mesh with a weave of 238 meshes by 5 cm^2 was used). The sheet closed in the mesh was stapled to prevent the samples from slipping out. Three sets of sheets were prepared for each test: A—new test, not exposed to UV radiation, B—test exposed to UV rays for 20 h, C—test exposed to UV rays for 50 h. The 36 W UV lamp (OSRAM, Munich, Germany) was used for irradiation. The samples were irradiated from a distance of 0.7 m. The irradiation time was supposed to correspond to the irradiation during the storage of waste in the composting plant for 2 or 5 days. Irradiation with UV rays was to initiate the process of decomposition of oxo-biodegradable polymers. The composting of the samples prepared in this way was conducted in an industrial composting plant (Jarocin, Greater Poland Voivodeship, Poland). Initially, composting took place in closed reactors with active aeration, and then on a heap where the compost matured. The entire process took about 5 months. After composting had finished, the sample sheets were removed from the mesh and rinsed thoroughly with distilled water. After washing, a visual assessment of the condition of the samples was carried out. A detailed description of the method of conducting the research is presented in the work from 2019 [31]. All samples were composted at the same time and under the same conditions. Only three out of the 12 samples had completely decomposed, and one was significantly defragmented.

After conducting a visual inspection, all samples were analysed using FTIR spectroscopy. A Nicolet iN10 MX spectrometer (Thermo Scientific, Waltham, MA, USA) equipped with an adapter was used for the samples. The spectra were recorded in the range of 4000–500 cm^{-1}, with a resolution of 4 cm^{-1}. Interferograms were obtained from 32 scans. Before beginning scanning the samples, the background was irradiated in an empty transparency adapter. The spectra analysis was conducted for all samples from trials A, B and C after the composting process, as well as for new samples, not subjected to composting, constituting the initial spectra (NEW). All laboratory tests were performed in replicates (several fragments of each sample were analysed). Among the spectra obtained, those were selected that did not show any disturbances, caused e.g., by the presence of water. In the case of different coloured samples, scans of parts with different colours were made. In this way, four sets of spectra were obtained for each sample and analysed.

Based on the results of the FTIR analysis, the carbonyl index (CI) values were calculated, according to the Equation (1):

$$CI = \text{Absorbance at } 1713 \text{ cm}^{-1} / \text{Absorbance at } 1464 \text{ cm}^{-1} \tag{1}$$

CI—absorbance ratio of carbonyl and methylene groups. This allows us to determine the amount of carbonyl compounds formed during the photo-oxidation process [32].

The cluster analysis method was used to detect similarities between the spectra of the new samples. This is one of multivariate analyses, useful in cases of large amounts of data (the graph plot of each spectrum consists of 7209 points). Hierarchical cluster analysis permits group observations into clusters. Inside the clusters similar observations can be found, though the clusters differ from each other. Grouping is based on similarities or distance (dissimilarity). The clusters are aggregated according to a decreasing degree of similarity (or increasing degree of dissimilarity) into one single tree-like cluster, called a dendrogram [33].

Grouping of observations (agglomeration) was conducted by the Ward method (the minimum increase of the sum of squares, MISSQ), sometimes called the "minimum variance" method. This method is based on minimising the heterogeneity (variance) in the clusters and finding the greatest possible similarity between the observations. Many studies have demonstrated its accuracy and usefulness in recreating the original structure of clusters [34,35]. The distances between the objects were determined on the basis of the Euclidean distance.

3. Results

All of the samples selected for research were composted under identical conditions in an industrial composting plant. This was to reflect the actual conditions in which the waste bags are sent together with biodegradable waste for processing and to check whether their decomposition is possible during the processing of biodegradable waste.

Only three out of the 12 samples failed completely. These were samples 1, 2 and 11. They were certified in accordance with the EN 13432 norm, and as described by the producers, they were biodegradable and compostable. Sample no. 3 was significantly defragmented. Figure 1 shows the appearance of the sample before (a) and after the composting process (b). Thin threads remained between the two layers of the sample protection mesh, disintegrating into dust when touched, which made it impossible to make scans. As sample no. 3 also had a compostability certificate, it was considered biodegradable. According to the information provided by the manufacturer, the decomposition of the packaging in sample no. 3 should take place within 6 weeks to a year. Such a long time may not ensure complete decomposition of the material in an industrial composting plant, where the process usually takes several months.

Figure 2 shows the FTIR spectra of new samples 1, 2, 3 and 11, which were decomposed in the composting process. This was confirmed during a visual assessment. The greatest changes in peak intensity are visible in new sample no. 3. This sample was made from the starch or sweet corn and potatoes, but it was also one of the most intensely coloured. Its composition should have been similar to that in samples 1, 2 and 11, but it may have been made more difficult by the use of a large amount of dyes. Our 2019 research [36], in which we analysed the composition of the tested samples, indicated that sample no. 3 contained large amounts of copper, which is used in the production process to make green dyes.

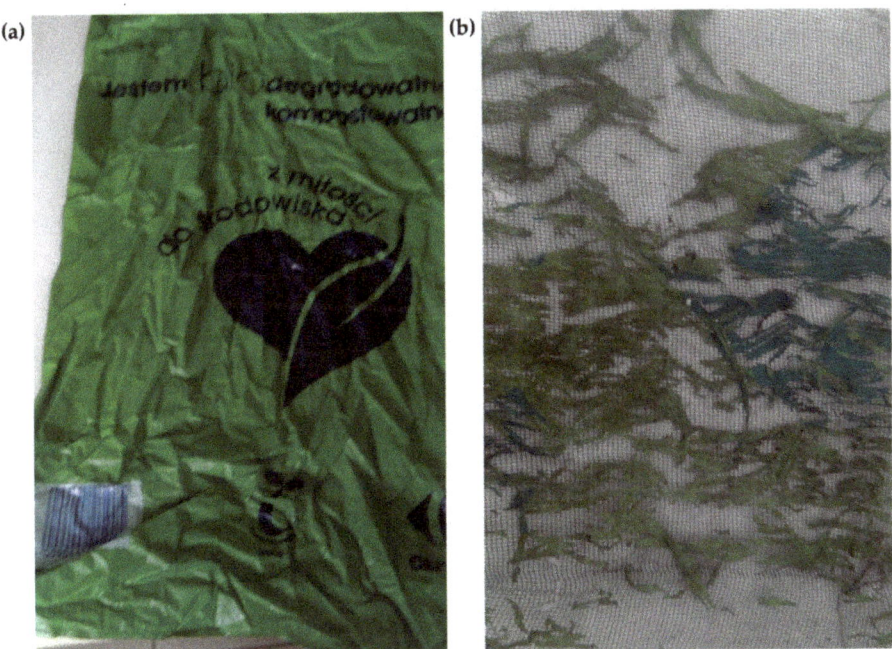

Figure 1. Appearance of sample no. 3: before (**a**) and after the composting process (**b**) [photo F. Markowicz].

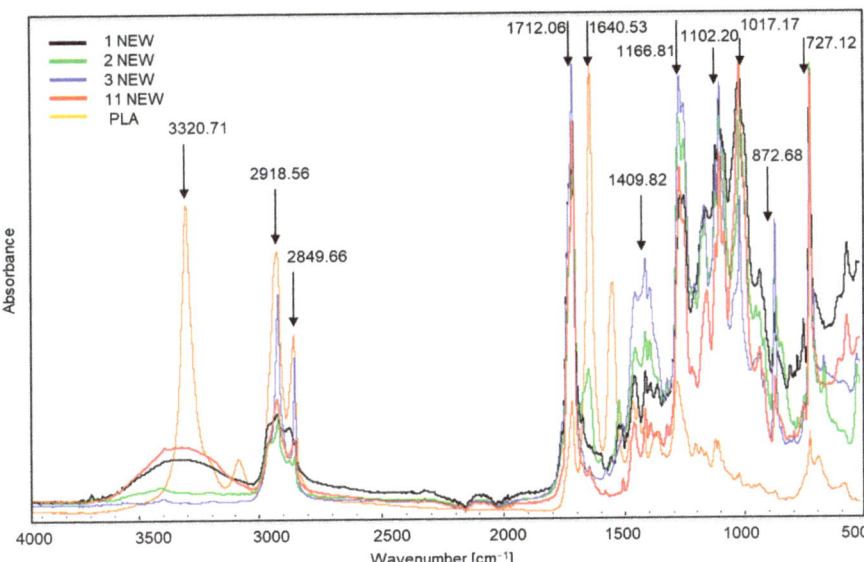

Figure 2. Spectra of new samples 1, 2, 3 and 11, which were decomposed in the composting process. The polylactide (PLA) spectrum was added for comparison.

The spectra of samples 1, 2, 3 and 11 (especially samples 2 and 11) are very similar to the spectra of biodegradable polymers, such as polylactide (PLA) [37]. We can observe characteristic peaks at approximately 3320 cm^{-1}, which corresponds to the OH bond, sharp peaks at approximately 1712 cm^{-1} are stretching vibrations of the C≡O ester groups C≡O [38]. Such changes may indicate the use of renewable raw materials in their production, such as corn meal. According to the information provided by the producers, samples 1, 2, 3 and 11 were made entirely of renewable, biodegradable materials.

Table 2 show the changes observed during the analysis of FTIR spectra of samples that did not decompose during composting.

Table 2. List of changes noticed during the analysis of the Fourier transform infrared (FTIR) spectra of the tested packages (samples 4, 5, 6, 7, 8, 9, 10, and 12).

Sample Number	Wave Number (cm^{-1})			
	>3001	2001–3000	1001–2000	500–1000
4	~3289.77 and 3298.11 (C=C double bonds)	~2915 and 2848 (stretching vibrations of the methylene C–H group)	~1618–1625 (deformation vibrations of the amino group N–H), ~1471 and 1461 (C–H bending vibrations), ~1320–1000 (C–O oxygen groups in carboxylic and ester bonds)	~874, 730 and 718 (C–Cl stretching vibrations of alkyl halides)
5		~2915 and 2847 (stretching vibrations of the C–H methylene group)	~1472 (C–H bending vibrations), ~1116 and 1111 (oxygen groups C–O in carboxylic and ester bonds)	~874, 730 and 718 (C–Cl stretching vibrations of alkyl halides)
6		~2915 and 2847 (stretching vibrations of the C–H methylene group)	~1651 (stretching vibrations of the C=C bonds), ~1525 (N–O nitro compounds), ~1461 and 1472 (C–H bending vibrations), ~1279, 1180 and 1068 (C–O oxygen groups of carboxylic and ester bonds)	~874, 841 and 718 (C–Cl stretching vibrations of alkyl halides)
7	~3394 (stretching vibrations of the amino group N–H)	~2915, 2909, 2844 and 2847 (vibrations of the C–H methylene group)	~1619 (bending vibrations of the N–H amino group), ~1461 and 1462 (C–H bending vibrations), ~1116 and 1107 (C–O oxygen groups of carboxylic and ester bonds)	~718 (C–Cl stretching vibrations), ~599 (stretching vibrations of alkyl halides C-Br bonds)
8	~3266 and 3394 (stretching vibration of the N–H amino group)	~2915 and 2847 (vibrations of the methylene C–H group—bond stretching and cleavage)	~1652 and 1648 (N–H amino group (bending vibrations), ~1461 and 1471 (C–H bending vibrations), ~1279, 1099 and 1077 (C–O oxygen groups in carboxylic and ester bonds)	~874, 844, 730 and 718 (C–Cl stretching vibrations), ~600 (C-Br stretching vibrations)

Table 2. Cont.

Sample Number	Wave Number (cm^{-1})			
	>3001	2001–3000	1001–2000	500–1000
9	~3266 (stretching vibrations of the amino group N–H)	~2916 and 2848 (vibrations of the methylene C–H group—bond stretching and cleavage)	~1651 (bending vibrations of the N–H amino group), ~1461 (C–H bending vibrations), ~1075 (C–O oxygen groups in carboxylic and ester bonds)	~718 (C–Cl stretching vibrations), ~617 (stretching, broad and strong vibrations of the C–H and –C≡C–H alkynes groups)
10		~2915 and 2847 (vibrations of the C–H methylene group)	~1472 and 1461 (C–H bending vibrations), ~1094 and 1075 (C–O oxygen groups in carboxylic and ester bonds)	~718 (C–Cl stretching vibrations of alkyl halides)
12	~3236 (stretching vibrations of the N–H amino group)	~2914 and 2847 (vibrations of the C–H methylene group—stretching and bond cleavage)	~1639 (bending vibrations of the N–H amino group), ~1460 (C–H bending vibrations), ~1101 and 1032 (C–O oxygen groups in carboxylic and ester bonds)	~874 and 718 (C–Cl stretching vibrations of alkyl halides)

In all samples that did not decompose during composting, was found the occurrence of C–O oxygen groups in carboxylic and ester bonds, C–H bending vibrations and C–Cl (or C–Br) stretching vibrations of alkyl halides. At wavenumbers above 3000 cm^{-1} they occurred stretching vibrations of the N–H amino group (samples no. 7, 8, 9, 12). The bending vibrations of the N–H amino group also occurred with wavenumbers between 1001 a 2000 cm^{-1} (samples no. 4, 7, 8, 9, 12). Moreover, in samples 4 and 6, vibrations of C=C bonds were found, and in sample 6 also vibrations of N–O nitro compounds. The starting decomposition of samples was mainly evidenced by visible discoloration, roughness and lower elasticity of the material.

Figures 3–5 show the spectra of new samples and three composted variants (Figure 3—sample no. 5, with the addition of TDPA; Figure 4—sample no. 7, with the addition of d2w®; Figure 5—the sample described by the manufacturer as a bag for organic waste—no. 12), that did not decompose. Figures showing the FTIR spectra of the remaining samples are included in the supplementary materials. The spectra were obtained during the X-raying of new samples (NEW), composted without irradiation (A), exposed to UV radiation for 20 h and composted (B) and UV irradiated for 50 h and composted (C). The Table of Characteristic IR Absorptions, published by the University of Colorado in Boulder, Department of Chemistry and Biochemistry, was used to analyse the spectra [39,40].

Figure 3 shows the spectra obtained during the X-ray of sample no. 5 (new and three composted variants). The bands visible at the wave numbers of approximately 2915 and 2847 cm^{-1} correspond to the stretching vibrations of the methylene C–H group. Initially, C–H stretching occurs, and then there is a marked drop in the peak, corresponding to bond cleavage. This applies to all of the samples, including the new ones. However, in the case of the coloured part of the material, the peaks are most intense for sample C, i.e., irradiated for 50 h and then composted. Further changes occur only at approximately 1472 cm^{-1} (occurrence of C–H bending vibrations). The presence of the bands in the wave number range 1320–1000 cm^{-1} is small. The growth of the bands around 1116 and 1111 cm^{-1} confirms the appearance of oxygen groups C–O in carboxylic and ester bonds. The bands at approximately 874, 730 and 718 cm^{-1} correspond to the C–Cl stretching vibrations. During visual inspection, no significant changes were noted for sample no. 5 after the composting process.

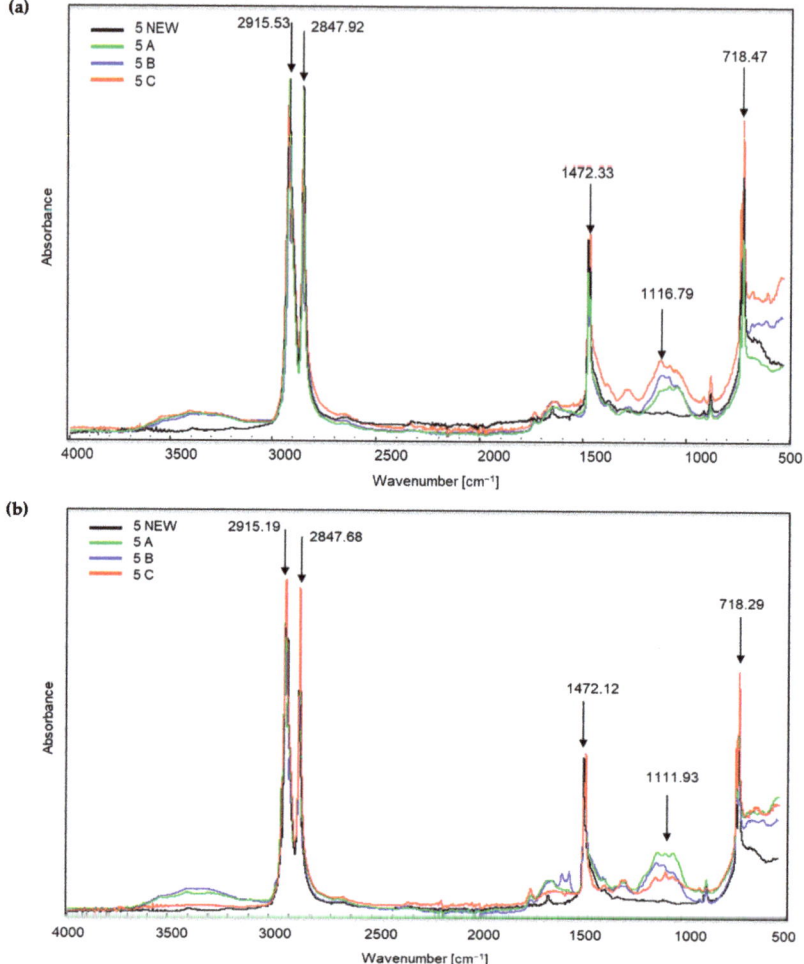

Figure 3. Sample no. 5 spectra: (**a**) spectra generated by the white part of the sample material, (**b**) spectra generated by the coloured part of the sample material.

In the case of sample no. 7 (Figure 4), the first significant changes appeared already in the range of wave numbers amounting to approximately 3394 cm^{-1}, which proves the occurrence of stretching vibrations of the amino group N–H. These changes are most intense for the coloured part of sample B, i.e., irradiated for 20 h and then composted. One can also see very intense peaks at approximately 2915, 2909, 2844 and 2847 cm^{-1}, which correspond to the vibrations of the C–H methylene group. Initially, C–H stretching occurs, followed by bond cleavage. This applies to all the samples, including the new ones. Both in the white and coloured parts of the material, peaks in this range are the most intense for trial C, i.e., irradiated for 50 h and then composted. Further changes occur in the spectra of coloured materials at the wave numbers of approximately 1619 cm^{-1}, which corresponds to the appearance of bending vibrations of the N–H amino group. The peaks at wave numbers around 1461 and 1462 cm^{-1} (for white and coloured material) are the reappearance of C–H bonds (bending vibrations). The next changes concern the wave numbers in the range from approximately 1116 and 1107 cm^{-1} and confirm the appearance of C–O oxygen groups of carboxylic and ester bonds. The bands at approximately 718 cm^{-1} are C–Cl stretching vibrations. Finally, in the coloured part of sample no. 7B, a very intense band appears at a wave number of approximately 599 cm^{-1}. According to the authors of the tables [39,40] these are stretching vibrations of C–Br bonds, alkyl halides, which are very stable compounds. The presented changes can be related to the visual assessment. After the composting process, we observed clear changes in the coloured part of sample no. 7. The sample material was also less flexible.

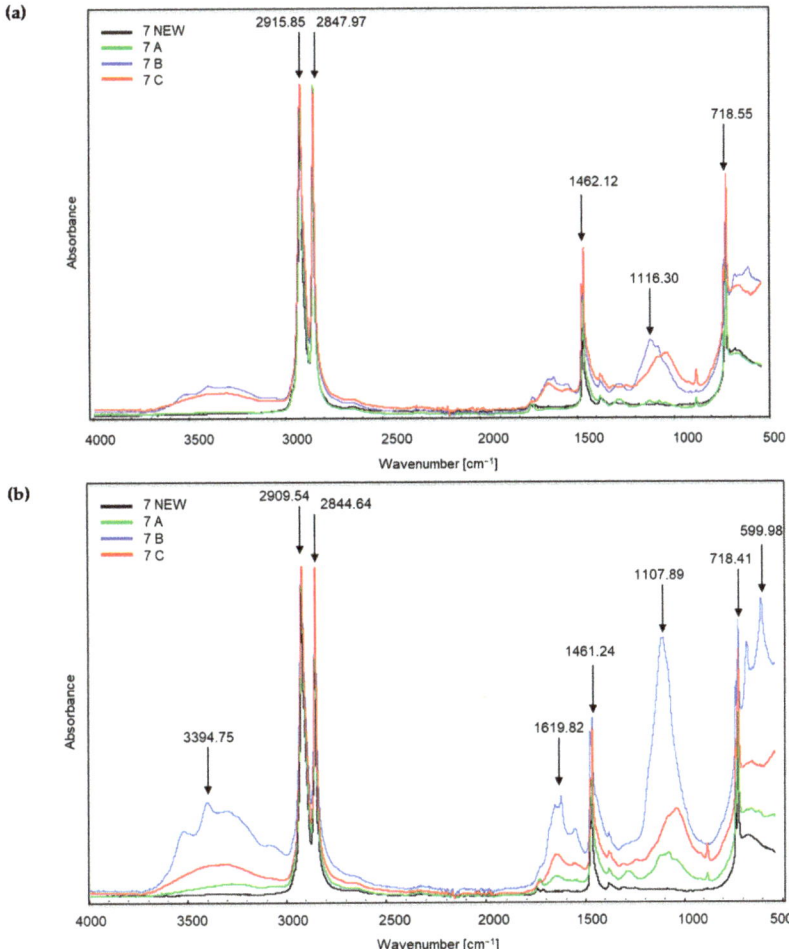

Figure 4. Sample no. 7 spectra: (**a**) spectra generated by the white part of the sample material, (**b**) spectra generated by the coloured part of the sample material.

In the case of sample no. 12 (Figure 5) the first changes appear in all samples in the range of wave numbers amounting to approximately 3236 cm^{-1}, and this is the occurrence of stretching vibrations of the amino group N–H. The bands visible at the wave numbers of approximately 2914 and 2847 cm^{-1} correspond to the vibrations of the methylene C–H group (stretching and bond cleavage). Changes at approximately 1639 cm^{-1} correspond to the appearance of bending vibrations of the N–H amino group. Further changes occur at approximately 1460 cm^{-1} (occurrence of C–H bending vibrations). The growth of the bands around 1101 and 1032 cm^{-1} is the appearance of oxygen groups C–O in carboxylic and ester bonds. The bands at approximately 874 and 718 cm^{-1} are C–Cl stretching vibrations. Sample no. 12 was made entirely of a brown-coloured material which did not change its structure after the composting process, and the colour changes were minor. The material was still flexible and durable. The producers suggested choosing this type of bag for collecting organic waste (as indicated by the description on the packaging), but even a visual assessment aroused doubts regarding its purpose. The structure, appearance and flexibility of the bag were almost identical to those made of polyethylene.

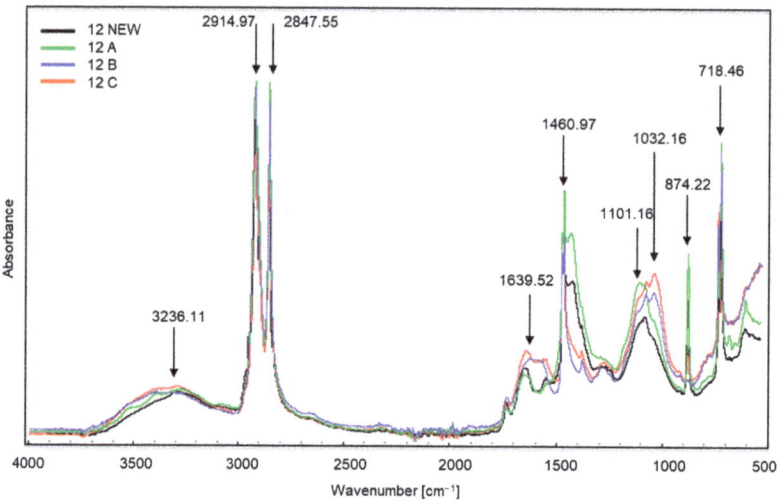

Figure 5. Spectra of sample no. 12 (the entire sample was made up of brown material).

Figure 6 shows the carbonyl index values for the samples after composting. In most cases, the indexes achieved higher values for samples that were exposed to UV radiation for 20 h before composting (B).

This does not confirm the results of the spectra analysis, which showed some changes (indicating the ongoing degradation process) in some of the samples irradiated for 50 h. The changes were not clear, they indicated that the decomposition of the samples was starting, even though the composting process was finished.

The use of hierarchical cluster analysis (supplementary materials, Figure S6) confirmed the existence of similar relationships between the spectra of the tested samples. Samples that were completely decomposed during composting (1, 2, 3, 11) ended up in the first main cluster (A), along with the color part of sample no. 6. According to the manufacturer's declaration, 85% of it was made from sugar cane and there were signs of a biodegradation process. Samples 1, 2, 3 and 11 could be made of materials containing similar components. The remaining samples were included in the second cluster (B). It includes all the samples that were not decomposed in the composting process (except for the coloured part of sample no. 6). This cluster is divided into two smaller ones: the first one (B') is dominated by coloured parts of samples (4, 10, 5, 8) and samples of uniform colour (9 and 12). It also includes the white parts of samples 4 and 5. Most of the mentioned samples showed an increase in the CI value after 20 h irradiation compared to the samples composted without irradiation. The second cluster (B'') has the most white samples (6, 8, 7, 10), and the coloured part of sample number 7 belongs to this.

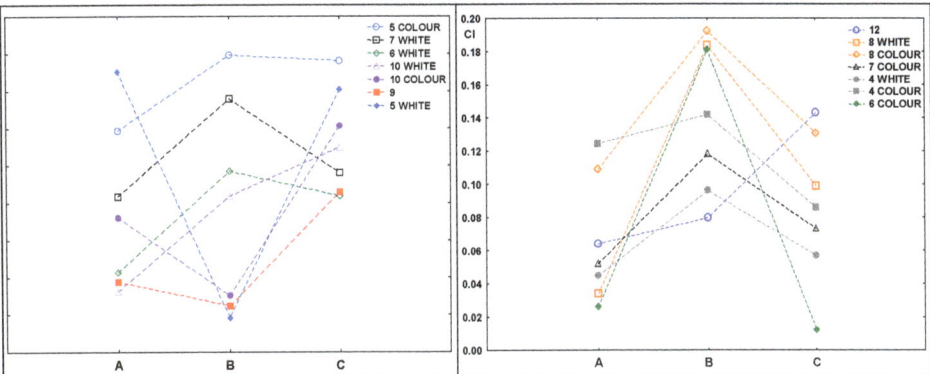

Figure 6. Values of the carbonyl indexes for samples after composting: A—composted without irradiation, B—exposed to UV radiation for 20 h and composted, C—irradiated for 50 h and composted.

4. Discussion

Synthetic polymer packaging is readily available, flexible and resistant to many factors. However, they are not renewable, non-biodegradable and pose a threat to the environment. Packages made of natural raw materials (biodegradable and ecological) are usually very light, thin and have low durability. They are not resistant to factors such as water, air and high temperature. It is very difficult to obtain a material that is both biodegradable, environmentally safe and durable [41]. Work on finding the golden mean in this area is ongoing. New types of material are appearing that may replace those previously used. There are many different packages available on the market for consumers described by producers as biodegradable and oxo-biodegradable. The information on these packages is not always accurate. Our research has shown that not all packaging decomposes, despite the fact that the research conditions corresponded to real-world conditions. Another scenario, e.g., longer exposure to UV rays, additional mechanical actions, heating, are not possible or available under real composting conditions.

All samples analysed with the use of FTIR spectroscopy were characterised by differences in the obtained spectra. The samples that did not decompose in the composting process were characterised by intense bands in the 3000–2800 cm^{-1} range, which correspond to the stretching and cleavage of the C–H group bonds. These bands are characteristic, for example, for polypropylene [42], they did not occur in samples 1, 2, 3 and 11, which completely decomposed in the composting process. All of the samples also had new peaks below 1500 cm^{-1}, known as the dactyloscopic range. Bands of stretching vibrations of single bonds appear (for example, C–H, C–Cl), corresponding to deformation vibrations, as well as groups of C–O oxygen carboxylic and ester bonds. The observed changes in the spectra indicate the occurrence of the oxidation and degradation process of the materials used in the samples. The spectra of samples 4, 7, 8 and 9, belonging to the films described as oxo-biodegradable, are similar to those presented by Benitez et al., who studied the degradation of polyethylenes with pro-oxidants added. The authors also found that these types of material oxidize the components and degrade the material [43,44]. Also in the case of sample no. 6, some changes appeared, indicating the occurrence of the sample decomposition process (especially the coloured part). In the case of samples no. 7 and 8, the analysis of the spectra revealed the presence of compounds from the group of alkyl halides (C–Br bonds). These compounds are used in the chemical industry and can pose a threat to the environment. Bromine is used in the production of dyes, and its compounds are persistent and stable in the environment. Some of the dyes can also be carcinogenic and mutagenic [45]. Perhaps due to the intense colour of the samples (prints, inscriptions), this compound appeared in them.

Registered changes were found for both white and coloured material. However, they prove abiotic degradation, i.e., the process of initial oxidation of the sample material. This is the stage before proper biodegradation with the use of microorganisms [46–48]. It should be noted that the results relate to samples that have already been treated in the composting process. Therefore, they cannot be considered biodegradable, nor can they be considered to degrade in the environment or in compost. In this case, the analysis of spectra obtained using FTIR spectroscopy cannot be treated as the only confirmation of biodegradability. They should be supplemented with other types of test. The carbonyl index allows us to determine the amount of carbonyl compounds formed during the photo-oxidation process. However, it has been shown that a significant part of the oxidation product is omitted, because it evaporates into the atmosphere. Therefore, this indicator cannot be considered as a reliable probe to measure the extent of oxidation and does not reflect the total degradation of the mechanical properties of the polymer [49]. Moreover, the conditions under which biodegradability was tested are important. If these were laboratory conditions, they also do not have to demonstrate the ability to be completely biodegradable, because such conditions are different from real ones and this information may mislead the consumer.

The conducted tests were aimed at assessing the degree and nature of changes occurring during the decomposition of selected packaging and confirming that at various stages of material biodegradation, harmful microplastics and substances contained in plastics may be released into the compost. The processing of plastics together with organic waste causes the pollution to enter the compost, which can then be used as fertiliser [50].

Analysis of the spectra also showed that earlier irradiation of samples did not bring the expected results. According to some authors, earlier exposure of plastics to UV radiation accelerates biodegradation [51] or increases growth of carbonyl indicators [52]. The results show only partial degradation of the samples. In samples no. 5, 6 and 7 there were greater differences between the spectra of samples irradiated for 50 h and the others. Visual inspection confirmed that the structure of some samples had been disturbed. An earlier publication presented the content of selected components in samples before and after the composting process [36]. Analyses have shown that elements such as zinc, copper, chromium and lead can get into the compost. During the decomposition of biodegradable and oxo-biodegradable plastics, microplastic particles may also form, which accumulate in the environment. This was also confirmed by other authors [53,54]. Microplastics are a potential threat to flora and fauna [55], and they affect the functions of soil and microbial communities [56]. The biodegradation of packaging used as waste bags or for storing food is very important, because it concerns many elements of the environment. Errors resulting from a too rash assessment of biodegradability potential may pose a threat to the environment and even to the health and life of animals.

5. Conclusions

The analysis of the results of the conducted research allowed for the formulation of the following conclusions:

- As a result of the research, only some of the composted samples decomposed. In the remaining samples, only slight traces of degradation were visible, despite the fact that all the bags selected for the tests were to undergo oxo- and biodegradation.
- FTIR spectroscopy is a method that can be used to confirm the degradation of biodegradable and oxo-biodegradable materials; however, the results obtained in this way may not give unequivocal information about the degree of actual degradation of a given material. It may, however, help in detecting similarities in the structure of the analysed plastics.
- The analysis of spectra obtained with the use of FTIR spectroscopy indicated the presence of compounds in the tested samples that may be a potential source of compost contamination. Apart from the observation of the compounds proving the degradation of the material (oxygen groups, changes in C–H bonds), groups of alkyl halides were detected.

- Alkyl halide groups were found in most of the samples that were not decomposed in the composting process. These were the following samples: oxo-biodegradable (4, 7, 8, 10, 12) and biodegradable (6).
- The analysis of the spectra of the samples subjected to irradiation with UV rays and non-irradiated ones shows that there are no clear differences between the spectra of the irradiated and non-irradiated samples.
- Plastics with certificates confirming their biodegradability and compostability should be completely biodegradable in real conditions (e.g., during composting). Each component used in their production should be similarly tested, i.e., the dyes used should be completely biodegradable and safe for the environment. An example is the intensely coloured biodegradable sample no. 3, the decomposition of which was difficult, and metals such as copper, zinc and chromium could have entered the environment along with the compost.

Supplementary Materials: The following are available online at https://www.mdpi.com/article/10.3390/ma14216449/s1, Figure S1: Sample no. 4 spectra: (a) spectra generated by the white part of the sample material, (b) spectra generated by the coloured part of the sample material, Figure S2: Sample no. 6 spectra: (a) spectra generated by the white part of the sample material, (b) spectra generated by the coloured part of the sample material, Figure S3: Sample no. 8 spectra: (a) spectra generated by the white part of the sample material, (b) spectra generated by the coloured part of the sample material, Figure S4: Spectra of sample no. 9 (the entire sample was made up of white material), Figure S5: Sample no. 10 spectra: (a) spectra generated by the white part of the sample material, (b) spectra generated by the coloured part of the sample material, Figure S6: Dendrogram showing the analyzed new samples. Agglomeration was carried out using the Ward method.

Author Contributions: Conceptualization, F.M. and A.S.-P.; methodology, F.M. and A.S.-P.; software, A.S.-P.; validation, F.M. and A.S.-P., formal analysis, F.M. and A.S.-P.; investigation, F.M.; resources, F.M.; data curation, A.S.-P.; writing—original draft preparation, F.M. and A.S.-P.; writing—review and editing, F.M. and A.S.-P.; visualization, F.M. and A.S.-P.; supervision, A.S.-P.; project administration, F.M.; funding acquisition, F.M. All authors have read and agreed to the published version of the manuscript.

Funding: The research was carried out as part of a targeted subsidy for research and development of young scientists and PhD students at the Faculty of Environmental Engineering and Geodesy of the Wrocław University of Environmental and Life Sciences—contract no. B030/0111/18.

Institutional Review Board Statement: Not applicable.

Informed Consent Statement: Not applicable.

Data Availability Statement: The data presented in this study are available on request from the corresponding author.

Conflicts of Interest: The authors declare no conflict of interest.

References

1. Geyer, R.; Jambeck, J.R.; Law, K.L. Production, use, and fate of all plastics ever made. *Sci. Adv.* **2017**, *3*, e1700782. [CrossRef] [PubMed]
2. Van Eygen, E.; Laner, D.; Fellner, J. Circular economy of plastic packaging: Current practice and perspectives in Austria. *Waste Manag.* **2018**, *72*, 55–64. [CrossRef] [PubMed]
3. Rujnić-Sokele, M.; Pilipović, A. Challenges and opportunities of biodegradable plastics: A mini review. *Waste Manag. Res.* **2017**, *35*, 132–140. [CrossRef] [PubMed]
4. Contat-Rodrigo, L. Thermal characterization of the oxo-degradation of polypropylene containing a pro-oxidant/pro-degradant additive. *Polym. Degrad. Stab.* **2013**, *98*, 2117–2124. [CrossRef]
5. Gewert, B.; Plassmann, M.M.; MacLeod, M. Pathways for degradation of plastic polymers floating in the marine environment. *Environ. Sci. Process. Impacts* **2015**, *17*, 1513–1521. [CrossRef] [PubMed]
6. Chinaglia, S.; Tosin, M.; Degli-Innocenti, F. Biodegradation rate of biodegradable plastics at molecular level. *Polym. Degrad. Stab.* **2018**, *147*, 237–244. [CrossRef]
7. Agboola, O.; Sadiku, R.; Mokrani, T.; Amer, I.; Imoru, O. Polyolefins and the environment. In *Polyolefin Fibres: Structure, Properties and Industrial Applications*, 2nd ed.; Ugbolue, S.C.O., Ed.; The Textile Institute Book Series; Woodhead Publishing: Cambridge, UK, 2017; pp. 89–133.

8. biodeg.org. Available online: https://www.biodeg.org/opa-briefing-note-3/ (accessed on 29 January 2021).
9. epi-global.com. Available online: https://epi-global.com/tdpa-oxo-biodegradable/how-tdpa-works/ (accessed on 29 January 2021).
10. tuv-at.be. Available online: https://www.tuv-at.be/green-marks/certifications/ (accessed on 29 January 2021).
11. Eubeler, J.P.; Zok, S.; Bernhard, M.; Knepper, T.P. Environmental biodegradation of synthetic polymers I. Test methodologies and procedures. *TrAC Trends Anal. Chem.* **2009**, *28*, 1057–1072. [CrossRef]
12. Ruggero, F.; Gori, R.; Lubello, C. Methodologies to assess biodegradation of bioplastics during aerobic composting and anaerobic digestion: A review. *Waste Manag. Res.* **2019**, *37*, 959–975. [CrossRef]
13. Harrison, J.P.; Boardman, C.; O'Callaghan, K.; Delort, A.-M.; Song, J. Biodegradability standards for carrier bags and plastic films in aquatic environments: A critical review. *R. Soc. Open Sci.* **2018**, *5*, 171792. [CrossRef]
14. Poonam, K.; Rajababu, V.; Yogeshwari, J.; Patel, H. Diversity of plastic degrading microorganisms and their appraisal on biodegradable plastic. *Appl. Ecol. Environ. Res.* **2013**, *11*, 441–449. [CrossRef]
15. Giacomucci, L.; Raddadi, N.; Soccio, M.; Lotti, N.; Fava, F. Polyvinyl chloride biodegradation by Pseudomonas citronellolis and Bacillus flexus. *New Biotechnol.* **2019**, *52*, 35–41. [CrossRef]
16. Eubeler, J.P.; Bernhard, M.; Knepper, T.P. Environmental biodegradation of synthetic polymers II. Biodegradation of different polymer groups. *TrAC Trends Anal. Chem.* **2010**, *29*, 84–100. [CrossRef]
17. Viera, J.S.C.; Marques, M.R.C.; Nazareth, M.C.; Jimenez, P.C.; Sanz-Lázaro, C.; Castro, I.I.B. Are biodegradable plastics an environmental rip off? *J. Hazard. Mater.* **2021**, *416*, 125957. [CrossRef]
18. Reddy, M.M.; Deighton, M.; Gupta, R.K.; Bhattacharya, S.N.; Parthasarathy, R. Biodegradation of oxo-biodegradable polyethylene. *J. Appl. Polym. Sci.* **2009**, *111*, 1426–1432. [CrossRef]
19. Yang, Y.; Yang, J.; Wu, W.-M.; Zhao, J.; Song, Y.; Gao, L.; Yang, R.; Jiang, L. Biodegradation and Mineralization of Polystyrene by Plastic-Eating Mealworms: Part 1. Chemical and Physical Characterization and Isotopic Tests. *Environ. Sci. Technol.* **2015**, *49*, 12080–12086. [CrossRef]
20. Wang, J.; Peng, J.; Tan, Z.; Gao, Y.; Zhan, Z.; Chen, Q.; Cai, L. Microplastics in the surface sediments from the Beijiang River littoral zone: Composition, abundance, surface textures and interaction with heavy metals. *Chemosphere* **2017**, *171*, 248–258. [CrossRef]
21. Nawong, C.; Umsakul, K.; Sermwittayawong, N. Rubber gloves biodegradation by a consortium, mixed culture and pure culture isolated from soil samples. *Braz. J. Microbiol.* **2018**, *49*, 481–488. [CrossRef]
22. Ruggero, F.; Carretti, E.; Gori, R.; Lotti, T.; Lubello, C. Monitoring of degradation of starch-based biopolymer film under different composting conditions, using TGA, FTIR and SEM analysis. *Chemosphere* **2020**, *246*, 125770. [CrossRef]
23. Sánchez-Monedero, M.A.; Cegarra, J.; García, D.; Roig, A. Chemical and structural evolution of humic acids during organic waste composting. *Biogeochemistry* **2002**, *13*, 361–371. [CrossRef]
24. Castaldi, P.; Alberti, G.; Merella, R.; Melis, P. Study of the organic matter evolution during municipal solid waste composting aimed at identifying suitable parameters for the evaluation of compost maturity. *Waste Manag.* **2005**, *25*, 209–213. [CrossRef]
25. Smidt, E.; Meissl, K. The applicability of Fourier transform infrared (FT-IR) spectroscopy in waste management. *Waste Manag.* **2007**, *27*, 268–276. [CrossRef]
26. Jung, M.R.; Horgen, F.D.; Orski, S.V.; C., V.R.; Beers, K.L.; Balazs, G.H.; Jones, T.T.; Work, T.; Brignac, K.C.; Royer, S.-J.; et al. Validation of ATR FT-IR to identify polymers of plastic marine debris, including those ingested by marine organisms. *Mar. Pollut. Bull.* **2018**, *127*, 704–716. [CrossRef]
27. Demestre, M.; Masó, M.; Fortuño, J.M.; De Juan, S. Microfouling communities from pelagic and benthic marine plastic debris sampled across Mediterranean coastal waters. *Sci. Mar.* **2016**, *80*, 117–127. [CrossRef]
28. Da Luz, J.M.R.; Paes, S.A.; Nunes, M.D.; Silva, M.D.C.S.D.; Kasuya, M.C.M. Degradation of Oxo-Biodegradable Plastic by Pleurotus ostreatus. *PLoS ONE* **2013**, *8*, e69386. [CrossRef]
29. Barbes, L.; Radulescu, C.; Stihi, C. ATR-FTIR spectrometry characterization of polymeric materials. *Rom. Rep. Phys.* **2014**, *66*, 765–777.
30. Hou, L.; Xi, J.; Chen, X.; Li, X.; Ma, W.; Lu, J.; Xu, J.; Lin, Y.B. Biodegradability and ecological impacts of polyethylene-based mulching film at agricultural environment. *J. Hazard. Mater.* **2019**, *378*, 120774. [CrossRef]
31. Markowicz, F.; Król, G.; Szymańska-Pulikowska, A. Biodegradable Package-Innovative Purpose or Source of the Problem. *J. Ecol. Eng.* **2019**, *20*, 228–237. [CrossRef]
32. Kumanayaka, T.O.; Parthasarathy, R.; Jollands, M. Accelerating effect of montmorillonite on oxidative degradation of polyethylene nanocomposites. *Polym. Degrad. Stabil.* **2010**, *95*, 672–676. [CrossRef]
33. Mongi, C.E.; Langi, Y.A.R.; Montolalu, C.E.J.C.; Nainggolan, N. Comparison of hierarchical clustering methods (case study: Data on poverty influence in North Sulawesi). In IOP Conference Series: Materials Science and Engineering, Proceedings of the 3rd Indonesian Operations Research Association-International Conference on Operations Research, Manado, Indonesia, 20–21 September 2018; IOP Publishing Ltd.: Bristol, UK, 2019; p. 012048.
34. Majerová, I.; Nevima, J. The measurement of human development using the Ward method of cluster analysis. *J. Int. Stud.* **2017**, *10*, 239–257. [CrossRef]
35. Eszergár-Kiss, D.; Caesar, B. Definition of user groups applying Ward's method. *Transp. Res. Procedia* **2017**, *22*, 25–34. [CrossRef]
36. Markowicz, F.; Szymańska-Pulikowska, A. Analysis of the possibility of environmental pollution by composted biodegradable and oxo-biodegradable plastics. *Geosciences* **2019**, *9*, 460. [CrossRef]

37. Malinowski, R.; Moraczewski, K.; Raszkowska-Kaczor, A. Studies on the Uncrosslinked Fraction of PLA/PBAT Blends Modified by Electron Radiation. *Materials* **2020**, *13*, 1068. [CrossRef] [PubMed]
38. Vuković-Kwiatkowska, I.; Kaczmarek, H.; Dzwonkowski, J. Innovative composites of poly(lactic) acid for the production of packaging foils. *Chemik* **2014**, *68*, 135–140.
39. Organic Chemistry. Available online: https://orgchemboulder.com/ (accessed on 3 October 2020).
40. Chemistry LibreTexts. Available online: https://chem.libretexts.org/ (accessed on 3 October 2020).
41. Shanmugam, K.; Doosthosseini, H.; Varanasi, S.; Garnier, G.; Batchelor, W. Nanocellulose films as air and water vapour barriers: A recyclable and biodegradable alternative to polyolefin packaging. *Sustain. Mater. Technol.* **2019**, *22*, e00115. [CrossRef]
42. Sibeko, M.A.; Luyt, A.S. Preparation and characterisation of vinylsilane crosslinked low-density polyethylene composites filled with nano clays. *Polym. Bull.* **2014**, *71*, 637–657. [CrossRef]
43. Benitez, A.; Sánchez, J.J.; Arnal, M.L.; Müller, A.J.; Rodriguez, O.; Morales, G. Abiotic degradation of LDPE and LLDPE formulated with a pro-oxidant additive. *Polym. Degrad. Stabil.* **2013**, *98*, 490–501. [CrossRef]
44. Corti, A.; Sudhakar, M.; Chiellini, E. Assessment of the Whole Environmental Degradation of Oxo-Biodegradable Linear Low Density Polyethylene (LLDPE) Films Designed for Mulching Applications. *J. Polym. Environ.* **2012**, *20*, 1007–1018. [CrossRef]
45. Bromine-Containing Dyes Dwarf Flame Retardants in House Dust. *Chemical & Engineering News*. Available online: https://cen.acs.org/articles/94/web/2016/11/Bromine-containing-dyes-dwarf-flame.html (accessed on 16 November 2020).
46. Peng, B.-Y.; Chen, Z.; Chen, J.; Yu, H.; Zhou, X.; Criddle, C.S.; Wu, W.-M.; Zhang, Y. Biodegradation of Polyvinyl Chloride (PVC) in Tenebrio molitor (Coleoptera: Tenebrionidae) larvae. *Environ. Int.* **2020**, *145*, 106106. [CrossRef]
47. Jacquin, J.; Cheng, J.; Odobel, C.; Pandin, C.; Conan, P.; Pujo-Pay, M.; Ghiglione, J.F. Microbial ecotoxicology of marine plastic debris: A review on colonization and biodegradation by the "Plastisphere". *Front. Microbiol.* **2019**, *10*, 865. [CrossRef]
48. Sheik, S.; Chandrashekar, K.R.; Swaroop, K.; Somashekarappa, H.M. Biodegradation of gamma irradiated low density polyethylene and polypropylene by endophytic fungi. *Int. Biodeter. Biodegr.* **2015**, *105*, 21–29. [CrossRef]
49. Rouillon, C.; Bussiere, P.O.; Desnoux, E.; Collin, S.; Vial, C.; Therias, S.; Gardette, J.L. Is carbonyl index a quantitative probe to monitor polypropylene photodegradation? *Polym. Degrad. Stabil.* **2016**, *128*, 200–208. [CrossRef]
50. Weithmann, N.; Möller, J.N.; Löder, M.G.J.; Piehl, S.; Laforsch, C.; Freitag, R. Organic fertilizer as a vehicle for the entry of microplastic into the environment. *Sci. Adv.* **2018**, *4*, eaap8060. [CrossRef]
51. Al-Salem, S.; Al-Hazza'A, A.; Karam, H.; Al-Wadi, M.; Al-Dhafeeri, A.; Al-Rowaih, A. Insights into the evaluation of the abiotic and biotic degradation rate of commercial pro-oxidant filled polyethylene (PE) thin films. *J. Environ. Manag.* **2019**, *250*, 109475. [CrossRef]
52. Gupta, K.K.; Devi, D. Characteristics investigation on biofilm formation and biodegradation activities of *Pseudomonas aeruginosa* strain ISJ14 colonizing low density polyethylene (LDPE) surface. *Heliyon* **2020**, *6*, e04398. [CrossRef]
53. Shruti, V.; Kutralam-Muniasamy, G. Bioplastics: Missing link in the era of Microplastics. *Sci. Total Environ.* **2019**, *697*, 134139. [CrossRef]
54. De Oliveira, T.A.; Barbosa, R.; Mesquita, A.B.; Ferreira, J.H.; de Carvalho, L.H.; Alves, T.S. Fungal degradation of reprocessed PP/PBAT/thermoplastic starch blends. *J. Mater. Res. Technol.* **2020**, *9*, 2338–2349. [CrossRef]
55. Rummel, C.D.; Jahnke, A.; Gorokhova, E.; Kühnel, D.; Schmitt-Jansen, M. Impacts of biofilm formation on the fate and potential effects of microplastic in the aquatic environment. *Environ. Sci. Tech. Let.* **2017**, *4*, 258–267. [CrossRef]
56. Guo, J.-J.; Huang, X.-P.; Xiang, L.; Wang, Y.-Z.; Li, Y.-W.; Li, H.; Cai, Q.-Y.; Mo, C.-H.; Wong, M.-H. Source, migration and toxicology of microplastics in soil. *Environ. Int.* **2020**, *137*, 105263. [CrossRef]

Article

Whey Protein Concentrate/Isolate Biofunctional Films Modified with Melanin from Watermelon (*Citrullus lanatus*) Seeds

Łukasz Łopusiewicz [1,*], Emilia Drozłowska [1], Paulina Trocer [1], Mateusz Kostek [1], Mariusz Śliwiński [2], Marta H. F. Henriques [3,4], Artur Bartkowiak [1] and Peter Sobolewski [5]

1. Center of Bioimmobilisation and Innovative Packaging Materials, Faculty of Food Sciences and Fisheries, West Pomeranian University of Technology Szczecin, Janickiego 35, 71-270 Szczecin, Poland; emilia_drozlowska@zut.edu.pl (E.D.); p.trocer@gmail.com (P.T.); mkosa9406@gmail.com (M.K.); Artur-Bartkowiak@zut.edu.pl (A.B.)
2. Dairy Industry Innovation Institute Ltd., Kormoranów 1, 11-700 Mrągowo, Poland; mariusz.sliwinski@iipm.pl
3. Polytechnic Institute of Coimbra, College of Agriculture, Bencanta, PT-3045-601 Coimbra, Portugal; mhenriques@esac.pt
4. CERNAS—Research Center for Natural Resources, Environment and Society, Polytechnic Institute of Coimbra, Bencanta, PT-3045-601 Coimbra, Portugal
5. Department of Polymer and Biomaterials Science, Faculty of Chemical Technology and Engineering, West Pomeranian University of Technology Szczecin 45 Piastów Ave, 70-311 Szczecin, Poland; piotr.sobolewski@zut.edu.pl
* Correspondence: lukasz.lopusiewicz@zut.edu.pl; Tel.: +48-91-449-6135

Received: 6 August 2020; Accepted: 1 September 2020; Published: 2 September 2020

Abstract: Valorization of food industry waste and plant residues represents an attractive path towards obtaining biodegradable materials and achieving "zero waste" goals. Here, melanin was isolated from watermelon (*Citrullus lanatus*) seeds and used as a modifier for whey protein concentrate and isolate films (WPC and WPI) at two concentrations (0.1% and 0.5%). The modification with melanin enhanced the ultraviolet (UV) blocking, water vapor barrier, swelling, and mechanical properties of the WPC/WPI films, in addition to affecting the apparent color. The modified WPC/WPI films also exhibited high antioxidant activity, but no cytotoxicity. Overall, the effects were melanin concentration-dependent. Thus, melanin from watermelon seeds can be used as a functional modifier to develop bioactive biopolymer films with good potential to be exploited in food packaging and biomedical applications.

Keywords: melanin; watermelon seeds; whey protein; bioactive films; plant residues

1. Introduction

The food market represents a large part of the global economy and is growing every year. Hand-in-hand, this economic sector is now also responsible for approx. 1.3 billion tons of waste per annum [1]. This waste, from fruit, vegetable, and food, includes waste generated during all aspects of food production: cleaning, processing, cooking, and packaging. However, some of these waste products and/or by-products can be important sources of bioactive compounds, such as phenolic compounds, dietary fiber, polysaccharides, vitamins, carotenoids, pigments, and oils [2]. These compounds can be potentially used in the development of novel food products (food additives and functional foods) or food packaging materials. This is an attractive path towards waste valorization in line with current market trends connected with "zero waste" goals and the so-called circular economy [3,4].

Therefore, continuing research into both the characterization and utilization of compounds obtained from food-industry waste/by-products is important, because it may offer a path towards improved sustainability of the food industry. This could significantly mitigate environmental problems associated with this industry, as well as have a positive impact from the point of view of climate change [1,2,5].

Watermelon (*Citrullus lanatus*, clade: Rosids, order: Cucurbitales, family: *Cucurbitaceae*) is a very popular fruit, with the flesh both consumed, as well as processed into juice and juice concentrates, due to water content approaching 92% of total weight. However, watermelon seeds, which constitute about 1 to 4% of total fruit weight, are not routinely eaten with the pulp [6–9]. At the same time, these seeds do have economic value, particularly in countries where cultivation is increasing. They can be used to prepare snacks or be milled into flour and used in sauces. Watermelon seeds are reported to be a rich source of proteins, vitamins B and E, minerals (such as magnesium, potassium, phosphorous, sodium, iron, zinc, manganese and copper), polyunsaturated fatty acids such as omega-6 (linoleic acid), and monounsaturated fatty acids, such as omega-9 (oleic acid). They also consist of saturated fatty acids, such as palmitic acid and stearic acid, and were found to be rich in γ-sitosterol, β-sitosterol, and lupeol [6–9]. Further, they are a promising source of useful compounds with potential biofunctional properties such as polyphenols, saponins, alkaloids and flavonoids [10,11]. However, despite these applications, watermelon seeds are still typically discarded, with only the fruit being eaten [10,12].

Biodegradable edible films are defined as a thin layer of material, that can be consumed. They are typically used to extend the shelf life and/or to improve the quality of foods. For example, they can be used to act as barriers to mass transfer, carriers of specific ingredients, or for the improvement of mechanical/handling characteristics of the product [13–15]. Growing consumer demand for high-quality foods, along with increasing environmental concern regarding the disposal of non-renewable food packaging materials, has led to a great deal of interest in the development of novel, biodegradable edible films/coatings [1,15,16]. However, such films, which are typically composed of biopolymers, can be also used in biomedical applications i.e., as wound dressings [17]. Further, the functional properties of such biopolymer films can be improved by adding different biofunctional compounds (e.g., antioxidant and/or antimicrobial properties). In this fashion, one can obtain biodegradable, bioactive materials with properties suitable for a range of diverse applications, while reducing the use of synthetic chemical additives that may have negative on human health or the environment [1,15,16].

At present, the packaging industry is dominated by synthetic polymers (plastics), because they are very cheap and possess good mechanical and physical properties. The annual plastic production is estimated to be approx. 300 million tons, of which 40% is used in packaging. However, this wide use of synthetic packaging materials has caused serious concerns, due to their high environmental impact [1,18]. Synthetic packaging polymers are petroleum-based and thus non-renewable, while at the same time being typically non-biodegradable. As a result, packaging accounts for large amounts of waste materials and pollution in the environment [1,13,19]. As a result, there is a pressing need to develop new, more eco-friendly packaging materials. In this context, biopolymers are very promising, because, compared to petroleum-based synthetic plastics, they are derived from a biological origin, making them renewable, biodegradable, and non-toxic or biocompatible [1,14]. A wide range of carbohydrates, proteins, and lipids—all derived from renewable sources—are being investigated as biodegradable alternatives, to improve sustainability and recyclability [1,13,20,21]. In particular, protein-based films are promising, due to better mechanical attributes, barrier characteristics, and nutritional-promoting properties, as compared to polysaccharide and lipid-based materials [20,22]. Gradually, bio-sourced materials are likely to replace the commonly utilized petroleum-based polymers, as environmental and sustainability externalities become increasingly accounted for in their cost [23].

Among protein-based edible films, whey protein (WP) films have received increased interest, because they possess interesting sensorial, optical, and mechanical barrier properties [24,25]. Whey is a protein-rich, major by-product of the cheese manufacturing industry [1,24,26]. In fact, this industry generates large volumes of fluid whey, that need to be properly disposed of, in order to avoid potential environmental problems [25]. Thus, whey protein-based edible films and coatings are not only value-added products, but also offer a potential solution to the disposal problem [1]. Heat-denatured whey proteins, with the addition of a plasticizer, yield transparent, bland, and flexible films with very good resistance to oxygen, aroma, and lipid transfer at low humidity [15,23]. However, the hydrophilic nature of the proteins enables interactions with water, which leads to a reduction in the moisture barrier properties [25,27]. In addition to applications in edible films, whey protein concentrates and isolates (WPC, WPI) also have the potential to be used in the biomedical field, for example forming hydrogels as bioactive carriers or by leveraging their antioxidative properties [28–31].

Melanins are black and brown biopigments, consisting of high molecular weight heterogeneous polymers derived from the oxidation of monophenols and the subsequent polymerization of intermediate o-diphenols and their resulting quinones [32]. The molecular structure of melanins includes multiple different reactive functional groups (−OH, −NH, and −COOH) [21]. They can be obtained and have been characterized by a variety of natural sources, including animals, plants, bacteria, and fungi [10,11,33,34]. Importantly, melanins are multifunctional and biologically-active, natural macromolecules and can be characterized as antioxidant, radioprotective, thermo-regulative, chemoprotective, antitumor, antiviral, antimicrobial, immunostimulating and/or anti-inflammatory [10,32,34–36]. Potentially, melanins could be used to impart some of these important attributes to polymers. In the case of biopolymers, this could enhance performance, as well as sustainability credentials. Further, melanins could enable a wide range of applications, for example by facilitating cross-linking during polymerization, providing antioxidant or antimicrobial activity, altering light scattering ability, or improving other biological properties of the polymers [18,19,37]. Importantly, melanins, like biopolymers, are obtained from renewable resources and are non-toxic; these two features make their use "greener" than many existing commercial additives [35]. Importantly, large-scale production of melanins by microorganisms digesting food waste, as well as by sustainable extraction from natural plant-residues (e.g., watermelon seeds) have been demonstrated [10,11,34]. In fact, melanin from watermelon seeds has been shown to have antioxidant and UV-barrier properties [10]. However, compared to their potential, the use of melanins remains under-explored. The relatively few examples typically involve their blending/use with polymers (as chemical modifiers or nanofillers) to modify films and coatings, for example: gelatin [38,39], poly(lactic acid) [18], alginate [21], agar [19], carrageenan [13], cellulose [22], chitosan [14], poly(vinyl alcohol) [40], polypropylene/poly(butylene adipate-co-terephthalate) [37], polyhydroxybutyrate [41] and, ethylene-vinyl acetate copolymer [35].

In this study, our aim was to investigate the effect of adding melanin obtained from watermelon seed on the properties of whey protein concentrate/isolate (WPC/WPI) films. To the best of our knowledge, no reports have been published on the modification of WPC or WPI films with natural melanin to improve the functionality of the materials. We used UV-Vis and IR spectroscopy to examine the chemical composition of films after melanin addition. Additionally, we also assessed the influence of melanin on the color, hydrodynamic, and optical properties of the films. Finally, in order to evaluate the potential (bio) functionality of the obtained materials, we evaluated their mechanical, barrier, and antioxidant properties and screened for any potential cytotoxicity in vitro.

2. Materials and Methods

2.1. Materials and Reagents

Whey protein concentrate (WPC, 85% protein content) and whey protein isolate (WPI, 90% protein content) manufactured from sweet cheese whey using cross-flow membrane filtration were purchased from Volac International Ltd. (Hertfordshire, UK). Calcium chloride, hydrogen peroxide, disodium phosphate, monosodium phosphate, 2,2-diphenyl-1-picrylhydrazyl (DPPH), 2,2′-azino-bis(3-ethylbenzothiazoline-6-sulfonic acid) (ABTS), potassium persulphate, potassium ferricyanide, trichloroacetic acid, ferric chloride, iron sulphate, tris(hydroxymethyl)aminomethane, pyrogallol, ortophenantroline, L929 murine fibroblasts, Dulbecco's Modified Eagle Medium (DMEM), fetal bovine serum (FBS), resazurin, L-glutamine, penicillin, streptomycin, and all other cell culture reagents were purchased from Sigma Aldrich (Darmstad, Germany). Glycerol, ammonia water, hydrochloric acid, sodium hydroxide, chloroform, ethyl acetate, ethanol and methanol were supplied from Chempur (Piekary Śląskie, Poland). Cell culture plasticware was purchased from VWR International (Radnor, PA, USA). All chemicals were of analytical grade.

2.2. Isolation, Purification and Preparation of Melanin Powder

Fresh Crimson Sweet watermelons (*Citrullus lanatus*) were purchased at a local market (Szczecin, Poland). Melanin isolation and purification were performed as described previously [10]. Briefly, watermelon seeds were first manually removed, then rinsed three times with distilled water, and finally dried at room temperature. Then, melanin was extracted by soaking 5 g of seeds in 50 mL of 1 M NaOH on an orbital shaker (150 rpm, 50 °C, 24 h), followed by centrifugation (6000× g rpm, 10 min) to remove plant tissue. Next, in order to precipitate the melanin, 1 M HCl was added to the alkaline mixture until the pH was 2.0, followed by centrifugation (6000× g rpm, 10 min). Then, the resultant pellet was first hydrolyzed in 6 M HCl (90 °C, 2 h), centrifuged (6000× g rpm, 10 min), and washed with distilled water five times to remove acid. After this procedure, in order to remove lipids and other residues, the pellet was washed with chloroform, ethyl acetate, and ethanol three times. Thus obtained, the purified melanin was dried and ground to a fine powder in a mortar.

2.3. Preparation of WPC and WPI Films

WPC/WPI-based films were prepared based on the methodology of Catarino et al. with minor modifications [24]. Briefly, film-forming solutions with a protein concentration of 10% (*w/w*) WPC or WPI were prepared in distilled water, at room temperature under continuous stirring. Once completely dissolved, ammonia water was added to adjust the pH to 8.0. Next, melanin was added to obtain concentrations of 0.1% and 0.5% (*w/w*) and stirred (250 rpm) for 1 h, until the melanin was completely dissolved. This mixture was then heated for 10 min in a water bath at 90 °C, until a uniform appearance was observed. Next, the mixture was cooled to room temperature and 5% (*w/w*) of glycerol (on a film-forming solution basis) was added, followed by homogenization. As reference materials, neat WPC/WPI films, without melanin addition, were also produced following the same procedure. All film samples were prepared in 10 repetitions. The film-forming solutions were cast on square (120 mm × 120 mm) polystyrene plates and dried at 40 °C for 48 h. Then, the dry films were carefully peeled off of the plates and conditioned at 25 °C and 50% RH in the clean room, prior to any tests.

2.4. Determination of Moisture Content, Water Solubility and Swelling Ratio

The moisture content (MC), water solubility (WS), and swelling ratio (SR) of obtained films were analyzed following the methodology of Roy et al. [14]. In brief, MC was determined as the weight change of the films after drying at 105 °C for 24 h. To determine the water solubility (WS), film specimens (2.5 cm × 2.5 cm) were first dried at 60 °C overnight and then weighed. The dried films were then dipped in 30 mL of distilled water for 24 h with occasional shaking at 25 °C, then carefully removed with a tweezer, and dried at 105 °C for 24 h, and finally re-weighed. The WS of the films was then calculated using the following formula:

$$WS\ (\%) = \frac{W_1 - W_2}{W_1} \times 100 \quad (1)$$

where W_1 is the initial and W_2 is the final weight of the films, respectively.

To determine the SR of the films, pre-weighed samples were submerged in 30 mL of distilled water for 1 h. Then, surface water was carefully removed using filter paper and the samples were re-weighed. The following formula was used to calculate SR:

$$SR\ (\%) = \frac{W_2 - W_1}{W_1} \times 100 \quad (2)$$

where W_1 is initial and W_2 is the final weight of the films, respectively.

2.5. Thickness, Mechanical, and Thermal Properties of WPC/WPI Films

The thickness of all obtained films was measured using a hand-held micrometer (Dial Thickness Gauge 7301, Mitoyuto Corporation, Kangagawa, Japan) with an accuracy of 0.001 mm. Each film was measured in five random points and the results were averaged.

The mechanical properties of the obtained films were tested using a Zwick/Roell 2,5 Z universal testing machine (Ulm, Germany). Static tensile testing was carried out to assess tensile strength and elongation at break (The gap between tensile clamps was 25 mm and crosshead speed was 100 mm/min).

Differential scanning calorimetry (DSC) measurements to assess thermal properties were carried out using a DSC calorimeter (DSC 3, Mettler-Toledo LLC, Columbus, OH, USA) over a temperature range from 30 to 300 at $\varphi = 10°$/min and under nitrogen flow (50 mL/min), performing two heating and one cooling scans.

2.6. The Water Vapour Transmission Rate (WVTR) of the Films

A gravimetric method was used to determine the Water Vapour Transmission Rate (WVTR) of the obtained films, as described previously [18]. This method relies on the sorption of humidity by calcium chloride. Briefly, 9 g of dry $CaCl_2$ was placed inside a container and sealed with 8.9 cm^2 samples of each film. Over the course of four days, the containers were weighed daily and the increase in mass indicated that water vapor passed through the films. For each film type, 10 film samples were tested, and average values for each day were calculated and used to express WVTR in g/(m^2 × day).

2.7. The Water Contact Angle (WCA)

The water contact angle of all obtained films was measured using a Haas μL goniometer (Poznań, Poland). Briefly, for each film, a microsyringe was used to deposit a drop of water on the surface. Three drops were analyzed and the contact angles were averaged.

2.8. Spectral Analysis

The UV-Vis spectra (300–700 nm) of the film samples were measured using a UV-Vis Thermo Scientific Evolution 220 spectrophotometer (Waltham, MA, USA).

Infrared spectroscopy was used in order to assess the chemical composition of obtained films, as described previously [18]. Briefly, 4 cm^2 squares of each film were placed directly on the ray-exposing stage of the ATR accessory of a Perkin Elmer Spectrum 100 FT-IR spectrometer (Waltham, MA, USA) operating in ATR mode. Spectra (64 scans) were recorded over a wavenumber range of 650–4000 cm^{-1}, at a resolution of 4 cm^{-1}. For analysis, spectra were baseline corrected and normalized using SPECTRUM software [18].

2.9. Color Analysis

The effect of melanin on the color of the films was measured using a colorimeter (CR-5, Konica Minolta, Tokyo, Japan). For each film type, five samples were analyzed, by making three measurements on both sides of each sample. The results (mean ± standard deviation) were expressed as L* (lightness), a* (red to green), and b* (yellow to blue). Additionally, ΔE (color difference) and YI (yellowness index), compared to unmodified WPC/WPI films, were also calculated as follows:

$$\Delta E = [(L_{standard} - L_{sample})^2 + (a_{standard} - a_{sample})^2 + (b_{standard} - b_{sample})]^{0.5} \quad (3)$$

$$YI = 142.86 b^* L^{-1} \quad (4)$$

2.10. Antioxidant Potential of the Films

2.10.1. Reducing Power

The reducing power of the films was determined based on the previously described methodology [42] with our own modification. Briefly, film samples (100 mg) were placed in 1.25 mL of phosphate buffer (0.2 M, pH 6.6), followed by the addition of 1.25 mL of 1% potassium ferricyanide solution. Samples were then incubated for 20 min at 50 °C followed by the addition of 1.25 mL of trichloroacetic acid. Next, the test tubes were centrifuged at 3000× g rpm for 10 min and 1.25 mL of obtained supernatant was diluted with 1.25 mL of deionized water. Finally, 0.25 mL of 0.1% ferric chloride solution was added and the absorbance was measured at 700 nm.

2.10.2. Free Radical Scavenging Activity

The free radical scavenging activity of WPC/WPI films was assessed towards ABTS, DPPH, superoxide (O_2^-), and hydroxyl ($\cdot$OH) radicals. ABTS and DPPH tests were performed according to Bishai et al. with a slight modification [43]. For the ABTS test, 5 mL of 7 mM ABTS was mixed with 5 mL of 2.45 mM potassium persulfate to obtain the radical 2,2'-azino-bis(3-ethylbenzothiazoline)-6-sulphonic acid (ABTS$^+$). After 16 h of incubation at room temperature protected from light, the solution was diluted to an absorbance maximum of 1.00 at 734 nm using water. To 25 mL of this ABTS$^+$ solution, samples of each film (100 mg) were added and incubated up to 1 h at room temperature. As a control, tubes of ABTS$^+$ solution were incubated under identical conditions, but without films. Finally, absorbance was measured and ABTS scavenging was calculated using the equation:

$$\textit{Free radical scavenging activity } (\%) = \frac{A_{sample} - A_{control}}{A_{sample}} \times 100, \qquad (5)$$

where A_{sample} is the absorbance of ABTS$^+$ solution with the film sample and $A_{control}$ is the absorbance of ABTS$^+$ solution without sample.

To determine DPPH radical scavenging activity, 100 mg of each film was placed in 25 mL of 0.01 mM DPPH methanolic solution, incubated for 30 min at room temperature, and absorbance at 517 nm was measured. As a control, the same solution was measured but without any film samples. The DPPH radical scavenging activity was calculated using Equation (5).

Superoxide (O_2^-) scavenging activity was assessed using the pyrogallol oxidation inhibition assay, following the methodology of Ye et al. with some modification [44]. Briefly, 100 mg of each film was incubated for 5 min in 3 mL of 50 mmol/L (pH 8.2) Tris-HCl buffer with gentle stirring. Then, 0.3 mL of 7 mM pyrogallol solution that was preheated to 25 °C was added and the mixture was allowed to react for exactly 4 min. To terminate the reaction 1 mL of 10 mM HCl was added and the absorbance was measured at 318 nm. The O_2^- scavenging rate was calculated from the formula:

$$O_2^- \textit{ inhibition } (\%) = \left[1 - \frac{(A_1 - A_1')}{A_0}\right] \times 100, \qquad (6)$$

where A_1 is the absorbance of the mixture in the presence of the sample, A_1' is the absorbance of water instead of the reaction agent, and A_0 is the absorbance without the sample.

Hydroxyl ($\cdot$OH) scavenging was assessed using the method of Ye et al. [44] with some modification. Film specimens (100 mg) were placed in a mixture of 1.5 mL of 5 μM ortophenantroline solution and 2 mL of phosphate buffer (pH 7.4, 0.05 M). Then, 1 mL of 7.5 mM FeSO$_4$ solution was added, followed by 1 mL of 0.1% H$_2$O$_2$, and, finally, distilled water was added to bring the total volume to 10 mL. The reaction solution was incubated at 37 °C for 1 h, protected from light, and the absorbance was measured at 536 nm. Ortophenatroline solution without H_2O_2 (replaced by 1 mL of methanol) served as a blank. The following formula was used to calculate hydroxyl scavenging:

$$\cdot OH \text{ inhibition } (\%) = \left[\frac{A_2 - A_1}{A_0 - A_1}\right] \times 100, \qquad (7)$$

where A_0 is the ortophenatroline solution without H_2O_2 addition, A_1 is the absorbance without the sample, and A_2 is the absorbance in the presence of the sample.

2.11. Evaluation of Cytotoxicity

In order to screen for potential cytotoxicity, extract tests and direct contact tests were carried out based on ISO 10993-5 using L929 murine fibroblasts [45]. Cells (passage 20–25) were maintained in complete growth medium: DMEM containing 10% FBS, 2 mM L-glutamine, 100 U/mL penicillin, and 100 μg/mL streptomycin. For each material, 8-mm discs were cut using a steel punch and sterilized using 20 min exposure to UV lamp in BSL-2 safety cabinet (Telstar Bio II Advance, Barcelona, Spain). Extracts were prepared by placing six 8-mm diameter discs of each material in a tissue culture plate and soaking in 1 mL of growth media (ratio: 3 cm^2/mL) for 24 h at 37 °C. Medical grade PCL (CAPA6340) was used as a negative control, nitrile glove (Mercator Nitrylex Classic, Kraków, Poland) served as a positive (toxic) control, and as a sham extract, 1 mL of media was added to an empty well. In parallel, 10,000 L929 cells were seeded per well of a 96-well plate and allowed to adhere and spread for 24 h. Next, the media was aspirated and replaced with 100 μL of extract, six technical replicates per material. After a further 24 h of culture, cell viability was assessed using an inverted light microscope (Delta Optical IB-100, Mińsk Mazowiecki, Poland) and resazurin viability assay [46] using a fluorescent plate reader (Biotek Synergy HTX, Winooski, VT, USA) (excitation 540 nm, emission 590 nm).

For the direct contact assay, 30,000 L929 cells were seeded per well of a 48-well plate and allowed to adhere and spread for 24 h. Then, discs of each material (8 mm diameter) were placed directly on top of the cell monolayer ($n = 5$ discs per material). Cells were then maintained for another 24 h of culture and viability was assessed using an inverted light microscope and resazurin viability assay—without removal of discs—as described previously.

For both experiments, viability data obtained from resazurin assay was analyzed by subtracting blank signal (growth media only, no cells) from all other measurements and normalizing to sham-treated cells. Both the extract and direct contact assay were performed twice.

2.12. Statistical Analyses

Statistical comparisons were performed using Statistica version 10 (StatSoft Polska, Kraków, Poland). Differences between means were determined using analysis of variance (ANOVA) followed by Fisher's LSD post-hoc testing with a significance threshold of $p < 0.05$. All measurements were carried out in at least triplicate.

3. Results and Discussion

3.1. Hydrodynamic Properties (Moisture Content, Water Solubility and Swelling Ratio)

The MC, WS, and SR of the WPC/WPI films are summarized in Table 1. The MC of neat WPC film was 27.16 ± 1.31%, whereas the MC of neat WPI film was 24.31 ± 0.51%. Both increased significantly as melanin content increased. This observation is in line with the findings of other authors who used melanins to modify biopolymer-based films, such as chitosan and agar [14,19]. In contrast, Yang et al. [21] noticed that the MC of alginate/poly(vinyl alcohol) films decreased with increased melanin content, similarly as observed by Roy et al. for the case of cellulose/melanin films [22]. According to Roy et al., a significant increase in the MC of polymer/melanin films may be the result of reduced polymer network interactions and, as a consequence, greater accessibility of free hydroxyl groups to absorb more water molecules [14,19].

Table 1. Moisture content (MC), water solubility (WS) and swelling ratio (SR) of neat whey protein concentrate/isolate films (WPC/WPI) and melanin-modified films.

Sample	MC (%)	WS (%)	SR (%)
WPC-C	27.16 ± 1.31 [b]	65.90 ± 6.12 [a]	324.30 ± 68.92 [a]
WPC-0.1	27.55 ± 0.25 [b]	55.31 ± 4.29 [b]	331.47 ± 18.62 [a]
WPC-0.5	31.60 ± 0.56 [a]	45.02 ± 2.98 [c]	469.47 ± 21.69 [b]
WPI-C	24.31 ± 0.51 [a]	70.57 ± 5.24 [a]	75.85 ± 8.58 [a]
WPI-0.1	26.78 ± 0.38 [b]	64.25 ± 3.54 [a,b]	111.94 ± 11.69 [b]
WPI-0.5	27.92 ± 0.06 [b]	57.08 ± 6.37 [b]	147.58 ± 18.04 [c]

Values are means ± standard deviation. Means with different lowercase are significantly different at $p < 0.05$.

In terms of WS, melanin addition resulted in a significant decrease in WS for both types of films ($p < 0.05$). The WS of WPC-0.5% and WPI-0.5% was reduced by approximately 20.87% and 13.50%, respectively, as compared to the neat WPC and WPI films ($p < 0.05$). Previously, a WS reduction for gelatin films modified with fungal melanin was reported [38]. Finally, for the case of SR, we observed that the addition of melanin markedly increased the SR of films ($p < 0.05$). Further, for the case of WPC films, they had a much higher SR than WPI films: the SR values of WPC-0.5% and WPI-0.5% were 147.58 ± 18.04% and 469.47 ± 21.69%, respectively and are lower than those reported previously for agar/melanin films [19]. Interestingly, Roy et al. observed that SR of chitosan films was improved by the reinforcement of melanin nanoparticles [14], but in contrast, SR was reported to be decreased in the case of agar/melanin nanocomposite films [19]. In biopolymer-based films, SR primarily depends on the cross-linking, porosity, and nature of the materials [14,47]. Thus, in the present case, the cross-linking density of the melanin-modified films might be higher and, consequently, the porosity and SR of the samples may have increased [14].

3.2. The Thickness, Mechanical and Thermal Properties

The thickness, mechanical, and thermal properties of the samples are presented in Table 2. The modification with melanin did not affect the thickness of WPC or WPI films ($p > 0.05$), probably because of the small amount of melanin used. However, in studies by other authors who used melanin particles at higher concentrations, an increase of thickness was observed, due to higher dry mass content [19]. Here, the addition of melanin significantly enhanced the mechanical strength of WPC and WPI films ($p < 0.05$). The tensile strength (TS) of the WPI films was higher than the corresponding WPC films; the highest TS was observed for sample WPI-0.5% (6.13 ± 0.41 MPa). However, the TS of the WPC/WPI films was lower than that reported for agar/melanin [19], chitosan/melanin [14] and poly(lactic acid)/melanin films [18]. We also observed that the elongation at break (EB) of the modified films decreased in comparison to the control samples ($p < 0.05$). The increase in TS and decrease of EB can be attributed to strong hydrogen bonding (H-bonding) interactions between melanin and the polymer matrices, which improves the mechanical properties of the films. This observation is in good agreement with the results of other authors [14,19,22]. It has also been reported that reinforcement with a low percentage (0.025% and 0.05%) of fungal melanin improved the mechanical properties of poly(lactic acid)/melanin composite films; however, at higher melanin concentration (0.2%), the TS of the films decreased [18]. These results clearly indicate that melanins (synthetic or natural) can improve the mechanical properties of polymer-based films, although the effect of melanin is highly dependent on the type of polymer matrix used and melanin concentration, which play pivotal roles in the inter- and intramolecular interactions between the polymer chains in the film [14,18,19].

Table 2. Thickness, mechanical, and thermal characteristics of neat WPC/WPI and melanin-modified films.

Sample	Thickness (mm)	TS (MPa)	EB (%)	T_m (°C)	ΔH_m (J/g)
WPC-C	0.18 ± 0.02 [a]	4.11 ± 0.36 [a]	18.32 ± 1.28 [a]	98.17	−68.83
WPC-0.1	0.18 ± 0.04 [a]	4.61 ± 0.45 [b]	12.66 ± 1.28 [b]	99.57	−62.50
WPC-0.5	0.18 ± 0.01 [a]	4.87 ± 1.04 [b]	11.14 ± 1.08 [b]	106.90	−59.30
WPI-C	0.14 ± 0.02 [a]	4.50 ± 1.02 [a]	16.22 ± 0.62 [a]	103.71	−162.24
WPI-0.1	0.16 ± 0.01 [a]	5.51 ± 1.34 [a]	14.88 ± 2.12 [a,b]	106.35	−125.92
WPI-0.5	0.16 ± 0.04 [a]	6.13 ± 0.41 [b]	13.96 ± 1.08 [b]	108.21	−96.91

Values are means ± standard deviation. Means with different lowercase are significantly different at $p < 0.05$.

In the present study, DCS measurements were also carried out to investigate the thermal characteristics of the films. Table 3 presents the melting temperatures (T_m) and melting enthalpies (ΔH_m) of the samples. Compared to the neat WPC/WPI films, the addition of melanin markedly increased the T_m and ΔH_m of the modified films. These changes may be due to different polymer-water interactions as a result of modification with melanin [14]. For example, Dong et al. reported that the addition of synthetic and natural melanin to PVOH improved the thermal stability of the PVOH films [40]. However, there are also several reports that modification with melanin did not affect the thermal stability of agar [19], cellulose [22], and poly(lactic acid) films [18]. At the same time, in those reports, melanins were used as fillers. Thus, it can be reasonably concluded that the different effects of melanins on thermal stability may be due to different melanin sources, differences in the polymer matrix, as well as different methods of preparation of the films [13].

Table 3. Color parameters (L*, a*, b*), total color difference (ΔE), and yellowness index (YI) of neat WPC/WPI and melanin-modified films.

Sample	L*	a*	b*	ΔE	YI
WPC-C	86.42 ± 0.91 [a]	−0.51 ± 0.10 [a]	11.74 ± 1.73 [a]	used as standard	19.41 ± 3.08 [a]
WPC-0.1	85.84 ± 1.18 [a]	−0.44 ± 0.17 [a]	13.96 ± 2.24 [a]	2.29 ± 0.85 [a]	23.23 ± 4.10 [a]
WPC-0.5	80.83 ± 1.27 [b]	0.90 ± 0.56 [b]	25.73 ± 3.26 [b]	15.13 ± 0.13 [b]	45.48 ± 7.18 [b]
WPI-C	90.57 ± 0.11 [a]	−0.85 ± 0.07 [a]	4.64 ± 0.47 [a]	used as standard	7.32 ± 0.74 [a]
WPI-0.1	89.38 ± 0.61 [b]	−0.72 ± 0.08 [a,b]	7.77 ± 1.22 [b]	3.35 ± 1.53 [a]	12.42 ± 1.87 [b]
WPI-0.5	86.41 ± 0.64 [c]	−0.66 ± 0.10 [b]	16.14 ± 1.78 [c]	12.23 ± 1.08 [b]	26.68 ± 3.14 [c]

Values are means ± standard deviation. Means with different lowercase are significantly different at $p < 0.05$.

3.3. Color

The apparent color, the total color difference (ΔE), and the yellowness index of the WPC/WPI films are presented in Table 3 and Figure 1. The neat WPI film was transparent and colorless, whereas neat WPC film showed a light yellowish color. Generally, WPC films were darker (lower L values) when compared with WPI films. The films with 0.1% and 0.5% melanin showed a yellowish-brown color and became darker as the concentration of melanin increased ($p < 0.05$). The redness and yellowness of the modified films also increased, due to the red-brown color of melanin. The yellowness index of WPC/WPI films increased significantly with increased melanin concentration ($p < 0.05$). As a result, the ΔE of the films was remarkably increased ($p < 0.05$) and ranged from 2.29 (WPC-0.1%) to 15.13 (WPC-0.5%). $\Delta E > 1$ is considered perceptible to the human eye, so both concentrations of melanin in WPC/WPI films caused noticeable color changes [18]. Overall, the observed results are in good agreement with previous reports, where chitosan [14], agar [19], polypropylene/poly(butylene-co-terephthalate) [37], gelatin [38] and poly(lactic acid) [18] films were modified with various melanins.

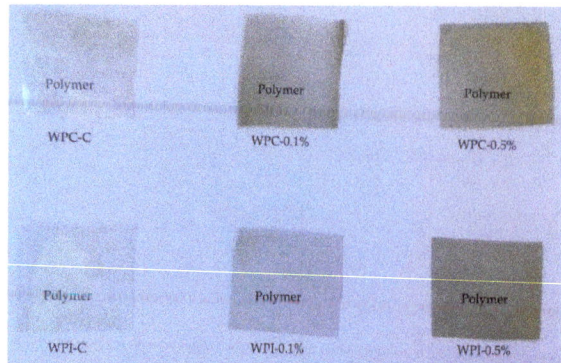

Figure 1. The visual appearance of neat and modified WPC/WPI films.

3.4. UV-Barrier Properties

UV-Vis light transmittance spectra of the neat and modified WPC/WPI films are shown in Figure 2. As can be seen, the neat WPC and WPI films were highly transparent to visible light at wavelengths greater than 380 nm and exhibited high transmittance for both UVA and UVB light. The transparency of WPI and WPC films decreased markedly by the addition of melanin and the decrease was concentration-dependent. These results clearly indicate that melanin, even at low concentration (0.1%) improves UV-light barrier properties of WPC, as well as WPI films. The decrease in the light transmittance of WPC and WPI films was mainly due to the absorption of UV light by melanin, similarly as in gelatin/melanin films [38]. However, when melanins were used as nanofillers, the UV barrier properties of materials were due to blocking of the light path, as was reported for poly(lactic acid)/melanin films [18] as well as for agar/melanin nanocomposite films [19]. In fact, in nature, melanins play a pivotal role in UV-protection, due to their strong UV-blocking properties resulting from the presence of phenolic and indole groups [22,37]. The high UV-barrier properties of melanin from watermelon seeds have also already been reported [10].

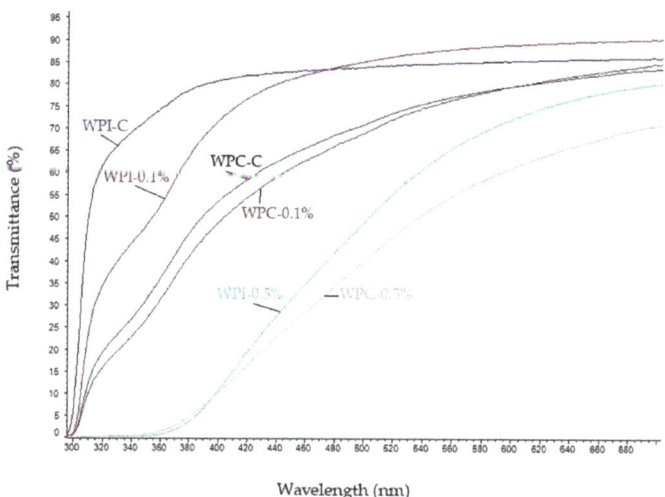

Figure 2. UV–Vis spectra of neat and modified WPC/WPI films.

3.5. Antioxidant Activity

The antioxidant activity of the films was determined by measuring the reducing power and radical (DPPH, ABTS, O_2^-, OH) scavenging activity, as presented in Table 4. It was noticed that the antioxidant activity of the films was melanin concentration-dependent and increased significantly with increased melanin concentration ($p < 0.05$). The highest reducing power (1.186 ± 0.06) was measured for sample WPI-0.5%. The ABTS scavenging efficiencies of the neat WPI and WPI films were 72.02 ± 5.78% and 60.93 ± 4.50%, respectively, and an increase to 85.18 ± 1.74% and 96.87 ± 0.48%, respectively, was noticed when melanin concentration was 0.5%. A similar trend, but lower values, were observed in the case of DPPH and O_2^- scavenging activity. On the other hand, the highest hydroxyl radical scavenging activity was observed for sample WPC-0.5% (39.08 ± 0.14%). In fact, the antioxidant activity of edible films is known to increase with the number of antioxidants [16,20,22,48]. Further, the antioxidant activity of the neat WPC/WPI films was expected, as they are reported to contain bioactive proteins, peptides, and amino acids with antioxidant capacity [28,31,49]. The antioxidant activity observed in the present report is also in good agreement with previous findings regarding agar [19], carrageenan [13], cellulose [22] and poly(lactic acid) [18] composite films modified with various melanins as fillers. Further, the antioxidant activity of melanin/gelatin films has been shown to increase when melanin is dissolved in a film-forming alkaline solution [38]. Likewise, it has already been observed that melanin from watermelon seeds has antioxidant activity [10]. In fact, it is well known that melanins act as effective antioxidants, because of intramolecular non-covalent electrons that can easily interact with free radicals and other reactive species [19]. Thus, the WPC/WPI melanin-modified films obtained here can be potentially used in active antioxidant packaging, to prevent oxidation of lipid-containing food or to preserve oxidation-sensitive food, as well to increase the shelf life of food [18,22,39]. Indeed, gelatin-based coatings modified with fungal melanin have been shown to have preservative activity against pork lard rancidity [39]. However, it is possible that the materials obtained here could also be used in biomedical applications, such as wound healing, due to their strong radical scavenging activity [31,50,51]. Wound healing is a complex process requiring an optimal balance between oxidative stress and antioxidant status. Typically, proper wound healing requires low levels of reactive oxygen species and low oxidative stress, while excessive oxidative stress can to impaired wound healing. As a result, antioxidants that can help control wound oxidative stress can improve wound healing [50,51].

Table 4. Reducing power (RP) and radical (DPPH, ABTS, O_2^-, ·OH) scavenging activity of neat WPC/WPI and melanin-modified films.

Sample	RP (700 nm)	DPPH (%)	ABTS (%)	O_2^- (%)	·OH (%)
WPC-C	0.661 ± 0.245 [a]	37.70 ± 1.67 [a]	72.03 ± 5.78 [a]	26.80 ± 1.61 [a]	31.68 ± 3.21 [a]
WPC-0.1	0.863 ± 0.122 [a]	49.87 ± 0.13 [b]	76.86 ± 6.13 [a,b]	29.07 ± 1.17 [a,b]	33.45 ± 0.30 [a]
WPC-0.5	0.924 ± 0.069 [b]	63.68 ± 2.69 [c]	85.18 ± 1.74 [b]	38.11 ± 0.31 [b]	39.08 ± 0.14 [b]
WPI-C	0.740 ± 0.192 [a]	58.76 ± 0.10 [a]	60.94 ± 4.50 [a]	27.96 ± 3.45 [a]	32.33 ± 0.69 [a]
WPI-0.1	1.110 ± 0.111 [b]	65.45 ± 0.20 [b]	95.83 ± 1.52 [b]	33.48 ± 5.78 [b]	33.91 ± 0.27 [b]
WPI-0.5	1.186 ± 0.060 [b]	72.70 ± 0.43 [c]	96.87 ± 0.48 [b]	38.19 ± 1.85 [c]	37.47 ± 0.05 [c]

Values are means ± standard deviation. Means with different lowercase are significantly different at $p < 0.05$.

3.6. FT-IR

FT-IR is a commonly used technique to assess the miscibility and compatibility of biopolymers and additives, owing to its rapid and nondestructive nature [3,18,22]. Absorbance spectra of WPC-based and WPI-based films are presented in Figure 3. As can be seen the films have characteristic peaks at 3280 cm^{-1} attributed to C–H, N–H and/or O–H of WPC/WPI and melanin [10,27]. Characteristic CH_3 and CH_2 stretching peaks at approximately 2940 cm^{-1} and 2880 cm^{-1} were also observed. The peak centered at 1630 cm^{-1} can be attributed to Amide-I associated with C=O stretch and C–N vibrations; however, in the case of melanin-modified films, it also corresponds to the vibration of aromatic C=C of melanin [10]. Peaks in the region 1400–1550 cm^{-1} and 1200–1350 cm^{-1} can be assigned to Amide-II (groups N-H) and to Amide III (with N–H bonds and C–N stretching), respectively [27]. The region from 750 to 1200 cm^{-1} was assigned to the absorption of glycerol (plasticizer), vibrations C–O and C–C bonds [27]. In the case of WPC and WPI films modified with melanin, slight variations in peak intensities and positions were observed, particularly at 3280 cm^{-1}, 2940 cm^{-1}, 2880 cm^{-1} and 1630 cm^{-1}. Overall, no substantial variations were noted in the functional groups of WPC/WPI modified films. The observed findings suggest that there were no structural changes in WPC and WPI films due to the addition of melanin. The small changes in intensities and minor shifts in the peaks are likely due to physical interactions (H-bonding, van der Waals force) between WPC/WPI and melanin, which is consistent with the findings of other authors [14,19,21].

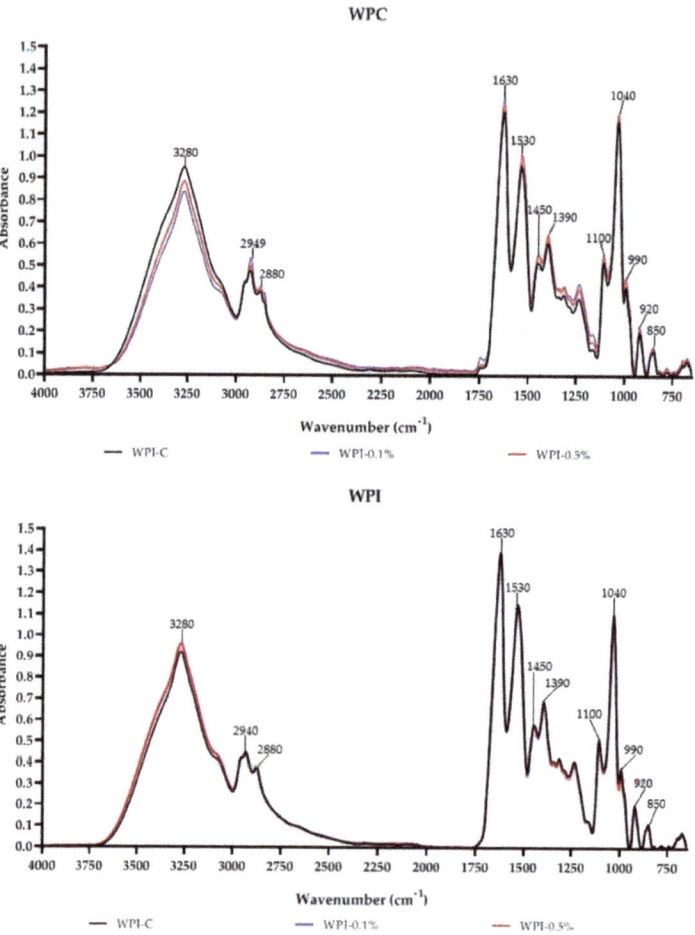

Figure 3. FT-IR spectra of neat WPC/WPI and melanin-modified films.

3.7. Water Contact Angle and WVTR

The results of water contact angle (WCA) measurements are presented in Table 5. The WCA of neat WPC and WPI films was 27.67 ± 0.47° and 45 ± 0.00°, respectively. Generally, biopolymer films with a WCA of less than 65° are considered hydrophilic [14]. Concerning the effect of melanin addition, it was interestingly observed that the WCA of the films decreased significantly with increasing melanin content to 14.33 ± 0.47° and 31 ± 0.00° for samples WPC-0.5% and WPI-0.5%, respectively. This observation is in contrast to the other reports that melanins generally increase the WCA of polymeric films due to their hydrophobic nature [14,18]. However, the difference might be attributed to the use of melanin particles in those studies, whereas in the present study melanin was used to modify the alkaline film-forming solution, which could result in other interactions in the polymer matrix.

Table 5. Water Contact Angle (WCA) and Water Vapor Transmission Ratio (WVTR) of neat WPC/WPI and melanin-modified films.

Sample	WCA (°)	WVTR (g/(m^2 × Day))
WPC-C	27.67 ± 0.47 [a]	1712.64 ± 7.46 [a]
WPC-0.1	18.00 ± 0.00 [b]	1599.23 ± 5.01 [b]
WPC-0.5	14.33 ± 0.47 [c]	1483.53 ± 5.49 [c]
WPI-C	45.00 ± 0.00 [a]	1618.57 ± 6.23 [a]
WPI-0.1	33.00 ± 0.00 [b]	1566.70 ± 7.14 [b]
WPI-0.5	31.00 ± 0.00 [c]	1490.49 ± 5.37 [b]

Values are means ± standard deviation. Means with different lowercase are significantly different at $p < 0.05$.

The water vapor permeability of the samples is also displayed in Table 5. It was observed that in the case of WPC, as well as WPI, the WVTR decreased as a result of modification with melanin ($p < 0.05$). The observed results are in good agreement with previous reports, where various melanins were used for the modification of alginate/poly(vinyl alcohol) [21], cellulose [22] and gelatin [38] films. However, there are also reports that the effect of melanin on WVTR depends on the concentration used. For instance, in the case of melanin added to poly(lactic acid) composite films, it was observed that WVTR decreased in the presence of low melanin content, but increased at high melanin content [18]. Likewise, in the case of agar/melanin nanoparticles composite films, the incorporation of low melanin-particle content did not result in WVTR changes, but at higher content, the WVTR increased. Whey protein-based materials are generally permeable or semipermeable to moisture penetration [25,52]. Although the WVTR of the melanin-modified films prepared here decreased (WVTR of WPC-0.5% and WPI-0.5%, 1483.53 ± 5.49 and 1490.49 ± 5.37 (g/(m^2 × Day)) respectively), as compared to unmodified samples (WPC-C and WPI-C -1712.64 ± 7.46 and 1618.57 ± 6.23 (g/(m^2 × Day)), respectively), they still exhibited weak water vapor barrier properties, lower than those reported in other studies [14,18,19,37].

3.8. Cytotoxicity

The cytotoxicity studies confirmed the non-toxic nature of the samples. As can be seen in Figure 4, after 24 h of exposure to extracts, robust cell growth was observed for all tested materials, similar to sham. Further, no marked differences in morphology were observed that could indicate a negative effect. The results of the resazurin cell viability assay, presented in Figure 5, were in good agreement with the microscopic observations. In all cases, cell viability was similar, near 90%—higher than the 70% threshold for cytotoxicity specified by the ISO 10993 norm.

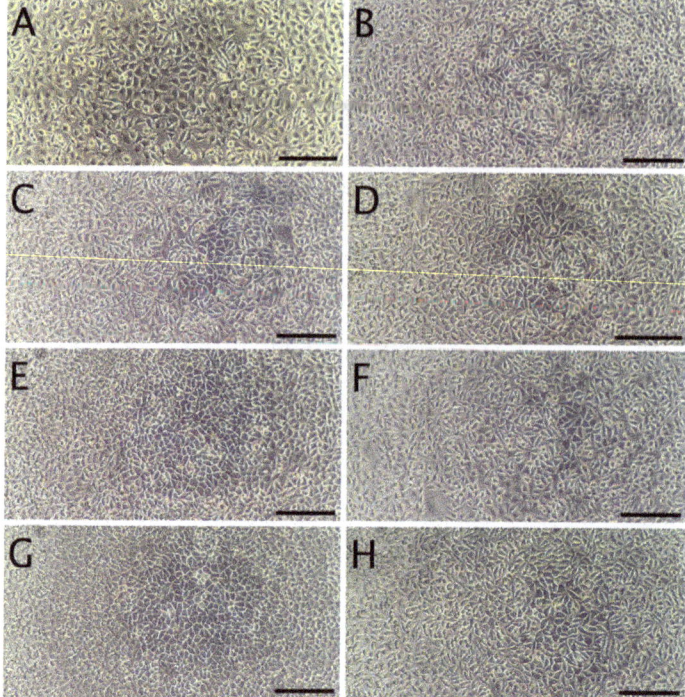

Figure 4. Representative micrographs of L929 fibroblasts. Panel (**A**): cells 24 h after seeding, but prior to the addition of extracts. Panel (**B**): Cells incubated for 24 h with the sham extract. Panels (**C,E,G**) present cells incubated with WPC extracts (WPC; WPC-0.1% and WPC-0.5%) for 24 h. Panels (**D,F,H**) present cells incubated with WPI extracts (WPI; WPI-0.1% and WPI-0.5%) for 24 h. The scale bar represents 200 µm.

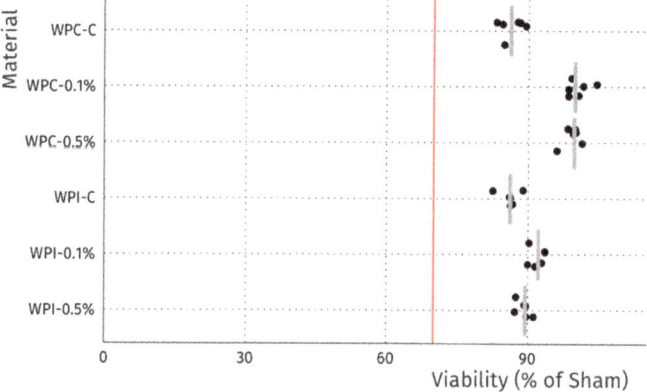

Figure 5. Viability of L929 fibroblasts after 24 h incubation with extracts. Data is normalized to sham extract. Dots represent technical replicates, grey bars represent median, and the red line indicates 70% viability, the threshold for cytotoxicity.

In terms of the direct contact assay, again robust cell growth was observed beneath the discs of each material, visually similar to wells without discs (Figure 6). Likewise, no marked differences in morphology were observed that could indicate a negative effect. Thus, the reactivity is graded as 0, according to ISO 10993. The resazurin viability assay indicated a modest reduction in viability for 4 of the 6 tested materials (Figure 7), which is likely due to mechanical damage to the monolayer caused by the physical presence of the disc on top of the cells. Meanwhile, the values above 100% may be due to an interaction between melanin and the resazurin reagent, as the assay was performed with the discs remaining in place to avoid the risk of further mechanical damage. To compare the viability data to the 70% viability threshold set by the ISO10993 standard, we fitted a linear model (estimated using ordinary least squares, using R (RStudio)) to predict (Normalized Viability (%)-Threshold (70%)) with Material. The model explains a significant and substantial proportion of variance ($R^2 = 0.96$, $F(6, 24) = 94.64$, $p < 0.01$, adj. $R^2 = 0.95$). For each material, the effect was positive, indicating viability values were above the threshold, and significant ($p < 0.001$). Both cell culture experiments were repeated and similar results were observed. Overall, the non-toxic nature of the films was expected, as no cytotoxic effects of whey proteins [29] nor melanins [14,53] have been reported.

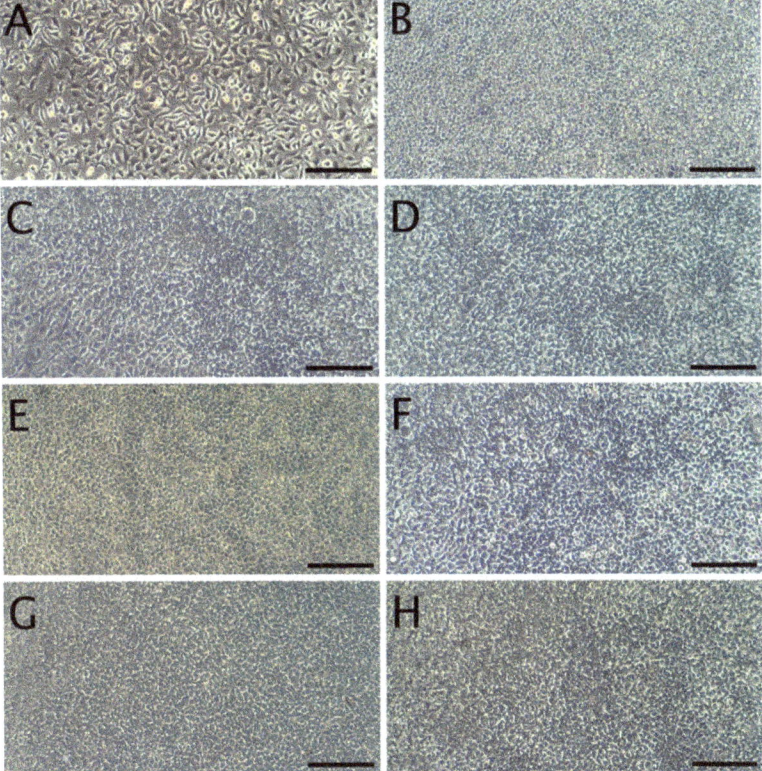

Figure 6. Representative micrographs of L929 fibroblasts. Panel (**A**): Cells 24 h after seeding, but prior to the addition of discs. Panel (**B**): cells incubated for 24 h without disc. Panels (**C,E,G**) present cells beneath WPC discs (WPC; WPC-0.1% and WPC-0.5%) after 24 h of culture. Panels (**D,F,H**) present cells beneath WPI discs (WPI; WPI-0.1% and WPI-0.5%) after 24 h of culture. The scale bar represents 200 µm.

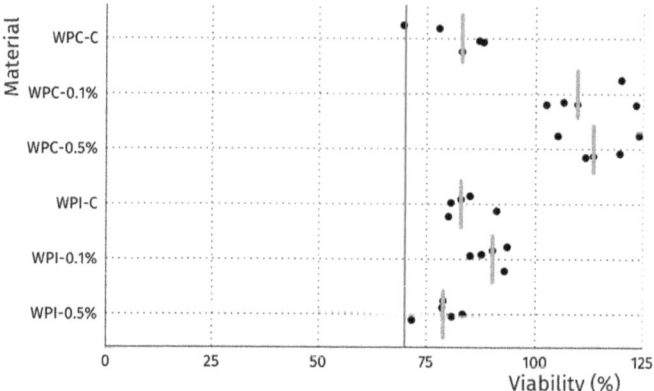

Figure 7. Viability of L929 fibroblasts after 24 h incubation in direct contact with discs. Data is normalized to wells without discs. Dots represent individual discs, grey bars represent median, and the red line indicates 70% viability, the threshold for cytotoxicity.

4. Conclusions

This article explored the properties of WPC and WPI films modified with plant melanin derived from agro-industrial by-product: watermelon seeds. The properties of melanin-modified WPC/WPI films were compared to neat WPC and WPI films. The modification with melanin had a marked effect on the mechanical, antioxidant, hydrodynamic and barrier properties, but did not introduce cytotoxicity. We conclude that the obtained biofunctional WPC/WPI melanin-modified films with improved mechanical, UV-barrier, and water vapor barrier properties, along with strong antioxidant activity could be used for active food packaging or, potentially, biomedical applications, such as wound dressings.

Author Contributions: Ł.Ł.: conceptualization, formal analysis, investigation, methodology, supervision, visualization, writing—original draft, and writing—review and editing; E.D., P.S.: visualization, investigation, methodology, formal analysis, and writing—review and editing; P.T., M.K.: investigation; M.Ś.: funding acquisition, investigation; M.H.F.H.: methodology; A.B.: funding acquisition, formal analysis. All authors have read and agreed to the published version of the manuscript.

Funding: This research received no external funding.

Acknowledgments: The authors thank Karol Fijałkowski (Faculty of Biotechnology and Animal Husbandry, ZUT) for access to the multi-functional plate reader.

Conflicts of Interest: The authors declare no conflict of interest.

References

1. Avramescu, S.M.; Butean, C.; Popa, C.V.; Ortan, A.; Moraru, I.; Temocico, G. Edible and functionalized films/coatings—Performances and perspectives. *Coatings* **2020**, *10*, 687. [CrossRef]
2. Dueñas, M.; García-Estévez, I. Agricultural and food waste: Analysis, characterization and extraction of bioactive compounds and their possible utilization. *Foods* **2020**, *9*, 817. [CrossRef] [PubMed]
3. Drozłowska, E.; Łopusiewicz, Ł.; Mężyńska, M.; Bartkowiak, A. Valorization of flaxseed oil cake residual from cold-press oil production as a material for preparation of spray-dried functional powders for food applications as emulsion stabilizers. *Biomolecules* **2020**, *10*, 153. [CrossRef] [PubMed]
4. Turon, X.; Venus, J.; Arshadi, M.; Koutinas, M.; Lin, C.S.K.; Koutinas, A. Food waste and byproduct valorization through bio-processing: Opportunities and challenges. *BioResources* **2014**, *9*, 5774–5777. [CrossRef]
5. Maina, S.; Kachrimanidou, V.; Koutinas, A. A roadmap towards a circular and sustainable bioeconomy through waste valorization. *Curr. Opin. Green Sustain. Chem.* **2017**, *8*, 18–23. [CrossRef]

6. Deshmukh, C.D.; Jain, A.; Tambe, M.S. Phytochemical and Pharmacological profile of *Citrullus lanatus* (THUNB). *Biolife* **2015**, 483–488. [CrossRef]
7. Mehra, M.; Pasricha, V.; Gupta, R.K. Estimation of nutritional, phytochemical and antioxidant activity of seeds of musk melon (*Cucumis melo*) and water melon (*Citrullus lanatus*) and nutritional analysis of their respective oils. *J. Pharmacogn. Phytochem.* **2015**, *3*, 98–102.
8. Tabiri, B. Watermelon seeds as food: Nutrient composition, phytochemicals and antioxidant activity. *Int. J. Nutr. Food Sci.* **2016**, *5*, 139. [CrossRef]
9. Seidu, K.T.; Otutu, O.L. Phytochemical composition and radical scavenging activities of watermelon (*Citrullus lanatus*) seed constituents. *Croat. J. Food Sci. Technol.* **2016**, *8*, 83–89. [CrossRef]
10. Łopusiewicz, Ł. Antioxidant, antibacterial properties and the light barrier assessment of raw and purified melanins isolated from *Citrullus lanatus* (watermelon) seeds. *Herba Pol.* **2018**, *64*, 25–36. [CrossRef]
11. Glagoleva, A.Y.; Shoeva, O.Y.; Khlestkina, E.K. Melanin pigment in plants: Current knowledge and future perspectives. *Front. Plant Sci.* **2020**, *11*. [CrossRef]
12. Wani, A.A.; Sogi, D.S.; Singh, P.; Shivhare, U.S. Characterization and functional properties of watermelon (*Citrullus lanatus*) seed protein isolates and salt assisted protein concentrates. *Food Sci. Biotechnol.* **2011**, *20*, 877–887. [CrossRef]
13. Roy, S.; Rhim, J.W. Carrageenan-based antimicrobial bionanocomposite films incorporated with ZnO nanoparticles stabilized by melanin. *Food Hydrocoll.* **2019**, *90*, 500–507. [CrossRef]
14. Roy, S.; Van Hai, L.; Kim, H.C.; Zhai, L.; Kim, J. Preparation and characterization of synthetic melanin-like nanoparticles reinforced chitosan nanocomposite films. *Carbohydr. Polym.* **2020**, *231*, 115729. [CrossRef] [PubMed]
15. Soazo, M.; Rubiolo, A.C.; Verdini, R.A. Effect of drying temperature and beeswax content on physical properties of whey protein emulsion films. *Food Hydrocoll.* **2011**, *25*, 1251–1255. [CrossRef]
16. Łupina, K.; Kowalczyk, D.; Zięba, E.; Kazimierczak, W.; Mężyńska, M.; Basiura-Cembala, M.; Wiącek, A.E. Edible films made from blends of gelatin and polysaccharide-based emulsifiers—A comparative study. *Food Hydrocoll.* **2019**. [CrossRef]
17. Szymańska, M.; Karakulska, J.; Sobolewski, P.; Kowalska, U.; Grygorcewicz, B.; Böttcher, D.; Bornscheuer, U.T.; Drozd, R. Glycoside hydrolase (PelAh) immobilization prevents Pseudomonas aeruginosa biofilm formation on cellulose-based wound dressing. *Carbohydr. Polym.* **2020**, *246*. [CrossRef]
18. Łopusiewicz, Ł.; Jędra, F.; Mizielińska, M. New poly(lactic acid) active packaging composite films incorporated with fungal melanin. *Polymers* **2018**, *10*, 386. [CrossRef]
19. Roy, S.; Rhim, J.W. Agar-based antioxidant composite films incorporated with melanin nanoparticles. *Food Hydrocoll.* **2019**, *94*, 391–398. [CrossRef]
20. Moghadam, M.; Salami, M.; Mohammadian, M.; Khodadadi, M.; Emam-Djomeh, Z. Development of antioxidant edible films based on mung bean protein enriched with pomegranate peel. *Food Hydrocoll.* **2020**, *104*, 105735. [CrossRef]
21. Yang, M.; Li, L.; Yu, S.; Liu, J.; Shi, J. High performance of alginate/polyvinyl alcohol composite film based on natural original melanin nanoparticles used as food thermal insulating and UV–vis block. *Carbohydr. Polym.* **2020**, *233*, 115884. [CrossRef] [PubMed]
22. Roy, S.; Kim, H.C.; Kim, J.W.; Zhai, L.; Zhu, Q.Y.; Kim, J. Incorporation of melanin nanoparticles improves UV-shielding, mechanical and antioxidant properties of cellulose nanofiber based nanocomposite films. *Mater. Today Commun.* **2020**, *24*, 100984. [CrossRef]
23. Schmid, M.; Merzbacher, S.; Müller, K. Time-dependent crosslinking of whey protein based films during storage. *Mater. Lett.* **2018**, *215*, 8–10. [CrossRef]
24. Catarino, M.D.; Alves-Silva, J.M.; Fernandes, R.P.; Gonçalves, M.J.; Salgueiro, L.R.; Henriques, M.F.; Cardoso, S.M. Development and performance of whey protein active coatings with *Origanum virens* essential oils in the quality and shelf life improvement of processed meat products. *Food Control* **2017**, *80*, 273–280. [CrossRef]
25. Yoshida, C.M.P.; Antunes, A.C.B.; Antunes, L.J.; Antunes, A.J. An analysis of water vapour diffusion in whey protein films. *Int. J. Food Sci. Technol.* **2003**, *38*, 595–601. [CrossRef]
26. Pires, A.F.; Marnotes, N.G.; Bella, A.; Viegas, J.; Gomes, D.M.; Henriques, M.H.F.; Pereira, C.J.D. Use of ultrafiltrated cow's whey for the production of whey cheese with Kefir or probiotics. *J. Sci. Food Agric.* **2020**. [CrossRef]

27. Agudelo-Cuartas, C.; Granda-Restrepo, D.; Sobral, P.J.A.; Castro, W. Determination of mechanical properties of whey protein films during accelerated aging: Application of FTIR profiles and chemometric tools. *J. Food Process Eng.* **2020**. [CrossRef]
28. Xu, R.; Liu, N.; Xu, X.; Kong, B. Antioxidative effects of whey protein on peroxide-induced cytotoxicity. *J. Dairy Sci.* **2011**, *94*, 3739–3746. [CrossRef]
29. Owonubi, S.J.; Mukwevho, E.; Aderibigbe, B.A.; Revaprasadu, N.; Sadiku, E.R. Cytotoxicity and in vitro evaluation of whey protein-based hydrogels for diabetes mellitus treatment. *Int. J. Ind. Chem.* **2019**, *10*, 213–223. [CrossRef]
30. Gunasekaran, S.; Xiao, L.; Ould Eleya, M.M. Whey protein concentrate hydrogels as bioactive carriers. *J. Appl. Polym. Sci.* **2006**, *99*, 2470–2476. [CrossRef]
31. Kerasioti, E.; Stagos, D.; Priftis, A.; Aivazidis, S.; Tsatsakis, A.M.; Hayes, A.W.; Kouretas, D. Antioxidant effects of whey protein on muscle C2C12 cells. *Food Chem.* **2014**, *155*, 271–278. [CrossRef] [PubMed]
32. Solano, F. Melanins: Skin pigments and much more—Types, structural models, biological functions, and formation routes. *New J. Sci.* **2014**, *2014*, 498276. [CrossRef]
33. Xu, C.; Chen, T.; Li, J.; Jin, M.; Ye, M. The structural analysis and its hepatoprotective activity of melanin isolated from *Lachnum* sp. *Process Biochem.* **2020**, *90*, 249–256. [CrossRef]
34. Ghadge, V.; Kumar, P.; Singh, S.; Mathew, D.E.; Bhattacharya, S.; Nimse, S.B.; Shinde, P.B. Natural melanin produced by the endophytic *Bacillus subtilis* 4NP-BL Associated with the Halophyte *Salicornia brachiata*. *J. Agric. Food Chem.* **2020**, *68*, 6854–6863. [CrossRef]
35. Di Mauro, E.; Camaggi, M.; Vandooren, N.; Bayard, C.; De Angelis, J.; Pezzella, A.; Baloukas, B.; Silverwood, R.; Ajji, A.; Pellerin, C.; et al. Eumelanin for nature-inspired UV-absorption enhancement of plastics. *Polym. Int.* **2019**, *68*, 984–991. [CrossRef]
36. Caldas, M.; Santos, A.C.; Veiga, F.; Rebelo, R.; Reis, R.L.; Correlo, V.M. Melanin nanoparticles as a promising tool for biomedical applications—A review. *Acta Biomater.* **2020**, *105*, 26–43. [CrossRef]
37. Bang, Y.J.; Shankar, S.; Rhim, J.W. Preparation of polypropylene/poly (butylene adipate-co-terephthalate) composite films incorporated with melanin for prevention of greening of potatoes. *Packag. Technol. Sci.* **2020**, 1–9. [CrossRef]
38. Łopusiewicz, Ł.; Jędra, F.; Bartkowiak, A. New active packaging films made from gelatin modified with fungal melanin. *World Sci. News* **2018**, *101*, 1–30.
39. Łopusiewicz, Ł.; Jędra, F.; Bartkowiak, A. The application of melanin modified gelatin coatings for packaging and the oxidative stability of pork lard. *World Sci. News* **2018**, *101*, 108–119.
40. Dong, W.; Wang, Y.; Huang, C.; Xiang, S.; Ma, P.; Ni, Z.; Chen, M. Enhanced thermal stability of poly(vinyl alcohol) in presence of melanin. *J. Therm. Anal. Calorim.* **2014**, *115*. [CrossRef]
41. Kiran, G.S.; Jackson, S.A.; Priyadharsini, S.; Dobson, A.D.W.; Selvin, J. Synthesis of Nm-PHB (nanomelanin-polyhydroxy butyrate) nanocomposite film and its protective effect against biofilm-forming multi drug resistant Staphylococcus aureus. *Sci. Rep.* **2017**, *7*, 1–13. [CrossRef]
42. Łopusiewicz, Ł.; Drozłowska, E.; Siedlecka, P.; Mężyńska, M.; Bartkowiak, A.; Sienkiewicz, M.; Zielińska-Bliźniewska, H.; Kwiatkowski, P. Development, characterization, and bioactivity of non-dairy kefir-like fermented beverage based on flaxseed oil cake. *Foods* **2019**, *8*, 544. [CrossRef] [PubMed]
43. Bishai, M.; De, S.; Adhikari, B.; Banerjee, R. A comprehensive study on enhanced characteristics of modified polylactic acid based versatile biopolymer. *Eur. Polym. J.* **2014**, *54*, 52–61. [CrossRef]
44. Ye, M.; Wang, Y.; Guo, G.Y.; He, Y.L.; Lu, Y.; Ye, Y.W.; Yang, Q.H.; Yang, P.Z. Physicochemical characteristics and antioxidant activity of arginine-modified melanin from *Lachnum* YM-346. *Food Chem.* **2012**, *135*, 2490–2497. [CrossRef] [PubMed]
45. ISO 10993-5. *Biological Evaluation of Medical Devices—Part 5: Tests for In Vitro Cytotoxicity*; International Organization for Standardization: Geneva, Switzerland, 2009.
46. Riss, T.L.; Moravec, R.A.; Niles, A.L.; Duellman, S.; Benink, H.A.; Worzella, T.J.; Minor, L. *Cell Viability Assays*; Sittampalam, G.S., Grossman, A., Brimacombe, K., Arkin, M., Auld, D., Austin, C.P., Baell, J., Bejcek, B., Caaveiro, J.M.M., Chung, T.D.Y., et al., Eds.; Bethesda: Rockville, MD, USA, 2004.
47. Jayaramudu, T.; Varaprasad, K.; Kim, H.C.; Kafy, A.; Kim, J.W.; Kim, J. Calcinated tea and cellulose composite films and its dielectric and lead adsorption properties. *Carbohydr. Polym.* **2017**, *171*, 183–192. [CrossRef]
48. Łupina, K.; Kowalczyk, D.; Drozłowska, E. Polysaccharide/gelatin blend films as carriers of ascorbyl palmitate—A comparative study. *Food Chem.* **2020**, *333*. [CrossRef]

49. Ebaid, H.; Salem, A.; Sayed, A.; Metwalli, A. Whey protein enhances normal inflammatory responses during cutaneous wound healing in diabetic rats. *Lipids Health Dis.* **2011**, *10*, 235. [CrossRef]
50. Fitzmaurice, S.D.; Sivamani, R.K.; Isseroff, R.R. Antioxidant therapies for wound healing: A clinical guide to currently commercially available products. *Skin Pharmacol. Physiol.* **2011**, *24*, 113–126. [CrossRef]
51. Garraud, O.; Hozzein, W.N.; Badr, G. Wound healing: Time to look for intelligent, "natural" immunological approaches? *BMC Immunol.* **2017**, *18*. [CrossRef]
52. Coltelli, M.; Aliotta, L.; Gigante, V.; Bellusci, M.; Cinelli, P.; Bugnicourt, E.; Schmid, M.; Staebler, A.; Lazzeri, A. Preparation and compatibilization of PBS/Whey protein isolate based blends. *Molecules* **2020**, *25*, 3313. [CrossRef]
53. Al-Tayib, O.A.; Elbadwi, S.M.; Bakhiet, A.O. Cytotoxicity assay for herbal melanin derived from *Nigella sativa* seeds using in vitro cell lines. *IOSR J. Humanit. Soc. Sci.* **2017**, *22*, 43. [CrossRef]

© 2020 by the authors. Licensee MDPI, Basel, Switzerland. This article is an open access article distributed under the terms and conditions of the Creative Commons Attribution (CC BY) license (http://creativecommons.org/licenses/by/4.0/).

Article

Alginate Biofunctional Films Modified with Melanin from Watermelon Seeds and Zinc Oxide/Silver Nanoparticles

Łukasz Łopusiewicz [1,*], Szymon Macieja [1], Mariusz Śliwiński [2], Artur Bartkowiak [1], Swarup Roy [3] and Peter Sobolewski [4]

1 Center of Bioimmobilisation and Innovative Packaging Materials, Faculty of Food Sciences and Fisheries, West Pomeranian University of Technology Szczecin, Janickiego 35, 71-270 Szczecin, Poland; ms40205@zut.edu.pl (S.M.); artur-bartkowiak@zut.edu.pl (A.B.)
2 Dairy Industry Innovation Institute Ltd., Kormoranów 1, 11-700 Mrągowo, Poland; mariusz.sliwinski@iipm.pl
3 School of Bioengineering and Food Technology, Shoolini University, Solan 173229, Himachal Pradesh, India; swaruproy2013@gmail.com
4 Department of Polymer and Biomaterials Science, Faculty of Chemical Technology and Engineering, West Pomeranian University of Technology Szczecin 45 Piastów Ave, 70-311 Szczecin, Poland; piotr.sobolewski@zut.edu.pl
* Correspondence: lukasz.lopusiewicz@zut.edu.pl; Tel.: +48-91-449-6135

Abstract: Bioactive films find more and more applications in various industries, including packaging and biomedicine. This work describes the preparation, characterization and physicochemical, antioxidant and antimicrobial properties of alginate films modified with melanin from watermelon (*Citrullus lanatus*) seeds at concentrations of 0.10%, 0.25% and 0.50% w/w and with silver and zinc oxide nanoparticles (10 mM film casting solutions for both metal nanoparticles). Melanin served as the active ingredient of the film and as a nanoparticle stabilizer. The additives affected the color, antioxidant (~90% ABTS and DPPH radicals scavenging for all melanin modified films) and antimicrobial activity (up to 4 mm grow inhibition zones of *E. coli* and *S. aureus* for both zinc oxide and silver nanoparticles), mechanical (silver nanoparticles addition effected two-fold higher tensile strength), thermal and barrier properties for water and UV-vis radiation. The addition of ZnONP resulted in improved UV barrier properties while maintaining good visible light transmittance, whereas AgNP resulted in almost complete UV barrier and reduced visible light transmittance of the obtained films. What is more, the obtained films did not have an adverse effect on cell viability in cytotoxicity screening. These films may have potential applications in food packaging or biomedical applications.

Keywords: melanin; watermelon; alginate; bioactive films; nanoparticles

Citation: Łopusiewicz, Ł.; Macieja, S.; Śliwiński, M.; Bartkowiak, A.; Roy, S.; Sobolewski, P. Alginate Biofunctional Films Modified with Melanin from Watermelon Seeds and Zinc Oxide/Silver Nanoparticles. *Materials* 2022, *15*, 2381. https://doi.org/10.3390/ma15072381

Academic Editor: Sandra Maria Fernandes Carvalho

Received: 17 February 2022
Accepted: 18 March 2022
Published: 23 March 2022

Publisher's Note: MDPI stays neutral with regard to jurisdictional claims in published maps and institutional affiliations.

Copyright: © 2022 by the authors. Licensee MDPI, Basel, Switzerland. This article is an open access article distributed under the terms and conditions of the Creative Commons Attribution (CC BY) license (https://creativecommons.org/licenses/by/4.0/).

1. Introduction

During the transport and storage of food products, it is important to ensure that the product is protected from any negative physical, chemical, or microbiological factors. In this way, packaging should delay food spoilage and slow down the loss of beneficial properties of the processed product, including both visual and nutritional qualities [1]. For example, wax coatings applied to the surface of fruits to extend their shelf life were the first materials used as a coating on food products [2]. In low-income countries, the main share of the total waste is those of organic origin, while in high-income countries, inorganic waste is the most common [3]. Potentially, some of these organic wastes (including agro-industrial by-products) can be used as a source of bioactive compounds such as phenolic compounds, dietary fiber, polysaccharides, vitamins, carotenoids, pigments and oils [4].

Plastics, due to their ease of processing and relatively low cost of raw materials, have become one of the most widely used materials in virtually all industries. Their annual production exceeds 300 million tons worldwide [5]. However, plastics present a problem with regard to waste management: much of plastic waste consists of packaging waste, of which up to 60% is food packaging. Furthermore, the majority of plastic waste (approximately 80%) ends up in landfills or pollutes land or water reservoirs, thus affecting entire ecosystems, from plants through to animals and humans [6–8]. Additionally, plastic production and disposal generate large amounts of greenhouse gases, which also has a significant negative impact on the environment. Therefore, finding less harmful alternatives to conventional plastics, such as renewable or biodegradable materials, is important to reduce the harmful impact of packaging on ecosystems [9].

One promising strategy involves biopolymers, polymers of natural origin: usually proteins or polysaccharides obtained from plants, bacteria, fungi, animals or seaweed. As biopolymers have extremely diverse properties, they can find applications in many different industries, including use in packaging, for example, of medical equipment, agricultural products, and other commodities [10,11].

Polysaccharides are a class of biopolymers that are particular abundant, including cellulose, chitin, xanthan gum, starch, carrageenan, and alginate [12]. Alginates are naturally occurring polymers consisting of linear copolymers of D-mannuronic and L-guluronic acids, linked by β-1,4-glycosidic bonds. Depending on the source, the two monomers can differ in proportion and be arranged in different, irregular patterns—these structural aspects affect the properties of alginate-based materials [13]. Example sources include seaweeds, mainly algae (*Laminaria hyperborea, L. digitata, L. japonica, Ascophyllum nodosum,* or *Macrocystis pyrifera*), whose alginic acid is a component of cell walls, or bacterial biosynthesis using *Azotobacter* and *Pseudomonas* bacteria [14]. Alginates have found applications in the textile, food, pharmaceutical, and chemical industries [15]. Importantly, the alginate backbone includes many free hydroxyl and carboxyl groups, making them highly amenable to modification and functionalization, using techniques such as oxidation, esterification and amidation. In this way it is possible to significantly influence the material properties, both physicochemical (such as solubility and affinity for water), as well as biological [16].

Melanins are a group of high molecular weight pigments, formed by the oxidative polymerization of indole and phenolic compounds [17]. In addition to imparting color to living organisms, melanins are also responsible for free radicals scavenging, immunostimulation, thermoregulation, protection from ultraviolet radiation, and have antiviral and antimicrobial properties [18–20]. Due to their antioxidant activity, melanins can be used to reduce metal compounds and thus synthesize metal nanoparticles. In contrast to traditional techniques for obtaining nanomaterials, the use of melanins can enable processes that are relatively fast and inexpensive, making them environmentally friendly [21,22]. Due to the aforementioned properties of melanins, they can carry out many active functions, which have resulted in numerous works describing the use of these pigments in combination with various polymers, including, although not limited to, bioactive food packaging [23].

Zinc oxide nanoparticles (ZnONP) are among the most commonly produced nanomaterials, next to titanium oxide and silicon dioxide NPs. As with other nanomaterials, the size and shape (and thus properties) of ZnONP depends on the preparation method. ZnONP, similar to titanium dioxide NP have a high ability to absorb ultraviolet radiation with a high band gap, making them transparent to visible light [24]. As a result, nano ZnO is commonly used in cosmetics, such as sunscreens, for protecting the skin against ultraviolet radiation. Additionally, ZnONP possess antimicrobial activity, making them suitable for the purification of water from microbial contaminants [25] or as antibacterial additives to paints and coatings [26,27]. Importantly, ZnONP are regarded as being safe (GRAS) by the United States Food and Drug Administration (USFDA, 21CFR182.8991). Combined, these properties make them very attractive for applications in active packaging systems [28].

Silver nanoparticles (AgNP) also possess potent antimicrobial properties, making them highly valued in the pharmaceutical and medical industries. As shown in numerous studies, AgNP are able to induce pore formation in bacterial cell membranes, most likely through the interaction of silver with sulfur present in membrane proteins. Further, when silver penetrates into the cell, it can interact with genetic material, causing its condensation inhibiting replication and gene expression [29–32]. Due to these properties, AgNP are used in various biomedical applications, such as in burn wound dressings and as catheter coatings. However, AgNP are also used in electronics, photonics, cosmetics, cleaning agents, and as disinfectants [26,33].

The antimicrobial and UV barrier properties of ZnONP and AgNP described above, along with potential other effects on the polymer matrix, provide prospective applications for these nanoparticles in active packaging materials. Nevertheless, the amount of nanoparticles loaded into the polymer matrix is crucial for the properties of the polymer, which as a consequence may deteriorate the quality of the packaging. As it has been shown for chitosan nanoparticles, the use of other than optimal amounts (smaller or larger) can affect the thermal, mechanical, barrier or optical properties [34].

In this work, our aim is to investigate the effect of the addition of melanin obtained from watermelon seeds, a food industry waste, on the properties of alginate films. The melanin was used as an additive directly or was used for synthesis of ZnONP and AgNP. To the best of our knowledge, there have been no reports on the modification of alginate films with either natural melanin or in combination with nanoparticles to improve the functionality of these materials. The chemical composition of the films after incorporation of melanin and nanoparticles was examined by FT-IR spectroscopy. In addition, the effect of melanin on the color, hydrodynamic and optical properties of the films was evaluated. Moreover, their potential bio(functionality) was assessed by evaluating their mechanical and barrier properties, as well as antioxidant and potential cytotoxicity in vitro.

2. Materials and Methods

2.1. Materials and Reagents

Calcium chloride, hydrogen peroxide, disodium phosphate, monosodium phosphate, 1,1-diphenyl-2-(2,4,6-trinitrophenyl) hydrazyl (DPPH), 2,2'-azino-bis(3-ethylbenzothiazoline-6-acid) sulfonic acid (ABTS), potassium persulfate, potassium hexacyanoferrate (III), trichloroacetic acid, iron (III) chloride, iron (II) sulfate, tris (hydroxymethyl)aminomethane, pyrogallol, sodium alginate (M_w = 1,450,000), and o-phenanthroline were purchased from Sigma-Aldrich (Darmstad, Germany). Glycerol, ammonia water, hydrochloric acid, sodium hydroxide, chloroform, ethyl acetate, ethanol and methanol were from Chempur (Piekary Śląskie, Poland). Zinc (II) nitrate, peptone water and MacConkey medium were from Scharlau Chemie (Barcelona, Spain). Silver (III) nitrate was from POCh (Gliwice, Poland). Chapman's medium was from Merck (Dramstadt, Germany). All reagents were of analytical grade.

The microorganisms used to evaluate the antimicrobial properties of the films were obtained from the American Type Culture Collection (ATCC). The strains used were *Escherichia coli* ATCC8739 and *Staphylococcus aureus* ATCC12600.

For in vitro cytotoxicity studies, murine fibroblast cell line (L929), as well as all cell culture reagents: Dulbecco's Modified Eagle Medium (DMEM), fetal bovine serum (FBS), L-glutamine, penicillin, streptomycin, and all other cell culture reagents were purchased from Sigma-Aldrich (Poznań, Poland). All sterile, single-use cell culture plasticware was purchased from VWR (Gdańsk, Poland).

2.2. Preparation of Alginate Films

Fresh Crimson Sweet watermelons (*Citrullus lanatus*) were purchased from a local market (Szczecin, Poland). Melanin isolation and purification was performed as previously described [35]. Briefly, watermelon seeds (5 g) washed in distilled water and dried were immersed in 50 mL of 1 M NaOH for 24 h with shaking (150 rpm, 50 °C). The resulting mixture was then centrifuged (6000× g rpm, 10 min) and the separated supernatant was brought to pH 2.0 by adding 1 M HCl and followed by centrifugation (6000× g rpm, 10 min). The resulting precipitate was hydrolyzed with 6 M HCl (90 °C, 2 h), centrifuged again (6000× g rpm, 10 min) and washed five times with distilled water. Finally, the obtained precipitate was washed with chloroform, ethyl acetate and ethanol, dried and ground in a mortar. To prepare melanin containing films, distilled water (400 mL) was poured into

500 mL bottles and melanin was added in appropriate amounts (0.008 g, 0.02 g and 0.04 g) in order to obtain melanin concentrations of 0.10%, 0.25% and 0.50% (w/w), respectively. Ammonia water (2 mL) was added to all samples to create an alkaline environment that facilitated melanin dissolution. The bottles were placed on a shaker overnight to completely dissolve the melanin. The resulting solutions were pressure filtered to separate insoluble residues. The solution bottles were then placed on magnetic stirrers and 8 g of alginate was slowly added to each bottle. They were then placed back on the shaker overnight at 60 °C to completely dissolve the alginate. Glycerol, a plasticizing agent, was added (30% (w/w) relative to the amount of alginate used) to the solutions and stirred on a magnetic stirrer for 10 min. A control sample without melanin was prepared in the same fashion. To prepare films, alginate solutions were poured into square polystyrene plates (120 mm × 120 mm) at 40 g of solution per plate and dried at 40 °C for 48 h. All assays were performed in eight replicates.

Samples of AgNP or ZnONP were prepared using the same general procedure, however, with the addition of silver nitrate or zinc nitrate in an aqueous solution of melanin. For this purpose, aqueous solutions of silver nitrate or zinc nitrate were slowly added dropwise to the melanin solutions using a pipette until 10 mM was obtained in the film-forming solution. The resulting solutions were then incubated at 90 °C for 1 h. The solutions were cooled to room temperature before adding alginate (8 g). To prepare the films, the alginate solutions were poured onto square polystyrene plates (120 mm × 120 mm) at 40 g of solution per plate and dried at 40 °C for 48 h. All experiments were performed in eight replicates.

2.3. Characterization on Nanoparticles

NP purification was performed prior to performing the analyses. For this purpose, the mixtures were centrifuged (ZnONP mixtures at 5000 rpm and AgNP at 14,000 rpm) for 5 min and then the resulting precipitate was washed with distilled water. The procedure was repeated three times.

Samples of mixtures of melanin with AgNP or ZnONP were filtered using syringe filters (pore size 0.45 µm) and 1 mL of the obtained filtrates analyzed using UV-Vis light absorption spectroscopy (Thermo Scientific (Waltham, MA, USA) Evolution 220 UV-Vis spectrophotometer). The spectra were collected over the wavelength range of 300–800 nm, with a resolution of 1 nm. Melanin-nanoparticles mixture were also dried overnight at 40 °C and the resulting powder was collected for chemical composition analysis using a Perkin Elmer Spectrum 100 FT-IR spectrophotometer (Waltham, MA, USA). The obtained powders were measured directly using attenuated total reflection (ATR). The spectra were recorded over a wavelength range of 650–4000 cm^{-1}, with a resolution of 1 cm^{-1}.

Dynamic light scattering (DLS) measurements were performed using Malvern Nanosizer ZS instrument (Worcestershire, UK) with a He-Ne laser source (633 nm). For each sample, three measurements were performed at each of two sample dilutions. For ZnONP, after purification, each pellet was re-dispersed in 2 mL of saline, yielding homogenous latté-colored suspensions. These suspensions were then further diluted 1:80 and 1:200 for measurements. For AgNP, after purification, pellets were dispersed in 1 mL of water, yielding pale dispersions, ranging from pink (0.5% MEL) to purple (0.1% MEL). These dispersions were measured neat, as well as after further 1:1 dilution with water.

Each DLS measurement was performed at 25 °C after 2 min of temperature equilibration and consisted of 10–15 runs, at 173° scattering angle, and using automatic attenuation to ensure count rates <500 kcps. Data presented consist of hydrodynamic diameter (intensity-weighted average, "z-average" diameter) and polydispersity index (PDI) obtained from cumulant analysis using Malvern Zetasizer software v3.30.

2.4. Biocomposite Film Characterisation

2.4.1. Determination of Moisture Content and Water Solubility

Moisture content (MC) was determined as the change in film (2 cm × 2 cm) weight after drying at 105 °C for 24 h. Water solubility (WS) was tested by adding pre-weighed film fragments (2 cm × 2 cm) to 30 mL of distilled water in conical tubes and stirring overnight. The samples were then centrifuged and the water was removed from the precipitate using a pipette. The samples were dried overnight at 60 °C and finally weighed again. Each material was tested in three replicates and mean values were calculated. The solubility of the films in water was calculated using the following formula:

$$S\ (\%) = \frac{W_1 - W_2}{W_1} \times 100 \tag{1}$$

where W_1 is the initial and W_2 the final weight of the films.

2.4.2. Thickness, Mechanical, and Thermal properties of Alginate Films

Film thickness was measured using an electronic thickness gauge (Dial Thickness Gauge 7301, Mitoyuto Corporation, Kangagawa, Japan) with an accuracy of 0.001 mm. Each sample was measured 10 times at randomly selected points and the results were averaged.

The tensile strength and elongation at break of the films were assessed using a Zwick/Roell 2.5 Z static testing machine (Ulm, Germany). The tensile clamp spacing was 25 mm and the head travel speed was 100 mm/min.

Differential scanning calorimetry (DSC) was used to assess thermal properties (DSC 3, Mettler-Toledo LLC, Columbus, OH, USA) using sequential heat-cool-heat cycles over a temperature range of 30–300 °C at $\phi = 10$ °/min, under nitrogen flow (50 mL/min).

2.4.3. Water Vapor Transmission Rate (WVTR) of Films

The water vapor permeability test (WVTR) was carried out using a gravimetric method: moisture sorption by pre-dried calcium chloride (9 g) was examined in containers tightly covered with test films (8.9 cm^2). The containers were weighed daily over a period of three days (starting at day zero every 24 h until day three of the start of the analysis) to monitor the weight gain of calcium chloride and thus the water vapor permeability of the films. Each material was tested in four replicates and mean values for each day were calculated to express WVTR in g/(m^2 × day) [36].

2.4.4. Spectral Analysis of Films

UV-Vis spectra of films were measured using a Thermo Scientific (Waltham, MA, USA) Evolution 220 UV-vis spectrophotometer. Strips of films, with a surface area matching the quartz cuvette (5.5 cm × 1 cm), were placed along with a cuvette in the apparatus and spectra were recorded over a wavelength range of 300–800 nm, with a resolution of 1 nm.

The chemical composition of the obtained films was also evaluated using a Perkin Elmer Spectrum 100 FT-IR spectrophotometer (Waltham, MA, USA). Pieces of the films were measured directly, in ATR mode (32 scans per sample), and spectra were recorded over a wavelength range of 650–4000 cm^{-1}, with a resolution of 1 cm^{-1}.

2.4.5. Film Color Analysis

The effect of melanin and silver or zinc nanoparticles on the color of the obtained films was assessed using a colorimeter (CR-5, Konica Minolta, Tokyo, Japan). Each sample was tested 10 times, at randomly selected points. The results (mean ± standard deviation) were expressed as L*, a* and b* parameters. In addition, ΔE (color difference) and YI (yellowing index) were calculated for comparison with unmodified alginate films, using the following equations:

$$\Delta E = \left[\left(L_{standard} - L_{sample} \right)^2 + \left(a_{standard} - a_{sample} \right)^2 + \left(b_{standard} - b_{sample} \right)^2 \right]^{0.5} \tag{2}$$

$$YI = 142.86 b \cdot L^{-1} \tag{3}$$

2.4.6. Antioxidant Potential of Films

The antioxidant activity was measured by reducing power and free radical scavenging activity (ABTS$^+$, DPPH, and superoxide (O_2^-) radicals) techniques as previously described [37], with minor modifications. For each film, two samples were tested and each measurement was performed in triplicate.

To assess reducing power, 100 mg film samples cut into pieces were placed in 1.25 mL of phosphate buffer (0.2 M, pH 6.6) and then 1.25 mL of 1% potassium ferricyanide solution was added. The samples were incubated at 50 °C for 20 min, after which 1.25 mL of trichloroacetic acid was added. The samples were then centrifuged at 1107 RCF for 10 min. From each sample, 1.25 mL of supernatant was taken, diluted with an equal volume of distilled water, and 0.25 mL of 0.1% ferric chloride solution was added. Finally, the absorbance of the samples was measured at 700 nm.

Free radical scavenging activity assay was performed against ABTS$^+$, DPPH, and superoxide (O_2^-) radicals. The ABTS$^+$ radical was induced prior to the addition of the film by mixing 5 mL of 7 mM ABTS with 2.45 mL of potassium persulfate and allowing it to stand overnight at room temperature, in the dark. ABTS$^+$ solutions were then prepared by diluting with ethanol to an absorbance of 0.7. For each test, 100 mg of film was added to 10 mL of ABTS$^+$ solution and incubated for six minutes in the dark. As negative control, polypropylene film with the same size as the samples was used, whereas the prepared ABTS$^+$ solution without any film was used as a blank control. The mixtures were incubated under identical conditions. The absorbance of the samples was measured at 734 nm and the free radical scavenging activity was calculated using the following equation:

$$\text{Free radical scavenging activity (\%)} = \frac{A_{sample} - A_{control}}{A_{sample}} \times 100 \tag{4}$$

where A_{sample} is the absorbance of the ABTS$^+$ solution with the addition of the tested films, and $A_{control}$ is the absorbance of the blank ABTS$^+$ solution.

For DPPH radical scavenging activity tests, 100 mg of each film was placed in 10 mL of 0.01 mM DPPH in methanol and incubated in the dark for 30 min. As negative control, polypropylene film with the same size as the sample was used, whereas the prepared DPPH solution without any film was used as a blank control. The mixtures were incubated under identical conditions. Absorbance was measured at 517 nm and free radical scavenging activity was calculated using the same equation as for ABTS.

Superoxide (O_2^-) radical scavenging activity was evaluated using the pyrogallol oxidation inhibition assay. For this purpose, 100 mg of each film was incubated for 5 min in 3 mL of 50 mmol/L Tris-HCl buffer (pH 8.2), with gentle stirring every minute. Then, 0.3 mL of pyrogallol was added. After exactly four minutes, the reaction was stopped by adding 1 mL of 10 mM HCl and the absorbance was immediately measured at 318 nm. As negative control polypropylene film was used. The radical scavenging activity was calculated using the following equation:

$$O_2^- \text{ inhibition (\%)} = \left[1 - \frac{A_1 - A'_1}{A_0}\right] \times 100 \tag{5}$$

where A_1 is the absorbance of the reaction mixture with the addition of the sample, A'_1 is the absorbance of the reaction mixture with water instead of pyrogallol and A_0 is the absorbance of the reaction mixture without addition of test samples.

2.4.7. Antimicrobial Activity

The antimicrobial properties were investigated by measuring the inhibition of microbial growth on plate count agar (PCA) in the presence of discs made of test films. For *Escherichia coli*, MacConkey agar was used, while for *Staphylococcus aureus*, Chapman agar was used. After a day of culture, single colonies were picked from the media and added to sterile peptone water prepared according to the manufacturer's instructions until an optical density of 0.5 McFarland was achieved. The microorganisms were then seeded onto PCA plates and two discs (2 cm in diameter) of each film were placed on top. After 24 h of culture, the zone of growth inhibition was measured.

2.4.8. Evaluation of Cytotoxicity

Cytotoxicity screening was performed using a direct contact assay, as described previously [37]. L929 cells were maintained in culture using DMEM containing 10% FBS, 2 mM l-glutamine, 100 U/mL penicillin, and 100 µg/mL streptomycin. For each test film, discs (8 mm in diameter) were cut using a steel punch, followed by sterilization with 20-min UV exposure in a BSL-2 safety cabinet (Telstar Bio II Advance, Barcelona, Spain).

For each experiment, L929 cells (passage 7–16) were trypsinized, counted using a hemocytometer, and seeded in the wells of 48-well plates (30,000 cells per well). After 24 h of incubation to permit the cells to adhere and spread, the media was aspirated and discs of each test film (5–6 samples each) were placed directly onto the cell monolayer—one well at a time. Manipulation of the thin discs was performed using a sterile 21 G needle: discs were "speared" vertically (from above), lifted, positioned in place in the well, and released by a gentle twist of the needle. Afterwards, 0.25 mL of fresh media was gently pipetted on top. As a sham control for normalization, wells with cells were aspirated, although no disc was placed, prior to adding 0.25 mL of fresh media. After a further 24 h of culture, cell viability was assessed by inverted light microscopy (Delta Optical IB-100, Minsk Mazowiecki, Poland) and using the resazurin assay [38]. Note that it was not possible to remove the discs, as they partially dissolved and in the case of samples containing nanoparticles, the nanoparticles interfered with the resazurin assay. Fluorescence plate reader measurements (Biotek Synergy HTX, Winooski, VT, USA, excitation 540 nm, emission 590 nm) were converted to normalized cell viability as percentage of sham by first subtracting the mean signal of six blank wells (no cells) from all values, followed by dividing by the mean signal of the sham wells (n = 6).

2.5. Statistical Analyses

All analyses were made at least in triplicate. Statistical analysis was performed using Statistica version 13 software (StatSoft Poland, Krakow, Poland). Differences between means were determined by analysis of variance (ANOVA) followed by Fisher's post hoc LSD test at a significance threshold of $p < 0.05$.

3. Results and Discussion

3.1. Characterization of Nanoparticles

In order to investigate the quality of obtained nanoparticles, UV-Vis analysis was carried out. UV-Vis spectra of ZnONP are shown in Figure 1A. The absorption maximum was at 307 nm. This value is shifted towards the UV-B radiation, as compared to the results obtained by Roy and Rhim [22], who noted a maximum at 365 nm for ZnO nanoparticles prepared using melanin.

UV-Vis spectra of AgNP are shown in Figure 1B. The maximum absorption peak can be observed at 420, 421, and 361 nm for 0.10% MEL + nAg, 0.25% MEL 0.25% + nAg, and 0.50% MEL + nAg, respectively. These results are similar to data obtained by Shankar and Rhim [39], as well as Vilmala et al. [40].

In order to investigate the presence of characteristic bonds in the samples FTIR analysis was performed. Both silver and ZnONP samples may contain melanin used in the synthesis of these nanoparticles, which was assessed using FTIR. Figure 2 shows the FTIR spectra of the AgNP obtained from the 0.50% MEL + nAg sample. Several distinct peaks can be seen. The stretching vibration of –OH bonds (phenols) at 3200 cm^{-1} are likely from the melanin present in the sample, along with the peaks at 1275 cm^{-1} and 1033 cm^{-1} corresponding to

C–O stretching, and 800 cm^{-1}, which can be attributed to aromatic groups. These results are similar to measurements of AgNP biosynthesized using plant extracts [41].

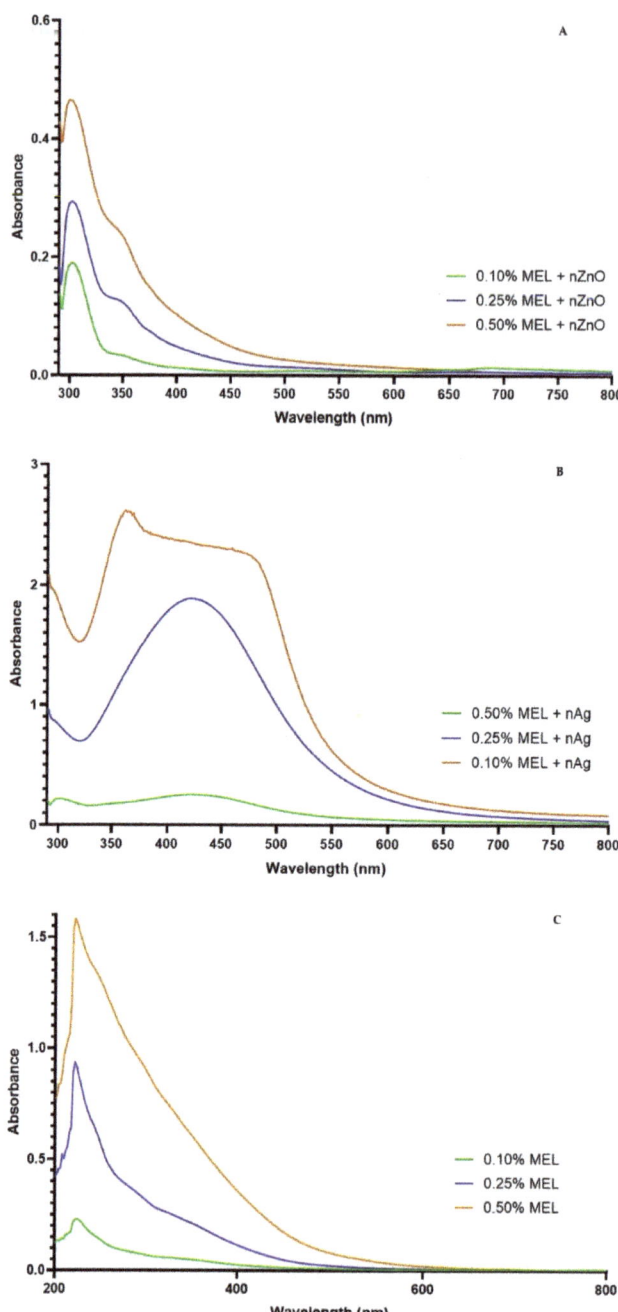

Figure 1. UV-Vis spectra of obtained zinc oxide nanoparticles (**A**), silver nanoparticles (**B**) samples, UV-Vis spectra of *C. lanatus* melanin at different concentrations for comparison (**C**).

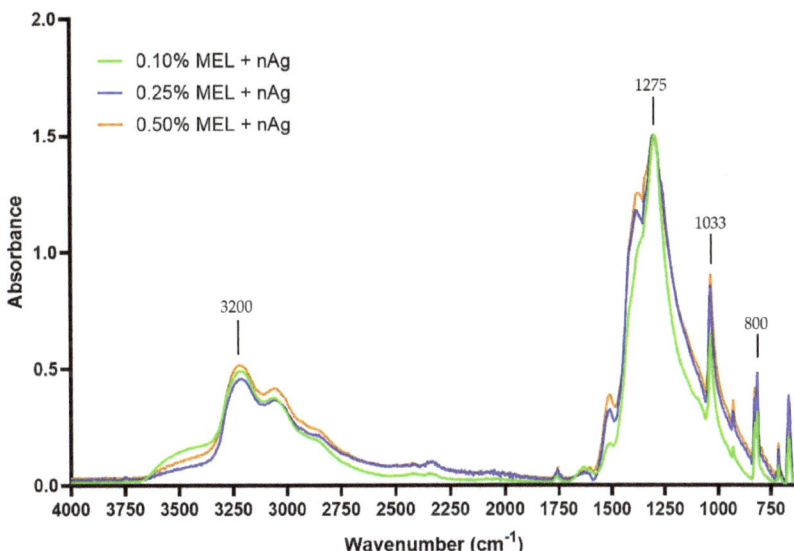

Figure 2. FTIR spectra of obtained silver nanoparticles samples.

Figure 3 shows the FTIR spectra of ZnONP. Considerable differences in the intensity and position of the FTIR peaks can be noted for the individual samples. This is likely due to the different amounts of residual melanin due to the different concentrations used during preparation. The primary peaks occur in the spectral regions: 3500–3250 cm^{-1}, 1600–1550 cm^{-1}, 1400–1300 cm^{-1}, 1100–950 cm^{-1}, and 900–750 cm^{-1}. These results are in good agreement with data obtained by Roy and Rhim [22], who also used melanin for ZnONP synthesis.

In order to investigate average particle size, dynamic light scattering was used. Dynamic light scattering results of the obtained particles are presented in Table 1. For all ZnONP samples, unimodal, yet broad distributions were obtained, with a trend towards smaller particles with higher MEL content. At 0.50% MEL, the particles were the smallest, sub-1-micron in diameter and with the narrowest distribution (PDI = 0.266 ± 0.061). For the case of AgNP, particle sizes were markedly smaller, all <300 nm, and the effect of MEL concentration was different: the highest MEL concentration (0.50%) yielded the largest particles, although again with the narrowest distribution (PDI = 0.288 ± 0.037). Additionally, for the 0.10% MEL AgNP, a minor shoulder was observed (~10% of area) at ~60 nm diameter.

Table 1. Dynamic light scattering results: hydrodynamic diameter (z-avg) and polydispersity index (PDI). Data are expressed as mean ± standard deviation (SD) of six measurements, 10–15 runs each.

Sample	Hydrodynamic Diameter (nm)	PDI
0.10% MEL + nZnO	1523 ± 201 [a]	0.476 ± 0.172 [a]
0.25% MEL + nZnO	1167 ± 74 [b]	0.312 ± 0.104 [a]
0.50% MEL + nZnO	988 ± 42 [b]	0.266 ± 0.061 [a]
0.10% MEL + nAg	230 ± 6 [c]	0.412 ± 0.023 [a]
0.25% MEL + nAg	216 ± 6 [c]	0.359 ± 0.037 [a]
0.50% MEL + nAg	265 ± 16 [c]	0.288 ± 0.037 [a]

Values are means ± standard deviation. Means with different letters (a–c) are significantly different at $p < 0.05$ (compared to other variants within groups).

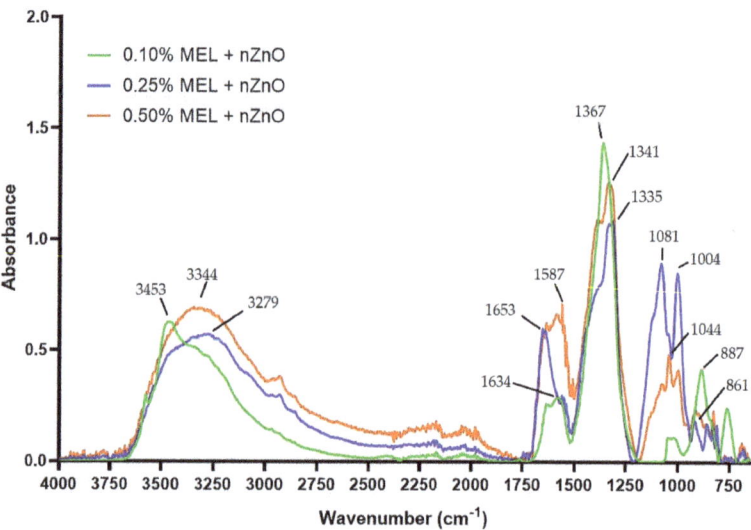

Figure 3. FTIR spectra of obtained zinc oxide nanoparticles samples.

3.2. Hydrodynamic Properties (Moisture Content and Water Solubility) and Water Vapor Transmission Rate (WVTR)

In order to investigate the influence on hydrodynamic properties of the addition of melanin, ZnONP and AgNP, moisture content, water solubility and water vapor transmission rate have been examined. The moisture content and water solubility of the unmodified and modified alginate films are shown in Table 2. The moisture content of the unmodified alginate film was 11.61 ± 0.22%. The addition of melanin caused a significant increase to 14.93 ± 3.14% for ALG + 0.10% MEL film ($p < 0.05$). These values are similar to those reported for agar films with melanin [42]. While the effect of melanin addition on the moisture content of the films was consistent with the results of other researchers [22,37,43,44], there are also reports of the opposite effect of melanin [45,46]. These differences can be attributed to the different properties of the biopolymers used and the resulting different interactions between their functional groups with melanin, leading to an increase or decrease in the overall availability of free hydroxyl groups able to bind water molecules [46].

Table 2. Moisture content (MC), water solubility (WS) and Water Vapor Transmission Ratio (WVTR) of unmodified alginate films and films modified with melanin and melanin with nanoparticles.

Sample	MC (%)	WS (%)	WVTR (g/(m² × Day))
ALG	11.61 ± 0.22 [a]	79.70 ± 11.26 [a]	1589.70 ± 54.96 [a]
ALG + 0.10% MEL	14.93 ± 3.14 [bA]	73.54 ± 14.05 [bA]	1240.92 ± 145.23 [bA]
ALG + 0.25% MEL	12.20 ± 1.27 [aB]	73.65 ± 5.93 [bA]	1160.92 ± 284.22 [bA]
ALG + 0.50% MEL	11.34 ± 0.31 [aB]	75.68 ± 9.64 [bA]	1117.23 ± 285.85 [bA]
ALG + 0.10% MEL + nZnO	13.20 ± 0.24 [aA]	69.37 ± 3.86 [bA]	1549.44 ± 129.52 [aA]
ALG + 0.25% MEL + nZnO	12.78 ± 0.62 [aA]	72.28 ± 1.52 [bA]	1503.18 ± 92.62 [aA]
ALG + 0.50% MEL + nZnO	12.64 ± 0.40 [aA]	73.23 ± 2.50 [bA]	1463.20 ± 116.17 [aA]
ALG + 0.10% MEL + nAg	10.96 ± 1.42 [aA]	61.92 ± 1.68 [bA]	1393.13 ± 89.11 [bA]
ALG + 0.25% MEL + nAg	10.97 ± 0.03 [aA]	66.66 ± 3.97 [bA]	1341.85 ± 173.89 [bA]
ALG + 0.50% MEL + nAg	11.61 ± 0.29 [aA]	69.79 ± 2.19 [bA]	1229.34 ± 57.05 [bA]

Values are means ± standard deviation. Means with different lowercase (a–c) are significantly different at $p < 0.05$ (compared to control), means with different uppercase (A–B) are significantly different at $p < 0.05$ (compared to other variants within groups).

In contrast to melanin, the addition of AgNP caused a decrease in MC, to a range from 10.96 ± 1.42 to 11.61 ± 0.29%, which is comparable to the results of Rhim et al. [47]. However, the addition of ZnONP to the films did not affect the MC of the obtained films. Interestingly, Kotharangannagari and Krishnan [48] observed a decrease in moisture content with increasing amount of zinc oxide added to films made of starch with lysine and various concentrations of ZnONP. This again indicates that the specific interactions between the biopolymer and the nanoparticle drive film affinity for moisture.

The water vapor barrier properties of the tested films are presented in Table 2. Compared to control alginate films, the WVTR of all of the modified alginate films had lower WVTR, with the effect being more pronounced with increasing concentration in the melanin used. For MEL addition alone, the WVTR drops from 1589.70 ± 54.96 g/(m^2 × Day) for neat alginate films to 1117.23 ± 285.85 g/(m^2 × Day) for the highest melanin concentration. As reported by Bang et al. [49], the effects of melanin on water vapor barrier properties can be attributed to the interactions of melanin with the polymer matrix, especially with free hydrophilic chains. As a result, previous studies have found both positive effects of melanin on WVTR [49], as well as negative effects [36,37] and a variable effect dependent on the melanin concentration [44].

Compared to the addition of melanin, the presence ZnONP and AgNP resulted in a smaller change in WVTR through the films. At the highest concentration of melanin, the WVTR was 1463.20 ± 116.17 and 1229.34 ± 57.05 g/(m^2 × Day), for ZnONP and AgNP, respectively. However, both of these values remain lower than control alginate. For the case of ZnONP, the reduced effect may be attributed to their much larger size (~1 µm). In previous work, Roy and Rhim [22] studied carrageenan films with the addition of ZnONP synthesized with melanin, similar to this work. They observed that the addition of ZnONP nanoparticles increased the barrier properties of the film. Likewise, Kanmani and Rhim [50] as well as Shankar et al. [51] obtained similar results. However, for PLA, Chu et al. [52] observed an increase in water permeability through films after adding ZnONP, although no significant effect of the addition of AgNP. Further, Shankar and Rhim [39] noted a slight increase in water vapor permeability after adding AgNP to agar films, whereas Rhim et al. [53], in a similar test system, noted a decrease in this parameter with increasing silver nanoparticle concentration. Overall, the water vapor barrier properties are highly dependent on the properties of the materials involved, their interactions, and the preparation, and dispersion of the nano-additives used.

In terms of water solubility, the addition of melanin reduced solubility from 79.70 ± 11.26% to 73.54 ± 14.05% for ALG + 0.10% MEL and 75.68 ± 9.64% for ALG + 0.50% MEL, although the differences were not significant ($p > 0.05$). A similar effect of melanin was previously described for whey protein isolate and concentrate films, as well as for gelatin films [33,47]. However, Roy et al. [43] noted an inverse relationship for chitosan films, attributed to interactions of melanin with polymer chains.

Meanwhile, films with AgNP had lower solubility (61.92 ± 1.68% for ALG + 0.10% MEL + nAg) than the controls, however, as the amount of melanin in the film increased, the solubility significantly increased, to 69.79 ± 2.19% ($p < 0.05$). The same relationship was also observed for the samples with ZnONP: a trend of increasing water solubility from 69.37 ± 3.86% to 73.23 ± 2.50% with increasing melanin concentration. Previously, Rhim et al. [45] reported an increase in the solubility of chitosan films after the addition of nanosilver, although the differences were not statistically significant.

3.3. The Thickness, Mechanical and Thermal Properties

The thickness, mechanical and thermal properties of the tested films are summarized in Table 3. In general, the addition of melanin did not have a significant effect on the thickness of the films. This is in good agreement with prior work with films made of whey protein concentrate/isolate [37], although not with those made of agar [42], gelatin [54], carrageenan [22] or polybutylene adipate terephthalate (PBAT) [49] where melanin addition tended to increase film thickness. Meanwhile, for carboxymethyl cellulose (CMC) films, melanin had no effect on the film thickness [36].

Table 3. Thickness, mechanical, and thermal characteristics of unmodified and modified alginate films.

Sample	Thickness (mm)	TS (MPa)	EB (%)	T_m (°C)	ΔH_m (J/g)
ALG	0.041 ± 0.020 [a]	7.59 ± 2.69 [a]	2.16 ± 0.96 [a]	127.5 ± 1.9 [a]	103.0 ± 13.2 [a]
ALG + 0.10% MEL	0.045 ± 0.028 [aA]	5.97 ± 1.12 [aA]	2.12 ± 1.12 [aA]	128.1 ± 1.9 [aA]	106.0 ± 13.6 [aA]
ALG + 0.25% MEL	0.036 ± 0.020 [aA]	7.26 ± 6.80 [aA]	3.44 ± 1.97 [bB]	128.0 ± 1.9 [aA]	73.5 ± 9.4 [bB]
ALG + 0.50% MEL	0.036 ± 0.014 [aA]	17.59 ± 3.91 [bB]	5.21 ± 1.40 [cC]	99.8 ± 1.5 [bB]	66.7 ± 8.5 [bB]
ALG + 0.10% MEL + nZnO	0.047 ± 0.007 [bA]	5.95 ± 2.84 [aA]	1.98 ± 1.04 [aA]	97.5 ± 1.5 [bA]	129.3 ± 16.6 [aA]
ALG + 0.25% MEL + nZnO	0.044 ± 0.010 [aA]	5.45 ± 1.85 [aA]	1.39 ± 0.67 [aA]	100.8 ± 1.5 [cB]	126.2 ± 16.2 [aA]
ALG + 0.50% MEL + nZnO	0.041 ± 0.008 [aA]	6.10 ± 3.06 [aA]	1.60 ± 0.32 [aA]	96.9 ± 1.5 [bA]	122.6 ± 15.7 [aA]
ALG + 0.10% MEL + nAg	0.033 ± 0.003 [aA]	14.99 ± 3.16 [bA]	1.69 ± 0.25 [aA]	105.1 ± 1.6 [bA]	158.4 ± 20.3 [bA]
ALG + 0.25% MEL + nAg	0.030 ± 0.006 [aA]	15.64 ± 8.27 [bA]	2.19 ± 0.36 [aA]	95.3 ± 1.4 [cB]	147.7 ± 18.9 [bA]
ALG + 0.50% MEL + nAg	0.037 ± 0.012 [cA]	15.78 ± 6.47 [bA]	2.40 ± 1.06 [aA]	97.5 ± 1.5 [cB]	151.7 ± 19.4 [bA]

Values are means ± standard deviation. Means with different lowercase (a–c) are significantly different at $p < 0.05$ (compared to control), means with different uppercase (A–C) are significantly different at $p < 0.05$ (compared to other variants within groups).

Regarding mechanical properties of the films, the tensile strength (TS) increased for films with increasing melanin content from 5.97 ± 1.12 MPa for films with 0.10% (w/w) melanin content to 17.59 ± 3.91 MPa for films with the highest content. An upward trend was also be observed for the films with the addition of nanoparticles, although the effect was not as pronounced. The results are broadly similar, however lower in magnitude than those obtained by other authors [22,36,42,44,45,49,54,55]. Elongation at break (EB) also increased with increasing melanin content. However, previous studies suggest that the value of the elongation at break may increase with increasing melanin concentration up to a point, beyond which it begins to decline; the same relationship may also exist for the tensile strength [43,49,55].

Surprisingly, the addition of ZnONP decreased the values of both tensile strength (TS) and elongation at break (EB) to 6.10 ± 3.06% and 1.60 ± 0.32%, respectively. Previously, Roy and Rhim [22] had found no significant changes for carrageenan films after the addition of these nanoparticles. For non-biopolymers the literature is mixed: Mania et al. [56] reported an increase in both of these parameters for polyethylene films due to the action of nano ZnO, whereas Chu et al. [52] recorded a decrease in TS and an increase in EB for poly (lactic acid) (PLA) films.

In contrast to ZnONP, AgNP caused an increase in tensile strength to approximately 15 MPa and a decrease in elongation at break to about 2% for all samples. In contrast, Rhim et al. [53], did not observe significant changes in EB after adding these nanoparticles to agar films, however, TS values were lower at low concentrations of nanoparticles used and then increased with the increase in the amount of AgNP. However, Shankar and Rhim [39] showed a decrease in TS for agar films, with a simultaneous increase in EB value. Again, the literature is mixed for non-biopolymer PLA: Chu et al. [52] observed that TS decreased slightly and EB slightly increased, while Fortunati et al. [57] found that both parameters tended to decrease with AgNP addition.

Overall, the results obtained for alginate films and the results reported in the literature for various polymer matrices indicate that the addition of both melanin and silver and ZnONP can have different effects, i.e., positive, negative or no change in the described mechanical parameters, depending on the type of polymer used, the amount of melanin/nanoparticles used, as well as the methods and conditions of nanocomposites preparation.

Table 3 presents the melting temperatures (T_m) and melting enthalpies (ΔH_m) of the films as measured by DSC. Compared to the neat alginate films, the addition of melanin decreased the T_m for films containing the highest concentration of melanin (from 127.5 ± 1.9 °C to 99.8 ± 1.5 °C), with ΔH_m also decreasing with increasing melanin concentration (from 106.0 ± 13.6 J/g to 66.7 ± 8.5 J/g). For both ZnONP and AgNP, T_m decreased, while ΔH_m increased, although the values of these parameters remained similar for individual trials. This is in contrast to previous work with WPC/WPI films [37], where modification with melanin increased their thermal stability. However, addition of melanin to agar [42], cellulose [55] or PLA [44] films did not affect these parameters.

3.4. UV Barrier Properties

To evaluate the UV-Vis barrier properties, spectrophotometric spectra of the obtained films were analyzed. Figure 4 presents the UV-Vis spectra of the unmodified alginate film and the alginate films modified with increasing concentrations of melanin. The unmodified alginate film exhibited very poor (almost no) barrier properties against visible wavelengths, UVA, and part of the UVB range (above 300 nm). The addition of melanin caused a similar decrease in the transparency for visible light, regardless of melanin concentration. However, in the range of blue light, UVA, and UVB, the barrier properties increased with increasing amounts of melanin. Even the lowest melanin concentration (0.10%) had a significant effect on the transmittance of light at specific wavelengths. As mentioned earlier, melanins are known to have very good light absorption in the UV range. As a result, similar effects of melanins on light transmission have been reported for films of gelatin [54], agar [42], carrageenan [43], PLA [44], cellulose [55], carboxymethylcellulose [36], PBAT [49], and chitosan [45].

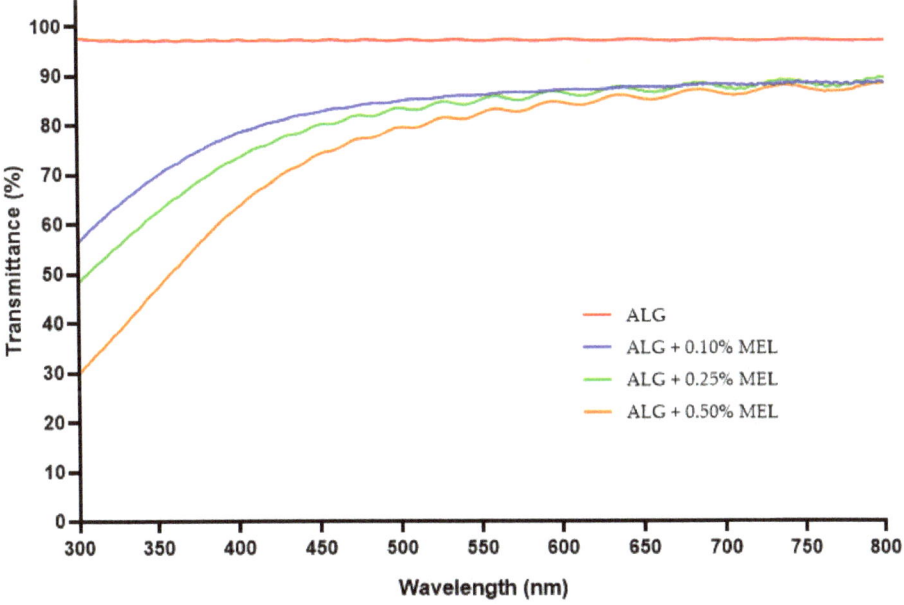

Figure 4. UV-Vis spectra of unmodified and melanin-modified alginate films.

Figure 5 presents a comparison of UV-Vis spectra of melanin-modified alginate films without and with the addition of ZnONP or AgNP. The ZnONP reduced the transmittance of the films by few to several percent in the visible and UV range. Metal oxide nanoparticles are commonly used in sunscreens due to their UV blocking properties [26,27]. Rashimi et al. [58] also observed an increased absorbance of UV-Vis radiation for polyvinyl alcohol films with the addition of ZnONP. A very significant effect can be observed with melanin concentrations of 0.25% and 0.50% resulting in essentially complete blocking of radiation with a wavelength shorter than 525 nm. For the sample with 0.10% melanin concentration, this parameter was also very low, on the order of 5 ± 0.5% transmittance.

Very similar results were obtained by other researchers for films modified with AgNP, such as agar and chitosan films [39,40,53].

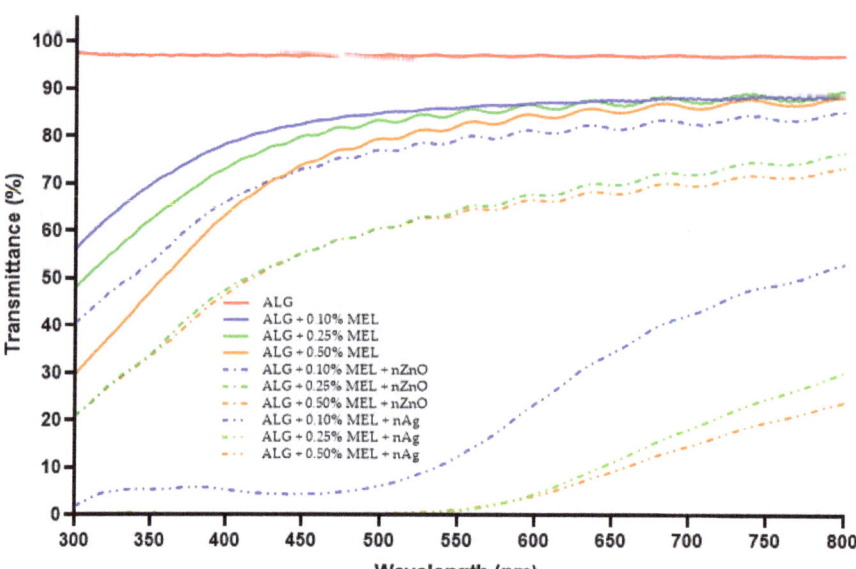

Figure 5. UV-Vis spectra of melanin-modified alginate film without or with the addition of zinc oxide silver nanoparticles.

3.5. FT-IR Analysis

In order to investigate the presence of characteristic bonds in the samples FT-IR assay was performed. Figure 6 summarizes the FT-IR spectra of unmodified alginate film and melanin-modified alginate films without nanoparticles. The graph shows numerous bands in the wavenumber range between 600 cm^{-1} and 1750 cm^{-1} and bands in the range of 3700–3000 cm^{-1} corresponding to the stretching vibrations of the –OH bonds [59], and between 3000 and 2850 cm^{-1} caused by –CH stretching vibration [60]. The bands present in the range of 1750–600 cm^{-1} are characteristic of alginate and they are as follows: at 1598 cm^{-1} it is the result of asymmetric vibrations stretching the –CO bond in the –COO– group [59,60], at 1407 cm^{-1} caused by symmetrical stretching vibrations –CO in the –COO– group [61], at 1026 cm^{-1} associated with asymmetric vibrations –COC [60] and at 816 cm^{-1} being characteristic of the presence of mannuronic acid residues [61].

The addition of melanin did not affect the formation of new bonds or the disappearance of existing ones. The slight shifts in bands and changes in their intensity are probably due to interactions between alginate and melanin in the form of hydrogen bonds and van der Waals forces, which is consistent with the observations of other researchers [36,37,42,54].

The addition of ZnONP (Figure 7) weakened the band at 1598 cm^{-1} (asymmetric vibrations stretching the –CO bond in the –COO– group) and increased the intensity of the band at 1355 cm^{-1} (asymmetric bond vibrations –CH– [52]. For the remaining bands, the differences in the measured intensity were small. Similar observations were noted for films made with the use of carrageenan [22] and PLA [52], while in the case of film made of polyvinyl alcohol, researchers reported a weakening of the peak intensity with an increase in the amount of added ZnONP [58].

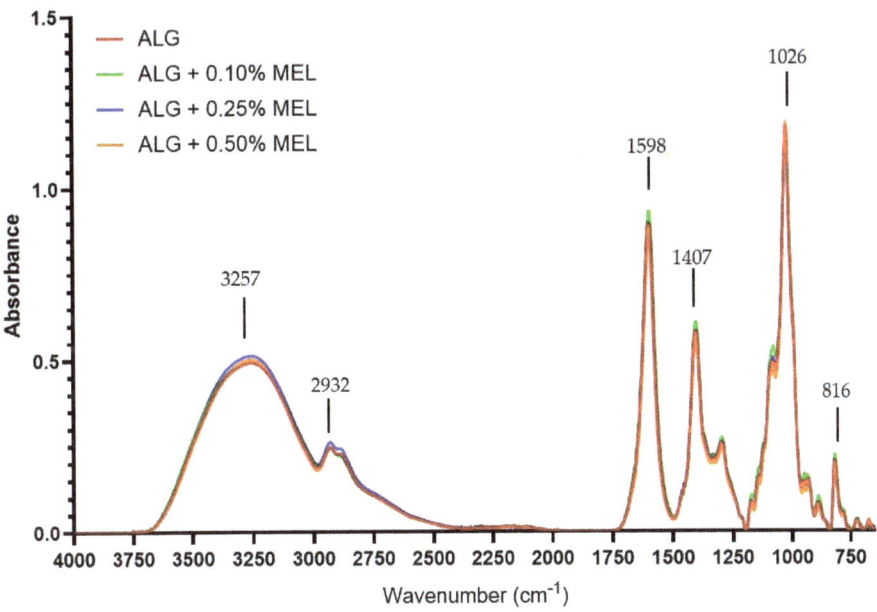

Figure 6. FT-IR spectra of unmodified and modified with melanin alginate films.

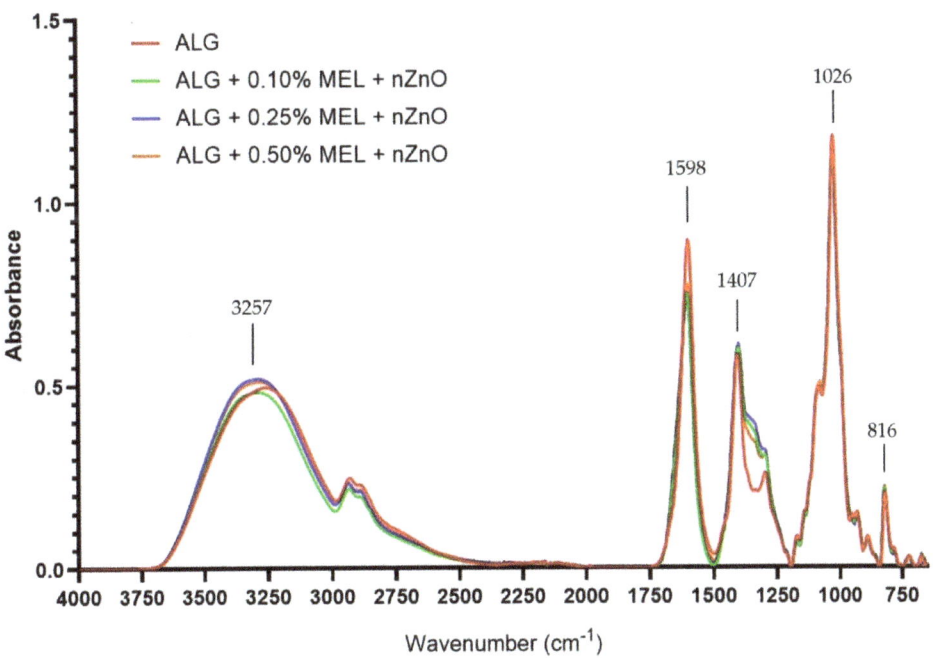

Figure 7. FT-IR spectra of unmodified and modified with melanin and zinc oxide nanoparticles alginate films.

In the case of AgNP (Figure 8), the intensity of the bands in the 1400–600 cm^{-1} range was enhanced. No additional peaks are visible, which indicates that the AgNP had no effect on the structure of the obtained alginate film. Shankar and Rhim [39], modifying the agar film with AgNP, also did not notice any changes in the chemical structure of the obtained product and their results show an increase in the intensity of most of the bands. However, to make their films, they used a concentration of silver nitrate that was twenty times lower, so the final content of nanoparticles was significantly lower. Vilmala et al. [40], on the other hand, noted a more pronounced effect of AgNP on films made of chitosan: AgNP addition caused not only a change in the intensity of some bands and their shift; it also resulted in the disappearance of some of bands and the appearance of new ones.

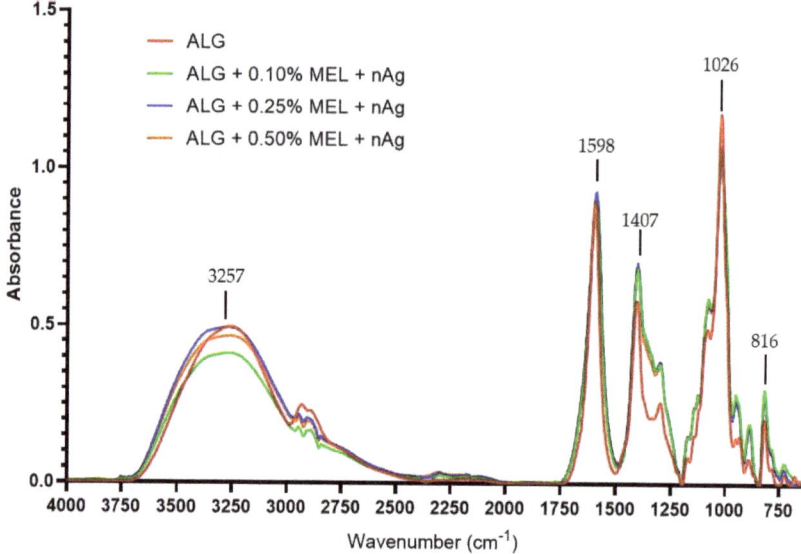

Figure 8. FT-IR spectra of unmodified and modified with melanin and silver nanoparticles alginate films.

3.6. Color

In order to investigate the influence of melanin and ZnONP and AgNP on color, chromatic parameter analyses were performed. Table 4 lists the chromatic parameters, the total color difference and the yellowness index of the unmodified alginate film and the film modified with melanin and with the addition of AgNP or zinc oxide, whereas Figure 9 shows the apparent color. As the melanin content increased, the parameter L* decreases slightly (the films become darker), while the parameters a* and b* increase (the red and yellow components of the overall color tone increase, respectively ($p < 0.05$)). The yellowness index also increases as the amount of melanin used increases ($p < 0.05$). The total color difference, for films containing only melanin, is noticeably different for each sample and is greater than unity, which is considered to be a difference noticeable to the human eye [44]. Similar relationships have been described in the literature for the effect of melanin addition on films made of gelatin [54], agar [42], carrageenan [43], chitosan [45], PLA [44], WPI/WPC [37], carboxymethylcellulose [36], and PBAT [49].

Table 4. Color (L*, a*, b*), total color difference (ΔE), yellowness index (YI) and transmittance of unmodified and modified alginate films.

Sample	L*	a*	b*	ΔE	YI	T_{280} (%)	T_{660} (%)
ALG	89.06 ± 0.76 [a]	−0.64 ± 0.04 [a]	6.75 ± 1.58 [a]	Used as standard	11.07 ± 2.84 [ab]	96.95	97.01
ALG + 0.10% MEL	88.87 ± 1.10 [aA]	−0.62 ± 0.04 [aA]	6.93 ± 1.67 [aA]	1.60 ± 1.04 [A]	11.59 ± 2.82 [ab]	61.22	87.65
ALG + 0.25% MEL	88.21 ± 0.54 [abA]	−0.60 ± 0.02 [aA]	8.56 ± 1.03 [abAB]	2.03 ± 0.77 [B]	13.84 ± 1.94 [ab]	62.38	86.73
ALG + 0.50% MEL	87.56 ± 2.21 [bA]	−0.48 ± 0.20 [bA]	10.19 ± 5.12 [bB]	4.64 ± 4.70 [C]	15.48 ± 4.82 [a]	43.67	85.24
ALG + 0.10% MEL + nZnO	87.96 ± 0.57 [ab]	−0.63 ± 0.03 [aA]	9.02 ± 1.03 [bA]	2.37 ± 1.16 [A]	14.65 ± 1.77 [ab]	61.19	71.76
ALG + 0.25% MEL + nZnO	86.73 ± 1.04 [b]	−0.44 ± 0.11 [bB]	11.03 ± 1.58 [cB]	4.72 ± 1.88 [B]	18.19 ± 2.82 [ac]	42.6	68.21
ALG + 0.50% MEL + nZnO	84.32 ± 0.84 [c]	0.07 ± 0.20 [cC]	15.07 ± 1.38 [dC]	9.46 ± 1.61 [C]	25.56 ± 2.61 [c]	39.39	70.14
ALG + 0.10% MEL + nAg	24.94 ± 4.92 [bA]	24.35 ± 2.81 [bA]	26.65 ± 6.30 [bA]	72.21 ± 1.15 [A]	129.86 ± 17.10 [b]	24.28	34.07
ALG + 0.25% MEL + nAg	24.39 ± 2.53 [bA]	25.57 ± 2.24 [bA]	23.60 ± 4.36 [bcA]	71.94 ± 0.65 [A]	137.25 ± 11.89 [c]	10.1	12.57
ALG + 0.50% MEL + nAg	22.61 ± 2.99 [cA]	24.48 ± 4.53 [bA]	20.94 ± 6.04 [cA]	72.82 ± 0.68 [A]	149.47 ± 18.47 [d]	4.19	10.29

Values are means ± standard deviation. Means with different lowercase (a–d) are significantly different at $p < 0.05$ (compared to control), means with different uppercase (A–C) are significantly different at $p < 0.05$ (compared to other variants within groups).

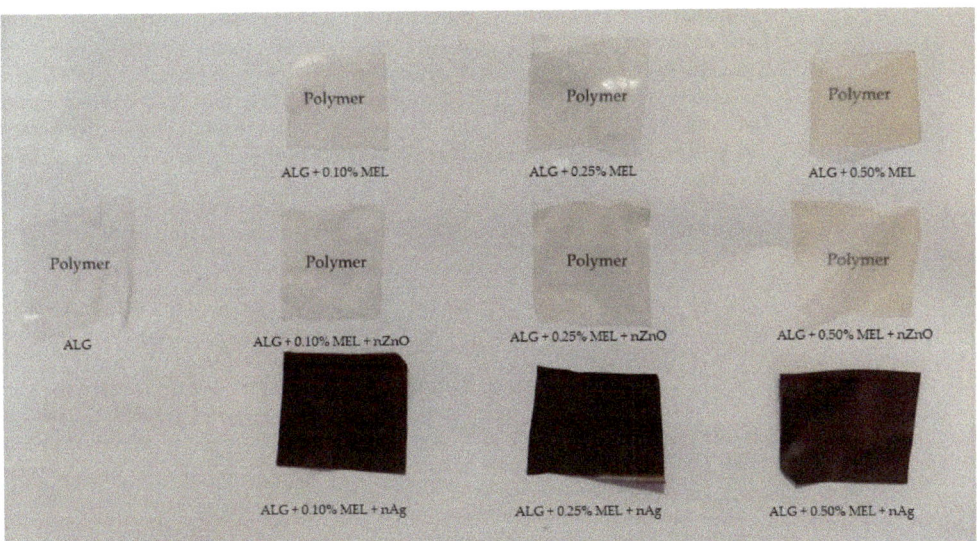

Figure 9. The visual appearance of neat and modified alginate films.

For the case of ZnONP, comparing the values of individual parameters for films with the same melanin content, both with or without nanoparticles, analogous changes can be observed, i.e., a decrease in the L* parameter and an increase in the other parameters. Roy and Rhim [22] obtained similar relationships for carrageenan film modified with ZnONP.

However, when making similar comparisons for films with and without AgNP, it can be seen that the L* parameter is over three-fold lower than for other types of films, which results from its much darker color. The parameters a* and b* have increased significantly, which indicates an increase in the red and yellow coordinates in the overall color tone of the film. This is due to the dark brown color imparted by the addition of nanosilver. The total color difference for these films has similar values, while the yellowness index is ten times higher than the results obtained for films with melanin alone. A similar effect of the addition of AgNP was described for agar film [39].

UV-barrier property (T_{280}) and transparency (T_{660}) of neat and modified films represented as transmittance at 280 nm and 660 nm, respectively, are shown in Table 4. Unmodified alginate films was transparent for both UV (96.95%) and visible (97.01%) light. As expected, based on melanin properties and visual appearance of the films, melanin-modified films were less transparent for both visual and UV light. UV light barrier properties increased as the amount of melanin used increased (from T_{280} equal to 61.22% for ALG + 0.10% MEL film to 43.67% for ALG + 0.50% MEL). ZnONP addition effected a further increase in visual light barrier properties and slight increase in UV light barrier properties. Addition of AgNP resulted in nearly no transmittance for both UV and visible light, especially for ALG + 0.50% MEL + nAg sample (4.19% and 10.29%, respectively). Those results are in line with findings for alginate films modified with copper sulfide nanoparticles [62].

3.7. Antioxidant Activity

In order to investigate the antioxidant properties of the obtained films reducing power (RP) and the free radicals scavenging ability were studied. Table 5 presents the antioxidant properties of the films (RP), and the free radicals scavenging ability (DPPH, ABTS, O_2^-). As expected, the addition of melanin improved the properties of free radicals scavenging activity by the tested films from $8.60 \pm 0.00\%$ for DPPH and $7.27 \pm 2.39\%$ for ABTS to $90.62 \pm 0.00\%$ and $90.87 \pm 0.02\%$, respectively, for the lowest melanin concentration used. Due to its insolubility in water and high antioxidant activity, all melanin-modified films exhibited ~90% DPPH and ABTS radicals scavenging regardless of the melanin concentration. For the reducing power, an upward trend was obtained with an increase in melanin concentration, while the ability to scavenge free radicals remains similar for all MEL concentrations. These results are consistent with previous studies that have shown that melanin has a significant effect on the scavenging of these radicals by various modified films [36,37,42–45,54,55].

Table 5. Reducing power (RP) and radical (DPPH, ABTS, O_2^-) scavenging activity of unmodified and modified alginate films.

Sample	RP (700 nm)	DPPH (%)	ABTS (%)	O_2^- (%)
ALG	0.023 ± 0.002 [a]	8.60 ± 0.00 [a]	7.27 ± 2.39 [a]	7.47 ± 1.12 [a]
ALG + 0.10% MEL	0.026 ± 0.007 [aA]	90.62 ± 0.00 [bA]	90.87 ± 0.02 [bA]	83.79 ± 3.37 [bA]
ALG + 0.25% MEL	0.027 ± 0.004 [aA]	89.38 ± 0.02 [cB]	90.86 ± 0.11 [bA]	80.50 ± 1.39 [bA]
ALG + 0.50% MEL	0.035 ± 0.011 [bA]	89.30 ± 0.09 [dC]	90.78 ± 0.12 [bA]	81.67 ± 1.21 [bA]
ALG + 0.10% MEL + nZnO	0.016 ± 0.002 [bA]	5.15 ± 0.00 [bA]	35.24 ± 0.42 [bA]	24.02 ± 1.43 [bA]
ALG + 0.25% MEL + nZnO	0.024 ± 0.010 [aAB]	5.69 ± 0.07 [cB]	40.39 ± 0.57 [cB]	47.74 ± 1.17 [cB]
ALG + 0.50% MEL + nZnO	0.025 ± 0.003 [aB]	6.20 ± 0.00 [dC]	47.51 ± 0.23 [dC]	59.25 ± 4.14 [dC]
ALG + 0.10% MEL + nAg	0.030 ± 0.022 [bA]	0.04 ± 0.00 [bA]	22.70 ± 0.13 [bcA]	68.39 ± 3.65 [bA]
ALG + 0.25% MEL + nAg	0.056 ± 0.018 [cAB]	0.19 ± 0.07 [cB]	21.56 ± 1.12 [bB]	68.30 ± 7.78 [bA]
ALG + 0.50% MEL + nAg	0.084 ± 0.046 [dB]	1.36 ± 0.07 [dC]	23.92 ± 0.48 [cC]	78.41 ± 6.33 [cB]

Values are means $\pm$ standard deviation. Means with different lowercase (a–d) are significantly different at $p < 0.05$ (compared to control), means with different uppercase (A–C) are significantly different at $p < 0.05$ (compared to other variants within groups).

The addition of ZnONP caused a decrease in RP and DPPH values to 0.016 ± 0.002 and $5.15 \pm 0.00\%$, respectively, for the 0.10% melanin concentration. These values were lower than those for unmodified alginate films. Meanwhile, the ABTS and superoxide scavenging results were noticeably lower than for the film with melanin alone, however still several times higher than for the control alginate. For AgNP, on the other hand, the reduction power markedly increased. However, radical scavenging ability was the lowest of all modified films. To the best of our knowledge, there are few if any similar studies on the synergic effect of melanin and metal nanoparticles on the antioxidant properties of films based on polymer matrices. Therefore, further studies should be conducted to determine the relationship between these additives and the antioxidant activity of different polymer matrix films.

3.8. Antimicrobial Activity

In order to test the antimicrobial properties of the obtained films, the ability to inhibit the growth of selected microorganisms was examined. Neither the unmodified alginate film nor the alginate films with the addition of melanin at various concentrations resulted in the formation of inhibition zones in any of the two tested microorganism strains. Previously, the lack of antimicrobial effect of melanin from the common mushroom (*Agaricus bisporus*) against the microorganisms used here was reported for PLA films [44]; however, the authors did note that melanin-modified materials were effective against *Enterococcus faecalis, Pseudomonas aeruginosa* and *P. putida*. On the other hand, melanin from *A. bisporus* did result in antimicrobial properties against *E. coli, S. aureus* and *Candida albicans* for the carboxymethylcellulose films. This again indicates that in this case the interactions of melanin with the polymer matrix have a significant influence on the properties of modified films.

For the films containing nanoparticles of silver or zinc oxide, the formation of zones of growth inhibition around the discs was observed. As illustrated in the pictures (supplementary materials), the nanoparticles migrate from the films into the medium. This effect was observed for all of the metal nanoparticle modified films tested. The sizes of the zones of growth inhibition for the tested microorganisms are presented in Table 6 and representative photos are shown in Supplementary Materials. The effects were similar for both types of nanoparticles against both microorganisms. The inhibitory effect of AgNP on the growth of *E. coli, Bacillus* and *Klebsiella pneumoniae* has been demonstrated for chitosan films [40], *E. coli* and *S. aureus* for PLA films [52,57], *E. coli* and *L. monocytogenes* for agar agar [39,53]. While chitosan film with the addition of cellulose dialdehyde nanocrystals showed inhibition for various strains of Gram positive, Gram negative bacteria and fungi [30]. Likewise, polymer films enriched with ZnONP have an antimicrobial effect well described in the literature; films with the addition of this nanomaterial showed antimicrobial activity for *E. coli* [22,52,56], *S. aureus* [56] and *L. monocytogenes* [22].

Table 6. Summary of the size of the zones of growth inhibition (expressed in mm) for *E. coli* (EC) and *S. aureus* (SA) around discs made of films modified with AgNP and ZnONP.

Sample	EC (mm)	SA (mm)
ALG + 0.10% MEL + nZnO	2.53 ± 0.38 [a]	3.02 ± 0.84 [a]
ALG + 0.25% MEL + nZnO	3.08 ± 1.21 [a]	3.08 ± 0.72 [a]
ALG + 0.50% MEL + nZnO	3.12 ± 0.93 [a]	3.54 ± 0.66 [a]
ALG + 0.10% MEL + nAg	1.44 ± 0.47 [b]	2.53 ± 0.54 [a]
ALG + 0.25% MEL + nAg	2.59 ± 0.79 [ab]	3.08 ± 1.12 [a]
ALG + 0.50% MEL + nAg	2.74 ± 0.41 [a]	3.52 ± 0.60 [a]

Values are means ± standard deviation. Means with a different letters (a-b) are significantly different at $p < 0.05$ (compared to other variants within groups).

The genotoxic effects of ZnONP are based on the following mechanisms: generation of reactive oxygen species (ROS) in cells; attachment directly to DNA or during cell division; influence on chromosome disorders. These phenomena occur even at ZnONP concentrations of several μg/mL [63]. AgNP also exhibit genotoxic effects, among others by binding to genetic material and thus affecting its replication and gene expression, however the smaller the nanoparticles are, the more toxic they are [64].

3.9. Cytotoxicity Screening

In order to screen for potential cytotoxicity of the prepared films, we performed direct contact assays where discs of materials were placed directly on top of pre-seeded fibroblast cells (L929 cell line). As the samples partially dissolved it was not possible to remove them prior to microscopic observation or viability assay. However, for the case of alginate films with different amounts of MEL, no interference was observed and after 24-h incubation, no cytotoxicity was detected by resazurin viability assay (Figure 10) or microscopy (see Supplemental Information for representative micrographs). Viability was well in excess of the 70% threshold guideline of ISO10993-5 and no marked changes in morphology were observed. The experiment was repeated with similar results. These results are in good agreement with our previous work with similar films obtained from whey protein concentrate/isolate [37]. The minor ~10% reduction in viability compared to sham may be due to some mechanical damage to the monolayer while discs are placed, as well as possible diffusional limitations, which should be minor due to swelling and dissolution of the discs.

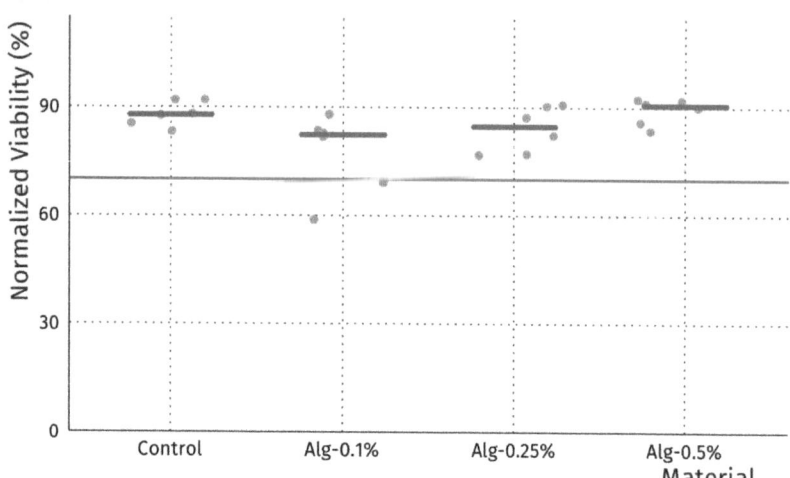

Figure 10. Cell viability data normalized to sham control (no disc) for alginate films with MEL and control film without MEL. Grey dots represent individual samples (n = 6), while blue bars indicate median values. Red line marks the 70% viability threshold of ISO10993-5.

For the case of alginate films with ZnONP and AgNP, the presence of particles interfered with the resazurin viability assay and significantly impeded microscopy, as compared to the alginate films without particles—particularly for the case of AgNP. This is likely due to the differences in reducing power as well as optical properties of the films noted previously. Unfortunately, due to swelling and partial dissolution of the films during the 24 h incubation, it was not possible to remove the samples to improve assay and imaging conditions. However, despite significantly reduced contrast and resolution, we were able to confirm robust cell growth using inverted light microscopy, similar to that for cells exposed to films without nanoparticles (see Supplemental Information for representative micrographs). The experiment was repeated with similar results. As a result, we conclude that the addition of ZnONP or AgNP did not have an adverse effect on cell viability, which is consistent with the fact that these materials are commonly used in both the pharmaceutical and cosmetics industries. However, further studies may be needed to rule out any negative effects.

4. Conclusions

This paper studied the properties of alginate films modified with melanin obtained from watermelon seeds as a by-product of agricultural industry and with Ag and ZnO NP. The properties of melanin-modified films and films with NP were compared to unmodified alginate films and to each other. Melanin modification had a clear effect on mechanical, antioxidant (~90% ABTS and DPPH radicals scavenging for all melanin modified films), hydrodynamic and barrier properties. The nanoparticles exhibited synergistic (silver nanoparticles addition effected two-fold higher tensile strength or both nanoparticles effected in increase in UV-Vis barrier properties) or antagonistic effects (decrease in some antioxidant properties compared to melanin-modified films) on the described melanin properties and they have developed an antimicrobial effect (up to four mm grow inhibition zones of *E. coli* and *S. aureus* for both zinc oxide and silver NP). It can be concluded that the obtained biofunctional melanin-modified alginate films, due to their antimicrobial and antioxidant activities, may find application for packaging of active food or potentially for biomedical applications.

Supplementary Materials: The following are available online at https://www.mdpi.com/article/10.3390/ma15072381/s1, Figure S1: Inverted light microscopy of L929 murine fibroblasts, Figure S2: Inverted light microscopy of L929 murine fibroblasts incubated for 24 h, Table S1: Example photographs of film samples and the inhibition zones they produced against E. coli and S. aureus microorganisms.

Author Contributions: Ł.Ł.—conceptualization, data curation, formal analysis, investigation, methodology, supervision, writing-original draft; S.M.—data curation, formal analysis, investigation, visualization, writing-original draft; M.Ś. —investigation; A.B. and S.R. —formal analysis; P.S. —investigation, methodology, formal analysis, writing-original draft. All authors have read and agreed to the published version of the manuscript.

Funding: This research received no external funding.

Institutional Review Board Statement: Not applicable.

Informed Consent Statement: Not applicable.

Conflicts of Interest: The authors declare no conflict of interest.

References

1. Marsh, K.; Bugusu, B. Food Packaging—Roles, Materials, and Environmental Issues: Scientific Status Summary. *J. Food Sci.* **2007**, *72*, R39–R55. [CrossRef] [PubMed]
2. Pavlath, A.E.; Orts, W. *Edible Films and Coatings for Food Applications*; Springer: Berlin/Heidelberg, Germany, 2009; pp. 1–23. [CrossRef]
3. Trang, P.T.T.; Dong, H.Q.; Toan, D.Q.; Hanh, N.T.X.; Thu, N.T. The Effects of Socio-Economic Factors on Household Solid Waste Generation and Composition: A Case Study in Thu Dau Mot, Vietnam. *Energy Procedia* **2017**, *107*, 253–258. [CrossRef]
4. Duenas, M.; Garciá-Estévez, I. Agricultural and Food Waste: Analysis, Characterization and Extraction of Bioactive Compounds and Their Possible Utilization. *Foods* **2020**, *9*, 817. [CrossRef] [PubMed]
5. Paletta, A.; Leal Filho, W.; Balogun, A.L.; Foschi, E.; Bonoli, A. Barriers and Challenges to Plastics Valorisation in the Context of a Circular Economy: Case Studies from Italy. *J. Clean. Prod.* **2019**, *241*, 118149. [CrossRef]
6. Thompson, R.C.; Moore, C.J.; Saal, F.S.V.; Swan, S.H. Plastics, the Environment and Human Health: Current Consensus and Future Trends. *Philos. Trans. R. Soc. B Biol. Sci.* **2009**, *364*, 2153–2166. [CrossRef]
7. Geyer, R.; Jambeck, J.R.; Law, K.L. Production, Use, and Fate of All Plastics Ever Made—Supplementary Information. *Sci. Adv.* **2017**, *3*, 19–24. [CrossRef]
8. Stafford, R.; Jones, P.J.S. Viewpoint—Ocean Plastic Pollution: A Convenient but Distracting Truth? *Mar. Policy* **2019**, *103*, 187–191. [CrossRef]
9. Jain, R.; Tiwari, A. Biosynthesis of Planet Friendly Bioplastics Using Renewable Carbon Source. *J. Environ. Health Sci. Eng.* **2015**, *13*, 11. [CrossRef]
10. Hassan, M.E.; Bai, J.; Dou, D.Q. Biopolymers; Definition, Classification and Applications. *Egypt. J. Chem.* **2019**, *62*, 1725–1737. [CrossRef]
11. George, A.; Sanjay, M.R.; Srisuk, R.; Parameswaranpillai, J.; Siengchin, S. A Comprehensive Review on Chemical Properties and Applications of Biopolymers and Their Composites. *Int. J. Biol. Macromol.* **2020**, *154*, 329–338. [CrossRef]
12. Luzi, F.; Torre, L.; Kenny, J.M.; Puglia, D. Bio- and Fossil-Based Polymeric Blends and Nanocomposites for Packaging: Structure-Property Relationship. *Materials* **2019**, *12*, 471. [CrossRef]
13. Matsumoto, T.; Mashiko, K. Viscoelastic Properties of Alginate Aqueous Solutions in the Presence of Salts. *Biopolymers* **1990**, *29*, 1707–1713. [CrossRef]
14. Lee, K.Y.; Mooney, D.J. Alginate: Properties and Biomedical Applications. *Prog. Polym. Sci.* **2012**, *37*, 106–126. [CrossRef] [PubMed]
15. Alihosseini, F. *Plant-Based Compounds for Antimicrobial Textiles*; Elsevier Ltd.: Amsterdam, The Netherlands, 2016; ISBN 9780081005859.
16. Yang, J.S.; Xie, Y.J.; He, W. Research Progress on Chemical Modification of Alginate: A Review. *Carbohydr. Polym.* **2011**, *84*, 33–39. [CrossRef]
17. Solano, F. Melanins: Skin Pigments and Much More—Types, Structural Models, Biological Functions, and Formation Routes. *New J. Sci.* **2014**, *2014*, 498276. [CrossRef]
18. Meredith, P.; Riesz, J. Radiative Relaxation Quantum Yields for Synthetic Eumelanin. *Photochem. Photobiol.* **2007**, *79*, 211–216. [CrossRef]
19. Zhang, M.; Xiao, G.; Thring, R.W.; Chen, W.; Zhou, H.; Yang, H. Production and Characterization of Melanin by Submerged Culture of Culinary and Medicinal Fungi *Auricularia auricula*. *Appl. Biochem. Biotechnol.* **2015**, *176*, 253–266. [CrossRef] [PubMed]

20. Cao, W.; Zhou, X.; McCallum, N.C.; Hu, Z.; Ni, Q.Z.; Kapoor, U.; Heil, C.M.; Cay, K.S.; Zand, T.; Mantanona, A.J.; et al. Unraveling the Structure and Function of Melanin through Synthesis. *J. Am. Chem. Soc.* **2021**, *143*, 2622–2637. [CrossRef] [PubMed]
21. Apte, M.; Girme, G.; Bankar, A.; RaviKumar, A.; Zinjarde, S. 3, 4-Dihydroxy-L-Phenylalanine-Derived Melanin from *Yarrowia lipolytica* Mediates the Synthesis of Silver and Gold Nanostructures. *J. Nanobiotechnol.* **2013**, *11*, 2. [CrossRef] [PubMed]
22. Roy, S.; Rhim, J.W. Carrageenan-Based Antimicrobial Bionanocomposite Films Incorporated with ZnO Nanoparticles Stabilized by Melanin. *Food Hydrocoll.* **2019**, *90*, 500–507. [CrossRef]
23. Roy, S.; Rhim, J. New Insight into Melanin for Food Packaging and Biotechnology Applications New Insight into Melanin for Food Packaging and Biotechnology Applications. *Crit. Rev. Food Sci. Nutr.* **2021**, 1–27. [CrossRef] [PubMed]
24. Kessler, R. Engineered Nanoparticles in Consumer Products: Understanding a New Ingredient. *Environ. Health Perspect.* **2011**, *119*, a120–a125. [CrossRef]
25. Dimapilis, E.A.S.; Hsu, C.S.; Mendoza, R.M.O.; Lu, M.C. Zinc Oxide Nanoparticles for Water Disinfection. *Sustain. Environ. Res.* **2018**, *28*, 47–56. [CrossRef]
26. Piccinno, F.; Gottschalk, F.; Seeger, S.; Nowack, B. Industrial Production Quantities and Uses of Ten Engineered Nanomaterials in Europe and the World. *J. Nanoparticle Res.* **2012**, *14*, 1109. [CrossRef]
27. Li, S.; Silvers, S.J.; El-Shall, M.S. Preparation, Characterization and Optical Properties of Zinc Oxide Nanoparticles. *Mater. Res. Soc. Symp.-Proc.* **1997**, *452*, 389–394. [CrossRef]
28. Mizielińska, M.; Łopusiewicz, Ł.; Mężyńska, M.; Bartkowiak, A. The Influence of Accelerated UV-A and Q-Sun Irradiation on the Antimicrobial Properties of Coatings Containing ZnO Nanoparticles. *Molecules* **2017**, *22*, 1556. [CrossRef] [PubMed]
29. Klasen, H.J. Historical Review of the Use of Silver in the Treatment of Burns.I. Early Uses. *Burns* **2000**, *26*, 117–130. [CrossRef]
30. Dong, F.; Li, S. Wound Dressings Based on Chitosan-Dialdehyde Cellulose Nanocrystals-Silver Nanoparticles: Mechanical Strength, Antibacterial Activity and Cytotoxicity. *Polymers* **2018**, *10*, 673. [CrossRef] [PubMed]
31. Hsueh, Y.-H.; Lin, K.-S.; Ke, W.-J.; Hsieh, C.-T.; Chiang, C.-L.; Tzou, D.-Y.; Liu, S.-T. The Antimicrobial Properties of Silver Nanoparticles in *Bacillus subtilis* Are Mediated by Released Ag^+ Ions. *PLoS ONE* **2015**, *10*, e0144306. [CrossRef]
32. Peiris, M.M.K.; Fernando, S.S.N.; Jayaweera, P.M.; Arachchi, N.D.H.; Guansekara, T.D.C.P. Comparison of Antimicrobial Properties of Silver Nanoparticles Synthesized from Selected Bacteria. *Indian J. Microbiol.* **2018**, *58*, 301–311. [CrossRef] [PubMed]
33. Janardhanan, R.; Karuppaiah, M.; Hebalkar, N.; Rao, T.N. Synthesis and Surface Chemistry of Nano Silver Particles. *Polyhedron* **2009**, *28*, 2522–2530. [CrossRef]
34. Garavand, F.; Cacciotti, I.; Vahedikia, N.; Rehman, A.; Tarhan, Ö.; Akbari-Alavijeh, S.; Shaddel, R.; Rashidinejad, A.; Nejatian, M.; Jafarzadeh, S.; et al. A Comprehensive Review on the Nanocomposites Loaded with Chitosan Nanoparticles for Food Packaging. *Crit. Rev. Food Sci. Nutr.* **2022**, *62*, 1383–1416. [CrossRef] [PubMed]
35. Łopusiewicz, Ł. Antioxidant, Antibacterial Properties and the Light Barrier Assessment of Raw and Purified Melanins Isolated from *Citrullus lanatus* (Watermelon) Seeds. *Herba Pol.* **2018**, *64*, 25–36. [CrossRef]
36. Łopusiewicz, Ł.; Kwiatkowski, P.; Drozłowska, E.; Trocer, P.; Kostek, M.; Śliwiński, M.; Polak-Śliwińska, M.; Kowalczyk, E.; Sienkiewicz, M. Preparation and Characterization of Carboxymethyl Cellulose-Based Bioactive Composite Films Modified with Fungal Melanin and Carvacrol. *Polymers* **2021**, *13*, 499. [CrossRef]
37. Łopusiewicz, Ł.; Drozłowska, E.; Trocer, P.; Kostek, M.; Śliwiński, M.; Henriques, M.H.F.; Bartkowiak, A.; Sobolewski, P. Whey Protein Concentrate/Isolate Biofunctional Films Modified with Melanin from Watermelon (*Citrullus lanatus*) Seeds. *Materials* **2020**, *13*, 3876. [CrossRef] [PubMed]
38. ISO 10993-5; Biological Evaluation of Medical Devices—Part 5: Tests for In Vitro Cytotoxicity. International Organization for Standardization: Geneva, Switzerland, 2009.
39. Shankar, S.; Rhim, J.W. Amino Acid Mediated Synthesis of Silver Nanoparticles and Preparation of Antimicrobial Agar/Silver Nanoparticles Composite Films. *Carbohydr. Polym.* **2015**, *130*, 353–363. [CrossRef]
40. Vimala, K.; Mohan, Y.M.; Sivudu, K.S.; Varaprasad, K.; Ravindra, S.; Reddy, N.N.; Padma, Y.; Sreedhar, B.; MohanaRaju, K. Fabrication of Porous Chitosan Films Impregnated with Silver Nanoparticles: A Facile Approach for Superior Antibacterial Application. *Colloids Surf. B Biointerfaces* **2010**, *76*, 248–258. [CrossRef] [PubMed]
41. Thirunavoukkarasu, M.; Balaji, U.; Behera, S.; Panda, P.K.; Mishra, B.K. Biosynthesis of Silver Nanoparticle from Leaf Extract of *Desmodium gangeticum* (L.) DC. and Its Biomedical Potential. *Spectrochim. Acta-Part A Mol. Biomol. Spectrosc.* **2013**, *116*, 424–427. [CrossRef] [PubMed]
42. Roy, S.; Rhim, J.W. Agar-Based Antioxidant Composite Films Incorporated with Melanin Nanoparticles. *Food Hydrocoll.* **2019**, *94*, 391–398. [CrossRef]
43. Roy, S.; Rhim, J.W. Preparation of Carrageenan-Based Functional Nanocomposite Films Incorporated with Melanin Nanoparticles. *Colloids Surf. B Biointerfaces* **2019**, *176*, 317–324. [CrossRef] [PubMed]
44. Łopusiewicz, Ł.; Jędra, F.; Mizielińska, M. New Poly(Lactic Acid) Active Packaging Composite Films Incorporated with Fungal Melanin. *Polymers* **2018**, *10*, 386. [CrossRef] [PubMed]
45. Roy, S.; Van Hai, L.; Kim, H.C.; Zhai, L.; Kim, J. Preparation and Characterization of Synthetic Melanin-like Nanoparticles Reinforced Chitosan Nanocomposite Films. *Carbohydr. Polym.* **2020**, *231*, 115729. [CrossRef] [PubMed]
46. Yang, M.; Li, L.; Yu, S.; Liu, J.; Shi, J. High Performance of Alginate/Polyvinyl Alcohol Composite Film Based on Natural Original Melanin Nanoparticles Used as Food Thermal Insulating and UV–Vis Block. *Carbohydr. Polym.* **2020**, *233*, 115884. [CrossRef]

47. Rhim, J.W.; Hong, S.I.; Park, H.M.; Ng, P.K.W. Preparation and Characterization of Chitosan-Based Nanocomposite Films with Antimicrobial Activity. *J. Agric. Food Chem.* **2006**, *54*, 5814–5822. [CrossRef]
48. Kotharangannagari, V.K.; Krishnan, K. Biodegradable Hybrid Nanocomposites of Starch/Lysine and ZnO Nanoparticles with Shape Memory Properties. *Mater. Des.* **2016**, *109*, 590–595. [CrossRef]
49. Bang, Y.J.; Shankar, S.; Rhim, J.W. Preparation of Polypropylene/Poly (Butylene Adipate-Co-Terephthalate) Composite Films Incorporated with Melanin for Prevention of Greening of Potatoes. *Packag. Technol. Sci.* **2020**, *33*, 433–441. [CrossRef]
50. Kanmani, P.; Rhim, J.W. Properties and Characterization of Bionanocomposite Films Prepared with Various Biopolymers and ZnO Nanoparticles. *Carbohydr. Polym.* **2014**, *106*, 190–199. [CrossRef]
51. Shankar, S.; Wang, L.F.; Rhim, J.W. Incorporation of Zinc Oxide Nanoparticles Improved the Mechanical, Water Vapor Barrier, UV-Light Barrier, and Antibacterial Properties of PLA-Based Nanocomposite Films. *Mater. Sci. Eng. C* **2018**, *93*, 289–298. [CrossRef] [PubMed]
52. Chu, Z.; Zhao, T.; Li, L.; Fan, J.; Qin, Y. Characterization of Antimicrobial Poly (Lactic Acid)/Nano-Composite Films with Silver and Zinc Oxide Nanoparticles. *Materials* **2017**, *10*, 659. [CrossRef] [PubMed]
53. Rhim, J.W.; Wang, L.F.; Hong, S.I. Preparation and Characterization of Agar/Silver Nanoparticles Composite Films with Antimicrobial Activity. *Food Hydrocoll.* **2013**, *33*, 327–335. [CrossRef]
54. Shankar, S.; Wang, L.F.; Rhim, J.W. Effect of Melanin Nanoparticles on the Mechanical, Water Vapor Barrier, and Antioxidant Properties of Gelatin-Based Films for Food Packaging Application. *Food Packag. Shelf Life* **2019**, *21*, 100363. [CrossRef]
55. Roy, S.; Kim, H.C.; Kim, J.W.; Zhai, L.; Zhu, Q.Y.; Kim, J. Incorporation of Melanin Nanoparticles Improves UV-Shielding, Mechanical and Antioxidant Properties of Cellulose Nanofiber Based Nanocomposite Films. *Mater. Today Commun.* **2020**, *24*, 100984. [CrossRef]
56. Mania, S.; Cieślik, M.; Konzorski, M.; Święcikowski, P.; Nelson, A.; Banach, A.; Tylingo, R. The Synergistic Microbiological Effects of Industrial Produced Packaging Polyethylene Films Incorporated with Zinc Nanoparticles. *Polymers* **2020**, *12*, 1198. [CrossRef]
57. Fortunati, E.; Armentano, I.; Zhou, Q.; Iannoni, A.; Saino, E.; Visai, L.; Berglund, L.A.; Kenny, J.M. Multifunctional Bionanocomposite Films of Poly(Lactic Acid), Cellulose Nanocrystals and Silver Nanoparticles. *Carbohydr. Polym.* **2012**, *87*, 1596–1605. [CrossRef]
58. Rashmi, S.H.; Raizada, A.; Madhu, G.M.; Kittur, A.A.; Suresh, R.; Sudhina, H.K. Influence of Zinc Oxide Nanoparticles on Structural and Electrical Properties of Polyvinyl Alcohol Films. *Plast. Rubber Compos.* **2015**, *44*, 33–39. [CrossRef]
59. Voo, W.P.; Lee, B.B.; Idris, A.; Islam, A.; Tey, B.T.; Chan, E.S. Production of Ultra-High Concentration Calcium Alginate Beads with Prolonged Dissolution Profile. *RSC Adv.* **2015**, *5*, 36687–36695. [CrossRef]
60. Lawrie, G.; Keen, I.; Drew, B.; Chandler-Temple, A.; Rintoul, L.; Fredericks, P.; Grøndahl, L. Interactions between Alginate and Chitosan Biopolymers Characterized Using FTIR and XPS. *Biomacromolecules* **2007**, *8*, 2533–2541. [CrossRef]
61. Fertah, M.; Belfkira, A.; Dahmane, E.m.; Taourirte, M.; Brouillette, F. Extraction and Characterization of Sodium Alginate from Moroccan *Laminaria digitata* Brown Seaweed. *Arab. J. Chem.* **2017**, *10*, S3707–S3714. [CrossRef]
62. Roy, S.; Rhim, J.W. Effect of CuS Reinforcement on the Mechanical, Water Vapor Barrier, UV-Light Barrier, and Antibacterial Properties of Alginate-Based Composite Films. *Int. J. Biol. Macromol.* **2020**, *164*, 37–44. [CrossRef] [PubMed]
63. Singh, S. Zinc Oxide Nanoparticles Impacts: Cytotoxicity, Genotoxicity, Developmental Toxicity, and Neurotoxicity. *Toxicol. Mech. Methods* **2019**, *29*, 300–311. [CrossRef] [PubMed]
64. Rodriguez-Garraus, A.; Azqueta, A.; Vettorazzi, A.; de Cerain, A.L. Genotoxicity of Silver Nanoparticles. *Nanomaterials* **2020**, *10*, 251. [CrossRef] [PubMed]

Article

Mechanical Testing of Epoxy Resin Modified with Eco-Additives

Agnieszka Derewonko [1,*], Wojciech Fabianowski [2] and Jerzy Siczek [2]

[1] Faculty of Mechanical Engineering, Institute of Mechanics and Computational Engineering, Military University of Technology, Sylwestra Kaliskiego 2, 00-908 Warsaw, Poland
[2] Military Institute of Chemistry and Radiation, gen. Antoniego Chruściela 105, 00-910 Warsaw, Poland
* Correspondence: agnieszka.derewonko@wat.edu.pl; Tel.: +48-261-837-906

Abstract: The future belongs to biodegradable epoxies. In order to improve epoxy biodegradability, it is crucial to select suitable organic additives. The additives should be selected so as to (maximally) accelerate the decomposition of crosslinked epoxies under normal environmental conditions. However, naturally, such rapid decomposition should not occur within the normal (expected) service life of a product. Consequently, it is desirable that the newly modified epoxy should exhibit at least some of the mechanical properties of the original material. Epoxies can be modified with different additives (such as inorganics with different water uptake, multiwalled carbon nanotubes, and thermoplastics) that can increase their mechanical strength but does not lead to their biodegradability. In this work, we present several mixtures of epoxy resins together with organic additives based on cellulose derivatives and modified soya oil. These additives are environmentally friendly and should increase the epoxy's biodegradability on the one hand without deteriorating its mechanical properties on the other. This paper concentrates mainly on the question of the tensile strength of various mixtures. Herein, we present the results of uniaxial stretching tests for both modified and unmodified resin. Based on statistical analysis, two mixtures were selected for further studies, namely the investigation of durability properties.

Keywords: epoxy resin; eco-additives; experimental tests

Citation: Derewonko, A.; Fabianowski, W.; Siczek, J. Mechanical Testing of Epoxy Resin Modified with Eco-Additives. *Materials* **2023**, *16*, 1854. https://doi.org/10.3390/ma16051854

Academic Editors: Krzysztof Moraczewski and Loic Hilliou

Received: 29 November 2022
Revised: 14 February 2023
Accepted: 21 February 2023
Published: 24 February 2023

Copyright: © 2023 by the authors. Licensee MDPI, Basel, Switzerland. This article is an open access article distributed under the terms and conditions of the Creative Commons Attribution (CC BY) license (https://creativecommons.org/licenses/by/4.0/).

1. Introduction

Composites are widely used in the automotive and aerospace industries. Their advantages include low weight and infinite durability. Unfortunately, infinite durability is also a disadvantage. A large amount of non-biodegradable waste is generated. Thus, a method is needed to recycle the waste. Pyrolysis is the most commonly used method of recycling in the case of carbon-fiber-reinforced composites with an epoxy matrix (CFRP) [1]. The recovered carbon fibers are used in thermoplastic and thermosetting coatings and nonwoven fabrics. On the other hand, the recycling of glass fiber reinforced polymers (GRP) is carried out via their simultaneous processing in the cement kilns with fuels from waste. Thermoplastic composites are also crushed and melted. Currently, popular methods for reducing environmental littering when using biodegradable or biocompostable polymers also have some disadvantages. One of them is the pollution of the aquatic environment by slowly degraded materials, mainly those that are hydrolyzed. For instance, in the case of epoxy–glass composites, there is a risk of releasing bisphenol A (BPA), which is highly toxic [2].

The degree of the composite degradation was determined using the methods utilized in the evaluation of the composite cracking resistance [3,4]. An effective and safe method for obtaining biodegradable or biocompostable composites also includes the issue of their mechanical strength in the assumed period of use (operation), which is often neglected. Therefore, the following question arises: How does an organic additive influence the mechanical strength of the epoxy resin?

Scientific and research literature on the use of organic additives is extensive and relates mainly to compounds modifying interactions at the interface between the glass (carbon) fiber and the liquid epoxy matrix [5–7]. Numerous additives—both organic and inorganic—were tested as epoxy modifying agents [8–10]. Note that even small amounts of added compounds, less than 1% w/w, can change the mechanical properties of epoxy composites in a significant way. The structural integrity of the composites and the interface of bonded layers, such as metal and composite, is ensured by adequate adhesion. The role of such issues in relation to the mechanical properties of structures was described by Bellini and his team in a 2019 paper [11]. However, the issue of modifying matrix polymers for CFRP composites in terms of strength was discussed, among others, in works [12–15].

One type of additive used in epoxy–glass composites is epoxidized natural oils that change the properties of the epoxy matrix, mainly by increasing their mechanical strength and impact strength. The effect of adding inedible oils of natural origins, e.g., karanja oil from Pongam tree seeds, to epoxy resins is described in [16]. In the case of composites, a series of works was carried out on the study, assessment and modeling of the effect of homophasic and heterophasic additives on mechanical properties, impact resistance and thermal resistance, e.g., [17]; however, they do not take into account their biodegradability.

Particularly interesting are the works on adding organic substances, including epoxidized edible and inedible oils known and used as additives for diesel fuels, e.g., biodiesel [18]. These substances, as rapidly biodegradable, cannot be used in composites for which their expected aging resistance is counted in years. Therefore, using organic substances that are more durable than biodiesel additives is proposed, namely, epoxidized cellulose derivatives, methyl cellulose, carboxymethyl cellulose and similar substances [19]. In work [20], it was shown that the chemical modification of cellulose pulp by epoxidation reactions has a positive effect on the rheological properties of the final product.

Environmental pollution and climate protection have intensified work on ecological additives that limit the extraction of fossil raw materials and enable the recycling of products after their use. A novel degradable and recyclable thermoset hyperbranched epoxy resin (EFTH-n) synthesized from bio-based 2,5-furandicarboxylic acid was described in work [21]. It has been demonstrated that EFTH-n is successfully used to improve the toughness, strength, modulus and elongation of DGEBA (Bisphenol A diglycidyl ether).

The issue of biodegradability has been raised frequently in recent times. This is due not only to massive and growing environmental pollution but also dwindling fossil resources. Among the many works from 2020, those in which the chemical aspect is related to the strength and durability of materials, such as [22–24], are included. As early as 2021, in paper [25], fully recyclable epoxy formulations using organic waste flour have been proposed. In contrast, paper [26] proposed environmentally friendly adhesives for aerospace applications. A comprehensive list of current works in the field of the application of clove oil in the production of composites can be found in Matykiewicz and Skórczewska, 2022 [27].

Although a new type of hardener was patented in 2021, Recylamine, which makes the epoxy resin biodegradable after curing, is not the only hardener used in the pyrolysis process; the Z-1 (triethylene tetramine) hardener is still used. The study investigated the effect of various organic additives and their content on the tensile strength of rowing specimens made from them.

2. Materials and Methods

The main objective of the study is to determine the tensile strength of specimens made of epoxy resin mixtures with natural additives.

A change in the crosslinking of a polymer is achieved by introducing an additive in a certain proportion. After a set finite time, such a change should cause the failure of the cured epoxy resin. A gradual degradation of the polymer crosslinking will allow for the separation of the epoxy matrix from fibers, which leads to the partial degradation of the composite and the segregation of its components. The types of analyzed mixtures are listed in Table 1. In Table 1, organic additives are as follows: ESO—epoxidized soybean oil (Boryszew S.A.); MC—methyl cellulose (C.T.S.); EC—ethyl cellulose (C.T.S.); CMC—carboxymethyl cellulose (C.T.S.). In Table 1, phr means parts by weight per 100 parts by weight of the resin. Epidian 601 is the epoxy resin (Zakłady Chemiczne Ciech Sarzyna).

Table 1. Composition of epoxy/additive mixtures for determining mechanical properties.

No.	Epoxy Resin	Additive	Phr	Remarks
1	Epidian 601	Without additive	-	-
2	Epidian 601	ESO	3	Ref. [28]
3	Epidian 601	ESO	10	Ref. [28]
4	Epidian 601	MC	3	Ref. [20]
5	Epidian 601	MC	10	Ref. [20]
6	Epidian 601	CMC	3	Ref. [20]
7	Epidian 601	CMC	10	Ref. [20]
8	Epidian 601	EC	3	Ref. [20]
9	Epidian 601	EC	10	Ref. [20]

Some basic properties of Epidian 601 are collected in Table 2.

Table 2. Basic properties of Epidian 601 resin (low viscosity liquid) [29].

Property	Unites	Value	Remarks
Colour	-	Yellow, slightly opaque	-
Boiling point	°C	210	Starts decomposition
Fire point	°C	180	-
Epoxide number	Mol/100 g	0.5–0.55	-
Density	g/cm^3 at 25 °C	1.14	-
Viscosity	mPas at 25 °C	700–1100	-
Gelling time	min	40	After addition of 13 phr Z1; RT

All mixtures were prepared using the method described below. In total, 100 g of Epoxy resin 601 was RT mixed up with 3 g or 10 g of organic additive, and 13 g of Z-1 crosslinking agent (triethylene tetramine from Zakłady Chemiczne Ciech Sarzyna) was added next, vigorously mixed for 5 min, and vacuum degassed for 10 min. All specimens of Epidian 601 with the organic additive and crosslinking agent were at first, right after addition, opaque and more viscous, but within 2–3 min after mixing, the specimens turned yellow again and were more transparent and less viscous, similarly to the starting Epidian 601 resin. Only after the addition of 3 phr or 10 phr of ethyl cellulose to the Epidian 601 resin did the specimens turn into an opaque white color and were more viscous; even after prolongated RT mixing for 10 min, they still remained unchanged. This observation suggests that EC, being more hydrophobic than other cellulose derivatives, was incompatible with Epidian 601 resin, forming a separate phase that weakly interacted with the surrounding resin. This was later confirmed by the poor mechanical properties of the Epidian/EC system. In preparation of Epidian/organic additive/crosslinking agent Z-1, it should be remembered that all mixing/degassing procedures must be completed within 30 min, because after 40 min, these systems start to gel, and no mixing or degassing is possible.

The described mixtures were used to create test specimens for uniaxial tensile tests. Polastosil AD4 (Zakłady Chemiczne Ciech Sarzyna) was used to prepare the silicone mold. There have to be holes in the mold that match the shape and dimensions of the specimens that will be created. The reference specimens shown in Figure 1a were used to create them. Both the mold and the reference specimens were made by 3D printing technology.

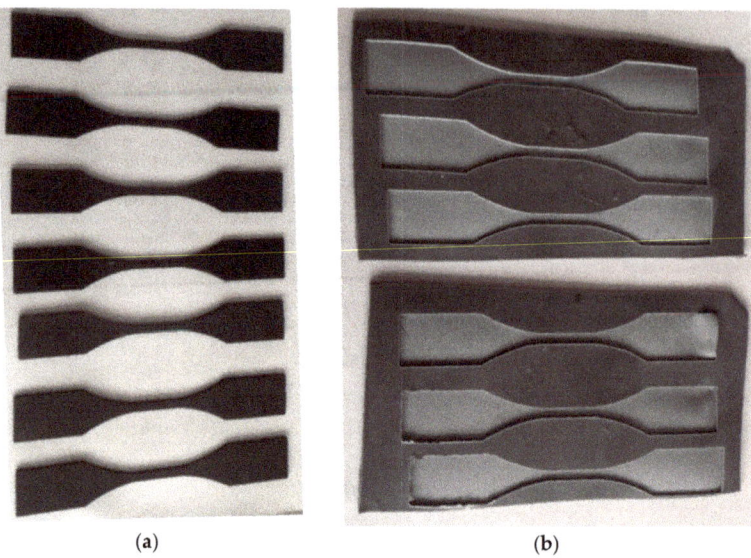

Figure 1. (a) Test reference specimens made by 3D printing and (b) silicone molds made from Polastosil.

The silicone molds with holes are shown in Figure 1b. Test specimens for the uniaxial tensile test were formed from mixtures 1 to 9 in Table 1.

In Figure 2a, the reference structure of the epoxy resin taken with the use of a digital microscope Keyence VHX 6000 with a magnification of 200× is shown. The effect of organic additives on the structure of the modified epoxy resin is shown in Figure 2b–i. The scale on each picture allows the determination of the size and distribution of additives. Inserting a scale is more practical than using a scale in pictures.

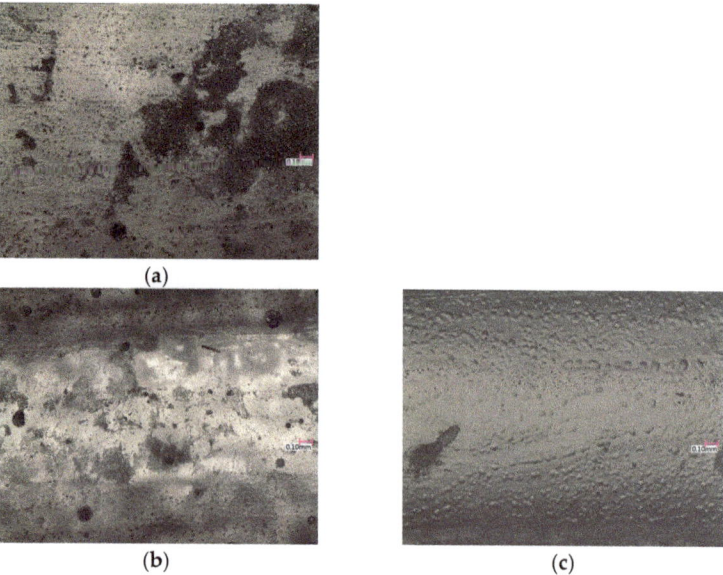

Figure 2. *Cont.*

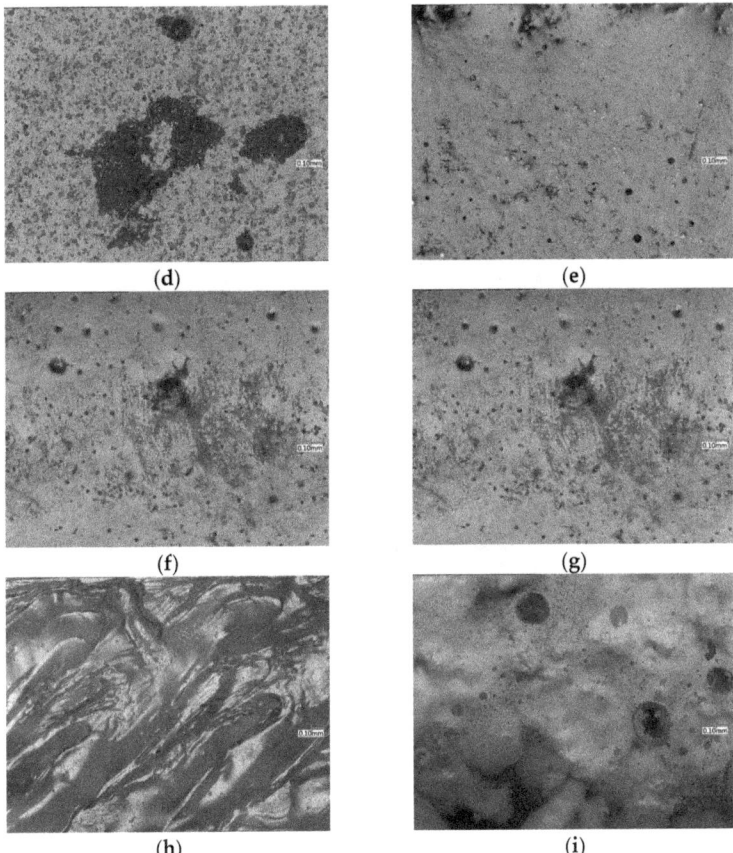

(d) (e) (f) (g) (h) (i)

Figure 2. Microscopic photos of the fragments of specimens made with mixtures of (**a**) epoxy resin E601 and additive, (**b**) ESO3, (**c**) ESO10, (**d**) MC3, (**e**) MC10, (**f**) CMC3, (**g**) CMC10, (**h**) EC3 and (**i**) EC10.

3. Uniaxial Stretching Tests

Biodegradable epoxy resins should indicate physical and chemical properties, as well as mechanical properties, that are similar to unmodified resins with practically unlimited disintegration time. Therefore, experimental uniaxial stretching tests according to ASTM638 were conducted on test specimens made of epoxy resins with an organic additive. A series of reference test specimens made of unmodified epoxy were also examined. The nominal dimension of the test specimen is shown in Figure 3.

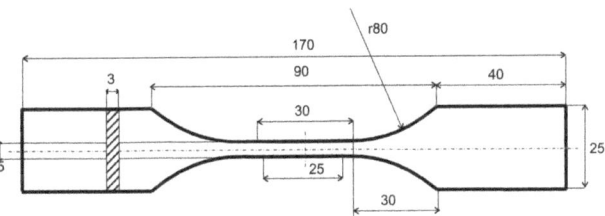

Figure 3. Nominal dimensions of the test specimen in mm.

Proper experimental testing requires additional specimen preparation. The specimens were cleaned and specially marked. The places of grips have been marked, as well as measurement points (black dots) spaced 30 mm apart. An exemplary set of test specimens prepared for testing that are made of epoxy resin E601 with the addition of ESO10 (epoxidized soybean oil) is shown in Figure 4.

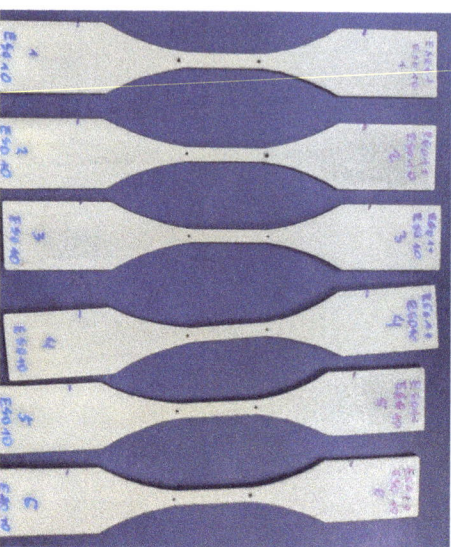

Figure 4. Exemplary set of test specimens prepared for testing that are made of epoxy resin E601 with the addition of ESO10.

The KAPPA 50 DS electromechanical loading system with a ZwickRoell video extensometer (Figure 5a) and specialized software was used for the tests, enabling the simultaneous measurement of the elongation and change in the width of the specimen in a given area. The machine ensured precise axial alignment to ASTM E292. ZwickRoell videoXtens uses image processing, allowing longitudinal and transverse strains to be determined with greater accuracy.

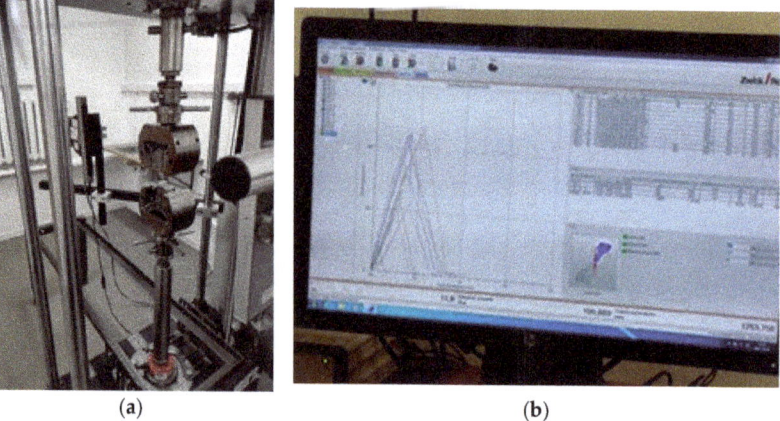

(a)   (b)

Figure 5. (a) KAPPA 50 DS electromechanical loading system with a ZwickRoell video extensometer and (b) specialized software testXpert ZwickRoell.

The specialized testXpert ZwickRoell (Figure 5b) software recorded measurement data such as time, distance, force, elongation and width change in the measurement area. The specimens were stretched at a speed of 2 mm/min until they were damaged. The frequency of data acquisition was set at 10 Hz due to the static nature of the load. Obtained in tensile test stress–strain curves are shown in Figure 6.

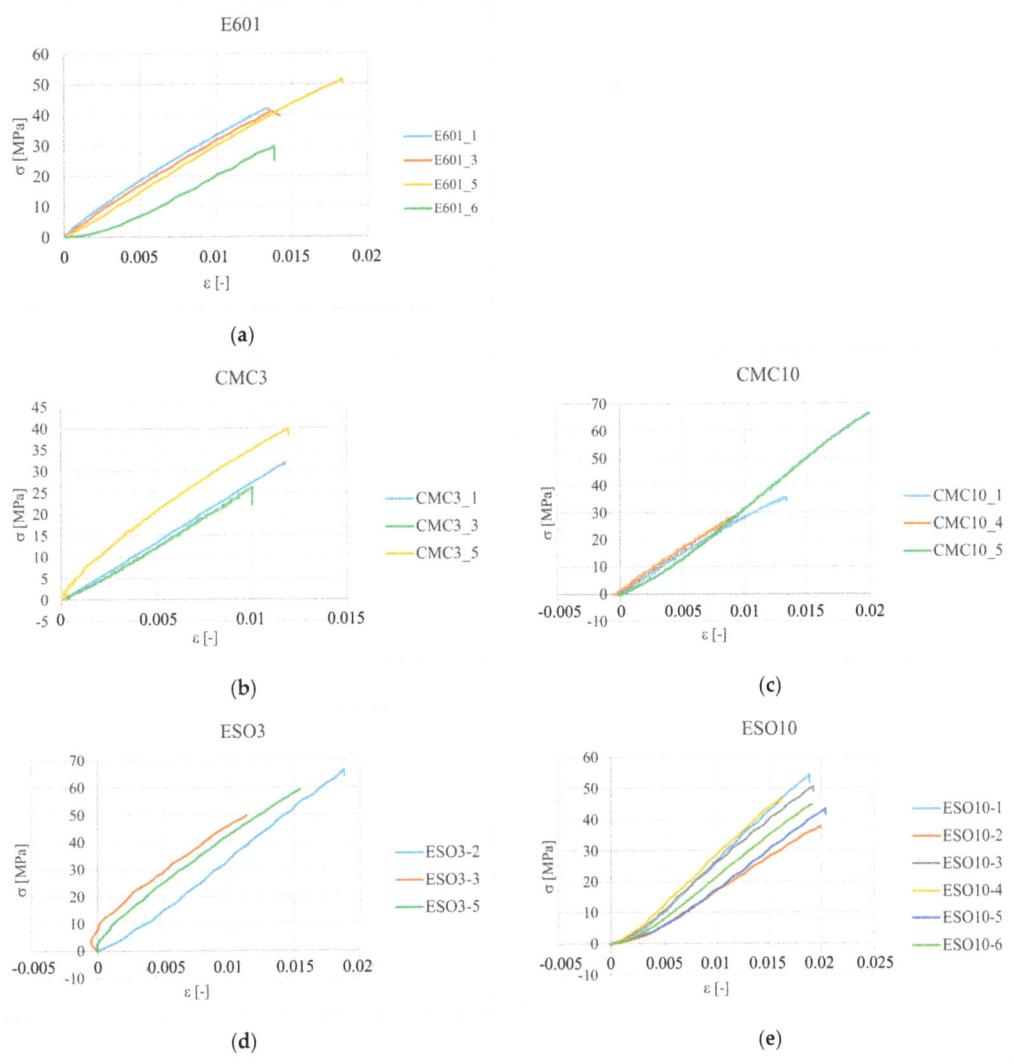

Figure 6. *Cont.*

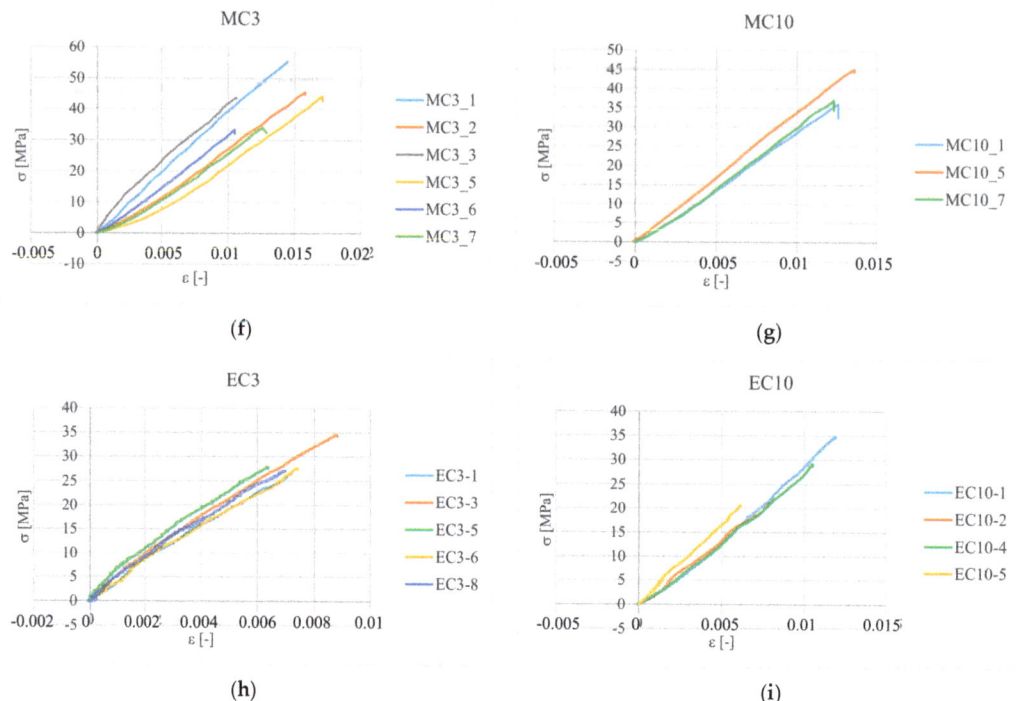

Figure 6. Sets of tensile stress–strain curve for (**a**) epoxy resin without fillers, (**b**) CMC3, (**c**) CMC10, (**d**) ESO3, (**e**) ESO10, (**f**) MC3, (**g**) MC10, (**h**) EC3 and (**i**) EC10.

4. Results and Discussion

The tensile test results were developed for each specimen separately. The tensile Young's modulus E, the ratio of the transverse strain to longitudinal strain in the uniaxial stress state (Poisson's ratio ν) and destructive stresses and strains, σ_f and ε_f, respectively, were determined for each group of specimens after rejecting the extreme results in the group. Young's modulus E was determined on the basis of the slope of the diagram of stresses as a function of strains. Poisson's ratio is a measure of deformation and has been defined as the slope angle of the transverse strain curve versus the longitudinal strain. The values of failure stresses and strains were determined as extreme stress and the corresponding strain of the stress function—longitudinal strain. The average values of Young's modulus, Poisson's ratio, stresses and failure strains together with the number of specimens are summarized in Table 3. The standard deviation (SD) values for each quantity are also included in Table 3.

The standard deviation, which is a measure of the width of the value scattering from the mean value, was determined using Excel according to

$$\sqrt{\frac{\sum(x-\bar{x})}{(n-1)}}, \qquad (1)$$

where $\bar{x}$ is the sample mean, and n is the sample size.

Figure 7 shows the average values of Young's modulus, with error bars showing the standard deviation values as a function of the specimen type.

Table 3. Average values of Young's modulus, Poisson's ratio, failure stresses and strains, and the number of specimens.

Name	E (MPa)	SD E	ν (-)	SD ν	σ_f (MPa)	SD σ_f	ε_f (MPa)	SD ε_f	No. of Specimens
E601	2991.5	296.72	0.42	0.2281	40.94	10.83	0.0152	0.0027	4
CMC3	2926.4	441.52	0.42	0.1457	32.49	20.17	0.0112	0.0054	3
CMC10	3091.7	378.46	0.33	0.1909	43.48	6.60	0.0141	0.0010	3
ESO3	3761.3	147.99	0.37	0.0635	58.50	8.36	0.0152	0.0038	3
ESO10	2692.6	415.15	0.37	0.0988	46.32	5.79	0.0189	0.0015	6
MC3	3297.3	499.24	0.39	0.1832	42.69	8.19	0.0135	0.0027	6
MC10	3152.1	201.14	0.2	0.0794	39.33	5.03	0.0128	0.0007	3
EC3	3845.5	284.94	0.35	0.22	28.81	3.25	0.0074	0.0009	5
EC10	2969.6	257.92	0.82	0.26	26.37	6.92	0.0091	0.0026	4

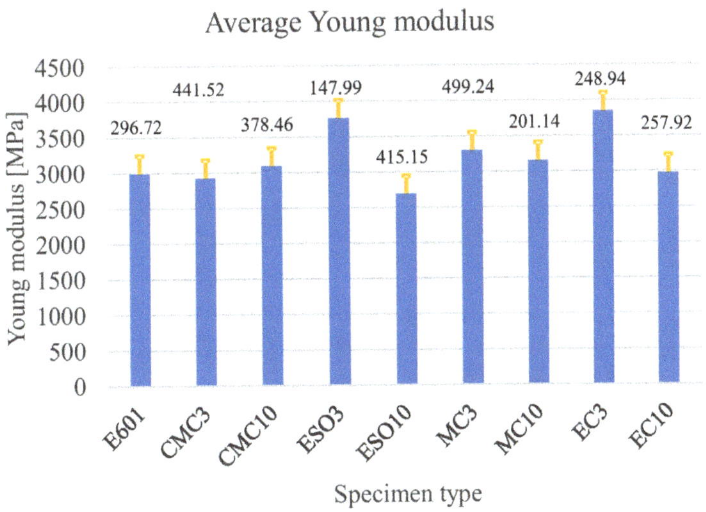

Figure 7. Average values of Young's modulus with standard deviation bars as a function of the specimen type.

Figure 8 places the average values of Poisson's ratio as well as standard deviation bars as a function of the specimen type. Poisson's ratio was determined as the angle of the slope of the transverse strain curve as a function of the longitudinal strain. Calculations were performed for each specimen tested using Excel. The value given in Table 3 is the average obtained after rejecting extreme values.

The obtained values of failure stresses (σ_f) (σ_f) were averaged for each test and presented in the form of diagrams (Figure 9) with standard deviation bars. The same method was used to visualize the longitudinal failure strains (ε_f) (ε_f) and their standard deviation, which are presented in Figure 10.

An example of a damaged specimen is shown in Figure 11a. A set of damaged specimens in the uniaxial tensile test specimens made of E601 epoxy resin with the addition of ESO3 is shown in Figure 11b. On the other hand, Figure 11c shows a set of damaged specimens made of E601 with the addition of MC10.

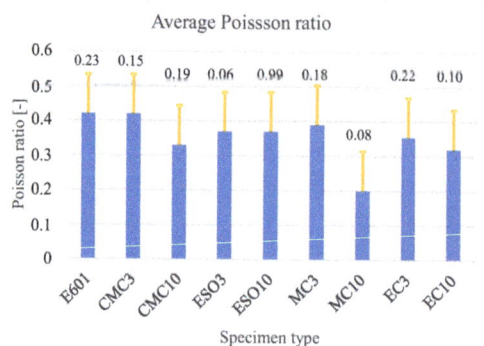

Figure 8. Average values of Poisson's ratio with standard deviation bars as a function of the specimen type.

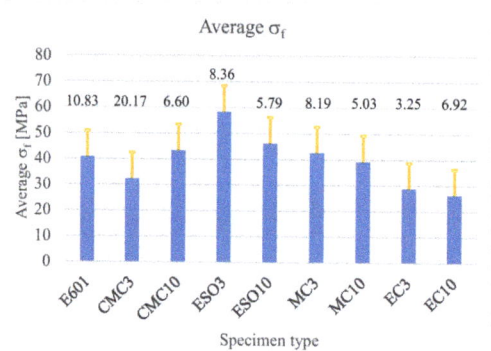

Figure 9. Average values of failure stress with standard deviation bars as a function of the specimen type.

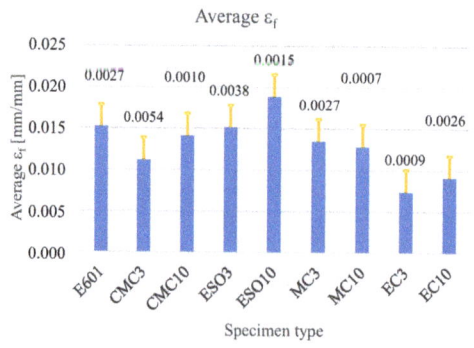

Figure 10. Average values of failure strain with standard deviation bars as a function of the specimen type.

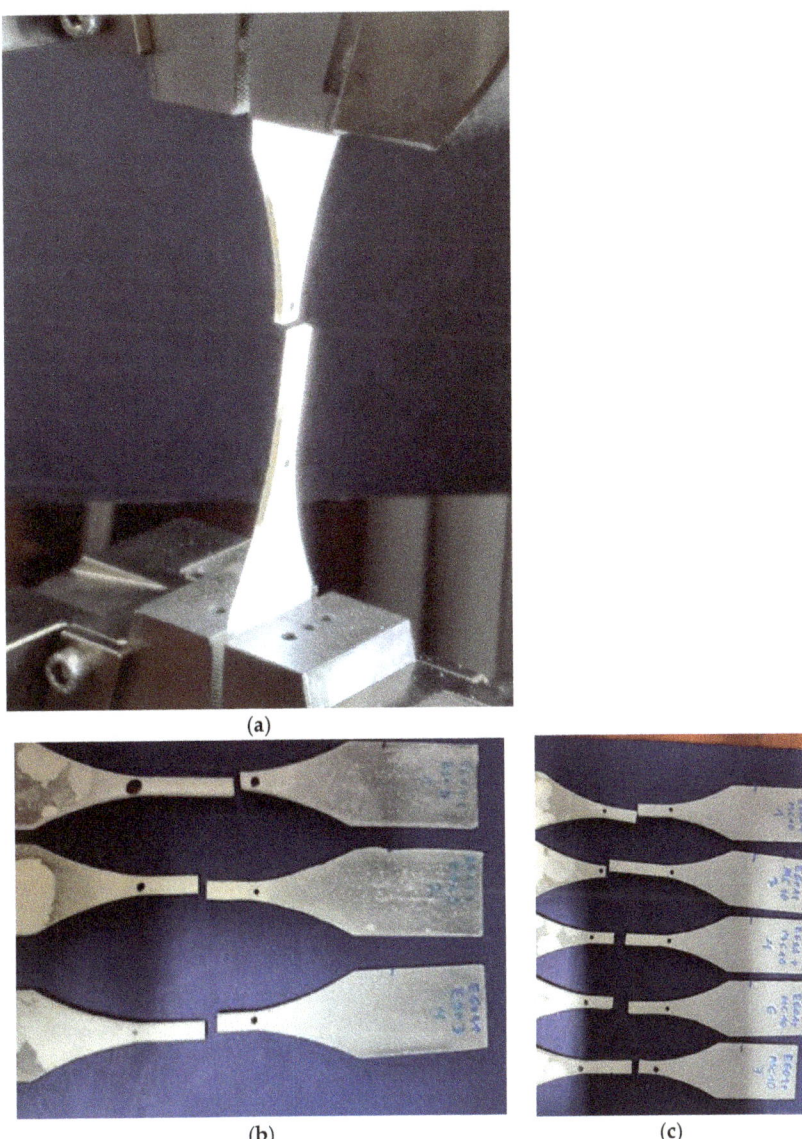

Figure 11. (**a**) An example of a damaged specimen. Set of damaged specimens in the uniaxial tensile test specimens made of (**b**) E601 + ESO3 and (**c**) E601 + MC10.

To determine the optimal content of the organic additive, the average values of Young's modulus, Poisson's ratio, failure stress and failure strain were determined for the E601 resin as reference values (E_{ref}, ν_{ref}, σ_{fref} and ε_{fref}). The average values of the same quantities obtained for specimens with additives were related to these values. The results in the form of ratios are shown as graphs in Figure 12.

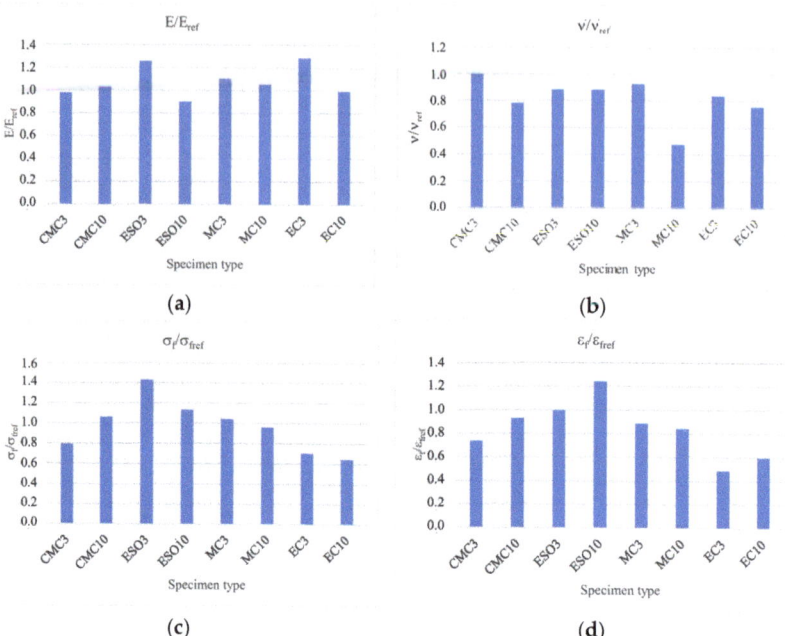

Figure 12. Ratios of (**a**) E/E_{ref}, (**b**) ν/ν_{ref}, (**c**) σ_f/σ_{fref} and (**d**) $\varepsilon_f/\varepsilon_{fref}$.

5. Conclusions

1. We do not recommend ethyl cellulose (EC) as an additive to the epoxy resin due to its poor miscibility. The best miscibility with epoxy resin and the best-looking test specimens were observed when using carboxymethyl cellulose (CMC) additives.
2. Both maximum values of strains and stresses in the stress–strain curves, higher than for epoxy resin, were observed for epoxy resins modified with epoxidized soya oil. The highest stresses were detected for epoxy E601 resin modified with 3 phr soya oil (ESO3), whereas the highest strains occurred for epoxy modified with 10 phr soya oil (ESO10) specimens.
3. The smallest standard deviation values of Young's modulus, Poisson's ratio and stress and strain values were observed for epoxy resins modified with 3 phr added epoxidized soya oil (ESO3).
4. Despite the imperfections of the prepared specimens and the small population of the tests carried out, the obtained results seem interesting and indicate the desirability of further extended research in biodegradation testing. We propose continuing research with epoxy resin modified with 3 phr epoxidized soya oil (ESO3) and 10 phr of methyl cellulose (MC10).

Author Contributions: Conceptualization, A.D.; methodology, A.D.; software, A.D.; validation, A.D.; formal analysis, A.D.; investigation, A.D., W.F. and J.S.; resources, A.D.; data curation, A.D.; writing—original draft preparation, A.D.; writing—review and editing, A.D. and W.F.; visualization, A.D.; supervision, A.D.; project administration, A.D.; funding acquisition, A.D. All authors have read and agreed to the published version of the manuscript.

Funding: This research study and APC were funded by Military University of Technology, grant number UGB 22-768.

Institutional Review Board Statement: Not applicable.

Informed Consent Statement: Not applicable.

Data Availability Statement: Not applicable.

Conflicts of Interest: The authors declare no conflict of interest.

References

1. Kramer, C.A.; Loloee, R.; Wichman, I.S.; Ghosh, R.N. Time Resolved Measurements of Pyrolysis Products from Thermoplastic Poly-Methyl-Methacrylate (PMMA). In Proceedings of the ASME 2009 International Mechanical Engineering Congress and Exposition, Lake Buena Vista, FL, USA, 13–19 November 2009; pp. 99–105. [CrossRef]
2. Żyłowska, A. Bisfenol A (BPA)—Gdzie Występuje, Jak Go Unikać? 2019. Available online: https://www.poradnikzdrowie.pl/diety-i-zywienie/zdrowe-odzywianie/bisfenol-a-bpa-gdzie-wystepuje-jak-go-unikac-aa-uTuv-sCmq-E8pN.html (accessed on 17 April 2020).
3. Fouad, H.; Mourad, A.-H.I.; ALshammari, B.A.; Hassan, M.K.; Abdallah, M.Y.; Hashem, M. Fracture toughness, vibration modal analysis and viscoelastic behavior of Kevlar, glass, and carbon fiber/epoxy composites for dental-post applications. *J. Mech. Behav. Biomed. Mater.* **2020**, *101*, 1034–1056. [CrossRef]
4. Di Boon, Y.; Joshi, S.C. A review of methods for improving interlaminar interfaces and fracture toughness of laminated composites. *Mater. Today Commun.* **2020**, *22*, 100830. [CrossRef]
5. Fu, Y.; Zhong, W.-H. Cure kinetics behavior of a functionalized graphitic nanofiber modified epoxy resin. *Thermochim. Acta* **2011**, *516*, 58–63. [CrossRef]
6. Wang, J.; Ma, C.; Chen, G.; Dai, P. Interlaminar fracture toughness and conductivity of carbon fiber/epoxy resin composite laminate modified by carbon black-loaded polypropylene non-woven fabric interleaves. *Compos. Struct.* **2020**, *234*, 111649. [CrossRef]
7. Wang, J.; Pozegica, T.R.; Xub, Z.; Nigmatullina, R.; Harnimanc, R.L.; Eichhorna, S.J. Cellulose nanocrystal-polyetherimide hybrid nanofibrous interleaves for enhanced interlaminar fracture toughness of carbon fibre/epoxy composites. *Compos. Sci. Technol.* **2019**, *182*, 107744. [CrossRef]
8. Sugiman, S.; Salman, S.; Maryudi, M. Effects of volume fraction on water uptake and tensile properties on epoxy filled with inorganic fillers having different reactivity to water. *Mater. Today Commun.* **2020**, *24*, 101360. [CrossRef]
9. Uthaman, A.; Lal, H.M.; Li, C.; Xian, G.; Thomas, S. Mechanical and Water Uptake Properties of Epoxy Nanocomposites with Surfactant-Modified Functionalized Multiwalled Carbon Nanotubes. *Nanomaterials* **2021**, *11*, 1234. [CrossRef]
10. Wu, J.; Li, C.; Hailatihan, B.; Mi, L.; Baheti, Y.; Yan, Y. Effect of the Addition of Thermoplastic Resin and Composite on Mechanical and Thermal Properties of Epoxy Resin. *Polymers* **2022**, *14*, 1087. [CrossRef]
11. Bellini, C.; Di Cocco, V.; Iacoviello, F.; Sorrentino, L. Influence of structural characteristics on the interlaminar shear strength of CFRP/Al fibre metal laminates. *Procedia Struct. Integr.* **2019**, *18*, 373–378. [CrossRef]
12. Liu, T.; Zhang, M.; Guo, X.; Liu, C.; Xin, J.; Zhan, J. Mild chemical recycling of aerospace fiber/epoxy composite wastes and utilization of the decomposed resin. *Polym. Degrad. Stab.* **2017**, *139*, 20–27. [CrossRef]
13. Kalita, D.J.; Tarnavchyk, I.; Chisholm, B.J.; Webster, D.C. Novel bio-based epoxy resins from eugenol as an alternative to BPA epoxy and high throughput screening of the cured coatings. *Polymer* **2021**, *233*, 124191. [CrossRef]
14. Kumar, S.; Krishnan, S.; Samal, S.K.; Mohanty, S.; Nayak, S.K. Toughening of Petroleum Based (DGEBA) Epoxy Resins with Various Renewable Resources Based Flexible Chains for High Performance Applications: A Review. *Ind. Eng. Chem. Res.* **2018**, *57*, 2711–2726. [CrossRef]
15. Jin, F.-L.; Park, S.-J. Impact-strength improvement of epoxy resins reinforced with a biodegradable polymer. *Mater. Sci. Eng.* **2008**, *478*, 402–405. [CrossRef]
16. Kadam, A.; Pawar, M.; Yemul, O.; Thamke, V.; Kodam, K. Biodegradable biobased epoxy resin from karanja oil. *Polymer* **2015**, *72*, 82–92. [CrossRef]
17. Mahnken, R.; Dammann, C. A three-scale framework for fibre-reinforced-polymer curing Part I: Microscopic modeling and mesoscopic effective properties. *Int. J. Solids Struct.* **2016**, *100–101*, 341–355. [CrossRef]
18. Skrzyńska, E.; Matyja, M. Porównanie właściwości fizykochemicznych wybranych tłuszczy naturalnych oraz ich estrów metylowych. *Chemik* **2011**, *65*, 923–935.
19. Ferenc, Z.; Pikoń, K. Przegląd rodzajów i ilości odpadów tłuszczowych i olejowych w Polsce. *Arch. Waste Manag. Environ. Prot.* **2005**, *2*, 69–80.
20. Cortés-Triviño, E.; Valencia, C.; Delgado, M.A.; Franco, J.M. Rheology of epoxidized cellulose pulp gel-like dispersions in castor oil: Influence of epoxidation degree and the epoxide chemical structure. *Carbohydr. Polym.* **2018**, *199*, 563–571. [CrossRef]
21. Chen, X.; Chen, S.; Xu, Z.; Zhang, J.; Miao, M.; Zhang, D. Degradable and recyclable bio-based thermoset epoxy resins. *Green Chem.* **2020**, *22*, 4187–4198. [CrossRef]
22. Zhang, T.; Wu, H.; Wang, H.; Sun, A.; Kan, Z. Creation of fully degradable poly(lactic acid) composite by using biosourced poly(4-hydroxybutyrate) as bioderived toughening additives. *Express Polym. Lett.* **2022**, *16*, 996–1010. [CrossRef]
23. Bardelli, T.; Marano, C.; Vangosa, F.B. Influence of curing thermal history on cross-linking degree of a polydimethylsiloxane: Swelling and mechanical analyses. *Express Polym. Lett.* **2022**, *16*, 924–932. [CrossRef]
24. Tanks, J.; Arao, Y.; Kubouchi, M. Network-level analysis of damage in amine-crosslinked diglycidyl ether resins degraded by acid. *Express Polym. Lett.* **2022**, *16*, 488–499. [CrossRef]

25. Ferrari, F.; Esposito Corcione, C.; Striani, R.; Saitta, L.; Cicala, G.; Greco, A. Fully Recyclable Bio-Based Epoxy Formulations Using Epoxidized Precursors from Waste Flour: Thermal and Mechanical Characterization. *Polymers* **2021**, *13*, 2768. [CrossRef]
26. Papanicolaou, G.C.; Anastasiou, D.E. Development of environmentally friendly epoxy and composite adhesives and applications in single and mixed-modulus joints. *J. Adhes. Sci. Technol.* **2021**, *35*, 1138–1153. [CrossRef]
27. Matykiewicz, D.; Skórczewska, K. Characteristics and Application of Eugenol in the Production of Epoxy and Thermosetting Resin Composites: A review. *Materials* **2022**, *15*, 4824. [CrossRef]
28. TeKrony, D.M. Accelerated Aging Test: Principles and Procedures. *Seed Technol.* **2005**, *27*, 135–146.
29. Farbyjachtowe pl. Epidian®601—Żywica Epoksydowa o Niskiej Lepkości 1kg. Available online: https://www.farbyjachtowe.pl/epidian-601-zywica-epoksydowa-o-niskiej-lepkosci-1kg-p-1365.html (accessed on 20 February 2023).

Disclaimer/Publisher's Note: The statements, opinions and data contained in all publications are solely those of the individual author(s) and contributor(s) and not of MDPI and/or the editor(s). MDPI and/or the editor(s) disclaim responsibility for any injury to people or property resulting from any ideas, methods, instructions or products referred to in the content.

Article

Analysis of Selected Properties of Injection Moulded Sustainable Biocomposites from Poly(butylene succinate) and Wheat Bran

Emil Sasimowski [1], Łukasz Majewski [1,*] and Marta Grochowicz [2]

[1] Department of Technology and Polymer Processing, Faculty of Mechanical Engineering, Lublin University of Technology, 20-618 Lublin, Poland; e.sasimowski@pollub.pl

[2] Department of Polymer Chemistry, Institute of Chemical Sciences, Faculty of Chemistry, Maria Curie-Sklodowska University, 20-400 Lublin, Poland; mgrochowicz@umcs.pl

* Correspondence: l.majewski@pollub.pl

Citation: Sasimowski, E.; Majewski, Ł.; Grochowicz, M. Analysis of Selected Properties of Injection Moulded Sustainable Biocomposites from Poly(butylene succinate) and Wheat Bran. *Materials* 2021, 14, 7049. https://doi.org/10.3390/ma14227049

Academic Editor: Krzysztof Moraczewski

Received: 25 October 2021
Accepted: 18 November 2021
Published: 20 November 2021

Publisher's Note: MDPI stays neutral with regard to jurisdictional claims in published maps and institutional affiliations.

Copyright: © 2021 by the authors. Licensee MDPI, Basel, Switzerland. This article is an open access article distributed under the terms and conditions of the Creative Commons Attribution (CC BY) license (https://creativecommons.org/licenses/by/4.0/).

Abstract: The paper presents a procedure of the manufacturing and complex analysis of the properties of injection mouldings made of polymeric composites based on the poly(butylene succinate) (PBS) matrix with the addition of a natural filler in the form of wheat bran (WB). The scope of the research included measurements of processing shrinkage and density, analysis of the chemical structure, measurements of the thermal and thermo-mechanical properties (Differential Scanning Calorimetry (DSC) and Thermogravimetric Analysis (TG), Heat Deflection Temperature (HDT), and Vicat Softening Temperature (VST)), and measurements of the mechanical properties (hardness, impact strength, and static tensile test). The measurements were performed using design of experiment (DOE) methods, which made it possible to determine the investigated relationships in the form of polynomials and response surfaces. The mass content of the filler and the extruder screw speed during the production of the biocomposite granulate, which was used for the injection moulding of the test samples, constituted the variable factors adopted in the DOE. The study showed significant differences in the processing, thermal, and mechanical properties studied for individual systems of the DOE.

Keywords: composite; injection moulding; biofiller; bioplastic; thermal properties; thermo-mechanical properties; mechanical properties; agro-waste materials; agro-flour filler

1. Introduction

Over the past several years, environmental issues have been increasingly raised, prompted by alarming reports of the environmental pollution caused by excessive use of petrochemical plastics [1–6]. One of the rapidly developing ways of prevention of the increasing pollution is the development and widespread use of biocomposites with natural fillers. In particular, the biocomposites are based on biodegradable or compostable polymers that are derived from natural sources or synthesized from substrates of natural origin [7,8]. Numerous complex compositions of multiple biodegradable polymers in various ratios are also used for that purpose [9–12]. Examples of such polymeric materials used to produce biocomposites include polylactide [13], polyvinyl alcohol [14], poly(hydroxyalkanoates) [15], polycaprolactone [16], and one of the more interesting—poly(butylene succinate) (PBS) [17]. PBS has very good functional properties that allow it to be widely used even in specific applications [18,19]. It is also characterized by very good mechanical and processing properties, which could classify this polymer as a structural material of common use [20–22]. However, PBS, like most of biodegradable polymers, has one significant disadvantage—a manifold higher price compared to traditional polyolefins of petrochemical origin, such as polypropylene or polyethylene [19,23,24]. This reduces its industrial popularity by excluding it from common use and marginalizing it to industries with high production costs [25,26]. The reason for the high costs consists mainly in the complex process of preparation and the high price of substrates but also the necessity of their drying, storage, and transportation in special conditions [27–31]. Therefore, the area of our current interest includes PBS-based polymer biocomposites with the addition of low-cost natural fillers, whose addition facilitates the possibility to reach the price competitiveness level, but often also provides a unique set of properties [21,32,33].

The literature includes numerous papers dealing with the manufacturing and properties of biocomposites on a PBS matrix with the addition of various natural fillers. Examples of such fillers are shredded wood shavings [34]; ground bran of cereals (wheat [35] and rice [36]); nut shells (pistachios [37], peanuts [38], and coconut [39]); and seeds (almonds [40]) but also dried pomace (apple [41] and grape [42]) or even wine lees [43]. The composition of all natural fillers of plant origin is based mainly on cellulose, hemicellulose, and lignin, but they differ in structure as well as in the proportion of their main components and the content of additional substances, such as simple and complex sugars, proteins, fats, and water [44–46]. Due to these differences, each lignocellulosic filler (LCF) will modify the properties of the polymer biocomposite in its own individual way. However, it can be generally assumed that the addition of the LCF positively affects the degradation rate and improves the stiffness of the composites [42,47,48] but also reduces the density and wear of the processing machine components compared to the mineral fillers [49–51]. However, the use of powdered byproducts of natural origin as fillers has some disadvantages and entails technological issues. Firstly, there is a decrease in the processability of composites due to the content of a significant amount of moisture and the increase of their viscosity and resistance during processing [52–55]. It is associated with an increased force on the drive system of the processing machine, a decrease in process efficiency and a risk of pore formation and hydrolysis during processing [52,53,56,57]. Secondly, the presence of LCF reduces the thermal resistance of the composites due to the low thermal decomposition temperatures of their structural components, which can be as high as approx. 150 °C. Therefore, PBS is suitable for the production of biocomposites with natural fillers because it has a low melting point (about 115 °C) [17,58–61]. Thirdly, the mechanical strength of LCF biocomposites is usually inversely proportional to the filler content [4,21,50,62]. The decrease in strength is usually related to the strength of the interfacial interactions at the polymer matrix/filler boundary. Due to their chemical structure, LCFs are hydrophilic in nature, whereas long polymer chains are hydrophobic or moderately hydrophilic due to the presence of local functional groups capable of forming hydrogen bonds [63,64]. Many authors indicate a significant decrease in the tensile strength of PBS biocomposites with the addition of powdered natural fillers. The decrease in strength is often higher the greater the filler content and can reach values of up to 50% [18,21,36,38,40–42]. The reduction in strength of biocomposites relative to unfilled polymeric materials is therefore an inherent aspect of the use of natural fillers. PBS, on the other hand, is hydrophilic in nature, and its water wetting angle is 70° [65] so that the level of interaction of PBS with the filler remains at a satisfactory level. This, combined with the good strength of neat PBS, makes it possible to efficiently produce biocomposites even with a high filling degree while maintaining satisfactory values of mechanical resistance parameters [66]. Nevertheless, many authors decided to use a compatibilizer during the manufacturing of the biocomposites, based on a PBS matrix with an addition of natural fillers. Due to the low popularity of PBS, there are no commercially available compatibilizers based on PBS, as is the case for polyethylene, where polyethylene grafted with maleic anhydride is widely available. In the case of PBS, maleinized, or epoxidized vegetable oils [38,40,67] or coupling agents based on, e.g., silanes [48,68,69] are used as compatibilizers in scientific research. Modified vegetable oils are most often ineffective and even cause a reduction in tensile strength, stiffness, and impact strength with respect to the PBS with filler but without oils [38,40,67]. Organosilane-based compatibilizers can improve the mechanical properties relative to biocomposites without a compatibilizer, but their cost is significant, and they exhibit significantly higher efficiencies over fibrous fillers than powdered ones [48,69]. Some authors choose to produce maleic anhydride-grafted PBS under laboratory conditions, such as by extrusion of reactive PBS and maleic anhydride in the presence of dicumyl peroxide. Despite the high efficiency of this compatibilizer, its use drastically increases the cost of manufacturing PBS matrix composites with natural filler [70–72]. The main purpose of using natural-waste fillers, which are most often technological waste from food or agricultural industry, is to reduce the cost of expensive polymeric materials such as PBS [49,73]. Therefore, it should be noted that the production of a compatibilizer or the use of a commercially available one significantly increases the cost of manufacturing the whole biocomposite, and the obtained strengthening effects are moderate or unsatisfactory. Thus, the use of compatibilizers during the manufacturing of PBS-based biocomposites is technically as well as economically unjustified [25,32,74].

The scientific literature abounds in papers dealing with the subject of biocomposites made of a PBS matrix with the addition of various fillers of natural origin, including fillers made of agricultural and food industry wastes. Despite the above, there is a shortage of works describing in detail the manufacturing process and characterizing the properties of compositions made of poly(butylene succinate) with the addition of ground wheat bran, which is a technological waste in the production of white flour. In the following work, an extensive and detailed analysis of selected properties of PBS injection mouldings filled with crushed wheat bran was carried out. The aim of this study was to evaluate the influence of wheat bran content and extruder screw speed during the extrusion of biocomposite pellets on the properties of the injection-moulded parts produced from them. The characteristics of the changes in the processing and the physical, structural, thermal, thermo-mechanical, and mechanical properties were determined as functions of variable factors; this was followed by an extensive analysis of the obtained results.

2. Experimental

2.1. Test Stand

Injection moulding of the biocomposite was carried out using an Arburg Allrounder 320C (Arburg, Lossburg, Germany) screw injection moulding machine equipped with a dual cavity mould to produce specimens for strength testing. The shape and dimensions of the samples were in accordance with ISO 294-1:2017-07 [75]. The specimens were dog-bone-shaped with a total length of 150 mm and a thickness of 4 mm; the width of the measuring part was 10 mm, and the grip part was 20 mm. Due to the danger of the thermal decomposition of the biocomposite components, low temperatures were applied during processing. The temperature of the plasticizing system was 30 °C in the feed zone, and in the individual heating zones it was: I–125 °C, II–145 °C, III–155 °C, and IV–160 °C; the injection nozzle temperature was 155 °C. The temperature of the thermostated mould was 25 °C. The injection of the biocomposition was performed at the following settings: maximum injection pressure 120 MPa, polymer flow rate 20 cm^3/s, packing pressure 110–80 MPa, packing time 15 s, and cooling time 20 s. In the case of the highest bran fraction of 50% (DOE layout 8), the injection pressure was increased by 130 MPa and the packing pressure to 120–80 MPa, which made it possible to eliminate the incomplete filling of mould cavities occurring at the lower values of these parameters.

2.2. Materials

The components of the studied biocomposition are: PBS constituting its matrix and filler in the form of wheat bran. A PBS designed for general-purpose injection moulding, trade name BioPBS FZ91 PB [76], was used to produce the biocomposition samples. This material is produced from bio-based succinic acid and 1,4-butanediol by PTT MCC BIOCHEM CO., Ltd., Bangkok, Thailand. The wheat grain husks, or wheat bran (WB), used in the biocomposition came from a local mill near Lublin (Poland). They are a waste product from the refining of white flour. The bran takes the form of thin flakes several millimetres thick, composed of fibrous substances, such as cellulose, lignin, and hemicellulose.

They also contain phytic acid, oligosaccharides, and non-starch polysaccharides, as well as fats and proteins, in their composition [77,78].

2.3. Research Programme and Methodology

Experimental tests were carried out according to the adopted DOE: central, composite, rotatable with star point distance = 1.414. The following independent variables—adjustable conditions of the process—were assumed: mass content of wheat bran introduced into poly(butylene succinate) $u = 10 \div 50\%$wt and extruder screw speed $n = 50 \div 200$ min^{-1} when obtaining the processed biocomposite pellets. A detailed analysis of the extrusion process and properties of the compositions obtained were presented in a previous paper [66]. The experimental design and test results obtained—the mean values of the dependent variables studied—are presented in Tables 1 and 2. Measurements were made in at least five replicates.

Table 1. Experimental design and experimental test results—mean values and standard deviation—part I.

Experimental Design Layout	n, min^{-1}	u, %	S_L, %	S_T, %	S_P, %	ρ, g/cm^3	HDT, °C	VST, °C
1	72	15.9	1.07 ± 0.01	1.83 ± 0.05	2.11 ± 0.02	1.3064 ± 0.0010	91.1 ± 0.8	109.1 ± 0.3
2	72	44.1	0.61 ± 0.01	1.37 ± 0.06	1.11 ± 0.01	1.3652 ± 0.0010	93.7 ± 0.4	108.4 ± 0.2
3	178	15.9	1.07 ± 0.01	1.88 ± 0.06	2.12 ± 0.02	1.3054 ± 0.0007	91.5 ± 0.8	109.3 ± 0.3
4	178	44.1	0.61 ± 0.01	1.37 ± 0.06	1.03 ± 0.03	1.3668 ± 0.0012	93.6 ± 0.5	108.3 ± 0.5
5	50	30.0	0.91 ± 0.01	1.66 ± 0.03	1.99 ± 0.02	1.3352 ± 0.0008	92.0 ± 0.5	108.7 ± 0.4
6	200	30.0	0.89 ± 0.01	1.58 ± 0.06	1.90 ± 0.03	1.3361 ± 0.0008	92.4 ± 0.4	108.9 ± 0.4
7	125	10.0	1.23 ± 0.01	2.03 ± 0.05	2.13 ± 0.02	1.2938 ± 0.0005	91.2 ± 0.2	109.7 ± 0.3
8	125	50.0	0.57 ± 0.01	1.11 ± 0.04	0.93 ± 0.02	1.3798 ± 0.0014	93.6 ± 0.5	108.5 ± 0.3
9 (C)	125	30.0	0.90 ± 0.01	1.58 ± 0.06	1.94 ± 0.02	1.3349 ± 0.0010	92.3 ± 0.4	108.8 ± 0.3

Table 2. Experimental design and experimental test results—mean values and standard deviation—part II.

Experimental Design Layout	n, min^{-1}	u, %	H, MPa	Impact Strength, kJ/m^2	σ, MPa	E, MPa	ε, %
1	72	15.9	55.9 ± 1.09	35.17 ± 1.53	30.20 ± 0.20	953 ± 5	22.1 ± 2.0
2	72	44.1	63.0 ± 0.47	11.70 ± 0.24	16.18 ± 0.30	1554 ± 27	9.1 ± 1.4
3	178	15.9	55.6 ± 1.19	38.96 ± 3.44	29.90 ± 0.27	929 ± 8	18.1 ± 1.7
4	178	44.1	62.2 ± 0.92	11.45 ± 0.34	16.58 ± 0.11	1530 ± 7	9.3 ± 1.1
5	50	30.0	59.8 ± 1.40	20.91 ± 1.43	23.00 ± 0.23	1208 ± 13	18.0 ± 3.4
6	200	30.0	59.7 ± 1.19	21.00 ± 2.09	22.34 ± 0.15	1206 ± 5	21.8 ± 1.9
7	125	10.0	53.3 ± 1.16	49.49 ± 1.48	31.96 ± 0.05	830 ± 3	26.6 ± 7.4
8	125	50.0	63.6 ± 0.47	8.88 ± 0.76	13.54 ± 0.13	1584 ± 9	8.1 ± 0.9
9 (C)	125	30.0	59.8 ± 1.40	19.44 ± 1.37	21.54 ± 0.11	1172 ± 8	16.9 ± 1.5

The following dependent (observed) variables were adopted in the experimental study: processing longitudinal S_L, transverse S_T and perpendicular S_P shrinkage [%], density ρ [g/cm^3], heat deflection temperature HDT [°C], Vicat softening temperature VST [°C], hardness H [MPa], impact strength [kJ/m^2], tensile strength σ [MPa], Young's modulus E [MPa], and elongation at break ε [%]. The measurements carried out according to the adopted design of the experiment made it possible to approximate the relationship between the mentioned dependent and independent variables by means of a polynomial value of many variables consisting of the following members: constant value, linear terms, quadratic terms, and a two-factor interaction term (Equation (1)) [66], where Y is the predicted response value (Y stands for H, S_L, S_T, S_P, HDT, VST, ρ, impact strength, σ, E, and ε), a_0 is a constant value, and a_x are the regression coefficients.

$$Y(n \cdot u) = a_0 + a_1 n + a_2 u + a_3 n^2 + a_4 u^2 + a_{12} nu \qquad (1)$$

Experimental tests of the injection mouldings made of the polymer compositions were carried out:

- Measurement of longitudinal shrinkage S_L and transverse shrinkage S_T and perpendicular shrinkage S_P of the samples with the use of a caliper as per ISO 294-4:2005 [79]. The measurement was made with an accuracy of 0.01 mm.
- Standard density was measured according to ISO 1183-1 A [80] using the immersion method. The mass of the samples in air and in water was measured. In order to obtain full soaking of the sample, the processed products were kept immersed in water for 24 h and then measured.
- FTIR analysis was performed using a TENSOR 27 FTIR spectrophotometer (Bruker, Billerica, MA, USA), with ATR (Attenuated Total Reflectance). The measurement was performed with a diamond crystal, recording 16 scans per spectrum in the range of 600–4000 cm^{-1} with a resolution of 4 cm^{-1}.
- Differential scanning calorimetry (DSC) measurements of biocomposite injection mouldings were performed according to ISO 11357-1:2016 [81] using a NETZSCH (Günzbung, Germany) model 204 F1 Phoenix DSC scanning calorimeter. Processing of the test data was carried out using the NETZSCH Proteus software. Measurements were made under the following conditions: heating cycle (I) with a heating rate of 10 K/min in the temperature range of −150 °C–140 °C; cooling cycle at a rate of 10 K/min within a temperature range of 140 °C–150 °C; heating cycle (II) at a rate of 10 K/min within the temperature range of −150 °C–140 °C; mass of measuring samples about 10 mg; and aluminium crucibles with pierced lids. On the basis of DSC curves obtained, the following findings were determined: crystallinity degree X_c, melting enthalpy ΔH_m, melting temperature T_m, crystallization temperature T_c, and glass transition temperature T_g of the investigated biocomposite samples. The adopted inflection point of the DSC curve in the glass transition region corresponded to the glass transition temperature. While determining the degree of crystallinity, the relation (Equation (2)):

$$X_c = \left(\frac{\Delta H}{(1-u) \times \Delta H_{100\%}} \right) \times 100\% \quad (2)$$

The adopted $\Delta H_{100\%}$ value for PBS in the calculation = 1103 J/g [82].

- Thermogravimetric (TG) measurements of the injection mouldings were carried out using a Jupiter STA 449 F1 thermal analyser (NETZSCH, Günzbung, Germany) in an oxidizing atmosphere. The gaseous products of the sample decomposition were analysed using an attached TENSOR 27 FTIR spectrophotometer from Bruker, (Germany). The measurements were carried out under the following conditions: temperature 40–800 °C, synthetic air flow rate 25 mL/min, sample mass about 12 mg, and measuring crucibles made of Al_2O_3.
- The heat deflection temperature (HDT) tests were performed using a Ceast HV3 apparatus manufactured by Instron (Turin, Italy) as per ISO 75-2:2013 [83]. Flat specimen alignment, B-measurement method (flexural stress 0.45 MPa), and a heating rate of 120 °C/h were used.
- Vicat softening temperature (VST) values were also determined on the Ceast HV3 apparatus by Instron (Turin, Italy) as per ISO 306:2013 [84]. The A120 measurement method was applied—10 N force and 120 °C/h heating rate.
- The hardness was measured employing a ball indentation method with the use of an HPK 8411 hardness tester with a ball-shaped indenter of 5 ± 0.025 mm diameter. Measurements were made in accordance with ISO 2039-1:2004 [85]
- Unnotched Charpy impact tests were carried out in accordance with ISO 179-2:2020 [86] on a Type 639F impact hammer by Cometech Testing Machines (Taizhong, Taiwan). The pendulum used had a maximum energy of 5093 J. The samples for impact tests were made by cutting the measuring part out of injection-moulded, dog-bone-shaped samples, obtaining rectangular samples with dimensions of 80 × 10 × 4 mm.

- The strength properties, such as tensile strength σ [MPa], elongation at break ε [%], and Young's modulus [MPa] were determined based on ISO 527-2 [87]. The tensile speed during the measurements was 50 mm/min. The measurements employed a Zwick Roell (Ulm, Germany) model Z010 testing machine.

3. Results

The collected results of the experimental investigations on the properties of the injection mouldings of poly(butylene succinate) (PBS) biocomposition filled with wheat bran are presented in Tables 1 and 2. The collected experimental results were used to determine empirical models describing the influence of adjustable process conditions (independent variables) on the examined properties of biocomposition (dependent variables). The models were adjusted using the backward stepwise regression method. The applied Pareto chart of standardized effects allowed us to illustrate the influence of the members of the regression equations on the studied quantity (dependent variable). Statistically significant are the members for which the absolute values of the standardized effects exceed the vertical line corresponding to the assumed significance level $p = 0.05$.

3.1. Physical and Structural Properties

3.1.1. Processing Shrinkage

Determined empirical models of the processing of longitudinal S_L, transverse S_T, and perpendicular S_P shrinkage were presented by means of polynomials (Equations (3)–(5)):

$$S_L = 1.362225 - 0.016295u \tag{3}$$

$$S_T = 2.205955 - 0.020187u \tag{4}$$

$$S_P = 1.915118 + 0.031844u - 0.001090u^2 \tag{5}$$

The results of the statistical analyses of the adopted processing shrinkage models are presented in Tables 3–5. It was observed that the wheat bran content u has a significant effect on the types of processing shrinkage S_L, S_T, and S_P, and that this relation is linear for the longitudinal and transverse shrinkage (Figures 1 and 2). For perpendicular shrinkage S_P, the quadratic term of the model with a negative effect is also statistically significant (Figure 3). Increasing the bran fraction results in a significant decrease in processing shrinkage (Figures 4–6). The greatest reductions in shrinkage values obtained by increasing the wheat bran content u in the composition from 10 to 50% (DOE layouts 7 and 8) were 0.66% for longitudinal shrinkage S_L (54% of the initial value), 0.92% for transverse shrinkage S_T (45% of the initial value), and 1.21% for perpendicular shrinkage S_P (57% of the initial value), respectively. No effect of the applied extruder screw speed n during the production of the polymeric composition and the occurrence of interactions between the variable factors studied on the processing shrinkage of the mouldings studied was observed.

In the case of mouldings made from the PBS alone, without the addition of bran, the processing shrinkage was higher and amounted to: longitudinal $S_L = 1.57 \pm 0.007\%$, transverse $S_T = 2.40 \pm 0.027\%$, and perpendicular $S_P = 2.22 \pm 0.014\%$.

The processing shrinkage of injection-moulded parts made of partially crystalline polymers depends on many factors, which include heat transfer during cooling, volume shrinkage due to thermal expansion, flow-induced residual stresses, orientation of macromolecules, and crystallization. The above factors are in turn dependent on the processing parameters and properties of the injected material [88]. All of the measurement series studied are characterized by partial crystallinity [66], by which the effect of shrinkage anisotropy was observed. It is recognized that for partially crystalline materials, the highest shrinkage values are observed in the flow direction due to the flow-induced orientation of the macromolecules [89], while in the analysed case the measured values of the longitudinal processing shrinkage were found to be smaller than the values of the transverse and perpendicular shrinkage. This is related to the elastic recovery effect generated by the amorphous phase [90], whose share is significant and varies in the range of 27–46%

depending on the filler share, as presented further in the description of the DSC studies. Moreover, the papers dealing with the shrinkage of semi-crystalline polymeric materials, including polyesters, demonstrated that when we take into account the additional factors affecting shrinkage, the processing shrinkage in this type of material occurs most intensively at the sample thickness [88,90]. This is consistent with the obtained shrinkage results for the tested PBS/WB biocomposites. The exceptions are the composites with the highest WB content (44% and 50%), for which the transverse shrinkage reached values higher than the perpendicular shrinkage. This is most likely to be due to a significant change in the material properties, such as viscosity and thermal conductivity, which leads to changes in cooling efficiency, the quality of pressure transmission in the flow system, and local flow rates. The value of shrinkage in the perpendicular direction occurring at the sample thickness is most susceptible to changes in the processing parameters during injection moulding [88–90].

The PBS/WB composites showed lower processing shrinkage values with respect to the unfilled PBS. This is due to the fact that the filler used is not subject to processing shrinkage. The effect of specific volume loss related to the crystallization effect during cooling is less pronounced with increasing filler content in the temperature and pressure ranges used during the injection moulding. This has been confirmed in previous work [66] via p-v-T tests.

Table 3. Model of longitudinal shrinkage S_L—ANOVA table, $R^2 = 0.98$; $R_{adj}^2 = 0.98$.

Source of Variation	SS	df	MS	F	p
u	2.123473	1	2.123473	2285.24	0.00000
Error	0.039956	43	0.000929		
Total SS	2.163429	44			

SS—sum of squares, df—number of the degrees of freedom, MS—mean sum of squares, F—values of the test statistic, p—value of probability corresponding to the test statistic value.

Table 4. Model of transverse shrinkage S_T—ANOVA table, $R^2 = 0.94$; $R_{adj}^2 = 0.94$.

Source of Variation	SS	df	MS	F	p
u	3.259054	1	3.259054	655.59	0.00000
Error	0.213760	43	0.004971		
Total SS	3.472814	44			

SS—sum of squares, df—number of the degrees of freedom, MS—mean sum of squares, F—values of the test statistic, p—value of probability corresponding to the test statistic value.

Table 5. Model of perpendicular shrinkage S_P—ANOVA table, $R^2 = 0.97$; $R_{adj}^2 = 0.96$.

Source of Variation	SS	df	MS	F	p
u	8.997936	1	8.997936	1063.73	0.00000
u^2	1.160503	1	1.160503	137.19	0.00000
Error	0.355273	42	0.008459		
Total SS	10.513711	44			

SS—sum of squares, df—number of the degrees of freedom, MS—mean sum of squares, F—values of the test statistic, p—value of probability corresponding to the test statistic value.

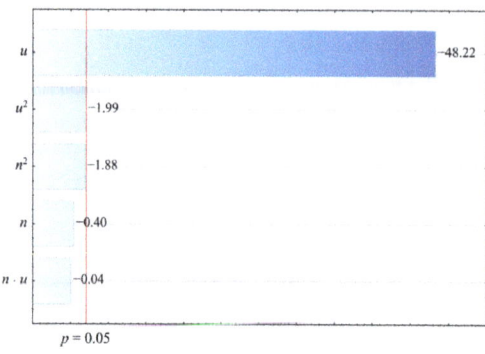

Figure 1. Pareto charts of the standardized effects for the empirical model of longitudinal shrinkage S_L; the vertical line in the plot corresponds to the arbitrarily chosen level of significance ($p = 0.05$).

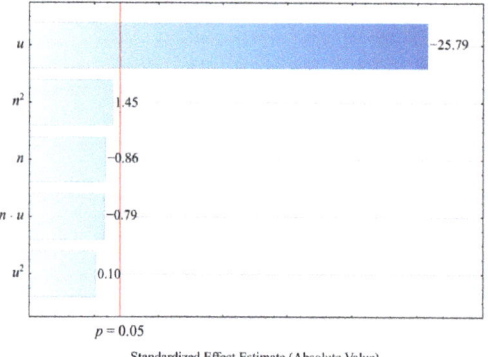

Figure 2. Pareto charts of the standardized effects for the empirical model of transverse shrinkage S_T; the vertical line in the plot corresponds to the arbitrarily chosen level of significance ($p = 0.05$).

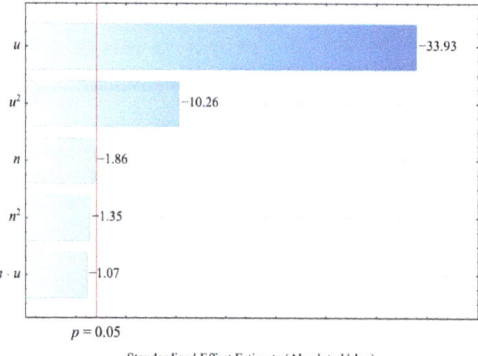

Figure 3. Pareto charts of the standardized effects for the empirical model of perpendicular shrinkage S_P; the vertical line in the plot corresponds to the arbitrarily chosen level of significance ($p = 0.05$).

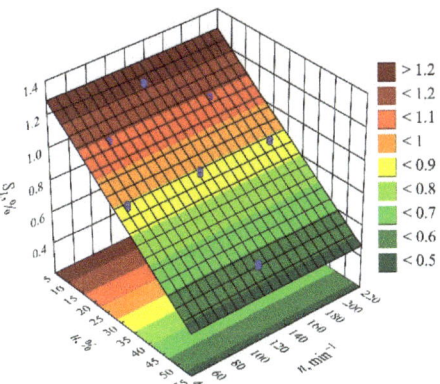

Figure 4. Response surface graph for the longitudinal shrinkage S_L versus wheat bran content u and screw speed n.

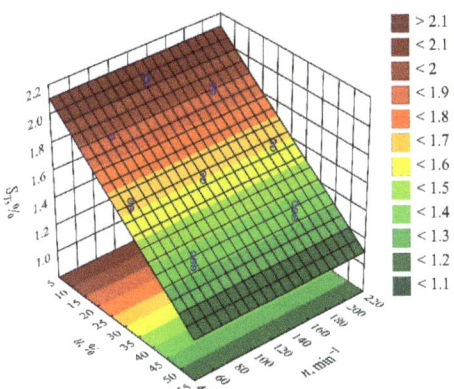

Figure 5. Response surface graph for the transverse shrinkage S_T versus wheat bran content u and screw speed n.

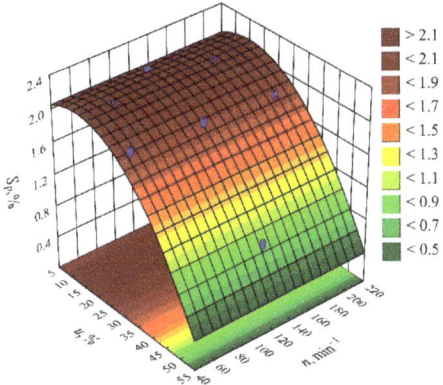

Figure 6. Response surface graph for the perpendicular shrinkage S_P versus wheat bran content u and screw speed n.

3.1.2. Density

The result of the performed modelling of the density ρ of the injection mouldings made of the compositions under study is an empirical model in the form of the polynomial (Equation (6)):

$$\rho = 1.274262 + 0.001905u + 0.000003u^2 + 0.0000002nu \qquad (6)$$

It has been observed that the wheat bran content u introduced into the composition has the strongest effect on the density of the obtained mouldings (linear and quadratic terms of the equation—Figure 7). Also statistically significant was the interaction between bran content u and extruder screw speed n during the production of the composites under study, but its influence is relatively very small. The results of the statistical analysis of the adopted model are presented in Table 6. The linear term in the model equation has the greatest influence on the density. Increasing the bran content, the density of which is $\rho = 1.5347 \pm 0.0084$ g/cm^3, causes an increase in the density of the mouldings obtained from the polymeric composition (Figure 8). The highest increase in the density of the mouldings, 0.0861 g/cm^3 (7%), was obtained by increasing the wheat bran content u in the composition from 10 to 50% (DOE layouts 7 and 8). The density of the samples made of PBS alone was significantly lower and amounted on average to $\rho = 1.275 \pm 0.0002$ g/cm^3, whereas the density of the filler was $\rho = 1.5347 \pm 0.0084$ g/cm^3. It should be noted that the density of the studied injection mouldings is in all cases clearly higher (from 0.0491 g/cm^3 to as much as 0.1927 g/cm^3) than the density of the pellets from which they were made. The relevant results of the density tests of the produced composition, in the form of pellets and microscopic pictures of its structure, were presented in a previous paper [66]. Reprocessing of the composition, this time by injection moulding, resulted in a decrease in the amount of pores present in the pellets as a result of the release of water vapour from the moisture contained in the bran and possibly partly from the products of the thermal decomposition of the composite components. During extrusion, there was a drastic reduction in pressure from 4–9 MPa to atmospheric pressure [66], which stimulated the formation of pores. During injection moulding, the material was pressed into the mould at 120 MPa and cooled at a gradient packing pressure of 120–80 MPa. This resulted in more packed material and prevented pore growth. The effect of pressure on the material packing was also confirmed in previous work with p-v-T T tests at 20 MPa and 110 MPa [66].

Table 6. Model of density ρ—ANOVA table, $R^2 = 0.99$; $R_{adj}^2 = 0.99$.

Source of Variation	SS	df	MS	F	p
u	0.036588	1	0.036588	37,269.04	0.000000
u^2	0.000011	1	0.000011	11.68	0.001441
nu	0.000008	1	0.000008	8.43	0.005907
Error	0.000040	41	0.000001		
Total SS	0.036648	44			

SS—sum of squares, df—number of the degrees of freedom, MS—mean sum of squares, F—values of the test statistic, p—value of probability corresponding to the test statistic value.

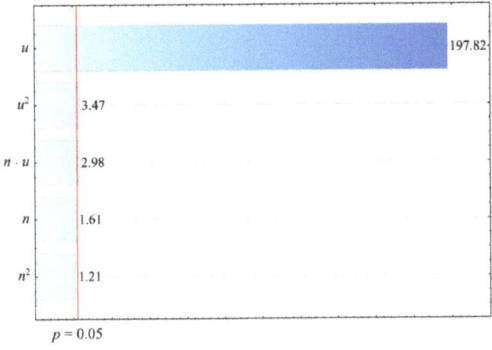

Figure 7. Pareto plots of the standardized effects of empirical full model density ρ, the vertical line in the plot corresponds to the arbitrarily chosen level of significance ($p = 0.05$).

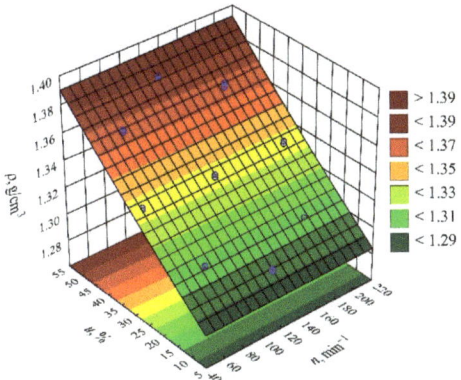

Figure 8. Response surface plot for the density ρ versus wheat bran content u and screw speed n.

3.1.3. Chemical Structure

The chemical structure of PBS and its biocomposites with wheat bran was confirmed using FTIR analysis. Characteristic absorption bands originating from the vibrations of C=O carbonyl groups (1714 cm^{-1}) and ester bonds -C-O-C and -O-(C=O) at 1262 cm^{-1}, 1173 cm^{-1}, and 1044 cm^{-1} are observed on PBS spectrum. The structure of wheat bran is mainly composed of polysaccharides (including cellulose, hemicellulose, and lignin), phenolic and lipid compounds, and proteins, as confirmed by the FTIR spectrum [66,91]. The spectra of PBS composites with bran (Figure 9) show absorption bands originating from functional groups that build both components. Figure 9 presents example FTIR spectra of the composites obtained at the same screw speed but with increasing bran content from 10% through 30% to 50%. An increase in the intensity of the absorption bands originating from the bran structure is observed along with an increase in the bran content in the composites. Particularly clear is the change in the maximum absorption band present on the PBS spectrum at 1173 cm^{-1} to 1156 cm^{-1} for composites 8 and 9, containing 50% and 30% bran, respectively. The shift of this band originating from the -C-O-C- vibration indicates the presence of a non-covalent interaction between PBS and the bran chemical components, which has also been previously reported [65,92]. It can be expected that this may be an interaction of the nature of hydrogen bonding, most likely between the -OH groups of the polysaccharides and lignin and the C=O groups in the polyester, but no shift in the absorption band of the carbonyl group was observed.

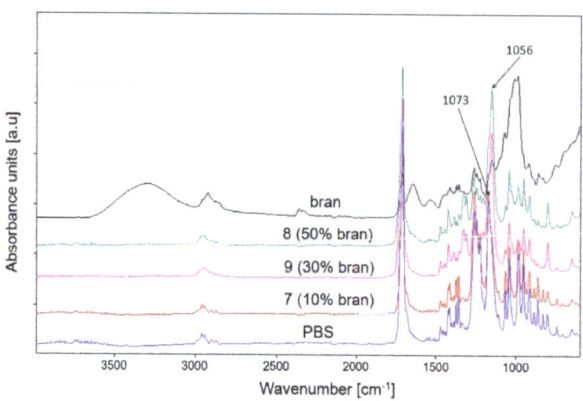

Figure 9. ATR-FTIR spectra of PBS, bran, and their composites 7, 8, 9 obtained with different bran content.

3.2. Thermal and Thermomechanical Properties

3.2.1. DSC

The DSC tests were performed in an inert gas atmosphere. The tests were performed in cycles: heating (I), cooling, and then heating (II). Table 7 presents the thermal parameters of the mouldings and their crystallinity degrees, determined on the basis of the DSC thermograms presented in Figure 10. The glass transition temperature (T_g) of PBS determined from the first heating cycle is about 5 °C higher than that of the composites. This difference almost disappears when T_g is read from the thermograms from the second heating cycle, and the secondary chain relaxation of PBS occurs at around −32 °C. The DSC thermograms show distinct endothermic peaks, which are due to the melting of the crystalline phase of the PBS present in the composites. The maximum of these endothermic transformations occurs at approx. 120 °C. However, the T_m of neat PBS is slightly higher than that of the composites, and in general, the T_m values determined from the first heating cycle are higher than those determined from the second heating cycle. Apart from the high PBS melting point, the DSC thermograms of the composites from the second heating cycle also show lower endothermic peaks with the maximum value at approx. 105 °C, while the curves from the first heating cycle show a broad peak extending from approx. 80 °C. The presence of this broad peak in the first heating cycle may be a result of the evaporation of a small amount of water absorbed within the structure of the composites, the presence of which was also confirmed by TG tests. The presence of a smaller endothermic peak on the thermograms from the second heating cycle has already been reported in our previous work [66], in which we presented DSC thermograms made for the pellets from which the injection mouldings under discussion were obtained. The course of the thermograms for the mouldings is different than that for the initial pellets for each composite, irrespective of the content of bran and screw speed; a clear first endothermic peak is present. If the mouldings with different bran content obtained with the same n, (1, 2 or 3, 4 or 7–9), are compared, a greater separation of endothermic peaks can be observed with the increasing bran content. As postulated earlier, the first endothermic peak is probably due to the melting of the less perfect PBS crystalline phase, which in this case was formed during the injection-moulding process. The values of the degree of crystallinity of the composite mouldings calculated from the first and second heating cycles are significantly different. Taking into account the fact that absorbed water is present in the structure of the composites, it can be concluded that the melting points of the crystalline phase and water desorption could overlap, which influenced the ΔH_m value, as well as the values of the crystallinity degree. Therefore, when considering the effect of the bran content on the degree of crystallinity of the composites, we refer to the values determined from the second heating cycle. In the case of the mouldings, a dependence of X_c on the content of bran is observed, which is analogous to the pellets. Neat PBS shows the highest crystallinity (60.5%) and increasing the content of bran in the composites decreases the X_c values (DOE layouts 1, 2 or 3, 4 or 7–9). In contrast to the previously described pellets, the screw speed at which the output pellets were obtained did not affect the degree of crystallinity of the mouldings obtained in the injection-moulding process.

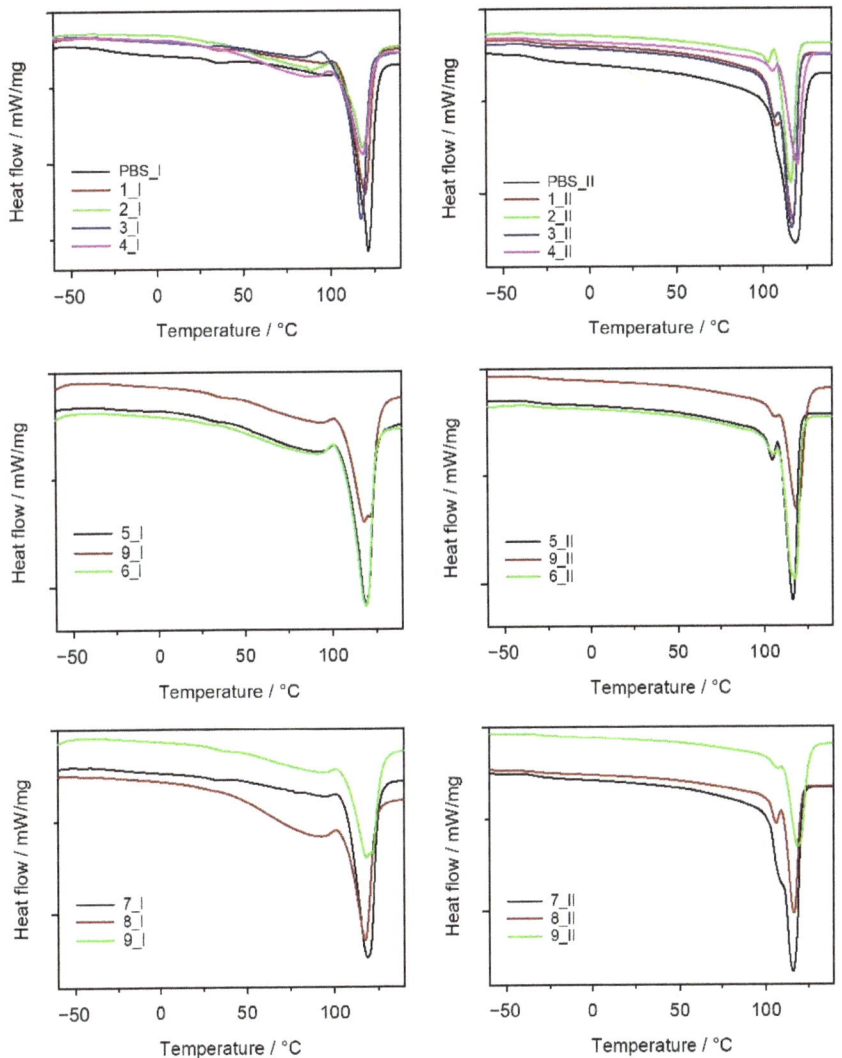

Figure 10. DSC thermograms of first (I) and second (II) heating scans of neat PBS and its composites with bran (exo up).

Table 7. Melting point (T_m), crystallization (T_c), and glass transition (T_g) temperatures; the enthalpy of melting (ΔH_m) and degree of crystallinity (X_c) of PBS and its composites with bran, based on differential scanning calorimetry (DSC) thermograms.

Sample	Heating I				Cooling	Heating II			
	T_g [°C]	T_m [°C]	ΔH_m [J/g]	X_c [%]	T_c [°C]	T_g [°C]	T_m [°C]	ΔH_m [J/g]	X_c [%]
PBS	−25.6	121.3	72.1	65.4	86.4	−31.7	118.5	66.7	60.5
1(16)	−32.5	119.6	58.0	62.6	85.3	−32.1	116.8	54.1	58.4
2(44)	−33.5	118.4	41.0	66.4	78.0	−33.8	115.5	35.4	57.3
3(16)	−30.2	117.3	67.5	72.9	84.3	−33.9	116.3	54.8	59.1
4(44)	−29.9	118.3	43.6	70.6	80.6	32.0	118.5	34.4	55.7
5(30)	−31.3	118.9	51.1	66.2	82.2	−32.1	116.3	38.2	49.5
6(30)	−31.5	118.9	51.7	67.0	81.3	32.4	117.2	39.4	51.0
7(10)	−33.3	118.8	63.3	63.8	86.4	−33.4	115.8	55.3	55.7
8(50)	−32.5	117.6	39.0	70.7	83.3	−32.5	116.6	30.6	54.5
9(30)	−31.4	118.0	41.9	54.3	79.6	−32.3	118.2	38.8	50.3

3.2.2. Thermal Stability

On the basis of a thermogravimetric analysis carried out in a synthetic air atmosphere, the thermal stability and thermal decomposition course of the obtained mouldings were determined. The TG curves presented in Figure 11 show a small mass loss, not exceeding 5%, related to the desorption of water present in the composite structure. The temperature at which 5% of the sample decomposed ($T_{5\%}$) was determined as the temperature of the onset of the mass loss of the samples. As can be seen from the data in Table 8, the $T_{5\%}$ value is the highest for neat PBS and decreases markedly with the increasing bran content in the composite structure. The lowest stability (261 °C) is shown by material 8, containing 50% bran. Comparing the thermal stability of the mouldings and the pellets, discussed in previous work, from which the mouldings were obtained, it can be stated that the injection moulding process did not affect the changes in the course of the thermal decomposition of the composites. The decomposition of neat PBS proceeds in a two-stage process; in the first stage, the hydrolysis of ester bonds occurs in parallel with the oxidation processes, whereas the second stage involves the final oxidation of the resulting deposit [66]. The decomposition of the composites proceeds in three stages; besides the mass loss associated with the decomposition of PBS, an additional first stage at about 303 °C, associated with the oxidative decomposition of the bran, is observed. The former mass loss is proportional to the bran content in the composites, which is also demonstrated by the TG and DTG curves of composites 1, 2 and 3, 4 and 7–9. An increase in the bran content causes a decrease in the thermal stability of the composites, which manifests itself in the values $T_{5\%}$ and $T_{50\%}$. Moreover, the R_m values for the composites and PBS, indicating the residual mass of the sample after the TG analysis, point that they decompose completely at 800 °C. Coming back to the comparison of the thermal stability of the discussed mouldings with that of the initial pellets, it is evident that in the case of mouldings 5, 9, and 6, the n parameter has no influence on the $T_{5\%}$ values. During the injection-moulding process, the initial pellets were heated above the value T_m, the internal structure of the composition was reorganized, and the effect of the conditions of the pellet preparation on the thermal stability of the moulded parts disappeared.

Table 8. Parameters characterizing the thermal stability of PBS, bran, and biocomposites, obtained based on thermogravimetry (TG) and derivative thermogravimetry (DTG) curves.

	$T_{5\%}$ [°C]	$T_{50\%}$ [°C]	T_{max1} [°C]	Δm_1 [%]	T_{max2} [°C]	Δm_2 [%]	T_{max3} [°C]	Δm_3 [%]	R_m [%]
bran	201	303	296	68.0	-	-	459	29.7	2.3
PBS	307	386	-	-	395	97.9	463	2.0	0.1
1(16)	288	380	303	11.9	389	82.0	474	6.0	0.1
2(44)	264	375	301	27.8	389	59.1	459	12.9	0.2
3(16)	293	384	303	10.7	394	82.3	478	6.9	0.1
4(44)	266	374	303	27.9	390	58.7	462	13.2	0.2
5(30)	274	380	303	20.5	392	70.3	476	9.1	0.1
6(30)	276	379	303	20.1	390	70.0	476	9.7	0.2
7(10)	299	383	303	8.5	391	87.0	475	4.5	0.3
8(50)	261	372	300	31.5	386	53.5	462	14.7	0.3
9(30)	274	381	303	19.8	391	71.1	476	9.0	0.1

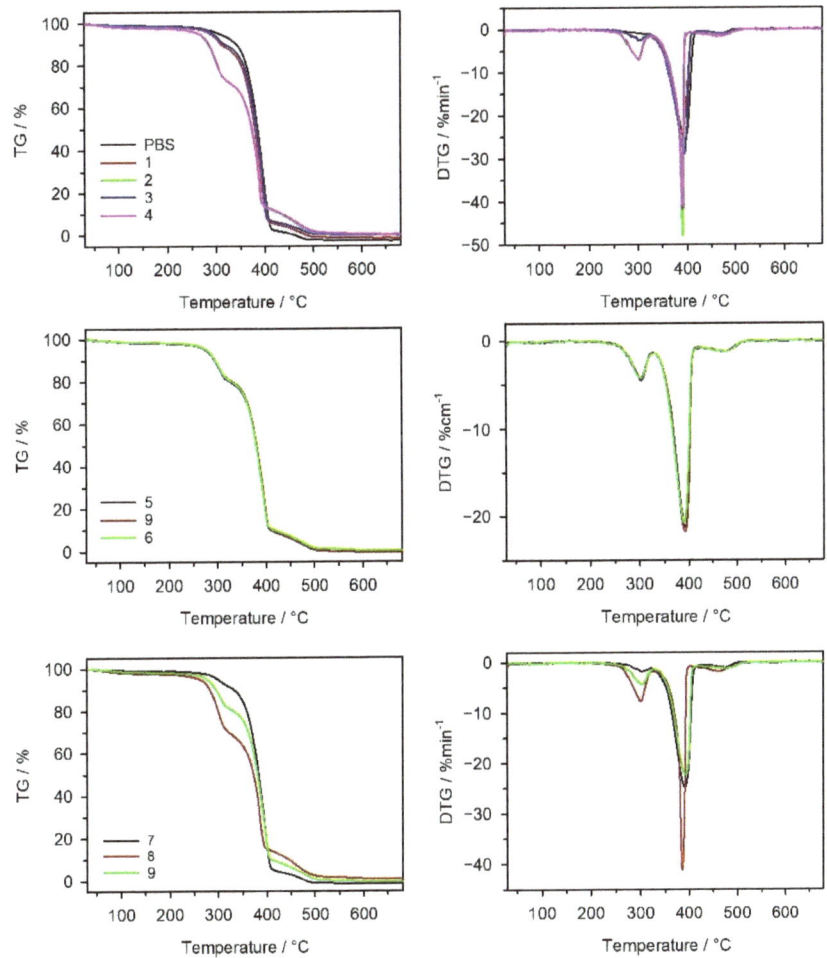

Figure 11. TG and DTG curves of PBS and its biocomposites with bran.

3.2.3. Heat Deflection Temperature

The determined relation describing the variation of the heat deflection temperature HDT was presented by means of a polynomial (Equation (7)):

$$HDT = 90.24372 + 0.07126u \quad (7)$$

The results of the statistical analysis of the adopted *HDT* model are presented in Table 9. Similarly, as in the case of the other studied quantities, a significant linear effect on the values of the heat deflection temperature *HDT* of the compositions studied is exerted by the wheat bran content *u* (Figure 12). Increasing bran content causes a significant increase in *HDT* in comparison with the samples of PBS alone, for which the determined *HDT* was significantly lower and amounted on average to HDT = 88.2 ± 0.4 °C (Figure 13). The highest increase in *HDT* during the tests, i.e., 2.4 °C (3%), was obtained by increasing the wheat bran content *u* in the composition from 10 to 50% (DOE layouts 7 and 8). There was no significant effect on the *HDT* of the applied extruder screw speed *n* during the production of the polymer composite.

Table 9. Model of heat deflection temperature HDT—ANOVA table, R^2 = 0.80; R_{adj}^2 = 0.79.

Source of Variation	SS	df	MS	F	p
u	24.36606	1	24.36606	98.33462	0.000000
Error	6.19468	25	0.24779	-	-
Total SS	30.56074	26	-	-	-

SS—sum of squares, df—number of the degrees of freedom, MS—mean sum of squares, F—values of the test statistic, *p*—value of probability corresponding to the test statistic value.

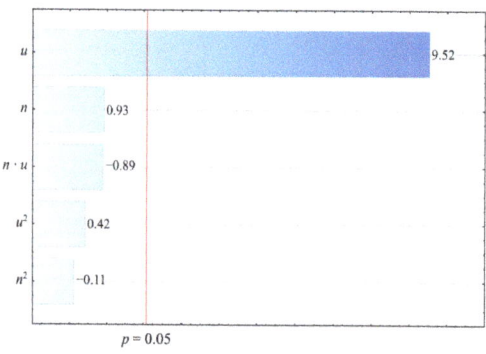

Figure 12. Pareto plots of the standardized effects of empirical full model of heat deflection temperature HDT, the vertical line in the plot corresponds to the arbitrarily chosen level of significance charts of the standardized effects for the empirical model *P* of the polymer blend pressure; the vertical line in the plot corresponds to the arbitrarily chosen level of significance (*p* = 0.05).

The increase in the HDT with the bran content is mainly due to the stiffening effect of the material. The presence of a dispersed, fine-grained natural filler in the PBS structure provides a mechanical barrier to the mobility of the macromolecules, improving the stiffness of the biocomposite and slowing down the deformation process. Therefore, it is necessary to reach temperatures in the higher range in order to obtain the preset bending deflection. Many authors obtain similar results, where the HDT value for the PBS matrix composites increases by several to over a dozen degrees relative to the unfilled PBS [43,76,93–95]. The literature also includes papers demonstrating the significant influence of interactions at the interfacial boundary of the PBS/natural filler, indicating the beneficial effect of compatibilizers on the HDT values [23,94]. Therefore, the non-covalent interactions between PBS and WB shown earlier in FTIR tests may also have a beneficial effect on the HDT values obtained for the PBS/WB composites under study.

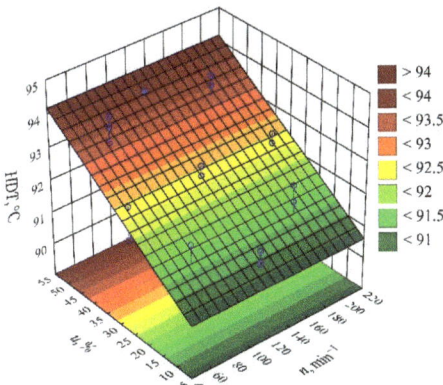

Figure 13. Response surface plot for the heat deflection temperature HDT versus wheat bran content u and screw speed n.

3.2.4. Vicat Softening Temperature

Based on the obtained measurement results, an empirical model describing the Vicat softening temperature VST (Equation (8)) was determined in the form of a polynomial:

$$VST = 109.7676 - 0.0307u \qquad (8)$$

Statistical analysis of the effects of the studied variable factors on the VST showed a statistically significant effect of only the mass content of bran u (Figure 14). The results of the statistical analysis of the adopted model are presented in Table 10. In contrast to the HDT, as the bran content of the composition increases, the value of the Vicat softening temperature VST decreases in a linear fashion (Figure 15). In the case of the samples made of PBS alone, the Vicat softening temperature was higher and amounted on average to $VST = 111.1 \pm 0.2$ °C. The observed changes, although statistically significant, are much smaller compared to those observed for the HDT. The largest observed decrease in the Vicat softening temperature VST, as a result of increasing the bran content in the composition from 10 to 50% (DOE layouts 7 and 8), was only 1.2 °C (change by 1%).

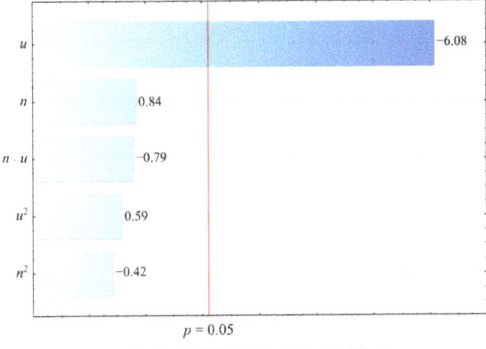

Figure 14. Pareto plots of the standardized effects of empirical full model Vicat softening temperature VST, the vertical line in the plot corresponds to the arbitrarily chosen level of significance (p = 0.05).

Table 10. Model of Vicat softening temperature VST—ANOVA table, $R^2 = 0.64$; $R_{adj}^2 = 0.62$.

Source of Variation	SS	df	MS	F	p
u	4.140601	1	4.140601	42.56	0.000001
Error	2.334784	24	0.097283	-	-
Total SS	6.475385	25	-	-	-

SS—sum of squares, df—number of the degrees of freedom, MS—mean sum of squares, F—values of the test statistic, p—value of probability corresponding to the test statistic value.

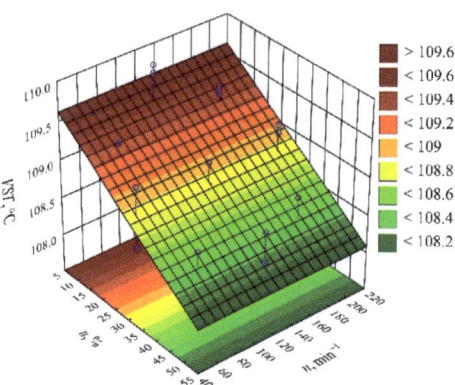

Figure 15. Response surface plot for the Vicat softening temperature VST versus wheat bran content u and screw speed n.

3.3. Mechanical Properties

3.3.1. Hardness

The result of the performed modelling of hardness H of the injection mouldings made of the compositions under study is an empirical model in the form of a polynomial (Equation (9)):

$$H = 49.32992 + 0.44949u - 0.00334u^2 \qquad (9)$$

The results of the statistical analysis of the adopted model are presented in Table 11. It was observed that the mass content of bran u introduced into the composite had a significant effect on the hardness of the obtained mouldings (Figure 16). The greatest influence is exerted by the linear term in the model equation, but the quadratic term is also statistically significant. However, its influence on hardness is many times smaller, and it has a negative effect. Increasing the bran content causes an increase in the hardness of the mouldings (Figure 17). The highest increase in hardness H, i.e., 10.3 MPa (19%), was obtained by increasing the wheat bran content u in the composition from 10 to 50% (DOE layouts 7 and 8). The statistical analysis, however, did not show any significant effect on the hardness of the mouldings of the extruder screw speed n applied during the production of the polymeric composition or the interaction between the bran content and the extruder screw speed. The hardness of the samples produced for comparison from PBS alone, without bran addition, was lower and averaged $H = 49.55 \pm 1.0$ MPa.

The hardness of the composites containing powder fillers depends on many factors, which include the mechanical and physical properties of the filler itself (stiffness, hardness, and fineness) but also on the uniformity of the filler distribution in the polymeric matrix and on the quality of the interfacial interactions. A slight increase in the hardness of the PBS matrix composites with the increasing content of the powdered natural fillers is a typical result reported by many authors [97,98]. However, only the adequate mixing and compatibilization between the natural filler and the PBS ensure an effective force transfer and a significant increase in hardness [99]. Concurrently, the use of, e.g., maleinized or epoxidized vegetable oils leads to the disruption of the interfacial interactions and a lower hardness relative to the composites without added oils [25,67].

Table 11. Model of hardness H—ANOVA table, $R^2 = 0.92$; $R_{adj}^2 = 0.91$.

Source of Variation	SS	df	MS	F	p
u	456.4350	1	456.4350	424.7162	0.000000
u^2	9.5844	1	9.5844	8.9184	0.004921
Error	40.8379	38	1.0747		
Total SS	500.6960	40			

SS—sum of squares, df—number of the degrees of freedom, MS—mean sum of squares, F—values of the test statistic, p—value of probability corresponding to the test statistic value.

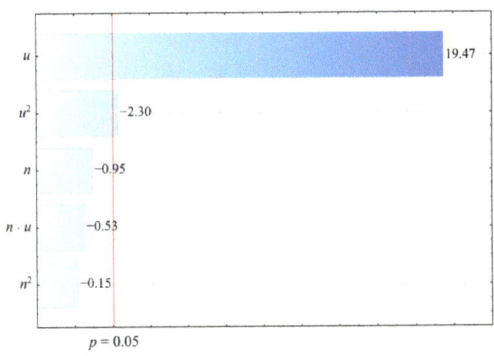

Figure 16. Pareto charts of the standardized effects for the empirical full model H; the vertical line in the plot corresponds to the arbitrarily chosen level of significance ($p = 0.05$).

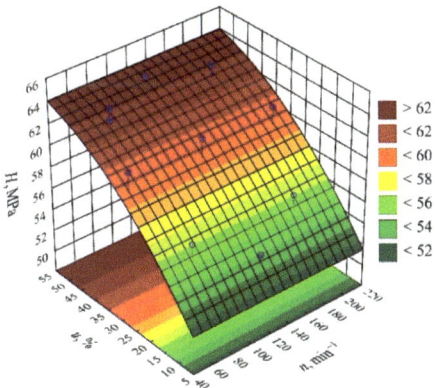

Figure 17. Response surface graph for the hardness H versus wheat bran content u and screw speed n.

3.3.2. Impact Strength

On the basis of the obtained measurement results, an empirical model was determined in the form of a polynomial describing the relation of the impact strength and the examined variable factors (Equation (10)):

$$Impact\ strength = 67.88949 - 2.21174u + 0.02094u^2 \tag{10}$$

The statistical analysis of the effects of the studied variable factors on *impact strength* showed a statistically significant effect of only the mass content of bran u (Figure 18). The results of the statistical analysis presented in Table 12 demonstrated the significance of the linear and quadratic terms in the adopted model. The *impact strength* values decrease sharply with the increase in wheat bran content u (Figure 19). As a result of a maximal increase in the wheat bran content u in the composite from 10 to 50% (DOE layouts 7 and 8), the value of the *impact strength* decreased by as much as 40.6 kJ/m^2 (82%). The unfilled PBS samples were the only ones that did not fracture under the test parameters used.

The nature of the fracture of the PBS/WB biocomposite samples changed with their content. For the two lowest WB contents (10% and 16%), the formation of a partial spall was observed in the middle part of the sample. No spalling was observed for the other contents, but the roughness of the resulting fracture increased along with the increasing filler content. The unfilled PBS samples were the only ones that did not fracture under the test parameters used. Taking into account the glass transition temperature (Table 7) and the VST values (Table 1) of the tested materials, it can be concluded that at room temperature, at which the impact test was conducted, the materials remain in an elastic state. For this reason, the unfilled PBS did not crack. At low bran contents, there occurs an initial accumulation of impact energy in the form of elastic deformation, followed by the initiation of a rapid crack of a brittle nature with spalling due to stress concentration on a random material defect, such as microcracks at the interfacial boundary. The energy is mostly used to initiate the crack. The samples with filler content from 30 to 50% have a significantly lower impact strength values due to numerous material defects, which are potential places for crack initiation. The fracture is brittle in nature and the energy is mainly utilized for crack propagation; so, the crack resistance decreases drastically with filler content [100,101]. Deterioration of the impact strength with the increasing natural filler content is a typical phenomenon for PBS-based biocomposites [17,21,36,67,102].

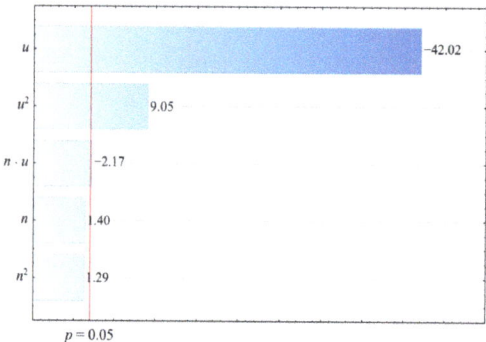

Figure 18. Pareto plots of the standardized effects of empirical full model impact strength; the vertical line in the plot corresponds to the arbitrarily chosen level of significance ($p = 0.05$).

Table 12. Model of impact strength—ANOVA table, $R^2 = 0.98$; $R_{adj}^2 = 0.97$.

Source of Variation	SS	df	MS	F	p
u	6223.385	1	6223.385	1574.945	0.000000
u^2	376.598	1	376.598	95.305	0.000000
Error	154.108	39	3.951		
Total SS	6468.503	41			

SS—sum of squares, df—number of the degrees of freedom, MS—mean sum of squares, F—values of the test statistic, p—value of probability corresponding to the test statistic value.

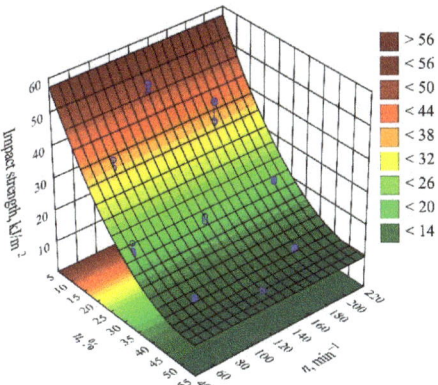

Figure 19. Response surface plot for the impact strength versus wheat bran content u and screw speed n.

3.3.3. Tensile Strength

The determined relation describing the variation of tensile strength was presented by means of a polynomial (Equation (11)):

$$\sigma = 37.94986 - 0.55714u + 0.00144u^2 \tag{11}$$

The initial analysis of the full model (Figure 20) indicated that the effect of both the wheat bran content u and the screw speed n was significant. However, the applied backward stepwise regression method of model building eventually demonstrated that, for this quantity, only the mass content of bran u (the linear and quadratic term) introduced into the composition has a significant effect on its values. The results of the statistical analysis of the adopted model σ are presented in Table 13. Moreover, in this case, increasing the bran content in the composition causes a significant decrease in the tensile strength (Figure 21). This is confirmed by the significantly higher tensile strength of the samples made of PBS alone, which was on average $\sigma = 40.68 \pm 0.52$ MPa. During the tests, following the maximum increase in wheat bran content u in the composition from 10 to 50% (DOE layouts 7 and 8), the greatest decrease in tensile strength σ of 18.4 MPa (58%) was recorded.

Tensile strength is strongly dependent on the quality of interactions at the matrix/filler interfaces and the quality of filler distribution in the matrix. Weak interaction forces cannot effectively transfer stresses between the filler grains and the polymer matrix, leading to the formation of microcracks and discontinuities. The lack of compatibilizer and the low strength of the bran itself leads to cavitation, i.e., the formation and enlargement of voids, which are the places where cracks initiate. A detailed description of the cavitation effect can be found in the work of Kim et al. [103]. As the WB content increases, so does the number of potential material defects that may become points of failure initiation during tension, especially for the hydrophilic WB and PBS matrix with a moderate affinity for water [21,43]. The description of the effect consisting in the decrease in tensile strength of the biocomposites based on the PBS matrix with the increase in natural filler content can be found in many works of other authors [21,36–38,40–43].

Moreover, during the processing PBS, unlike WB, is subject to the effect of processing shrinkage. This leads to a shrinkage of the plastic on the filler grains, exposing them to compressive stresses, while the matrix itself is then subjected to tensile stresses. The filler particles inside the PBS matrix consequently become stress concentration points, resulting in reduced tensile strength [104,105].

Table 13. Model of tensile strength—ANOVA table, $R^2 = 0.99$; $R_{adj}^2 = 0.99$.

Source of Variation	SS	df	MS	F	p
u	1726.567	1	1726.567	5428.771	0.000000
u^2	2.027	1	2.027	6.375	0.015538
Error	13.040	41	0.318		
Total SS	1741.222	43			

SS—sum of squares, df—number of the degrees of freedom, MS—mean sum of squares, F—values of the test statistic, p—value of probability corresponding to the test statistic value.

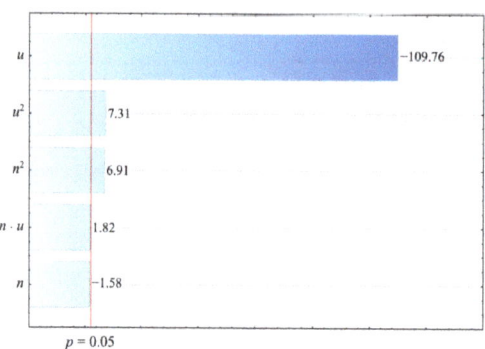

Figure 20. Pareto plots of the standardized effects of empirical full model tensile strength σ; the vertical line in the plot corresponds to the arbitrarily chosen level of significance ($p = 0.05$).

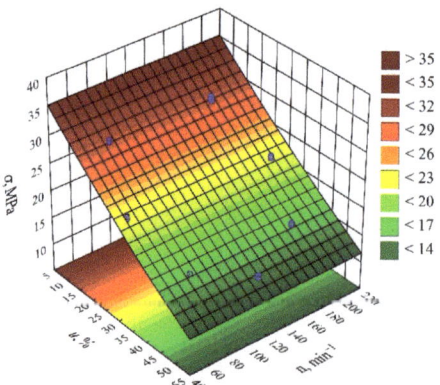

Figure 21. Response surface plot for the tensile strength σ versus wheat bran content u and screw speed n.

3.3.4. Young's Modulus

The result of the performed modelling of Young's modulus E of the injection mouldings made of the compositions under study is an empirical model in the form of a polynomial (Equation (12)):

$$E = 616.7247 + 20.0558u \qquad (12)$$

Once again, the preliminary analysis of the full model (Figure 22) indicated that the influence of both variable factors under study was significant. The results of the modelling performed (Table 14), as the most suitable model, indicated the linear dependence of taking into account only the effect of wheat bran content u. The highest increase in Young's modulus of the mouldings, i.e., 754 MPa (94%), was obtained by increasing the mass content of bran u in the composition from 10 to 50% (DOE layouts 7 and 8) (Figure 23). Young's modulus of the samples made from the PBS alone was significantly lower and averaged $E = 729 \pm 8$ MPa.

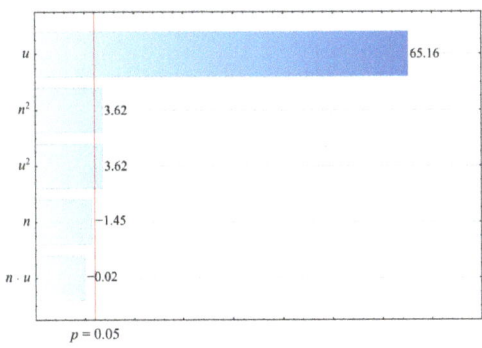

Figure 22. Pareto plots of the standardized effects of empirical full model Young's modulus E; the vertical line in the plot corresponds to the arbitrarily chosen level of significance ($p = 0.05$).

Table 14. Model of Young's modulus E—ANOVA table, $R^2 = 0.99$; $R_{adj}^2 = 0.99$.

Source of Variation	SS	df	MS	F	p
u	3,216,923	1	3,216,923	3198.350	0.000000
Error	43,250	43	1006		
Total SS	3,260,173	44			

SS—sum of squares, df—number of the degrees of freedom, MS—mean sum of squares, F—values of the test statistic, p—value of probability corresponding to the test statistic value.

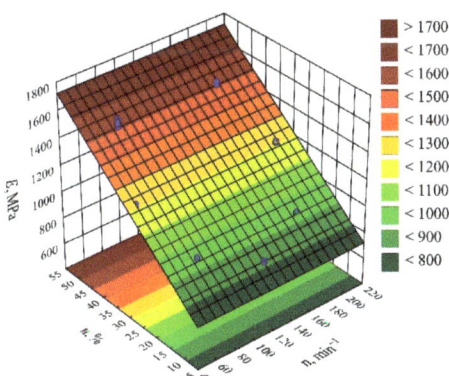

Figure 23. Response surface plot for the Young's modulus E versus wheat bran content u and screw speed n.

As mentioned earlier when describing the HDT results, the introduction of a powder filler into the polymer matrix results in an increase in stiffness by limiting the mobility of the macromolecules with the presence of a dispersed phase. The increase in stiffness is manifested by lower deformability, deterioration of the elastic and plastic properties, and an increase in brittleness, which clearly affects all the parameters related to the deformation of the samples. This is, of course, a common phenomenon occurring in polymer composites containing filler in powder form, either of natural or mineral origin [37,38,41–43,76].

3.3.5. Elongation at Break

On the basis of the obtained measurement results, an empirical model was determined in the form of a polynomial describing the relation of elongation at break ε and the examined variable factors (Equation (13)):

$$\varepsilon = 29.42612 - 0.42370u \quad (13)$$

The statistical analysis of the effects of the studied variable factors on elongation at break showed a statistically significant effect of only the mass content of bran u (Figure 24). The results of the statistical analysis presented in Table 15 demonstrated the significance of the linear and quadratic terms in the adopted model of elongation at break. As the bran content u increases, the values of elongation at break ε decrease in a linear fashion (Figure 25). As a result of a maximal increase in the mass content of bran u in the composite from 10 to 50% (DOE layouts 7 and 8), the value of elongation at break ε decreased by as much as 18.4% (69% of the initial value). Samples made from the PBS alone had a significantly higher elongation at break $\varepsilon = 218.6 \pm 14.6\%$, and a ductile fracture with a very long neck.

The obtained course of change in elongation at break is related to the aforementioned increase in the stiffness of the composition and, at the same time, its brittleness, which is manifested by a significant decrease in deformability. Analogous courses of changes in the maximum deformation of the PBS with the increasing natural filler content can be observed in other works and for other fillers [37,41–43,76]. The nature of the obtained fractures also changed from ductile to brittle, similarly to the case with impact strength. Even the addition of 10% WB resulted in a reduction in deformation by almost 200%, but the fracture was ductile, and numerous longitudinal pore-like structures could be clearly observed on the surface of the neck as a result of the cavitation effect. Each such structure represents a potential point of crack initiation. At a bran content of 16%, there was only a residual neck. For 30% and 44% bran content, the neck did not occur, and plastic deformation was manifested in the form of light discolouration on the measurement part of the tested samples. On the other hand, at the content of 50%, no plastic deformation indicators visible to the naked eye were observed, and a brittle fracture with high roughness was obtained.

Table 15. Model of elongation at break ε—ANOVA table, $R^2 = 0.72$; $R_{adj}^2 = 0.72$.

Source of Variation	SS	df	MS	F	p
u	1243.739	1	1243.739	104.1111	0.000000
Error	477.851	40	11.946		
Total SS	1721.590	41			

SS—sum of squares, df—number of the degrees of freedom, MS—mean sum of squares, F—values of the test statistic, p—value of probability corresponding to the test statistic value.

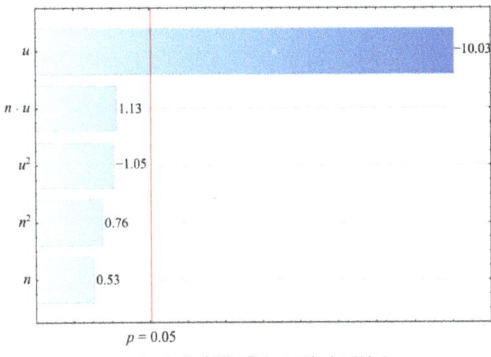

Figure 24. Pareto plots of the standardized effects of empirical full-model elongation at break ε; the vertical line in the plot corresponds to the arbitrarily chosen level of significance ($p = 0.05$).

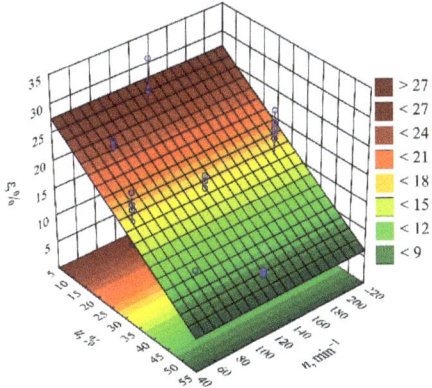

Figure 25. Response surface plot for the elongation at break ε versus wheat bran content u and screw speed n.

4. Conclusions

Analysing the presented work as a whole, it can be stated with certainty that there is no significant influence of the extruder screw speed during the production of pellets on all the properties of the injection-moulded parts made of those pellets. Thus, for the sake of efficiency and economy of production, high screw speed values are preferred during the extrusion of the composite pellets. As presented in a previous paper [66], this has a positive effect on the flow rate of the extrudate and minimizes the energy consumption due to the autothermal effect, while not exerting a negative effect on the properties of the finished injection-moulded products. Concurrently, a very significant effect of the filler content on the properties investigated was found.

A beneficial effect of WB on the processing shrinkage value was observed. As the filler content increases, the shrinkage values in all three directions decrease. Lower shrinkage values facilitate mould designing and help maintain the dimensional stability of the moulded parts. Moreover, the studied injection mouldings obtained higher density values in comparison with the pellets from which they were made. This is due to the high injection pressure and cooling under packing pressure.

Chemical structure tests showed the presence of structures and compounds typical for PBS and wheat bran. The possibility of non-covalent interactions between the matrix and the filler was also found.

The presence of WB in the structure of PBS affects its thermal properties. The DSC results demonstrated a significant effect of the presence of bran on the degree of crystallinity and the crystallization temperature of the PBS. The degree of crystallinity decreases relative to the unfilled PBS. This may affect the mechanical and thermal properties of the biocomposite to some extent. The melting point of the tested compositions is slightly lower than that for neat PBS. An analogous relationship was obtained for the VST. The crystallization temperature during cooling is also lower than for PBS alone. The HDT increased along with the increase in WB content, in spite of a general decrease in thermal stability shown in the TG tests.

As far as the mechanical properties are concerned, an increase in hardness and stiffness with the increasing WB content was demonstrated. The maximum tensile strength and impact strength deteriorated drastically, which is as expected and typical for PBS matrix composites containing powdered fillers of natural origin. In spite of the observed deterioration of some of the mechanical properties, their values still remain at a satisfactory level and can meet the design requirements of many objects of everyday use. The high filling degree of WB allows for an effective reduction in the cost of manufacturing PBS components, potentially contributing to the industrial popularity of this biodegradable material.

Author Contributions: Conceptualization, E.S. and Ł.M.; methodology, E.S., Ł.M. and M.G.; validation, E.S., Ł.M. and M.G.; formal analysis, E.S. and M.G.; investigation, E.S., Ł.M. and M.G.; resources, E.S. and Ł.M.; writing—original draft preparation, E.S. and Ł.M.; writing—review and editing, M.G.; visualization, E.S., M.G. and Ł.M.; supervision, E.S.; project administration, Ł.M. and E.S.; funding acquisition Ł.M. and E.S. All authors have read and agreed to the published version of the manuscript.

Funding: The research was financed in the framework of the project of Lublin University of Technology-Regional Excellence Initiative, funded by the Polish Ministry of Science and Higher Education (contract no. 030/RID/2018/19).

Institutional Review Board Statement: Not applicable.

Informed Consent Statement: Not applicable.

Data Availability Statement: The data presented in this study are available on request from the corresponding author.

Conflicts of Interest: The authors declare no conflict of interest. The funders had no role in the design of the study; in the collection, analyses, or interpretation of data; in the writing of the manuscript; or in the decision to publish the results.

List of Abbreviations and Symbols

PBS	Poly(butylene succinate)
WB	Wheat bran
DOE	Design of experiment
FTIR	Fourier transform infrared (spectroscopy)
DSC	Differential scanning calorimetry
X_c	Degree of crystallinity
ΔH_m	Melting enthalpy
T_m	Melting point
T_c	Crystallization temperature
T_g	Glass transition temperature
TG	Thermogravimetry
DTG	Derivative thermogravimetry
$T_{5\%}$, $T_{50\%}$	Temperature of 5% and 50% of mass loss
T_{max}	Temperature of the maximum rate of mass loss
Δm	Mass loss corresponding to T_{max}
R_m	Residual mass

U	Wheat bran content
N	Extruder screw speed
HDT	Heat deflection temperature
VST	Vicat softening temperature
LCF	Lignocellulosic filler
α	Star point distance in DOE
S_L	Longitudinal shrinkage
S_T	Transverse shrinkage
S_P	Perpendicular shrinkage
P	Density
p-v-T	Relationship between pressure p, specific volume v and temperature T
H	Hardness
σ	Tensile strength
E	Young's modulus
ε	Elongation at break

References

1. Vilaplana, F.; Strömberg, E.; Karlsson, S. Environmental and resource aspects of sustainable biocomposites. *Polym. Degrad. Stab.* **2010**, *95*, 2147–2161. [CrossRef]
2. Mohanty, A.K.; Misra, M.; Drzal, L.T. Sustainable bio-composites from renewable resources: Opportunities and challenges In the Green materials Word. *J. Polym. Environ.* **2002**, *10*, 19–26. [CrossRef]
3. Väisänen, T.; Das, O.; Tomppo, L. A review on new bio-based constituents for natural fiber-polymer composites. *J. Clean. Prod.* **2017**, *149*, 582–592. [CrossRef]
4. Babu, R.P.; O'Connor, K.; Seeram, R. Current progress on bio-based polymers and their future trends. *Prog. Biomater.* **2013**, *2*, 8. [CrossRef]
5. Nagarajan, V.; Mohanty, A.K.; Misra, M. Sustainable Green composites: Value addition to agricultural residues and parennial grasses. *ACS Sustain. Chem. Eng.* **2013**, *1*, 325–333. [CrossRef]
6. Korol, J.; Burchart-Korol, D.; Pichlak, M. Expansion of environmental impact assessment for eco-efficiency evaluation of biocomposites for industrial application. *J. Clean. Prod.* **2016**, *113*, 144–152. [CrossRef]
7. Saba, N.; Jawaid, M.; Sultan, M.T.H.; Alothman, O.Y. Green biocomposites for structural applications. In *Green Biocomposites. Green Energy and Technology*; Jawaid, M., Salit, M., Alothman, O.Y., Eds.; Springer: Cham, Switzerland, 2017; pp. 1–27. [CrossRef]
8. Wei, L.; McDonald, A.G. A review on grafting of biofibres for biocomposites. *Materials* **2016**, *9*, 303. [CrossRef]
9. Imre, B.; Pukánszky, B. Compatibilization in bio-based and biodegradable polymer blends. *Eur. Polym. J.* **2013**, *49*, 1215–1233. [CrossRef]
10. Hamad, K.; Kaseem, M.; Ko, Y.G.; Deri, F. Biodegradable polimer blends and composites: An overwiew. *Polym. Sci. Ser. A* **2014**, *56*, 812–829. [CrossRef]
11. Pivsa-Art, W.; Chaiyasat, A.; Pivsa-Art, S.; Yamane, H.; Ohara, H. Preparation of polymer blends between poly(lactic acid) and poly(butylene adipate-co-terephthalate) and biodegradable polymers as compatibilisers. *Energy Procedia* **2013**, *34*, 549–554. [CrossRef]
12. Takayama, T.; Todo, M.; Tsuji, H. Effect of annealing on the mechanical properties of PLA/PCL and PLA/PCL/LTI polimer blends. *J. Mech. Behav. Biomed. Mater.* **2011**, *4*, 255–260. [CrossRef] [PubMed]
13. Chun, K.S.; Husseinsyah, S.; Osman, H. Mechanical and thermal properties of coconut shell powder filled polylactic acid biocomposites: Effect of the filler content and silane coupling agent. *J. Polym. Res.* **2012**, *19*, 9859. [CrossRef]
14. Alias, N.F.; Ismail, H.; Wahab, M.K. Properties of polyvinyl alcohol/palm kernel shell powder biocomposites and their hybrid composites with halloysite nanotubes. *Bioresources* **2017**, *12*, 9103–9117. [CrossRef]
15. Cinelli, P.; Mallegni, N.; Gigante, V.; Montanari, A.; Seggiani, M.; Coltelli, M.B.; Bronco, S.; Lazzeri, A. Biocomposites based on polyhydroxyalkanoates and natural fibers from renewable byproducts. *Appl. Food Biotechnol.* **2019**, *6*, 35–43. [CrossRef]
16. Cocca, M.; Avolio, R.; Gentile, G.; Di Pace, E.; Errico, M.E.; Avella, M. Amorphized cellulose as filler in biocomposites based on poly(ε-caprolactone). *Carbohydr. Polym.* **2015**, *118*, 170–182. [CrossRef]
17. Terzopoulou, Z.N.; Papageorgiou, G.Z.; Papadopoulou, E.; Athanassiadou, E.; Reinders, M.; Bikiaris, D.N. Development and study of fully biodegradable composite materials based on poly(butylene succinate) and hemp fibers or hemp shives. *Polym. Compos.* **2016**, *37*, 407–421. [CrossRef]
18. Lin, N.; Yu, J.; Chang, P.R.; Li, J.; Huang, J. Poly(butylene succinate)-based biocomposites filled with polysaccharide nanocrystals: Structure and properties. *Polym. Compos.* **2011**, *32*, 472–482. [CrossRef]
19. Mohanty, A.K.; Vivekanandhan, S.; Pin, J.M.; Misra, M. Composites from renewable and sustainable resources: Challenges and innovations. *Science* **2018**, *362*, 536–542. [CrossRef]
20. Xu, J.; Guo, B.H. Poly(butylene succinate) and its copolymers: Research development and industrialization. *Biotechnol. J.* **2010**, *5*, 1149–1163. [CrossRef]

21. Kim, H.S.; Yang, H.S.; Kim, H.J. Biodegradability and mechanical properties of agro-flour-filled polybutylene succinate biocomposites. *J. Appl. Polym. Sci.* **2005**, *97*, 1513–1521. [CrossRef]
22. Anstey, A.; Muniyasamy, S.; Reddy, M.M.; Misra, M.; Mohanty, A. Processability and biodegradability evaluation of composites from poly(butylene succinate) (PBS) bioplastic and biofuel co-products from Ontario. *J. Polym. Environ.* **2014**, *22*, 209–218. [CrossRef]
23. Muthuraj, R.; Misra, M.; Mohanty, A.K. Injection molded sustainable biocomposites from poly(butylene succinate) bioplastic and perennial grass. *ACS Sustain. Chem. Eng.* **2015**, *3*, 2767–2776. [CrossRef]
24. Yun, I.S.; Hwang, S.; Shim, J.K.; Seo, K.H. A study on the thermal and mechanical properties of poly(butylene succinate) / thermoplastic starch binary blends. *Int. J. Precis. Eng. Manuf. Green Technol.* **2016**, *3*, 289–296. [CrossRef]
25. Liminana, P.; Garcia-Sanoguera, D.; Quiles-Carrillo, L.; Montanes, B.N. Development and characterization of environmentally friendly composites from poly(butylene succinate) (PBS) and almond shell flour with different compatibilizers. *Compos. Part B Eng.* **2018**, *144*, 153–162. [CrossRef]
26. Rafiqah, A.A.; Khalina, A.; Harmaen, A.S.; Tawakkal, I.A.; Zaman, K.; Asim, M.; Nurrazi, M.N.; Lee, C.H. A review on properties and application of bio-based poly(butylene succinate). *Polymers* **2021**, *13*, 1436. [CrossRef]
27. Cukalovic, A.; Stevens, C.V. Feasibility of production methods for succinic acid derivatives: A marriage of renewable resources and chemical technology. *Biofuels Bioprod. Biorefining* **2008**, *2*, 505–529. [CrossRef]
28. Bechthold, I.; Bretz, K.; Kabasci, S.; Kopitzky, R.; Springer, A. Succinic acid: A new platform chemical for biobased polymers from renewable sources. *Chem. Eng. Technol.* **2008**, *31*, 647–654. [CrossRef]
29. McKinlay, J.B.; Vielle, C.; Zeikus, J.G. Prospects for bio-based succinate industry. *Appl. Microbiol. Biotechnol.* **2007**, *76*, 727–740. [CrossRef]
30. Sheldon, R.A. Green and sustainable manufacture of chemicals from biomass: State of the art. *Green Chem.* **2014**, *16*, 950–963. [CrossRef]
31. Phua, Y.J.; Chow, W.S.; Ishak, Z.A.M. The hydrolytic effect of moisture and hygrothermal aging on po(butylene succinate)/organomontmorillonite nanocomposites. *Polym. Degrad. Stab.* **2011**, *96*, 1194–1203. [CrossRef]
32. Frollini, E.; Bartolucci, N.; Sisti, L.; Celli, A. Poly(butylene succinate) reinforced with different lignocellulosic fibers. *Ind. Crops Prod.* **2013**, *45*, 160–169. [CrossRef]
33. Mochane, M.J.; Magagula, S.I.; Sefadi, J.S.; Mekhena, T.C. A review on green composites based on natural fiber-reinforced polybutylene succinate (PBS). *Polymers* **2021**, *13*, 1200. [CrossRef]
34. Park, C.W.; Youe, W.J.; Han, S.Y.; Park, J.S.; Lee, E.A.; Park, J.Y.; Kwon, G.J.; Kim, S.J.; Lee, S.H. Influence of lignin and polymeric diphenylmethane diisocyante addition on the properties of poly(butylene succinate)/wood flour composite. *Polymers* **2019**, *11*, 1161. [CrossRef] [PubMed]
35. Sasimowski, E.; Majewski, Ł.; Jachowicz, T.; Sąsiadek, M. Experimental determination of coefficients for the Renner model of thermodynamic equation of state of the poly(butylene succinate) and wheat bran biocomposites. *Materials* **2021**, *14*, 5293. [CrossRef] [PubMed]
36. Kim, H.S.; Kim, H.J.; Lee, J.W.; Choi, I.G. Biodegradability of bio-flour filled biodegradable poly(butylene succinate) biocomposites in natural and compost soil. *Polym. Degrad. Stab.* **2006**, *91*, 1117–1127. [CrossRef]
37. Rojas-Lema, S.; Arevalo, J.; Gomez-Caturla, J.; Garcia-Garcia, D.; Torres-Giner, S. Peroxide-induced synthesis of maleic anhydride grafted poly(butylene succinate) and its compatibilizing effect on poly(butylene succinate)/pistachio shell flour composites. *Molecules* **2021**, *26*, 5927. [CrossRef]
38. Hongsriphan, N.; Kamsantia, P.; Sillapasangloed, P.; Loychuen, S. Bio-based composite from poly(butylene succinate) and peanut shell waste adding maleinized linseed oil. *IOP Conf. Ser. Mater. Sci. Eng.* **2020**, *773*, 012046. [CrossRef]
39. Akindoyo, J.O.; Husney, N.A.A.; Ismail, N.H.; Mariatti, M. Structure and performance of poly(lactic acid)/poly(butylene succinate-co-L-lactate) blend reinforced with rice husk and coconutshell filler. *Polym. Polym. Compos.* **2020**, *29*, 992–1002. [CrossRef]
40. Liminana, P.; Garcia-Sanoguera, D.; Quiles-Carrillo, L.; Balart, R.; Montanes, N. Optimization of maleinized linseed oil loading as a biobased compatibilizer in poly(butylene succinate) composites with almond shell flour. *Materials* **2019**, *12*, 685. [CrossRef]
41. Picard, M.C.; Rodriguez-Uribe, A.; Thimmanagari, M.; Misra, M.; Mohanty, A.K. Sustainable biocomposites from poly(butylenes succinate) and apple pomace: A study on compatibilization performance. *Waste Biomass Valorization* **2020**, *11*, 3775–3787. [CrossRef]
42. Gowman, A.; Wang, T.; Rodriguez-Uribe, A.; Mohanty, A.K.; Misra, M. Bio-poly(butylene succinate) and its composites with grape pomace: Mechanical performance and thermal properties. *ACS Omega* **2018**, *3*, 15205–15216. [CrossRef] [PubMed]
43. Nanni, A.; Messori, M. Thermo-mechanical properties and creep modeling of wine lees filled polyamide 11 (PA11) and polybutylene succinate (PBS) bio-composites. *Compos. Sci. Technol.* **2020**, *188*, 107974. [CrossRef]
44. Visakh, P.M.; Thomas, S. Preparation of bionanomaterials and their polymer nanocomposites from waste and biomas. *Waste Biomass Valorization* **2010**, *1*, 121–134. [CrossRef]
45. Paukszta, D.; Borysiak, S. The influence of processing and the polymorphism of lignocellulosic fillers on the structure and properties of composite materials—A review. *Materials* **2013**, *6*, 2747–2767. [CrossRef] [PubMed]
46. Avérous, L.; Le Digabel, F. Properties of biocomposites based on lignocellulosic fillers. *Carbohydr. Polym.* **2006**, *66*, 480–493. [CrossRef]
47. Yan, L.; Chouw, N.; Jayaraman, K. Flax fibre and its composites—A review. *Compos. Part B Eng.* **2014**, *56*, 296–317. [CrossRef]

48. Liu, L.; Yu, J.; Cheng, L.; Qu, W. Mechanical properties of poly(butylenes succinate) (PBS) biocomposites reinforced with surface modified jute fiber. *Compos. Part A* **2009**, *40*, 669–674. [CrossRef]
49. Zini, E.; Scandola, M. Green composites: An overview. *Polym. Compos.* **2011**, *32*, 1905–1915. [CrossRef]
50. Sasimowski, E.; Majewski, Ł.; Grochowicz, M. Influence of the design solutions of extruder screw mixing tip on selected properties of wheat bran-polyethylene biocomposite. *Polymers* **2019**, *11*, 2120. [CrossRef]
51. Mirmehdi, S.M.; Zeinaly, F.; Dabbagh, F. Date palm wood flour as a filler of linear low-density polyethylene. *Compos. Part B Eng.* **2014**, *56*, 137–141. [CrossRef]
52. Yan-Hong, F.; Yi-Jie, L.; Bai-Ping, X.; Da-Wei, Z.; Jin-Ping, Q.; He-Zhi, H. Effect of fiber morphology on rheological properties of plant fiber reinforced poly(butylene succinate) composites. *Compos. Part B Eng.* **2013**, *44*, 193–199.
53. Feng, Y.H.; Zhang, D.W.; Qu, J.P.; He, H.Z.; Xu, B.P. Rheological properties of sisal fiber/poly(butylene succinate) composites. *Polym. Test.* **2011**, *30*, 124–130. [CrossRef]
54. Bhattacharjee, S.K.; Chakraborty, G.; Kashyap, S.P.; Gupta, R.; Katiyar, V. Study of the thermal, mechanical and melt rheological properties of rice straw filled poly (butylene succinate) bio-composites through reactive extrusion process. *J. Polym. Environ.* **2020**, *29*, 1477–1788. [CrossRef]
55. Santi, C.R.; Hage, E.; Vlachopoulos, J.; Correa, C.A. Rheology and processing of HDPE/wood flour composites. *Int. Polym. Process.* **2009**, *24*, 346–353. [CrossRef]
56. Yeh, S.K.; Gupta, R.K. Improved wood-plastic composites through better processing. *Compos. Part A Appl. Sci. Manuf.* **2008**, *39*, 1694–1699. [CrossRef]
57. Tazi, M.; Erchiqui, F.; Godard, F.; Kaddami, H.; Ajji, A. Characterization of rheological and thermophysical properties of HDPE-wood composite. *J. Appl. Polym. Sci.* **2014**, *131*. [CrossRef]
58. Lee, S.M.; Cho, D.; Parl, W.H.; Lee, S.G.; Han, S.O.; Drzal, L.T. Novel silk/poly(butylene succinate) biocomposites: The effect of short fibre content on their mechanical and thermal poperties. *Compos. Sci. Technol.* **2005**, *65*, 647–657. [CrossRef]
59. Yang, H.S.; Wolcott, M.P.; Kim, H.S.; Kim, H.J. Thermal properties of lignocellulosic filler-thermoplastic polymer biocomposites. *J. Therm. Anal. Calorim.* **2005**, *82*, 157–160. [CrossRef]
60. Sahoo, S.; Misra, M.; Mohanty, A.K. Enhanced properties of lignin-based biodegradable polymer composites using injection moulding process. *Compos. Part A* **2011**, *42*, 1710–1718. [CrossRef]
61. Sahoo, S.; Misra, M.; Mohanty, A.K. Biocomposites from switchgrass and lignin hybrid and poly(butylene succinate) bioplastic: Studies on reactive compatibilization and performance evaluation. *Macromol. Mater. Eng.* **2014**, *299*, 178–189. [CrossRef]
62. Majewski, Ł.; Gaspar Cunha, A. Evaluation of suitability of wheat bran as a natural filler in polymer processing. *Bioresources* **2018**, *13*, 7037–7052. [CrossRef]
63. Abba, H.A.; Zahari, I.N.; Sapuan, S.M.; Leman, Z. Characterization of millet (Pennisetum glaucum) husk fiber (MHF) and its use as filler for high density polyethylene (HDPE) composites. *Bioresources* **2017**, *12*, 9287–9301.
64. Zharif, A.T.M.; Ishak, M.A.; Taib, R.M.; Sudin, R.; Leong, W.Y. Mechanical, water absorption and dimensional stability studies of kenaf bast fibre-filled poly(butylene succinate) composites. *Polym. Plast. Technol. Eng.* **2011**, *50*, 339–348. [CrossRef]
65. Domínguez-Robles, J.; Larrañeta, E.; Fong, M.L.; Martin, N.K.; Irwin, N.J.; Mutjé, P.; Tarrés, Q.; Delgado-Aguilar, M. Lignin/poly(butylene succinate) composites with antioxidant and antibacterial properties for potential biomedical applications. *Int. J. Biol. Macromol.* **2020**, *145*, 92–99. [CrossRef]
66. Sasimowski, E.; Majewski, Ł.; Grochowicz, M. Efficiency of twin-screw extrusion of biodegradable poly (butylene succinate)-wheat bran blend. *Materials* **2021**, *14*, 424. [CrossRef] [PubMed]
67. Liminana, P.; Quiles-Carrillo, L.; Boronat, T.; Balart, R.; Montanes, N. The effect of varying almond shell flour (ASF) loading in composites with poly(butylene succinate) (PBS) matrix compatibilized with maleinized linseed oil. *Materials* **2018**, *11*, 2179. [CrossRef] [PubMed]
68. Nam, T.H.; Ogihara, S.; Nakatani, H.; Kobayashi, S.; Song, J.I. Mechanical properties and water absorption of jute fiber reinforced poly(butylene succinate) biodegradable composites. *Adv. Compos. Mater.* **2012**, *21*, 241–258. [CrossRef]
69. Platnieks, O.; Gaidukovs, S.; Barkane, A.; Gaidukova, G.; Grase, L.; Thakur, V.K.; Filipova, I.; Fridrihsone, V.; Skute, M.; Laka, M. Highly loaded cellulose/poly (butylene succinate) sustainable composites for woody-like advanced materials application. *Molecules* **2020**, *25*, 121. [CrossRef]
70. Phua, Y.J.; Chow, W.S.; Ishak, Z.A.M. Reactive processing of maleic anhydride-grafted poly(butylene succinate) and the compatibilizing effect on poly(butylene succinate) nanocomposites. *Express Polym. Lett.* **2013**, *7*, 340–354. [CrossRef]
71. Wu, C.S.; Hsu, Y.C.; Liao, H.T.; Yen, F.S.; Wang, C.Y.; Hsu, C.T. Characterization and biocompatibility of chestnut shell fiber-based composites with polyester. *J. Appl. Polym. Sci.* **2014**, *131*, 40730. [CrossRef]
72. Lee, J.M.; Ishak, Z.A.M.; Taib, R.M.; Law, T.T.; Thirmizir, M.Z.A. Mechanical, thermal and water absorption properties of kenaf-fiber-based polypropylene and poly(butylene succinate) composites. *J. Polym. Environ.* **2013**, *21*, 293–302. [CrossRef]
73. Totaro, G.; Sisti, L.; Vannini, M.; Marchese, P.; Tassoni, A.; Lenucci, M.S.; Lamborghini, M.; Kalia, S.; Celli, A. A new route of valorization of rice endosperm by-product: Production of polymeric biocomposites. *Compos. Part B Eng.* **2018**, *139*, 195–202. [CrossRef]
74. Kim, H.S.; Lee, B.H.; Lee, S.; Kim, H.J.; Dorgan, J.R. Enhanced interfacial adhesion, mechanical and thermal properties of natural flour-filled biodegradable polymer bio-composites. *J. Therm. Anal. Calorim.* **2011**, *104*, 331–338. [CrossRef]

75. ISO 294-1:2017. *Plastics—Injection Moulding of Test Specimens of Thermoplastic Materials—Part 1: General Principles, and Moulding of Multipurpose and Bar Test Specimens*; ISO: Geneva, Switzerland, 2017.
76. BioPBS FZ91PB—Technical Data Sheet. Available online: https://www.mcpp-global.com/en/mcpp-asia/products/product/biopbsTM-general-properties/ (accessed on 5 October 2021).
77. Greffeuille, V.; Abecassis, J.; Lapierre, C.; Lullien-Pellerin, V. Bran size distribution at milling and mechanical and biochemical characterization of common wheat grain outer layers: A relationship assessment. *Cereal Chem.* **2006**, *83*, 641–646. [CrossRef]
78. Kamal-Eldin, A.; Lærke, H.N.; Knudsen, K.E.B.; Lampi, A.M.; Piironen, V.; Adlercreutz, H.; Katina, K.; Poutanen, K.; Åman, P. Physical, microscopic and chemical characterization of industrial rye and wheat brans from the Nordic countries. *Food Nutr. Res.* **2009**, *53*, 1912. [CrossRef]
79. ISO 294-4:2018. *Plastics—Injection Moulding of Test Specimens of Thermoplastic Materials—part 4: Determination of Moulding Shrinkage*; ISO: Geneva, Switzerland, 2018.
80. ISO 1183-1:2019. *Plastics—Methods for Determining the Density of Non-Cellular Plastics—Part 1: Immersion Method, Liquid Pycnometer Method and Titration Method*; ISO: Geneva, Switzerland, 2019.
81. ISO 11357-1:2016. *Plastics—Differential Scanning Calorimetry (DSC)—Part 1: General Principles*; ISO: Geneva, Switzerland, 2016.
82. Huang, Z.; Qian, L.; Yin, Q.; Yu, N.; Liu, T.; Tian, D. Biodegradability studies of poly(butylene succinate) composites filled with sugarcane rind fiber. *Polym. Test.* **2018**, *66*, 319–326. [CrossRef]
83. ISO 75-2:2013. *Plastics—Determination of Temperature of Deflection Under Load—Part 2: Plastics and Ebonite*; ISO: Geneva, Switzerland, 2013.
84. ISO 306:2013. *Plastics—Thermoplastic Materials—Determination of Vicat Softening Temperature (VST)*; ISO: Geneva, Switzerland, 2013.
85. ISO 2039-1:2001. *Plastics—Determination of Hardness—Part 1: Ball Indentation Method*; ISO: Geneva, Switzerland, 2001.
86. ISO 179-1:2020. *Plastics—Determination of Charpy Impact Properties—Part 2: Instrumented Impact Test*; ISO: Geneva, Switzerland, 2020.
87. ISO 527-2:2012. *Plastics—determination of Tensile Properties—part 2: Test Conditions for Moulding and Extrusion Plastics*; ISO: Geneva, Switzerland, 2012.
88. Kwon, K.; Isayev, A.I.; Kim, K.H. Theoretical and experimental studies of anisotropic shrinkage in injection molding of various polyesters. *J. Appl. Polym. Sci.* **2006**, *102*, 3526–3544. [CrossRef]
89. Chang, T.C. Shrinkage behavior and optimization of injection molded parts studied by the Taguchi method. *Polym. Eng. Sci.* **2001**, *41*, 703–710. [CrossRef]
90. Kwon, K.; Isayev, A.I.; Kim, K.H.; van Sweden, C. Theoretical and experimental studies of anisotropic shrinkage in injection moldings of semicrystalline polymers. *Polym. Eng. Sci.* **2006**, *46*, 712–728. [CrossRef]
91. Sasimowski, E.; Majewski, Ł.; Grochowicz, M. Analysis of Selected Properties of Biocomposites Based on Polyethylene with a Natural Origin Filler. *Materials* **2020**, *13*, 4182. [CrossRef]
92. Ahmad Saffian, H.; Hyun-Joong, K.; Md Tahir, P.; Ibrahim, N.A.; Lee, S.H.; Lee, C.H. Effect of lignin modification on properties of kenaf core fiber reinforced poly(butylene succinate) biocomposites. *Materials* **2019**, *12*, 4043. [CrossRef] [PubMed]
93. Wu, F.; Misra, M.; Mohanty, A.K. Sustainable Green composites from biodegradable plastics blend and natural fibre with balanced performance: Synergy of Nano-structured blend and reactive extrusion. *Compos. Sci. Technol.* **2020**, *200*, 108369. [CrossRef]
94. Sahoo, S.; Misra, M.; Mohanty, A.K. Effect of compatinilizer and fillers on the properties of injection molded lignin-based hybrid green composites. *J. Appl. Polym. Sci.* **2013**, *127*, 4110–4121. [CrossRef]
95. Khankrua, R.; Pivsa-Art, S.; Hiroyuki, H.; Suttiruengwong, S. Thermal and mechanical properties of biodegradable polyester/silica nanocomposites. *Energy Procedia* **2013**, *34*, 705–713. [CrossRef]
96. Ostrowska, J.; Sadurski, W.; Paluch, M.; Tyński, P.; Bogusz, J. The effect of poly(butylene succinate) content on the structure and thermal and mechanical properties of its blends with polylactide. *Polym. Int.* **2019**, *68*, 1271–1279. [CrossRef]
97. Karakehya, N. Comparison of the effects of various reinforcements on the mechanical, morphological, thermal and surface properties of poly(butylene succinate). *Int. J. Adhes. Adhes.* **2021**, *110*, 102949. [CrossRef]
98. Singsang, W.; Suetrong, J.; Choedsanthia, T.; Srakaew, N.L.; Jantrasee, S.; Prasoetsopha, N. Properties of biodegradable poly(butylene succinate) filled with activated carbon synthesized from waste coffee grounds. *J. Mater. Sci. Appl. Energy* **2021**, *10*, 87–95.
99. Huang, S.; Pan, B.; Wang, Q. Study on the hardness and wear behavior of eco-friendly poly(butylene succinate)-based bamboo carbon composites. *Arab. J. Sci. Eng.* **2019**, *44*, 7997–8003. [CrossRef]
100. Hertzberg, R.W.; Vinci, R.P.; Hertzberg, J.L. *Deformation and Fracture Mechanics of Engineering Materials*, 6th ed.; John Wiley & Sons: Hoboken, NJ, USA, 2020.
101. Wong, K.J.; Zahi, S.; Low, K.O.; Lim, C.C. Fracture characterization of short bamboo fibre reinforced polyester composites. *Mater. Des.* **2010**, *31*, 4147–4154. [CrossRef]
102. Pivsa-Art, S.; Pivsa-Art, W. Eco-friendly bamboo fiber reinforced poly(butylene succinate) biocomposites. *Polym. Compos.* **2021**, *42*, 1752–1759. [CrossRef]
103. Kim, G.M.; Michler, G.H. Micromechanical deformation processes in toughened and particle filled semicrystalline polymers: Part 2. Model representation for micromechanical deformation processes. *Polymer* **1998**, *39*, 5699–5703. [CrossRef]

104. Shahzad, A.; Isaac, D.H. Weathering of Lignocellulosic Polymer Composites. In *Lignocellulosic Polymer Composites: Processing, Characterisation and Properties*; Thakur, V.K., Ed.; John Willey & Sons: Hoboken, NJ, USA, 2014.
105. Huang, H.X.; Zhang, J.J. Effects of filler–filler and polymer–filler interactions on rheological and mechanical properties of HDPE–wood composites. *J. Appl. Polym. Sci.* **2009**, *111*, 2806–2812. [CrossRef]

Article

Artificial Ageing, Chemical Resistance, and Biodegradation of Biocomposites from Poly(Butylene Succinate) and Wheat Bran

Emil Sasimowski [1], Łukasz Majewski [1,*] and Marta Grochowicz [2]

[1] Department of Technology and Polymer Processing, Faculty of Mechanical Engineering, Lublin University of Technology, Nadbystrzycka 36, 20-618 Lublin, Poland; e.sasimowski@pollub.pl

[2] Department of Polymer Chemistry, Institute of Chemical Sciences, Faculty of Chemistry, Maria Curie-Sklodowska University, M. Curie-Sklodowska 3, 20-031 Lublin, Poland; mgrochowicz@umcs.pl

* Correspondence: l.majewski@pollub.pl

Citation: Sasimowski, E.; Majewski, Ł.; Grochowicz, M. Artificial Ageing, Chemical Resistance, and Biodegradation of Biocomposites from Poly(Butylene Succinate) and Wheat Bran. *Materials* 2021, *14*, 7580. https://doi.org/10.3390/ma14247580

Academic Editor: Krzysztof Moraczewski

Received: 10 November 2021
Accepted: 7 December 2021
Published: 9 December 2021

Publisher's Note: MDPI stays neutral with regard to jurisdictional claims in published maps and institutional affiliations.

Copyright: © 2021 by the authors. Licensee MDPI, Basel, Switzerland. This article is an open access article distributed under the terms and conditions of the Creative Commons Attribution (CC BY) license (https://creativecommons.org/licenses/by/4.0/).

Abstract: The results of comprehensive studies on accelerated (artificial) ageing and biodegradation of polymer biocomposites on PBS matrix filled with raw wheat bran (WB) are presented in this paper. These polymer biocomposites are intended for the manufacture of goods, in particular disposable packaging and disposable utensils, which decompose naturally under the influence of biological agents. The effects of wheat bran content within the range of 10–50 wt.% and extruder screw speed of 50–200 min^{-1} during the production of biocomposite pellets on the resistance of the products to physical, chemical, and biological factors were evaluated. The research included the determination of the effect of artificial ageing on the changes of structural and thermal properties by infrared spectra (FTIR), differential scanning calorimetry (DSC), and thermogravimetric analysis (TG). They showed structural changes—disruption of chains within the ester bond, which occurred in the composition with 50% bran content as early as after 250 h of accelerated ageing. An increase in the degree of crystallinity with ageing was also found to be as high as 48% in the composition with 10% bran content. The temperature taken at the beginning of weight loss of the compositions studied was also lowered, even by 30 °C at the highest bran content. The changes of mechanical properties of biocomposite samples were also investigated. These include: hardness, surface roughness, transverse shrinkage, weight loss, and optical properties: colour and gloss. The ageing hardness of the biocomposite increased by up to 12%, and the surface roughness (R_a) increased by as much as 2.4 µm at the highest bran content. It was also found that ageing causes significant colour changes of the biocomposition (ΔE = 7.8 already at 10% bran content), and that the ageing-induced weight loss of the biocomposition of 0.31–0.59% is lower than that of the samples produced from PBS alone (1.06%). On the other hand, the transverse shrinkage of moldings as a result of ageing turned out to be relatively small, at 0.05%–0.35%. The chemical resistance of biocomposites to NaOH and HCl as well as absorption of polar and non-polar liquids (oil and water) were also determined. Biodegradation studies were carried out under controlled conditions in compost and weight loss of the tested compositions was determined. The weight of samples made from PBS alone after 70 days of composting decreased only by 4.5%, while the biocomposition with 10% bran content decreased by 15.1%, and with 50% bran, by as much as 68.3%. The measurements carried out showed a significant influence of the content of the applied lignocellulosic fillers (LCF) in the form of raw wheat bran (WB) on the examined properties of the biocompositions and the course of their artificial ageing and biodegradation. Within the range under study, the screw speed of the extruder during the production of biocomposite pellets did not show any significant influence on most of the studied properties of the injection mouldings produced from it.

Keywords: accelerated ageing; biofiller; thermal properties; agro-waste materials; agro-flour filler; natural filler; thermal resistance; discolouration; lignocellulosic materials; biopolymer; composites

1. Introduction

Plastic is one of the most popular construction materials today. Its annual global production is counted in hundreds of million tons [1–4]. The scale of plastic use results in significant environmental pollution, which has been a large problem for many years [3,5–7]. Nevertheless, it is predicted that global plastic production may even double in the next 15–20 years, which may lead to inefficiencies in plastic waste disposal and recycling mechanisms [8–10]. Potential opportunities to reduce the scale of the problem are seen, inter alia, in the assumptions of a circular economy, mechanisms for reduction of plastic consumption, and in industrial scale popularization of biopolymers and biocomposites based on biopolymer matrix [8,11–17].

Biopolymers are mainly materials capable of biodegradation in environmental conditions with participation of microorganisms, as this term is also used for non-biodegradable polymers obtained from substrates of natural origin [18–20]. The kinetics of polymer degradation depends mainly on the chemical structure of macromolecules, which translates into their thermal resistance, chemical activity, or resistance to oxidation or action of acids, bases, and enzymes. The chemical structure of a polymer will therefore directly affect its susceptibility to degradation caused by various external factors [21–26]. Depending on the sensitivity of the material to the particular stimulating factors, other types of degradation are distinguished in addition to microbial biodegradation. These include thermal degradation at elevated temperatures, mechanical degradation caused by long-term stress, oxidative degradation in oxygen-containing atmospheres, photodegradation caused by light rays, hydrolytic degradation at high humidity, corrosion caused by the activity of chemical substances, and the effect of high-energy electromagnetic radiation (e.g., UV) [20,27–30]. The above-mentioned physical and chemical factors lead to irreversible changes in the material structure, such as a change in the proportion of the crystalline phase or decomposition of macromolecules into shorter chains, and consequently, also to changes in mechanical and physical properties [31–34]. These changes can be manifested, e.g., by an increase in the stiffness and brittleness leading to the fragmentation of material and an increase in the surface area affected by external factors, but also by a loss of mass or a change in the colour and roughness of the surface [35–39]. Therefore, the activity of physical and chemical agents significantly accelerates the degradation process of biopolymers increasing the specific surface area of their influence by fragmentation of the material. However, the actual process of degradation and mineralization of biopolymers to simple substances such as water, carbon dioxide, and inorganic compounds takes place primarily through the activity of microorganisms [28,29,37,40–42]. Different biopolymers have different susceptibility to degradation agents and different degradation rates [43]. Water-soluble biopolymers, such as poly(vinyl alcohol) or starch-based plastics, will degrade most rapidly in an aqueous environment [44–46]. Some polymers, with adequate moisture content, biodegrade relatively quickly at room temperature or slightly higher by incubation with mesophilic bacteria. The latter include polyhydroxyalkanoate and polyhydroxybutyrate [47,48]. There is also a group of biopolymers that require appropriate conditions for proper and quick biodegradation process. Such factors may include the appropriate level of humidity, the presence of specific strains of bacteria and the pH that is adequate for them, or an elevated temperature corresponding to thermophilic microorganisms. Examples of such biopolymers include polycaprolactone, polylactide, and poly(butylene succinate) (PBS) [37,49–53]. Resistance to external factors has a direct impact on the choice of biopolymer for specific applications and the life cycle of polymer products [54–56].

The life cycle encompasses all stages related to the presence of a given product on the market, from designing, manufacturing, exploitation, and management of waste generated during its use. The duration of each stage varies depending on the type of product and the material used [54–58]. In the case of polymeric materials of petrochemical origin, the landfilling stage of waste management, the process of natural decomposition of polyolefins, for example, takes up to several hundred years. This is a direct cause of the ever increasing environmental pollution from plastic waste [59–61]. The use of biodegradable polymers, especially those that degrade under conditions other than their operating conditions, allows the situation to be reversed. Under standard conditions, Poly(butylene succinate) can be exploited for a long time, and when directed to composting under conditions stimulating biodegradation, it decomposes within a few months [52,62]. This in combination with mechanical properties comparable to polyolefins as well as good processability and thermal resistance make PBS an attractive alternative for polymers of petrochemical origin [63–65]. However, PBS has a multi-stage and relatively complex manufacturing process, which at the same time is very expensive, drastically increasing the price of PBS as a raw material [64,66–69]. In addition, running such complex production processes on an industrial scale cannot be done without an environmental footprint [70,71]. Therefore, despite the biodegradability of PBS, it is justifiable to apply mechanisms to minimize the consumption of raw materials and energy during manufacturing, both ecologically and economically. This is done, among others, by filling PBS with lignocellulosic fillers (LCF) of natural origin, which are most often technological wastes from food and agricultural industry. This reduces the consumption of expensive plastic while using waste from other economic sectors [72,73]. Examples of such waste fillers used to produce PBS matrix biocomposites include ground rice husks [74], wheat bran [75], pistachio [76] and peanut shells [77], almond kernels [78], or even wine lees [79], as well as apple [80], and grape pomace [81]. The presence of LCF significantly modifies the physical, mechanical, thermal, and processing properties of such compositions [72,82–84]. In addition, due to their chemical structure, LCFs display hydrophilic character and are much less resistant to physical and chemical factors and microbial activity compared to PBS or other polymeric materials [85,86]. This may be a limiting factor, resulting in a shorter life cycle for products made from PBS/LCF biocomposites and a reduced spectrum of potential applications, but also imply simplified disposal through accelerated biodegradation. Therefore, when designing new composite systems and using new types of fillers, it is important to know the full characteristics of the material, taking into account processability, physical, mechanical and thermal properties, chemical resistance and resistance to ageing, as well as kinetics and course of biodegradation.

In spite of numerous papers dealing with biodegradable polymeric materials, the current literature lacks studies on the characterization of PBS-based biocomposites with a filler in the form of raw wheat bran (WB). A biocomposite with such a composition undergoing natural decomposition is the subject of a patent [87] and constitutes a completely new material. Consequently, the results of testing the resistance of PBS/wheat bran biocomposites to external factors presented in this paper are a scientific novelty and, at the same time, will be helpful in determining the spectrum of applications of the new biocomposite. The aim of this work was to evaluate the influence of wheat bran content in the range 10–50 wt.% and extruder screw speed in the range 50–200 min^{-1} during the production of composite pellets on the resistance of the manufactured biocomposite injection mouldings to chemical, biological, and physical factors. Changes of selected structural, thermal, mechanical, and optical properties of biocomposites, which occurred as a result of artificial ageing, were also evaluated.

2. Experimental Procedures

2.1. Test Stand

Injection moulding of the biocomposite was carried out using an Arburg Allrounder 320 C (Arburg, Lossburg, Germany) screw injection moulding machine equipped with a dual cavity mould to produce specimens for strength testing. The shape and dimensions of the samples were in accordance with ISO 294-1:2017-07 [88]. The specimens were dog-bone-shaped with a total length of 150 mm and a thickness of 4 mm; the width of the measuring part was 10 mm, and the grip part was 20 mm. Due to the danger of thermal decomposition of biocomposite components, low temperatures were applied during processing. The temperature of the plasticizing system was 30 °C in the feed zone, and in the individual heating zones: I—125 °C, II—145 °C, III—155 °C, IV—160 °C, and the injection nozzle temperature was 155 °C. The temperature of the thermostated mould was 25 °C. The injection of the biocomposition was performed at the following settings: maximum injection pressure 120 MPa, polymer flow rate 20 cm^3/s, packing pressure 110–80 MPa, packing time 15 s, and cooling time 20 s. In the case of the highest bran fraction of 50% (DOE layout 8), the injection pressure was increased to 130 MPa and the packing pressure to 120–80 MPa, which made it possible to eliminate incomplete filling of mould cavities, occurring at lower values of these parameters.

2.2. Materials

The prepared test samples were based on employing as matrix PBS with the trade name BioPBS FZ91 PB [89], manufactured in the form of pellets (PTT MCC BIOCHEM CO., LTD, Bangkok, Thailand). It was synthesized using bio-based succinic acid and 1,4-butanediol. This material is intended for manufacturing general-purpose products via injection moulding.

Wheat bran (WB), i.e., wheat grain shells, is a process waste product from the refining of white flour. It comes in the form of thin flakes with dimensions up to a few mm and was obtained from a local mill near the city of Lublin (Poland). WB is primarily composed of fibrous substances such as cellulose, lignin and hemicellulose, but includes phytic acid, oligosaccharides, non-starch polysaccharides, as well as fats and proteins [90,91].

2.3. Research Programme and Methodology

Experimental tests were carried out according to the adopted design of experiment (DOE), the experimental layouts of which are shown in Table 1. The following independent variables—adjustable conditions of the process—were assumed: mass content of wheat bran introduced into poly(butylene succinate) u = 10–50 wt.% and extruder screw speed n = 50–200 min^{-1} when obtaining processed biocomposite pellets. A detailed characterization and analysis of the twin-screw extrusion process of biocomposite pellets was presented in a previous paper [75]. Measurements were made in at least five replicates. PBS and its composites with bran containing 10% (3), 30% (5), and 50% (4) of biofiller obtained at the same screw speed and the composites containing 30% of biofiller, but obtained at n equal to 50 min^{-1} (1) and 200 min^{-1} (2) were selected for the tests. Such a choice of test samples allows us to consider the influence of the biofiller content and n parameter on the values of tested parameters.

Table 1. Experimental design.

Experimental Design Layout	n, min^{-1}	u, %
1	50	30.0
2	200	30.0
3	125	10.0
4	125	50.0
5	125	30.0
6 PBS	-	0

Experimental tests of injection mouldings made of polymer compositions, prepared according to DOE, were carried out and involved:

- Performing accelerated ageing test in Xenotest Alpha+ accelerated ageing chamber (Atlas, Chicago, IL, USA). The test lasted 1000 h, the samples were irradiated with a xenon lamp emitting radiation imitating solar radiation. The irradiance of 60 W/m^2 and the daylight filter system were used. The temperature in the chamber was kept at 38 °C, humidity 50%. During irradiation, the samples were sprayed with distilled water for 18 min every 102 min. The measurement conditions were in accordance with the standard: ISO 4892-2:2013 [92]. Before and after the test, and for selected tests also during the test, the following properties, which are the determinants of resistance to artificial ageing, were measured:

- The infrared spectra (FTIR) analysis of the samples. The accelerated ageing test was stopped for 10 min each time at 250, 500, and 750 h in order to obtain the FTIR spectra of the test samples. The used parameters of the ageing test allow us to assess the behaviour of composites in artificial weathering conditions. The FTIR of tested samples were taken using Tensor 27 spectrometer (Bruker, Germany) equipped with ATR (attenuated total reflectance) module with diamond crystal. The FTIR spectra were recorded from 600 to 4000 cm^{-1} with 32 scans per spectrum and the resolution of 4 cm^{-1}. The infrared carbonyl stretching region was deconvoluted into the Gaussian curves using OPUS 7.0 software;

- Differential scanning calorimetric (DSC) studies were performed on DSC 204 F1 Phoenix (NETZSCH, Günzburg, Germany) working with the NETZSCH Proteus software, in accordance with standard ISO 11357-1:2016 [93]. Each measurement was carried in three cycles: heating from −150 °C to 140 °C with heating rate of 10 K/min (I heating), cooling from 140 °C to −150 °C with cooling rate of 10 K/min, heating from −150 °C to 140 °C with heating rate of 10 K/min (II heating). Samples with mass about 10 mg were analysed in closed pierced aluminium pans in argon atmosphere with flow rate of 25 mL/min. Parameters such as melting enthalpy (ΔH_m), melting temperature (T_m), crystallization temperature (T_c), glass transition temperature (T_g), and crystallinity degree (X_c) were calculated based on the obtained thermograms. The T_g value was adopted as the inflection point of the DSC curve in the area of the glass transition. The X_c parameter was calculated from the equation:

$$X_c = \left(\frac{\Delta H}{(1-u) \times \Delta H_{100\%}} \right) \times 100\% \quad (1)$$

assuming that for pure PBS, $\Delta H_{100\%}$ = 110.3 J/g [94];

- Measurements of transverse shrinkage, determined as the percentage difference in linear dimension of the specimen before accelerated ageing and after the completed accelerated ageing cycle. Recommendations were applied from the standard ISO 294-4:2018 [95];

- Thermogravimetric analysis was performed in synthetic air with the use of STA 449 F1 Jupiter (Netzsch, Günzburg, Germany) coupled with FTIR TENSOR 27 spectrometer (Bruker, Germany). The measurement conditions were as follows: temperature range of 40–800 °C, heating rate 10 K/min gas flow 25 mL/min, and sample mass approximately 10 mg. Samples were analysed in Al_2O_3 opened crucibles;

- Colour measurement of samples according to ASTM E308 [96], for which X-Rite Ci4200 spectrophotometer was used. The colour is described in the CIELab system, where it is specified in L^*, a^*, b^* space. Parameter a describes the colour from green (negative values) to red (positive values), parameter b—the colour from blue (negative values) to yellow (positive values), and parameter L is the luminance—brightness, representing the grey scale from black to white (value 0 corresponds to black and 100 to white). The difference between two colours—two points in the three-dimensional space L^*, a^*, b^*—is described by the relation:

$$\Delta E = \sqrt{\Delta L^2 + \Delta a^2 + \Delta b^2} \quad (2)$$

in which: ΔL, Δa, and Δb denote the difference in colour parameters between the compared samples, respectively.

- The roughness parameters of samples before and after ageing were given on the basis of results obtained from the optical profiler Contur GT (Bruker, Karlsruhe,

Germany). Average roughness parameter (Ra) was calculated in accordance with ASME B46.1 [97], with the use of Vision 4.20 software. Measurements were performed at room temperature, the area of 156 μm × 117 μm was scanned in three different places for each sample. After the accelerated ageing test, the exact same sections of the surface of the samples were analysed.

- ○ Measurement of the surface gloss of the samples using an X-Rite Ci4200 spectrophotometer (X-Rite, Grand Rapids, MI, USA), performed in accordance with ISO 2813:2001 [98] at the 60° image slit opening angle for the light source and receiver.
- ○ Weight loss measurements, determined as the percentage difference between the initial weight of dry samples and the weight of dry samples after completion of the accelerated ageing cycle;
- ○ Hardness test using ball indentation method. The measurement was carried out in accordance with ISO 2039-1:2004 [99] using an HPK 8411 hardness tester with a ball-shaped indenter of 5 ± 0.025 mm diameter.
- ○ Shore D hardness test was performed as per ISO 868:2003 [100] with the use of Shore durometer model ART.13 by Affri System (Induno Olona, Italy) with a cone-shaped indenter with a rounded tip.
- Assessing the absorption of polar and non-polar fluids. The tested biocomposite samples were completely immersed in water and vegetable oil for 7 days. The percentage difference in mass and linear dimensions after immersion and before immersion in the fluid was then specified to determine the values of fluid absorption and swelling. The test procedure was in accordance with ISO 175:2010 [101];
- The chemical resistance of PBS and its biocomposites was assessed in 1 M solution of NaOH and 1 M solution of HCl at room temperature. The test consisted in placing the samples (10 mm × 10 mm × 4 mm) dried to constant weight in glass bottles containing the above-mentioned solutions. The samples were taken out from solutions at specified intervals, dried on blotting paper and then weighed. The percent change of the mass was calculated according to the equation:

$$mass\ change(\%) = \frac{(m - m_0)}{m_0} \times 100\% \qquad (3)$$

where m is the final mass of the sample; m_0—is the initial mass of the sample.
- Laboratory biodegradation tests conducted under controlled conditions as per ISO 20200:2015 [102]. Samples were placed in separate polypropylene bioreactors filled with commercially available compost from a local waste management facility. Then the bioreactors were placed in a climate chamber with a temperature of 58 °C and air humidity of 60%. At the intervals specified in the standard, water was replenished in the bioreactors and the compost was homogenized. After fixed incubation periods of 7, 14, 21, 28, 42, 56, and 70 days, respectively, the samples were extracted, washed, dried to constant weight, and weighed. On the basis of mass measurements, its loss was determined, which is an indicator of biodegradation rate.

3. Results

The results of the study were statistically processed in STATISTICA 13. The ANOVA variance analysis was used to determine if there were significant differences between the compared results. Before performing the above, the required assumptions such as normality of distribution of variables and homogeneity of variance were checked. Non-parametric test was applied where the variables did not meet the condition of normality of distribution. When these analyses confirmed the presence of statistically significant differences, Tukey's multiple comparisons test was then performed. This test, called the post-hoc test, facilitates grouping of means and separation of homogeneous groups. The significance level of $p = 0.05$ was adopted in the applied analyses. The results obtained are presented in the form of graphs, where the mean values are marked together with the standard error. For most of the properties under study, the influence of both variable parameters, i.e., bran content u, and extruder screw speed n, was presented separately on graphs.

3.1. Accelerated Artificial Ageing

3.1.1. Chemical Structure

The total residence time of the samples in the ageing chamber was 1000 h, but the samples were subjected to FTIR analysis several times during the study at specific time intervals. FTIR spectra of PBS before and after ageing are presented in Figure 1, whereas FTIR spectra performed in successive hours of the ageing test for the composite containing the highest percentage of bran—sample number 4—are shown in Figure 2. The course of the spectra obtained before and after ageing is different for both PBS and composite 4. These differences are due to the structural changes that occurred in the samples as a result of artificial weathering. Figure 2 shows that the structural changes in composite 4 occurred already after the first 250 h of the composite's stay in the ageing chamber, for the other tested materials the changes occurred in a similar way. Analysis of the FTIR spectra shows that the intensity of the absorption bands at 1262–1227 cm^{-1}, which originate from the asymmetric stretching vibration of the C–O–C group, and the band at 991 cm^{-1} (in the composite) and 986 cm^{-1} (in PBS) originating from the vibration of the chain backbone have clearly decreased. Additionally, as the accelerated ageing test progresses, an increase in the intensity of the absorption bands at 1153 cm^{-1} and 1329 cm^{-1} corresponding to the symmetric stretching vibrations of the C–O–C group is observed, as well as a marked decrease in the intensity of the band at 1713 cm^{-1} originating from the vibration of the C=O group compared to the band at 1153 cm^{-1}. On this basis, one may conclude that the effects of UV radiation, elevated temperature, and water result in structural changes in PBS within the ester bond. Several different mechanisms of PBS chain cleavage have been proposed in the literature; among them, α-hydrogen abstraction leading to the formation of carboxylic and aldehyde groups via hydroperoxides; the Norrish I of chain cleavage leading to carboxyl, aldehyde and ether groups; the hydroxyl end groups oxidation leading to carboxyl groups and β-hydrogen abstraction resulting in unsaturated compounds [103–105]. Analysis of the FTIR spectra of the composites after ageing indicates that the first three mentioned photodegradation mechanisms may have taken place. However, the lack of absorption bands from C=C unsaturated groups in the spectra excludes the possibility of β-hydrogen chain scission reaction. Moreover, considering the region of the spectra of PBS and composite 4 in the vibrational range of the C=O carbonyl group, it is apparent that this band consists of at least two overlapping bands: with a maximum at about 1714 cm^{-1} and with a maximum at about 1732 cm^{-1}. It is assumed that the structure of semicrystalline polymers can be described by a three-phase model including mobile amorphous fraction (MAF), rigid amorphous fraction (RAF), and crystalline phase [106,107]. The absorption bands around 1736 cm^{-1}, 1720 cm^{-1}, and 1714 cm^{-1} have been assigned to C=O in MAF, RAF, and crystalline phase of PBS, respectively [106]. Deconvolution of the carbonyl bands of PBS (Figure 3) as well as composite 4 (Figure 4) demonstrated that they consist of four bands. Assuming that the band at 1736 cm^{-1} originates from the amorphous phase, its percentage of the carbonyl band after PBS ageing increased by 15%, and the percentage of the band originating from the RAF phase increased by 3%. For composite 4, a 3% increase in the C=O band contribution of the MAF phase was calculated, but also a 1.5% increase in the RAF band contribution. For PBS and composite 4, the percentage of C=O bands at about 1718–1712 cm^{-1} (assuming that they originate from the crystalline phase) decreased after ageing by 14% and 6%, respectively. Such calculations may indicate a decrease in the crystallinity of the samples after the ageing test, but they also contradict the crystallinity tests performed with the use of the DSC method. It should be taken into account that in the case of FTIR spectra of the samples after ageing, the carbonyl band may have changed not only due to changes in the crystallinity of the samples but also due to decomposition processes of the ester bond. Such decomposition may have resulted in the formation of carbonyl groups, such as aldehyde groups or carboxyl groups, which may have affected the calculated percentage rates in the C=O band.

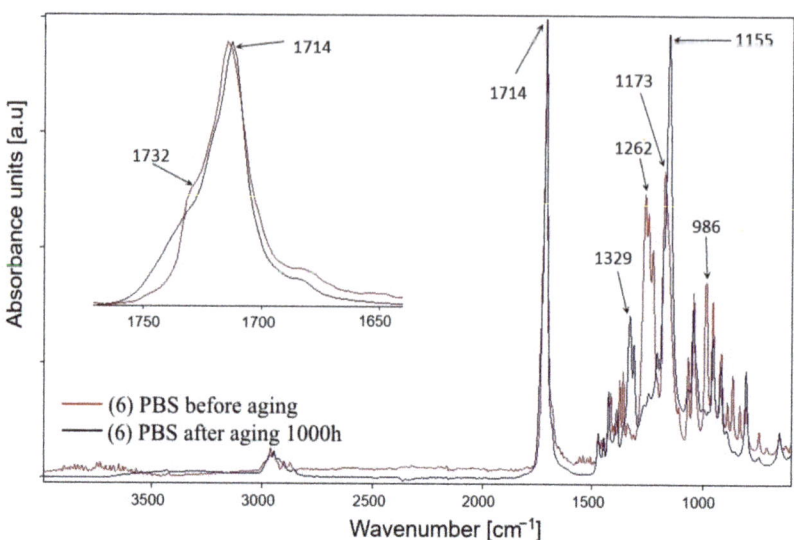

Figure 1. ATR-FTIR spectra of PBS acquired before and after ageing test.

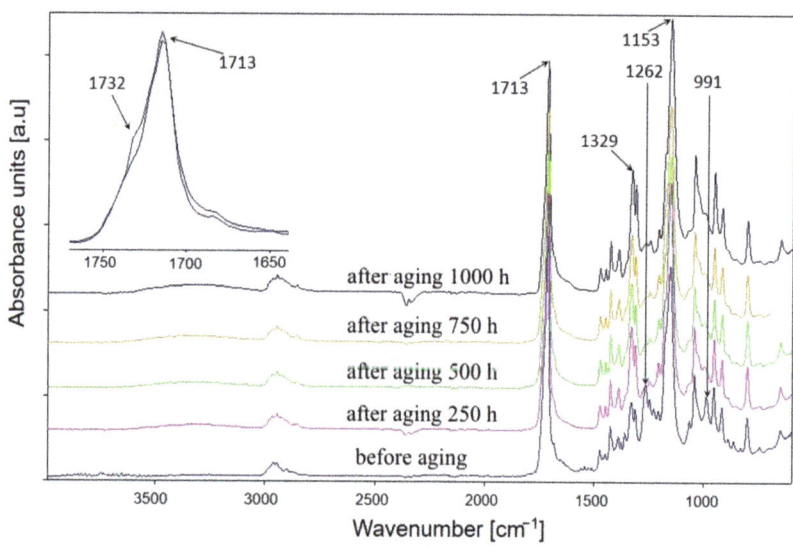

Figure 2. ATR-FTIR spectra of biocomposite 4 containing 50% of bran, in ageing test intervals.

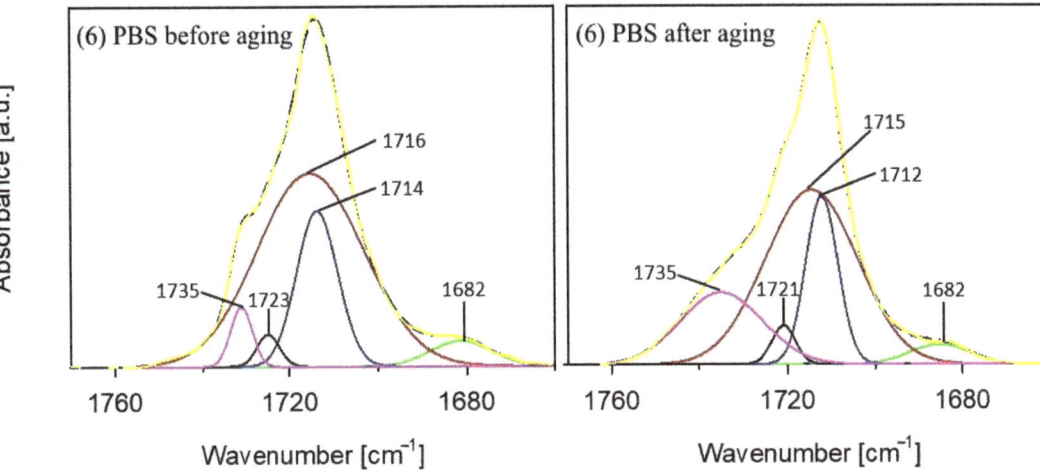

Figure 3. Deconvolution of the carbonyl stretching region of PBS before and after aging (dashed line: resolved peaks, solid line: recorded spectra).

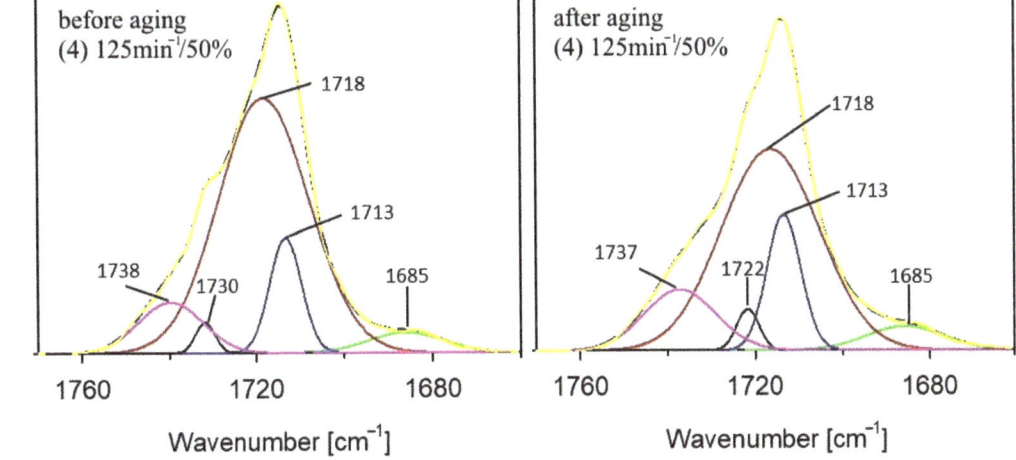

Figure 4. Deconvolution of the carbonyl stretching region of composite with 50% bran before and after aging (dashed line: resolved peaks, solid line: recorded spectra).

3.1.2. Differential Scanning Calorimetry

All results obtained from the DSC analysis before and after ageing the samples, including glass transition temperature (T_g), crystallization temperature (T_c), melting temperature (T_m), melting enthalpy (ΔH_m), as well as the calculated degree of crystallinity (X_c), are summarized in Table 2. As can be seen from Table 2, it was not possible to observe the glass transition temperature for all tested samples during the first heating scan. The structure of the samples could contain absorbed water and possibly low molecular weight organic compounds formed as a result of the degradation of composites in the ageing test conditions, which is evidenced by the broad endothermic peak on the DSC curves from the first heating scan. Therefore, when considering the effect of composition and conditions of preparation of composites on thermal properties, we relied on the results from the second heating cycle (Figure 5). The T_g values for neat PBS after ageing increased from −32 °C to −27 °C. On the other hand, no significant change in T_g values after ageing was observed for the composites. The change in T_g value for pure PBS is related to the increase in the degree of crystallinity of the polymer after ageing [108]. An increase in crystallinity degree was also observed for the composites, but due to the presence of bran particles, which separate the polymer chains, no change in T_g values was observed. Interestingly, the cooling curves show exothermic peaks originating from crystallization process (Figure 6). In case of PBS and sample 3, which contains only 10% of filler, two maxima on curves of approximately 77 °C and 85 °C are clearly visible, while before the ageing process, only one crystallization temperature (approximately 86 °C) was observed for these materials (Table 2) [109]. This is probably due to the existence of two types of crystals in the structure of the samples. On the basis of FTIR spectra, it can be expected that changes in the chemical structure of PBS took place during ageing, consisting in breaking of polymer chains within ester bonds. This led to the formation of fractions of the polymer with shorter chains, which have a greater capacity for mobility and thus increase their ability to form an ordered phase. As a result, this contributes to lower crystallization temperatures [103]. Moreover, after the ageing process, a clear increase in the degree of crystallinity is observed not only for PBS and composite 3, but also for all other composites. FTIR analysis of the chemical structure of the composites after ageing showed that the polymer chain disintegration reaction occurred in every tested sample. This resulted in the formation of PBS chains with smaller molar masses, which are more easily organized into a crystalline phase. On the other hand, amorphous regions of polymers are more susceptible to degradation than crystalline ones [104,110,111]. The decomposition of the chains present in the amorphous phase caused their release from the amorphous region and they could then arrange themselves into crystalline structures. This ultimately led to an increase in the degree of crystallinity of the samples after ageing, this increase reaching up to 48% for the composite containing 10% bran. With more bran, the increase in the degree of crystallinity was not as great due to limitations in chain mobility caused by the large presence of filler.

The X_c values clearly increased after the ageing process, while no significant changes were observed in T_m. Only for neat PBS is a decrease in T_m evident compared to T_m before ageing. In addition, the endothermic peak in the DSC curve originating from the melting of the crystalline phase after ageing is narrower than for PBS before ageing (Figure 5), without a broadened arm toward lower temperatures. This is due to the increased crystallinity of the sample. DSC curves from second heating scan for all tested materials after ageing show another difference in comparison with curves before ageing (Figure 5). A small endothermic peak preceding the actual melting peak was observed in the melting temperature region before ageing, which we attributed to the melting of less perfect PBS crystallites [75]. Previously reported studies on the melting process of semicrystalline PBS also showed a similar melting behaviour [112–115]. This complex behaviour is explained by melting recrystallization mechanism, whereby more defective crystals melt at a lower temperature, then further heating of the sample leads to their recrystallization, fusion and further melting at a higher temperature. After the composite ageing process, the first melting peak decreased significantly, which can be attributed to the increase in the degree of crystallinity of the materials.

Considering the changes in the parameters given in Table 2 before and after ageing for materials 1, 2, and 5 (obtained at different screw speeds and the same bran content), changes in the degree of crystallinity of the samples are evident. In each case, the degree increased; the increase of X_c was consistent with the decreasing value of n. It is clear that the breakdown of polymer chains into shorter units is responsible for the increase of X_c. In this case, it can be assumed that the homogeneity of bran distribution in the PBS network also affects the ageing process of the composites. Composite 2, obtained at the highest screw speed, shows the most homogeneous distribution of the biofiller in the PBS matrix among samples 1, 5, and 2 [75]. In its case, the increase of X_c after ageing is only 12%, while sample 1, obtained with the smallest n, showed an increase of X_c by 45%. The shorter polymer chains of PBS, formed by ageing of the samples, have a greater ability to rearrange into crystalline structures, and the bran acts as a nucleating agent. It can be claimed that heating the samples above the melting point and cooling again led to an increase in the homogeneity of the composite structure, and shorter PBS chains formed crystallites in the vicinity of the bran particles.

Table 2. DSC data for PBS and its composites obtained before (denoted as b) and after ageing (denoted as a).

Sample	Heating I				Cooling		Heating II			
	T_g (°C)	T_m (°C)	ΔH_m (J/g)	X_c (%)	T_c (°C)	T_g (°C)	T_m (°C)	ΔH_m (J/g)	X_c (%)
PBS b	−25.6	121.3	72.1	65.4	86.4	−31.7	118.5	66.7	60.5
PBS a	−25.7	116.3	94.9	86.0	77.6/88.9	−27.6	113.5	84.4	76.5
1(50/30) b	−31.3	118.9	51.1	66.2	82.2	−32.1	116.3	38.2	49.5
1(50/30) a	-	118.5	71.9	93.1	80.3	−32.0	115.3	55.6	72.0
2(200/30) b	−31.5	118.9	51.7	67.0	81.3	−32.4	117.2	39.4	51.0
2(200/30) a	-	117.7	49.5	64.1	81.8	−32.2	116.3	44.3	57.4
3(125/10) b	−33.3	118.8	63.3	63.8	86.4	−33.4	115.8	55.3	55.7
3(125/10) a	−26.6	118.8	98.2	98.9	77.4/85.3	−33.5	115.0	81.9	82.5
4(125/50) b	−32.5	117.6	39.0	70.7	83.3	−32.5	116.6	30.6	54.5
4(125/50) a	−15.3	118.8	48.1	87.2	83.8	−32.4	116.6	35.3	64.0
5(125/30) b	−31.4	118.0	41.9	54.3	79.6	−32.3	118.2	38.8	50.3
5(125/30) a	-	117.7	65.5	84.8	82.5	−31.5	114.7	49.8	64.5

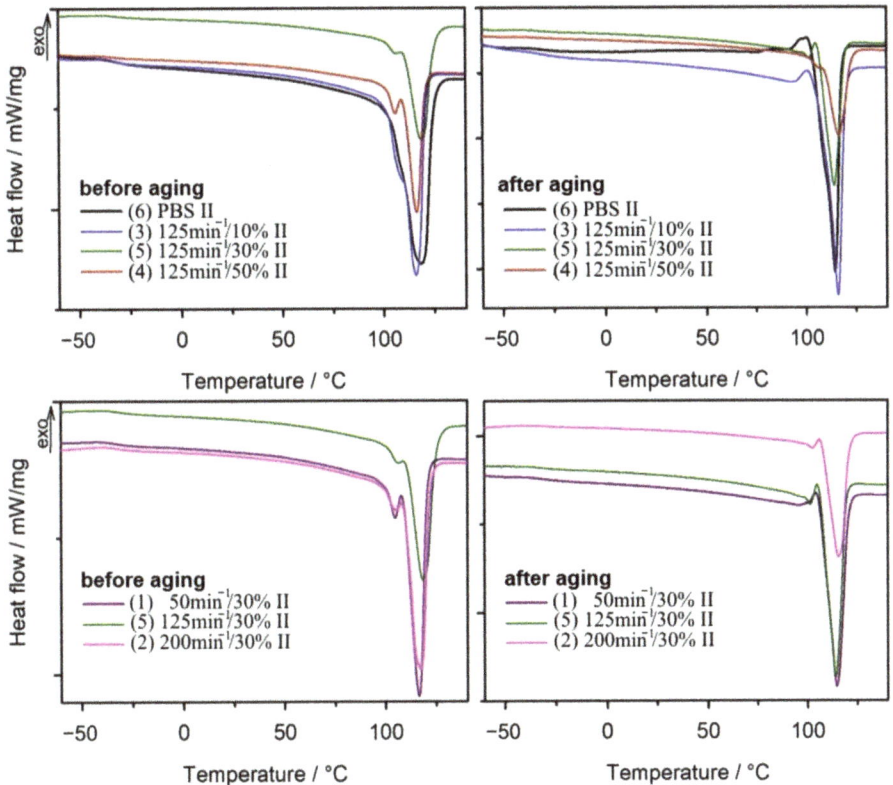

Figure 5. DSC thermograms of the second heating scans for PBS and composites obtained before and after ageing.

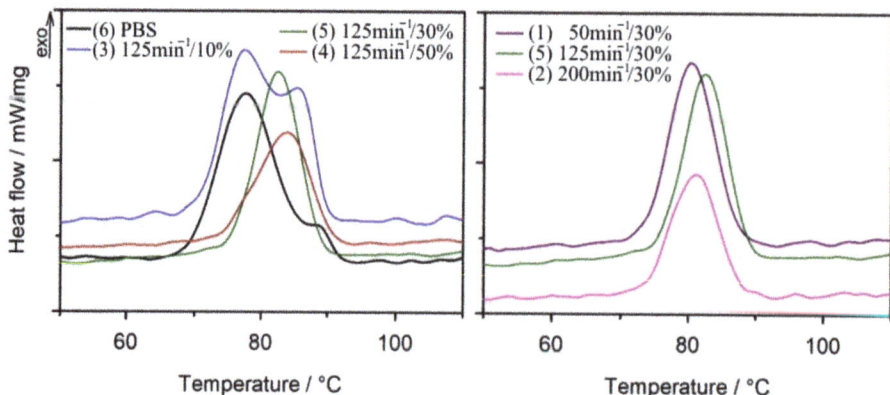

Figure 6. Cooling curves for PBS and its composites obtained after ageing.

3.1.3. Transverse Shrinkage

The effects of the bran content by weight in the polymeric composition and the screw speed during its manufacturing on the transverse shrinkage of the moulded parts as a result of ageing are shown in Figure 7. The values of transverse shrinkage of mouldings made of PBS and polymer compositions, resulting from ageing, were relatively small and ranged from 0.05% to 0.35%. Post-hoc tests of the obtained results showed that as a result of ageing the transverse shrinkage of mouldings with 10% and 30% bran content is comparable and also significantly higher than that of samples made of PBS alone and of compositions with 50% bran content, which also do not differ significantly from each other. The effect of screw speed on shrinkage proved significant when increased from 50 min^{-1} to 125 min^{-1}. When the screw speed was further increased to 200 min^{-1}, the shrinkage remained unchanged.

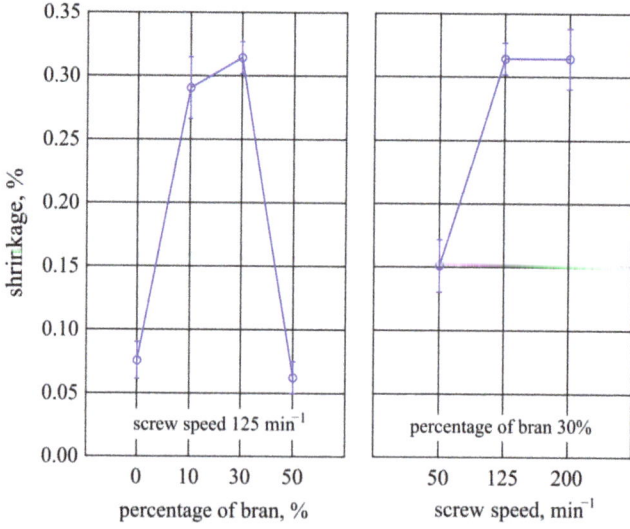

Figure 7. Transverse shrinkage of mouldings due to ageing as a function of bran mass content and extruder screw speed.

The observed changes in the linear dimensions of the samples are directly related to the effect of PBS degradation due to ageing and the increase in the degree of crystallinity.

During injection moulding, the samples undergo processing shrinkage, which is caused partly by thermal expansion effects, but mainly by the crystallization process. In a previous work [75], p-v-T studies showed that the crystalline phase of PBS has a significantly lower specific volume than its amorphous phase, which significantly affects the processing shrinkage values [109]. Arrangement of polymer chains into crystalline structures is a long-term process and although the largest increase in shrinkage is observed during cooling immediately after processing, small changes in linear dimensions can still be observed for several months [116]. The decrease in molecular weight of PBS due to ageing facilitates the rearrangement of macromolecules by increasing the degree of crystallinity, and the greater the degree of crystallinity, the greater the shrinkage of PBS [117]. Moreover, it was also shown that the filler particles act as a promoter of crystallization. This means that the arrangement of the crystalline phase occurs around the filler particles and the resulting volume loss exerts compressive stresses on them, which would further affect the shrinkage value [118,119]. At lower filler contents, the crystalline regions around the WB particles will be larger, whereas at 50% content, there will be more crystallization nuclei, but with smaller sizes, as the presence of filler will also limit the mobility of macromolecules [120,121].

3.1.4. Thermal Resistance

The thermal resistance of PBS and its composites after artificial ageing was tested in an oxidizing atmosphere. Figure 8 shows TG and DTG curves collected into two groups, depending on the amount of bran in the composite and depending on the extrusion process conditions. Table 3 shows the parameters characterizing the thermal resistance of the discussed materials examined before and after the ageing test. From the data presented, it is evident that artificial ageing altered the thermal resistance of PBS and composites. The course of TG and DTG curves obtained after ageing for composite samples is similar to the course of these curves before ageing [111]. The DTG curves still show three distinct peaks with maxima around 300 °C, 390 °C, and 480 °C. They suggest that the thermal decomposition of the composites proceeded in three stages. However, the T_{max1} and T_{max2} values are slightly shifted towards lower temperatures compared to the pre-ageing values. The situation is a little different for PBS. Before ageing, it underwent thermal decomposition in a two-stage process, while after ageing, an additional peak at about 300 °C appears on the DTG curve. This coincides with the first peak in the DTG curves for composites and corresponds to a mass loss of almost 9%. Therefore, it can be concluded that the structural fragments formed during the ageing of PBS are decomposed at about 300 °C; these are most likely short chains of PBS formed by photochemical degradation, terminated by carboxyl groups. The temperature taken as the onset of sample weight loss, $T_{5\%}$, decreased after ageing for all composites and PBS. The smallest decrease was observed for PBS, while the largest for composite 4 containing 50% bran. It is likely that artificial rain during the ageing process led to partial hydrolysis of the carbohydrate components of the bran. Similarly, the $T_{50\%}$ values after ageing also decreased slightly compared to the values before ageing. We have previously demonstrated that the composites experienced bran degradation in the first stage of thermal decomposition (at T_{max1}) [76,110]. After ageing, given the weight loss for pure PBS, one might expect that this step is also related to the breakdown of the polymer, not just the bran.

FTIR analysis of the gaseous degradation products of the samples allows us to infer the possible mechanism of their thermal degradation. Figure 9 shows 3D FTIR diagrams of the gas decomposition products of PBS and composite 4, while Figure 10 shows FTIR spectra at emission maxima. On the spectrum collected at the first stage of PBS decomposition at 300 °C, low intensity absorption bands are visible at 1812 cm^{-1}, originating from vibrations of C=O group, at 1055 cm^{-1} (vibrations of –C–O–C– groups) and at 907 cm^{-1} (vibrations of carboxyl group) as well as bands in the range 2980–2880 cm^{-1} characteristic for vibrations of methyl and methylene groups. These are absorption bands characteristic of succinic acid and butane-1,4-diol in the gas phase [122,123]. In addition, carbon dioxide (absorption bands at 2359–2310 cm^{-1} and 669 cm^{-1}) and water (broad bands at approximately

4000–3500 cm^{-1} and 1800–1300 cm^{-1}) were observed. These bands suggest that thermal decomposition of PBS after ageing starts with hydrolysis of ester bonds leading to release of succinic acid and butane-1,4-diol and is accompanied by oxidation process evidenced by the release of CO_2 and water. On the spectrum collected at 312 °C for composite 5, mainly absorption bands from carbon dioxide, carbon monoxide (2180 and 2114 cm^{-1}) and water are present. In the vibration region of the carbonyl group around 1812 cm^{-1}, a band of negligible intensity can be found. The FTIR spectra for all composites look similar. Thus, oxidation processes are the main ones that take place in their case. However, at the second stage of decomposition, at a temperature about 390 °C, absorption bands characteristic for succinic acid and butane-1,4-diol are observed on the spectra of the composites, indicating, besides the dominant oxidation process, also polyester hydrolysis. The spectrum of PBS in the second decomposition stage is similar to that of composites. On the other hand, only absorption bands from water and carbon dioxide resulting from oxidation processes are present in the spectra of both PBS and composites in the last stage of decomposition, at temperatures around 480 °C.

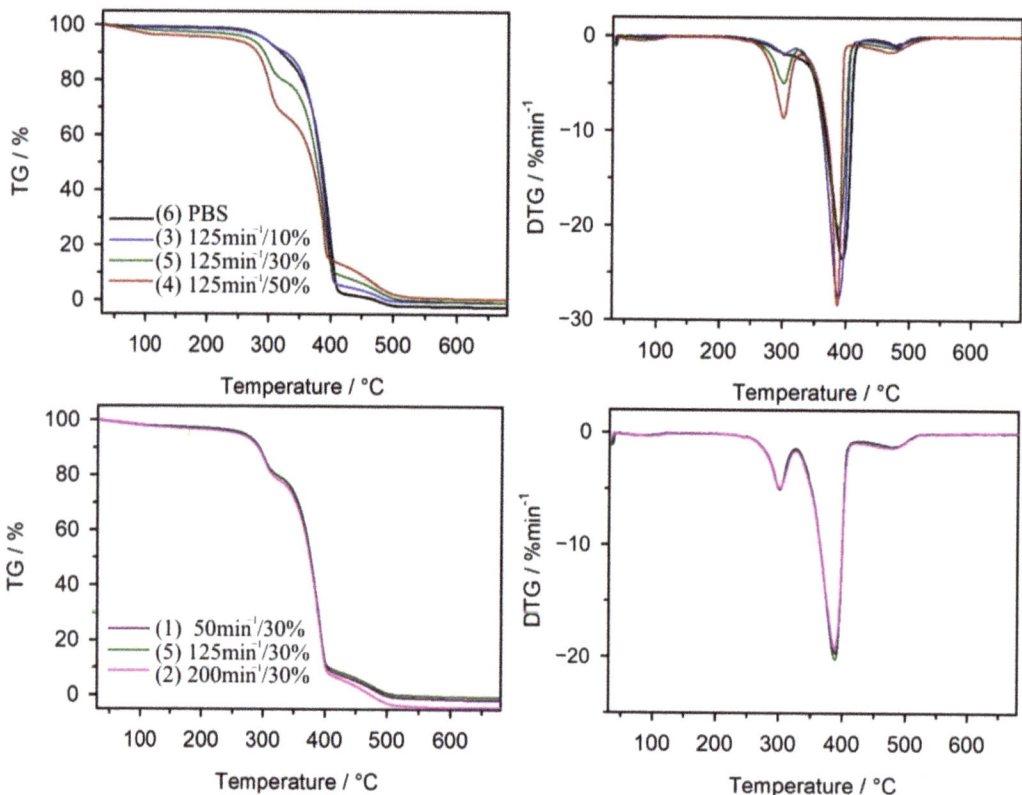

Figure 8. TG and DTG curves for PBS and its composites obtained after ageing.

Table 3. Parameters characterizing the thermal resistance of PBS and its biocomposites before (denoted as b) and after ageing (denoted as a), obtained based on thermogravimetry (TG) and derivative thermogravimetry (DTG) curves.

Sample	$T_{5\%}$ (°C)	$T_{50\%}$ (°C)	T_{max1} (°C)	Δm_1 (%)	T_{max2} (°C)	Δm_2 (%)	T_{max3} (°C)	Δm_3 (%)	R_m (%)
bran	201	303	296	68.0	-	-	459	29.7	2.3
PBS b	307	386	-	-	395	97.9	463	2.0	0.1
PBS a	297	384	300	8.9	394	88.6	477	2.4	0.1
1(50 30) b	274	380	303	20.5	392	70.3	476	9.1	0.1
1(50 30) a	260	376	300	20.5	389	71.1	476	8.2	0.2
2(200 30) b	276	379	303	20.1	390	70.0	476	9.7	0.2
2(200 30) a	259	375	301	21.8	388	71.1	476	7.1	0.0
3(125 10) b	299	383	303	8.5	391	87.0	475	4.5	0.3
3(125 10) a	288	382	301	9.5	388	85.1	478	5.3	0.1
4(125 50) b	261	372	300	31.5	386	53.5	462	14.7	0.3
4(125 50) a	231	369	302	33.5	387	52.0	469	14.2	0.3
5(125 30) b	274	381	303	19.8	391	71.1	476	9.0	0.1
5(125 30)	260	377	300	20.5	388	70.4	476	9.1	0

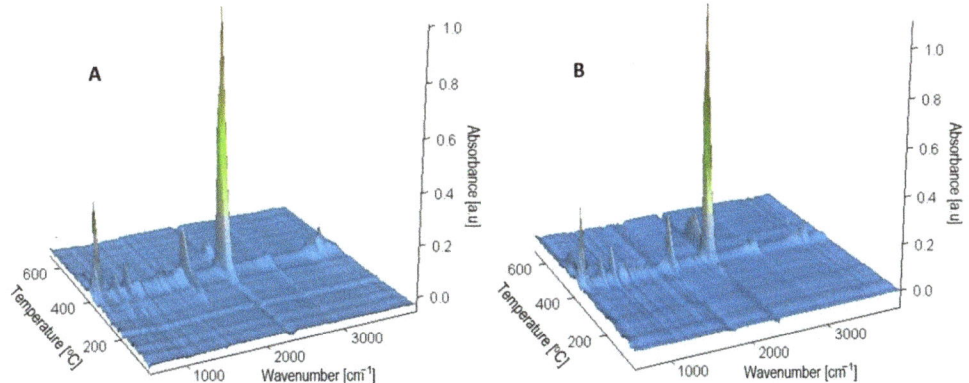

Figure 9. 3D-FTIR diagrams of gaseous degradation products of (A) PBS and (B) composite 4 (containing 50% w/w of bran).

3.1.5. Colour

The results of the measurement of L, a, b colour parameters before and after ageing of the mouldings made of the polymer compositions under study are shown in Figure 11. For comparative purposes, the Figure 11 presents also the results obtained for mouldings made of PBS alone without the addition of bran. As a result of ageing, the smallest colour change among the studied samples was observed in mouldings made of PBS alone. The calculated value of ΔE was 3.9, which allows us to define the colour change as distinct (3.5 < ΔE < 5). As a result of ageing, the colour of PBS mouldings changed toward blue shade—mainly decrease in b parameter, with negligible changes in a and L parameters. The effect of ageing on the change of the colour of mouldings produced from compositions containing bran was clearly greater. At bran content as low as u = 10%, the colour change amounted to ΔE = 7.8, which allows us to classify it as high (ΔE > 5). For samples of compositions with higher bran contents u = 30% and 50%, the colour change was even greater and amounted to ΔE = 23.3 and 36.9, respectively. It has been observed that as a result of the ageing process the colour of the mouldings made of compositions containing bran changes toward blue and green shades (parameter b and a decrease, respectively) and simultaneously the mouldings become lighter/faded, which is evidenced by an increase in the luminance value L. These changes intensify with increasing bran content in the composition. However, no significant influence of the extruder screw speed during the preparation of the composition pellets on the colour of the mouldings produced from it as well as on the colour difference between them after ageing was found.

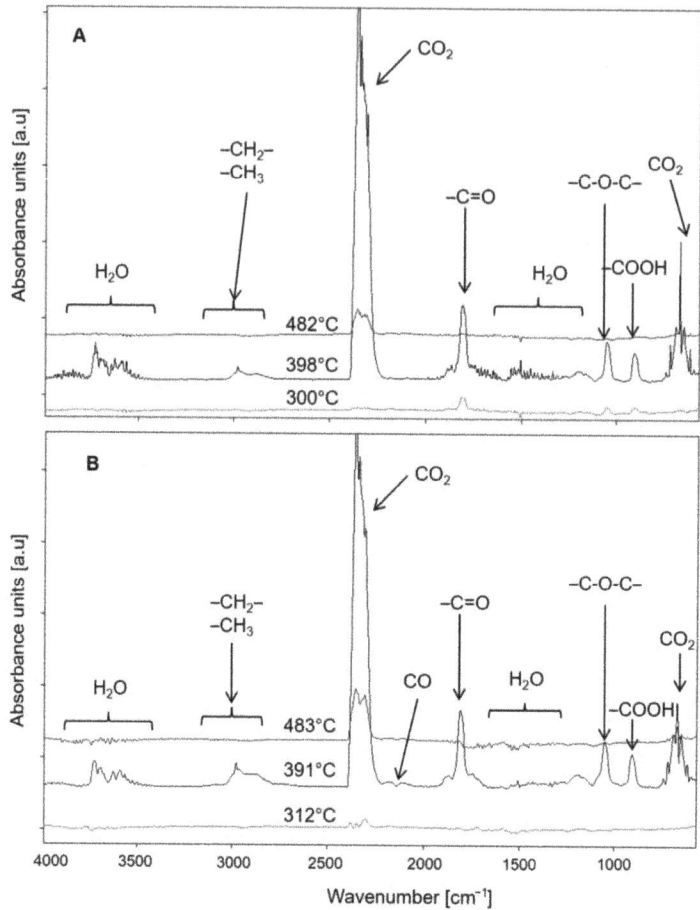

Figure 10. Extracted in maxima of emission FTIR spectra of gaseous degradation products of (**A**) PBS and (**B**) composite 4 (containing 50% w/w of bran).

Comparison of the mouldings before ageing showed that with increasing bran content their colour changes toward blue shades and darkens, the b parameter decreases and the luminance L decreases. These differences are large and in the case of increasing bran content from 10% to 30%, they are $\Delta E = 12.8$ and $\Delta E = 15.7$ when bran content is increased from 10% to 50%.

The obtained colour changes for PBS samples containing wheat bran are directly related to the degradation of the main WB building blocks, namely lignin and cellulose. The key to the course of their degradation is the simultaneous use of UV irradiation and water spraying in the chamber. Lignin itself shows hydrophobic nature and is quite resistant to degradation in water and solutions of weak concentrations [86,124]. However, it shows susceptibility to photodegradation by radiation activity, especially in the presence of water and at elevated temperatures. Depolymerisation of lignin exposes cellulose, which exhibits a hydrophilic nature and significantly increases surface wettability. The presence of water causes swelling of filler grains enabling deeper penetration of radiation and simultaneously accelerates oxidation reaction being the effect of photodegradation [125,126]. In addition, the vast majority of chromophores in lignocellulosic materials is located in lignin macromolecules; therefore, its decomposition will show a significant effect on colour loss [127]. Moreover, cyclic water spraying in the ageing chamber causes mechanical leaching of lignocellulose degradation products and exposure of PBS surfaces or deeper WB fragments [126]. Degradation of filler grains and their removal from the surface of samples as a result of ageing causes a shift in the colour parameters of biocomposites towards the colour of neat PBS. The above analysis is confirmed by the surface appearance of the samples before and after ageing. Figure 12 clearly shows that the fading is focal, and the number of focal spots increases with increasing filler content, covering virtually the entire surface of the composite containing 50% WB.

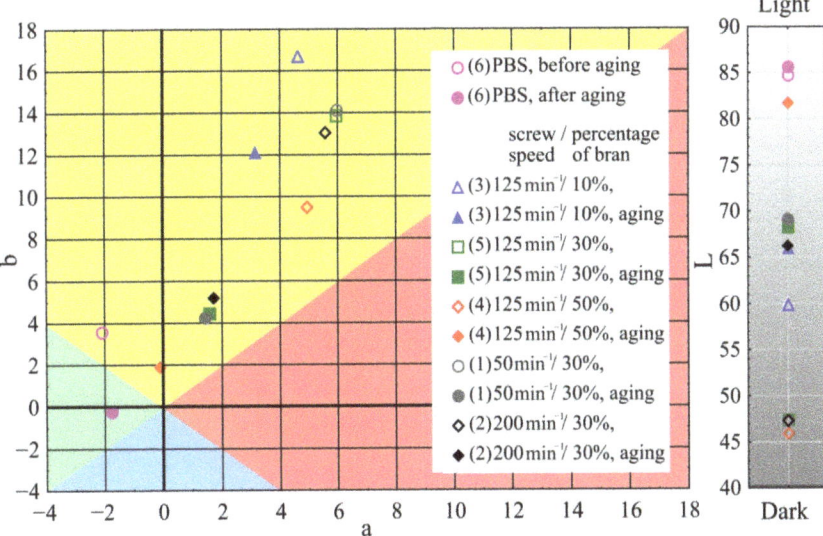

Figure 11. Variation of L, a, b colour parameters before and after ageing of mouldings made of polymer compositions depending on the bran mass fraction content and processing screw speed.

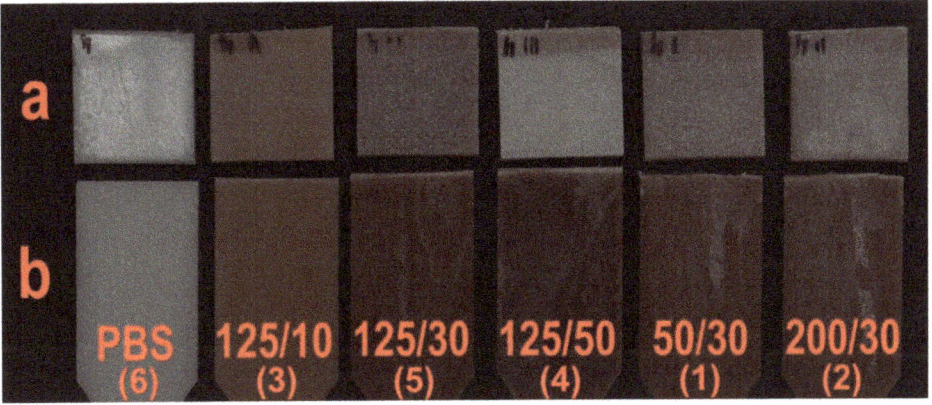

Figure 12. Surface appearance of tested samples after ageing (**a**) and before ageing (**b**).

3.1.6. Surface Roughness

Optical profilometer tests were employed to observe morphological changes in the samples before and after the accelerated ageing test. The 3D images of the surface topography of the PBS and the composite containing 50% bran before and after the ageing test are shown in Figure 13. In order to quantify the roughness changes of the materials, the R_a parameter was calculated, and the results are shown in Figure 14. Neat PBS has the lowest R_a value, both before and after ageing. As the amount of bran in the samples increases, the R_a value also increases. The bran particles, which are visible as blue dots in Figure 13, are responsible for the increase in surface roughness of the composites. Considering the composites containing 30% bran obtained at different values of n, no significant effect of this parameter on R_a value was found. A very good homogenization of the composite components occurred at that speed, thus reducing R_a. After the ageing test, the surface roughness of the composites increased significantly, and the trend of changes for samples containing different amounts of bran and obtained at different n is similar to that before ageing. Interestingly, as the amount of bran increased, the increase in R_a was steeper and the sample surfaces more heterogeneous, as indicated by the large confidence intervals. The situation was different for neat PBS. After the ageing test, the topography of its surface changed slightly.

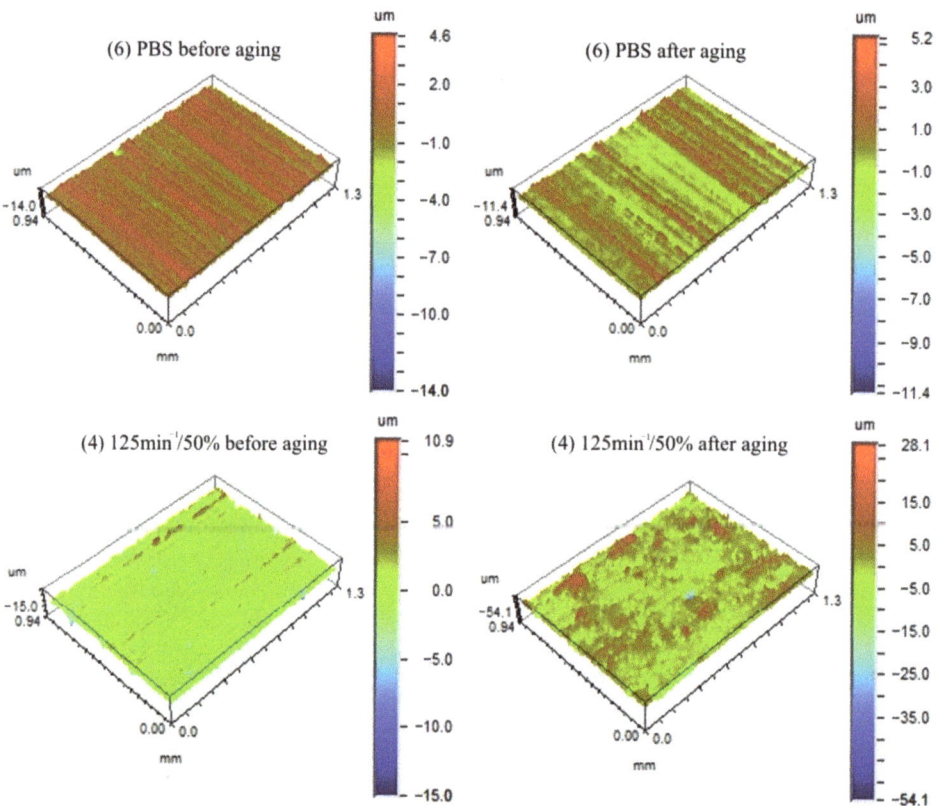

Figure 13. 3D images of the surface topography of the PBS and the composite containing 50% bran before and after the ageing test.

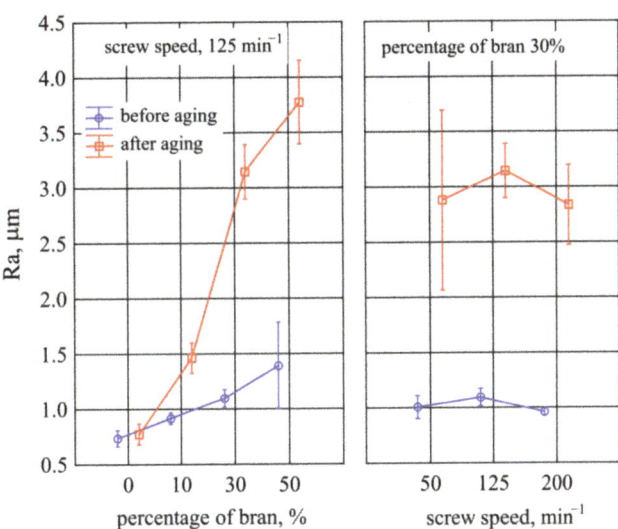

Figure 14. Graph of changes in Ra parameter for PBS and composite containing 50% bran (4) before and after ageing test.

The unfilled PBS showed the lowest roughness and its surface structure in the form of longitudinal parallel lines shown in Figure 13 constitutes a mapping of the injection mould surface structure obtained during machining. Biocomposites containing WB showed significantly higher roughness. As the filler content increases, the amount of filler particles localized directly on the sample surface statistically increases, which also increases the roughness. This is a typical result for PBS matrix composites with powder filler [128,129]. The obtained post-ageing roughness results are consistent with the above analysis of the colour results, and the obtained course of changes is directly related to the degradation of the lignocellulosic filler and the mechanical leaching of the products of this degradation [126].

3.1.7. Gloss

Figure 15 presents the results of gloss measurements of mouldings made of the polymer compositions under study and of PBS alone. Statistical analyses did not show any significant influence of the bran content and extruder screw speed during the preparation of the composite granules on the gloss of the mouldings before ageing. A post-hoc test indicated only a difference in gloss between the lowest value of 50 min^{-1} and the highest value of 200 min^{-1} (at constant content of 30%). One of the reasons for such outcome is a large scatter of obtained results especially visible for PBS alone and for compositions with 10% bran content. A statistically significant effect of ageing on the gloss of mouldings was observed for the 30% bran content. In the case of the lower content of 10% bran and PBS without its addition, the gloss before and after ageing did not differ significantly.

The gloss value indicates the ability and mechanism of a surface to reflect incident light in a specular direction. Whether a surface reflects, scatters, or absorbs light depends on the surface roughness, but also on the heterogeneity of the material and surface. The presence of the filler, and then also its composition, size, and quality of distribution in the matrix influence the increase of randomness of light reflection directions and decrease of gloss values [130,131]. Therefore, the introduction of bran powder, as expected, resulted in a decrease in gloss value. On the other hand, the decrease in gloss as a result of ageing is a consequence of the degradation of filler grains located on the surface of the samples, which increases the surface roughness. Consequently, no difference was observed in the gloss of unfilled PBS before and after ageing.

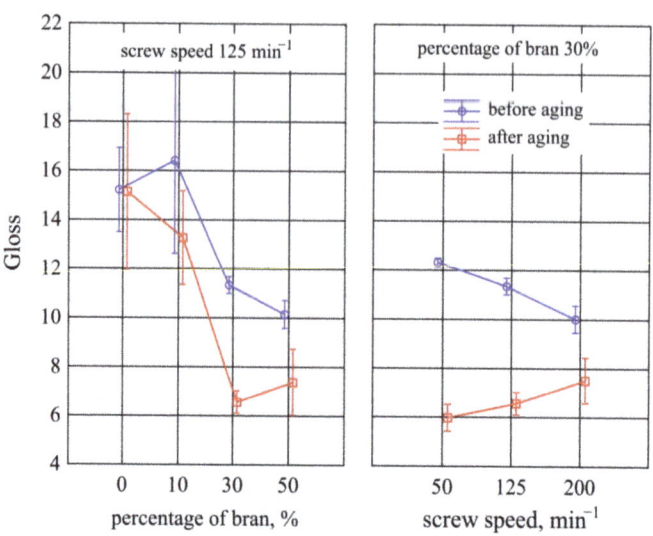

Figure 15. Variation of gloss value before and after ageing of injection mouldings made of polymer compositions depending on the bran mass fraction content and processing screw speed.

3.1.8. Weight Loss

The largest weight loss after ageing was observed for samples made from PBS alone and was 1.06% (Figure 16). Introduction of 10% bran into the composition resulted in a significant decrease in weight loss to 0.31%. Increasing their content caused a successive increase in the value of weight loss to a maximum of 0.59% at 50% bran content. At the same time, a significant decrease in weight loss of the samples prepared from the composition obtained at processing screw speeds higher than 50 min^{-1} was found.

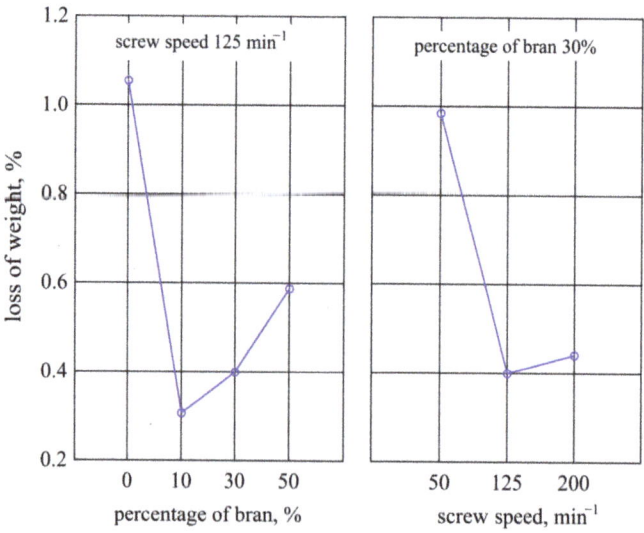

Figure 16. Weight loss resulting from ageing of injection mouldings made of polymer compositions depending on the bran mass fraction content and processing screw speed.

The obtained dependence of weight loss after ageing on bran content remains fully consistent with previous observations. As the bran content increases, the amount of WB particles located on the surface statistically increases, with the particles degraded and washed away during cyclic water spraying [126]. A surprising observation was that the largest weight loss due to ageing was recorded for unfilled PBS, although a weight loss of 1% due to accelerated ageing in the chamber for neat PBS was also obtained by other authors [62]. A potential reason could be that WB particles provided a physical barrier against radiation [125,126]. Due to the numerous chromophores in its macromolecules, lignin is responsible for absorbing about 80–95% of the light incident on lignocellulosic materials, thus it concentrates degradation on the filler present on the sample surface [132,133]. While in unfilled PBS, the entire surface is uniformly exposed to degrading agents. The profilometer tests did not show any difference in the PBS roughness before and after ageing, but some reduction in the height difference between the extreme points can be seen in Figure 13, which may suggest a uniform degradation across the surface. The effect of screw speed, on the other hand, may be related to intensive shearing and fragmentation of filler grains occurring at high speeds, which makes their distribution more homogeneous and the grains smaller [75]. Significantly higher uncertainty of roughness measurements for composites produced at $n = 50$ min^{-1} may suggest a larger dispersion of filler grain sizes.

3.1.9. Hardness

It was observed that the mass content of bran u introduced into the composite had a significant effect on the hardness of the mouldings, measured using ball indentation method (Figure 17). Even before the ageing process, 10% of their content caused significant increase of hardness by 3.9 MPa (8%) in comparison to samples with PBS without their addition. Increasing the bran content to 30% resulted in a further 24% increase in hardness compared to PBS. Further increase in bran content to 50% did not cause any significant changes in hardness. The statistical analyses carried out did not show any significant effect of the processing screw speed during the composite production on its hardness measured using the ball pressing method both before and after the ageing process.

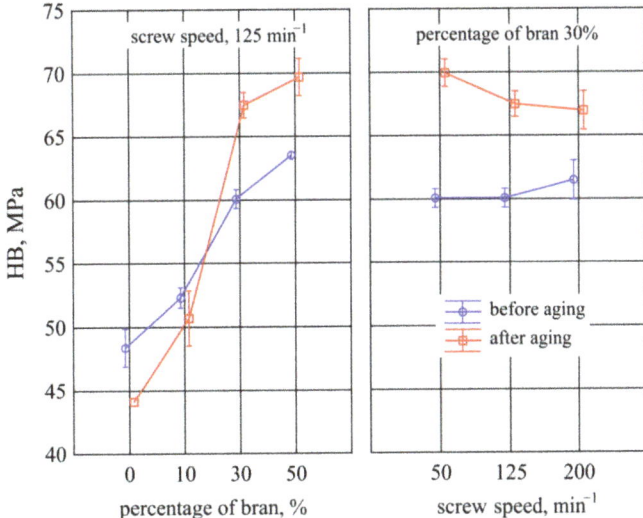

Figure 17. Variation of hardness HB through ball indentation method before and after ageing of injection mouldings made of polymer compositions depending on the bran mass fraction content and processing screw speed.

Ageing caused a significant decrease in hardness of samples made from PBS alone by an average of 4.3 MPa (9%). In the case of 10% bran content, no significant differences were found in the hardness of the samples before and after ageing. At higher bran contents of 30% and 50%, a significant increase in age hardness by 7.4 MPa (12%), and 6.2 MPa (10%), respectively, was observed.

The tested samples were also subjected to Shore D hardness tests and the results obtained are shown in Figure 18. The hardness results obtained before ageing showed no statistically significant differences for samples with PBS alone and with 10% bran content. Only 30% and 50% bran content caused a very slight but statistically significant decrease in hardness by an average of 0.55 °ShD (0.8%). As well, in the case of this hardness test method, no significant effect of the speed of processing screws during the manufacture of the composition on its hardness before ageing was observed and after ageing a very small increase in hardness occurred only at the highest screw speed.

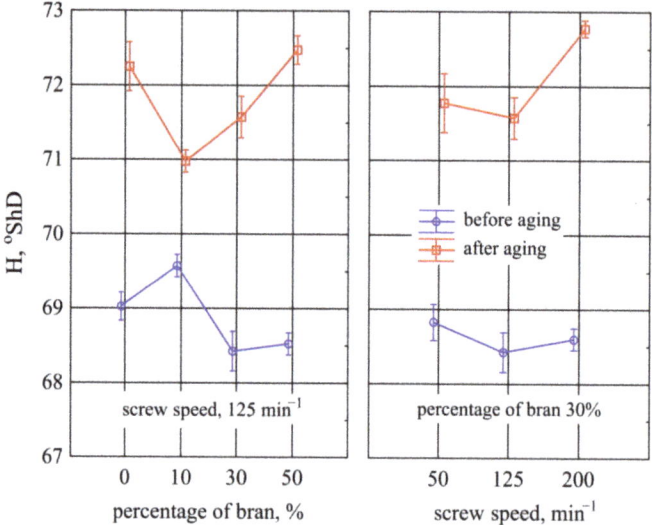

Figure 18. Variation of hardness H through Shore D method before and after ageing of injection mouldings made of polymer compositions depending on the bran mass fraction content and processing screw speed.

Ageing caused a significant increase in Shore hardness of all the samples tested. The bran content had a significant effect on hardness after ageing for only 10% of its content—a slight decrease of 1.3 °ShD (1.8%) compared to samples with PBS alone and 1.5 °ShD (2.1%) compared to samples with 50% bran content. The other samples were not significantly different from one another.

The different effect of bran content on pre-ageing hardness for each of the measurement methods used is due to their different characteristics. The ball indentation method involves deflecting a substantial area of the specimen surface with an indenter. The presence of the dispersed phase in the polymer matrix increases the stiffness and restricts the mobility of macromolecules making it difficult to deform [76,77,120]. In the Shore D method, the exerted load is concentrated over a small area, so that transferring the load to filler grains with a hardness less than that of the polymer matrix does not produce a pronounced strengthening effect. Hence, the decrease in hardness in the Shore method at high filler content.

The increase in hardness after ageing is directly related to the increase in degree of crystallinity shown by DSC studies. Due to the close arrangement of macromolecules, the crystalline phase is characterized by higher density, stiffness, and consequently hardness [75,109]. In the case of the ball indentation method, the strengthening effect after ageing is only visible from 30% bran content on. This is due to the penetration of the ball into the outer layer of PBS, which has been degraded by the sum effect of temperature, moisture, and radiation in the ageing chamber. At higher WB contents, the filler particles degrade and begin to cover a significant part of the surface, whose hardness is negligible anyway. In the Shore method, the indenter penetrates the sample to a much greater depth, where the material may not have degraded significantly, so that a strengthening effect due to an increase in crystallinity was observed for all samples differing in content.

3.2. Absorption of Polar and Non-Polar Fluids

3.2.1. Water Absorption

The results of the tests of water absorption X from the studied variable factors are presented in Figure 19. The mass content of bran u introduced in the compositions studied was found to have a significant effect on the water absorption X of the mouldings. Increasing the bran content causes a very large increase in water absorption, which is due to the hydrophilic properties of bran. According to literature data, water retention cap by wheat bran is at the level of 4–4.8 g H_2O/g, and it may be even higher at a high degree of fragmentation [134–136]. Additionally, with an increase in WB content, the distance between adjacent filler particles decreases and the amount of gases, mainly water vapour, emitted during the processing increases, causing the porosity of the composition. This creates opportunities for water to penetrate deeper into the sample structure. The greatest changes, similarly as in the case of the other examined properties, were observed between the compositions which differed most in the bran content of 10% and 50% (DOE layouts 3 and 4). Then, as a result of increasing the bran content, water absorption X increased by 7.3% (i.e., to 475% of the initial value). Measurements of samples made from PBS alone showed that their water absorption averaged only $X = 0.65 \pm 0.01\%$; thus, the bran is primarily responsible for water absorption. The results of statistical analyses showed no significant effect of screw speed on water absorption.

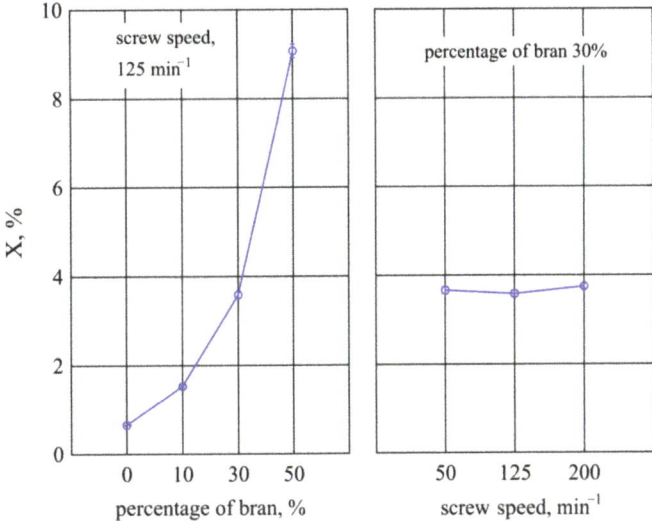

Figure 19. Dependence of water absorption X of injection mouldings made of polymer compositions on bran mass content and processing screw rotational speed.

A parameter directly related to water absorption is the change in linear dimensions of a sample due to absorption of a certain volume of fluid. The relationship describing the swelling variation SW depending on the bran content in the composition and processing screw speed during its production is shown in Figure 20. The mass content of bran u introduced into composition has the greatest influence on this quantity. In this case, the influence of the extruder screw speed n during the production of the polymeric compositions under study was clearly smaller, but also statistically significant. Increasing the bran content in the composition causes a significant increase in SW, which, similarly to water absorption X, is due to hydrophilic properties of bran. The previously mentioned ability of bran to retain about 4 g of water per g of bran must lead to a significant change in volume. The highest increase in SW during the tests, i.e., 1.6% (407% of the initial value), was obtained by increasing wheat bran content u in the composition from 10% to 50% (DOE layouts 3 and 4). In contrast, the maximum investigated increase in SW due to increased screw speed n was only 0.2% (i.e., 27% of the initial value). The probable reason for the increase in swelling with screw speed n is the fragmentation of soft bran grains due to the action of intense shear stress arising during extrusion of the composition [75]. The conducted analyses did not show the presence of interactions between the studied variable factors. Tests performed on samples made from PBS alone showed that their swelling averaged only SW = 0.15 ± 0.04%.

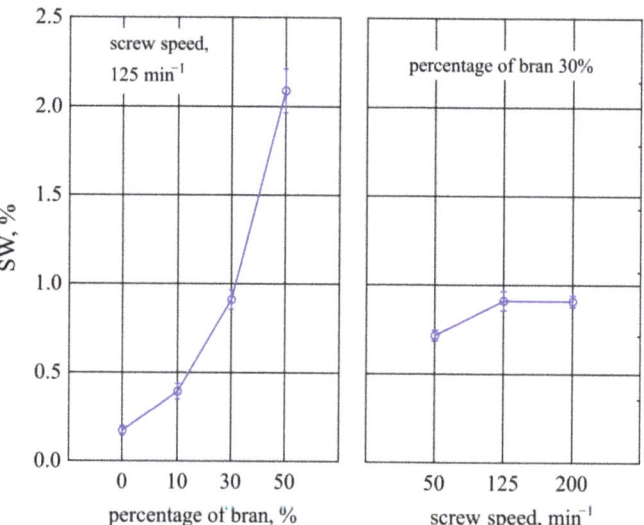

Figure 20. Dependence of swelling SW of injection mouldings made of polymer compositions on bran mass content and processing screw rotational speed.

3.2.2. Oil Absorption

Results the tests of oil absorption X_{oil} of injection mouldings made of the polymer compositions under study and of PBS alone are presented on Figure 21. It has been observed that the highest influence on the oil absorption of the obtained mouldings is exerted by the mass content of bran u introduced into the composition, but also by the extruder screw speed n during its production. The influence of the two mentioned variable factors is linear. The highest increase in oil absorption X_{oil} during the tests, i.e., 0.11% (87% of the initial value), was obtained by increasing wheat bran content u in the composition from 10% to 50% (DOE layouts 3 and 4). In contrast, the maximum investigated increase in X_{oil} due to increased screw speed n was 0.05% (i.e., 35% of the initial value). Comparative measurements of samples made from PBS alone showed that their oil absorption averages only X_{oil} = 0.06 ± 0.01%. This means that PBS is a material resistant to vegetable oils, which is also confirmed by the studies of other authors [137]. Therefore, it should be assumed that bran introduced into the composition is mainly responsible for oil absorption, and in particular, the hydrophobic lignin contained in the filler structure [124].

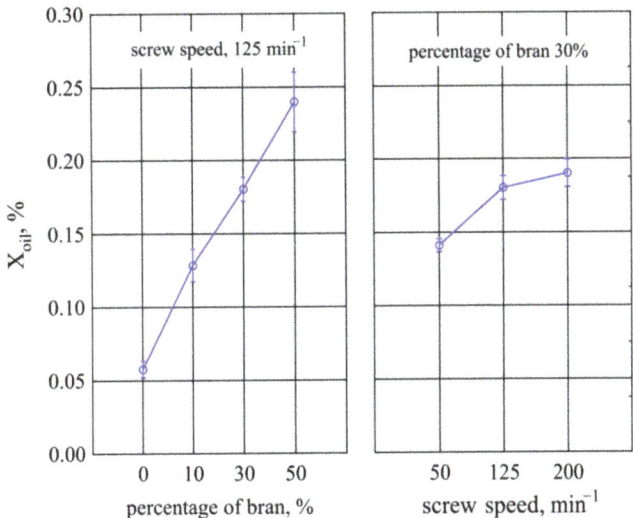

Figure 21. Dependence of oil absorption X_{oil} of injection mouldings made of polymer compositions on bran mass content and processing screw rotational speed.

Due to the low oil absorption values, the resulting swelling also took on very low values. The highest increase in linear dimensions was observed for composite 4, containing 50% bran, which was 0.25 ± 0.03%.

3.3. Chemical Resistance

3.3.1. Resistance to Acids

Figure 22 shows the weight changes of the samples during the acidic environment test in 1 M HCl aqueous solution. The observed increase in weight of the samples may seem surprising. It results from the adopted method of measurement. Usually, the mass of samples in chemical resistance tests is measured after they have been dried to a constant mass in an oven. This approach did not work for our study. Drying of the composite samples, after a one-week stay in acid solution, at 55 °C for 48 h, was not sufficient to achieve constant mass. On the other hand, after 48 h of drying, it was clearly visible that the bran was thermally decomposed. Therefore, decision was made for the samples to be dried in a paper towel before weighing them after being removed from the HCl solution. The mass increases observed in Figure 22 are related to the absorption of the acid solution by the PBS and composites. The highest weight gain is observed for the composite containing 50% bran, it decreases according to the decreasing amount of bran in the composites and reaches the lowest value for neat PBS. This relationship is consistent with the water absorption results. It results from the susceptibility of bran to swelling and additionally from the porosity of the samples. In the case of composites, an increase in weight is recorded up to day 35 of the test, except for the sample containing only 10% bran, which displays shows an increase in weight for as long as up to day 55. After 55 days of the samples staying in the acidic solution (after 90 days for sample 3), a progressive weight loss was noted. After 90 days of testing, the weight loss was the highest for the composite with 50% bran content, and the lowest—with 10% bran content. The change in weight of neat PBS over the 90 days of the test remained very similar (approximately 0.8%). Therefore, it can be concluded that the addition of bran to PBS affects the acceleration of chemical degradation in acidic solution. The decrease in weight of the composites is mainly related to the chemical degradation of the bran by acid hydrolysis. Admittedly, PBS belongs to the polyester group and is susceptible to hydrolysis of the ester bond even in pure water. This reaction, however, requires an elevated temperature (approximately 75 °C) [138,139]. Our experiment was conducted at room temperature and even the presence of a strong acid was not sufficient for hydrolysis of neat PBS to occur. The addition of bran susceptible to hydrolysis caused its decomposition, so that the acid solution could diffuse more easily into the internal structure of the composites and react with the PBS chains. Considering the influence of parameter n on the resistance of the composites 1, 2, and 5 in an acidic environment, it is apparent that it is not as pronounced as the percentage of bran content in the composites. Composite 5, obtained at a screw speed of 125 min^{-1}, showed the greatest weight loss between 35 and 90 days, by almost 2%, while for the other two speeds, the weight loss in this time interval was about 1.5%.

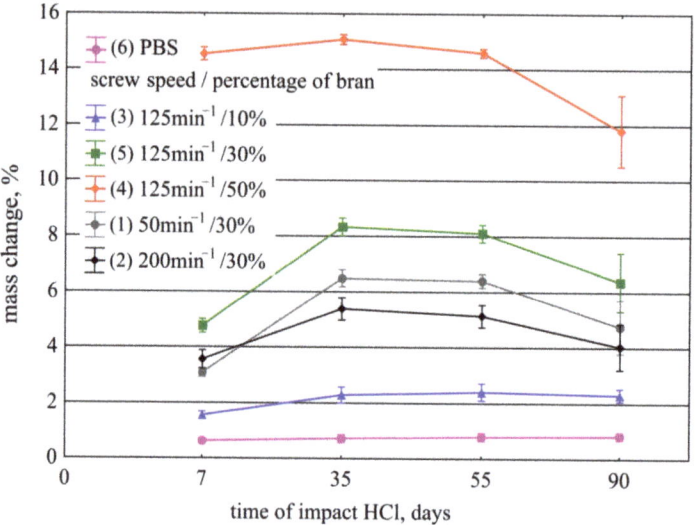

Figure 22. Variation of mass change with hydrolytic degradation time for neat PBS and its composites.

3.3.2. Resistance to Bases

Resistance test of the composites to basic environment in 1 M NaOH aqueous solution showed that the tested samples are not resistant to strong base solution. The composite samples containing 30% and 50% bran disintegrated visibly as soon as after the first day of the test, while the sample with 10% biofiller changed colour and structure, and after another day it also started to disintegrate visibly (Figure 23). After seven days, all composites had completely disintegrated in the base solution. Because the samples became sticky and began to disintegrate early in the test, it was impossible to weigh them. Only the neat PBS sample was integral for 7 days, after which it showed a 3.5% weight loss. The ester bonds in PBS are sensitive to base hydrolysis even at room temperature [140,141]. Bran composed of polysaccharides in the environment of strong bases is easily hydrolysed. As in the case of the action of the acidic solution, their chemical disintegration, enabled faster diffusion of the basic solution into the internal structure of the composite and accelerated the hydrolysis of PBS. In the case of degradation in base solution, only the effect of bran amount on the stability of the composites could be observed while, due to the fast degradation progress, it was impossible to notice the effect of the extruder screw speed.

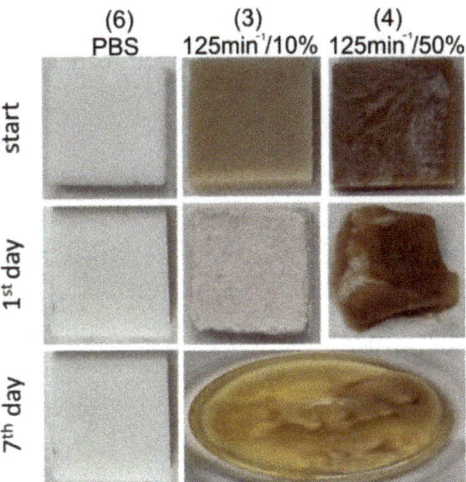

Figure 23. Images of the samples before and during degradation in base solution.

3.4. Biodegradation in Compost

The effects of the bran content by weight in the polymeric composition and the screw speed during its manufacturing on the weight loss of the mouldings under controlled biodegradation in compost are shown in Figure 24. The observed changes in the values of mass loss depending on the biodegradation time are linear. The lowest weight loss occurred in samples made with PBS alone and equalled 0.06% per day; after 70 days, the weight of the samples decreased by an average of 4.5%. The weight loss due to biodegradation of samples containing bran was significantly higher and increased with the bran content in the composition. At 10% bran content, it was 0.22% per day, and after 70 days, it averaged 15.1%. At 30% bran content, biodegradation resulted in an average daily weight loss of 0.41% for the three processing screw speeds used during composite production. Changes in weight loss associated with the speed of the processing screws in the production of composites were found to be negligibly small. The greatest weight loss occurred at 50% bran content and averaged 0.9% per day. After 70 days, the average weight loss of these samples was 68.3%.

The degradation kinetics are significantly different for the tested biocomposites and unfilled PBS. Degradation of PBS is relatively slow [62]; therefore, it can be expected that for biocomposites, at least in the initial phase, the degradation mainly involves bran. Due to its chemical structure, WB is more easily enzymatically hydrolysed than PBS and is preferred by microorganisms. As the WB content increases, the rate of weight loss increases significantly with time. This is related to the previously demonstrated high ability of bran to absorb and retain water due to the presence of numerous hydrophilic chemical structures. In addition, the bran grains swelling due to water absorption facilitate the penetration of microorganisms accelerating the degradation of WB. Thus, after the biological decomposition of the filler, the surface of active interaction of water and microorganisms on PBS macromolecules increases [4,142–144]. The significant increase in biodegradation kinetics of PBS in compost with increasing lignocellulosic filler content is a typical result also obtained by other authors [52,94,111,144].

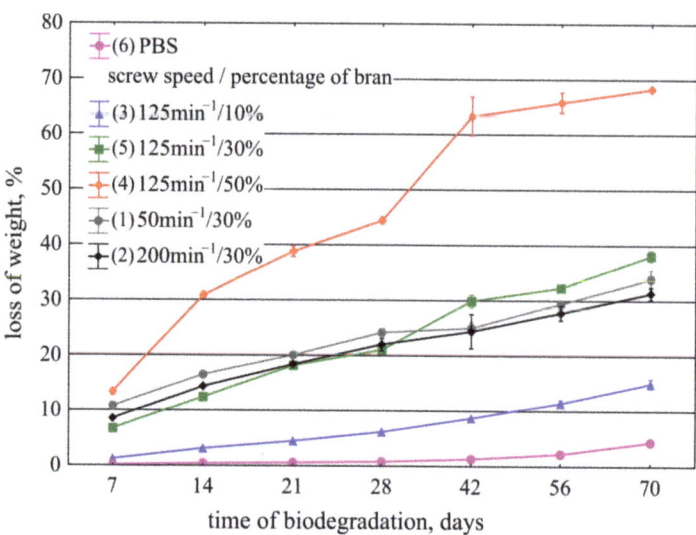

Figure 24. Weight loss resulting from biodegradation of injection mouldings made of polymer compositions depending on the bran mass fraction content and processing screw speed.

4. Conclusions

The results presented unambiguously show that the resistance of PBS-based polymer biocomposites filled with wheat bran to physical, chemical, and biological factors depends mainly on the mass content of the filler. The effect of manufacturing conditions of the biocomposite pellets subsequently used for injection moulding of the samples showed moderate or no effect on the properties under study. Thus, from the economic point of view, it will be preferable to produce composite pellets with the highest possible efficiency, as the screw speed in the investigated range does not determine the resistance of the injection mouldings made of them.

Accelerated ageing resulted in changes in the chemical structure of PBS within the ester bond, suggesting macromolecular disintegration. DSC tests performed after ageing showed a significant increase in the biocomposites and PBS degree of crystallinity up to 48%. This phenomenon significantly affects other properties of the tested materials, such as an increase of 0.05–0.35% in shrinkage and a growth of about 10% in hardness. Significant colour changes due to lignin photodegradation were also noted, leading to fading of the samples due to chromophore decomposition. The mentioned filler decomposition was in turn the reason for the changes in surface roughness, the R_a parameter increased of about 168% for biocomposition containing 50% of bran, and weight loss of about 0.31–0.59%. Moreover, the thermal stability of tested materials decreased after ageing of 10 °C for neat PBS and of 30 °C for the composition containing 50% bran.

The biocompositions under study were characterized by significant water absorption in comparison with neat PBS. Both X and SW parameters increased with increasing filler amount reaching for biocomposition containing 50% of bran of about 9% and 2.1%, respectively. The same relationship was observed for vegetable oil absorption. It was on moderate level, reaching almost 0.25%. It can be concluded that surface grains of lignocellulosic filler, which are composed of both polar and non-polar macromolecules, are responsible for the absorption of both polar and non-polar liquids.

Chemical resistance tests showed that PBS is resistant to acidic environment at room temperature, whereas biocompositions exhibited the mass loss of about 1.5–2% between the 35th and 90th day of test. Nevertheless, samples still retained their integrity after 90 days. The biocompositions, in contrast to neat PBS, exhibited negligible resistance to strong bases, as the samples containing the filler completely lost integrity within seven days. The lack of the resistance on the alkaline environment should be taken into account when the possible applications are planned for the compositions of PBS filled with wheat bran.

Biodegradation studies have shown a significant effect of wheat bran content on the kinetics of degradation of the analysed biocompositions. The most rapid degradation is, of course, in the bran itself, which is the preferred habitat of microorganisms. Unfilled PBS degrades relatively slowly under the test conditions; on the other hand, the composition containing 50% of WB reached 68% disintegrability after 70 days of incubation. Such high disintegration indicates that proposed biocompositions are compostable, which is very advantageous from an environmental point of view.

Author Contributions: Conceptualization, E.S., M.G. and Ł.M.; methodology, E.S., Ł.M. and M.G.; validation, E.S., Ł.M. and M.G.; formal analysis, E.S., M.G.; investigation, E.S., Ł.M. and M.G.; resources, E.S. and Ł.M.; writing—original draft preparation, E.S., Ł.M. and M.G.; writing—review and editing, M.G.; visualization, E.S., M.G. and Ł.M.; supervision, E.S.; project administration, Ł.M. and E.S.; funding acquisition, E.S. All authors have read and agreed to the published version of the manuscript.

Funding: The research was financed in the framework of the project Lublin University of Technology-Regional Excellence Initiative, funded by the Polish Ministry of Science and Higher Education (contract No. 030/RID/2018/19).

Institutional Review Board Statement: Not applicable.

Informed Consent Statement: Not applicable.

Data Availability Statement: The data presented in this study are available on request from the corresponding author.

Conflicts of Interest: The authors declare no conflict of interest. The funders had no role in the design of the study; in the collection, analyses, or interpretation of data; in the writing of the manuscript, or in the decision to publish the results.

Abbreviations

PBS	Poly(butylene succinate)
WB	Wheat bran
UV	Ultraviolet
LCF	Lignocellulosic filler
u	Wheat bran content
n	Extruder screw rotational speed
FTIR	Fourier transform infrared (spectroscopy)
DSC	Differential scanning calorimetry
X_c	Degree of crystallinity
ΔH_m	Melting enthalpy
T_m	Melting point
T_c	Crystallization temperature
T_g	Glass transition temperature
TG	Thermogravimetry
DTG	Derivative thermogravimetry
$T_{5\%}$, $T_{50\%}$	Temperature of 5% and 50% of mass loss
T_{max}	Temperature of the maximum rate of mass loss
Δm	Mass loss corresponding to T_{max}
R_m	Residual mass
L^*	Luminance/brightness
a^*	Colour from green to red
b^*	Colour from blue to yellow
ΔE	Difference between the two colors
R_a	Average roughness parameter

E_{jc} Specific energy consumption
p-v-T Relationship between pressure p, specific volume v and temperature T
HB Ball indentation hardness
H Shore D hardness
X Water absorption
SW Swelling
X_{Oil} Oil absorption

References

1. Babu, R.P.; O'Connor, K.; Seeram, R. Current progress on bio-based polymers and their future trends. *Prog. Biomater.* **2013**, *2*, 8. [CrossRef]
2. Hamad, K.; Kaseem, M.; Deri, F. Recycling of waste from polymer materials: An overview of the recent works. *Polym. Degrad. Stab.* **2013**, *98*, 2801–2812. [CrossRef]
3. Muniyadi, M.; Yit Siew Ng, T.; Munusamy, Y.; Xian Ooi, Z. Mimusops elengi seed Shell powder as a new bio-filler for polypropylene-based bio-composites. *Bioresources* **2018**, *13*, 272–289. [CrossRef]
4. Danso, D.; Chow, J.; Streit, W.R. Plastics: Environmental and biotechnological perspectives on microbial degradation. *Appl. Environ. Microbiol.* **2019**, *85*, e01095-19. [CrossRef] [PubMed]
5. Väisänen, T.; Das, O.; Tomppo, L. A review on new bio-based constituents for natural fiber-polymer composites. *J. Clean. Prod.* **2017**, *149*, 582–592. [CrossRef]
6. Vilaplana, F.; Strömberg, E.; Karlsson, S. Environmental and resource aspects of sustainable biocomposites. *Polym. Degrad. Stab.* **2010**, *95*, 2147–2161. [CrossRef]
7. Korol, J.; Burchart-Korol, D.; Pichlak, M. Expansion of environmental impact assessment for eco-efficiency evaluation of biocomposites for industrial application. *J. Clean. Prod.* **2016**, *113*, 144–152. [CrossRef]
8. Nielsen, T.D.; Hasselbalch, J.; Holmbetg, K.; Stripple, J. Politics and the plastic crisis: A review throughout the plastic life cycle. *WIREs Energy Environ.* **2020**, *9*, e360. [CrossRef]
9. De Oliveira, C.C.N.; Zotin, M.Z.; Rochedo, P.R.R.; Szklo, A. Achieving negative emissions in plastics life cycles through the conversion of biomass feedstock. *Biofuels Bioprod. Biorefin.* **2021**, *15*, 430–453. [CrossRef]
10. Aretoulaki, E.; Ponis, S.; Plakas, G.; Agalianos, K. Marine plastic littering: A review of socio economic impacts. *J. Sustain. Sci. Manag.* **2021**, *16*, 276–300. [CrossRef]
11. Syberg, K.; Nielsen, M.B.; Clausen, L.P.W.; van Calster, G.; van Wezel, A.; Rochman, C.; Koelmans, A.A.; Cronin, R.; Pahl, S.; Hanses, S.F. Regulation of plastic from a circular economy perspective. *Curr. Opin. Green Sustain. Chem.* **2021**, *29*, 100462. [CrossRef]
12. Sheldon, R.A.; Norton, M. Green chemistry, and the plastic pollution challenge: Towards a circular economy. *Green Chem.* **2020**, *22*, 6310–6322. [CrossRef]
13. Xanthos, D.; Walker, T.R. International policies to reduce plastic marine pollution from single-use plastics (plastic bags and microbeads): A review. *Mar. Pollut. Bull.* **2017**, *118*, 17–26. [CrossRef] [PubMed]
14. Jiang, J.Q. Occurrence of microplastics and its pollution in the environment: A review. *Sustain. Prod. Consum.* **2018**, *13*, 16–23. [CrossRef]
15. Wei, L.; McDonald, A.G. A review on grafting of biofibres for biocomposites. *Materials* **2016**, *9*, 303. [CrossRef] [PubMed]
16. Hamad, K.; Kaseem, M.; Ko, Y.G.; Deri, F. Biodegradable polimer blends and composites: An overwiew. *Polym. Sci. Ser. A* **2014**, *56*, 812–829. [CrossRef]
17. Pivsa-Art, W.; Chaiyasat, A.; Pivsa-Art, S.; Yamane, H.; Ohara, H. Preparation of polymer blends between poly(lactic acid) and poly(butylene adipate-co-terephthalate) and biodegradable polymers as compatibilisers. *Energy Procedia* **2013**, *34*, 549–554. [CrossRef]
18. Meeks, D.; Hottle, T.; Bilec, M.M.; Landis, A.E. Compostable biopolymer use in the real world: Stakeholder interviews to better understand the motivations and realities of use and disposal in US. *Resour. Conserv.* **2015**, *105*, 134–142. [CrossRef]
19. Thakur, S.; Chaudhary, J.; Sharma, B.; Verma, A.; Tamulevicius, S.; Thakur, V.K. Sustainability of bioplastics: Opportunities and challenges. *Curr. Opin. Green Sustain. Chem.* **2018**, *13*, 68–75. [CrossRef]
20. Tokiwa, Y.; Calabia, B.P.; Ugwu, C.U.; Aiba, S. Biodegradability of plastics. *Mol. Sci.* **2009**, *10*, 3722–3742. [CrossRef]
21. Artham, T.; Doble, M. Biodegradation of aliphatic and aromatic polycarbonates. *Macromol. Biosci.* **2007**, *8*, 14–24. [CrossRef]
22. Tnag, X.Z.; Kumar, P.; Alavi, S.; Sandeep, K.P. Recent advances in biopolymers and biopolymer-based nanocomposites for food packaging materials. *Crit. Rev. Food Sci. Nutr.* **2012**, *52*, 426–442. [CrossRef] [PubMed]
23. Noorunnisa Khanam, P.; Abdul Khalil, H.P.S.; Ramachandra Reddy, G.; Venkata Naidu, S. Tensile, flexural and chemical resistance properties of sisal fibre reinforced polymer composites: Effect of fibre surface treatment. *J. Polym. Environ.* **2011**, *19*, 115–119. [CrossRef]
24. Krishna Mohan, S.; Srivastava, T. Microbial deterioration and degradation of polymeric materials. *J. Biochem. Technol.* **2011**, *2*, 210–215.
25. Loredo-Treviño, A.; Gutiérrez-Sánchez, G.; Rodríguez-Herrera, R.; Aguilar, C.N. Microbial enzymes involved in polyurethane biodegradation: A review. *J. Polym. Environ.* **2012**, *20*, 258–265. [CrossRef]

26. Pathak, V.M. Review on the current status of polymer degradation: A microbial approach. *Bioresour. Bioprecess.* **2017**, *4*, 15. [CrossRef]
27. Brebu, M. Environmental degradation of plastic composites with natural fillers—A review. *Polymers* **2020**, *12*, 166. [CrossRef] [PubMed]
28. Siracusa, V. Microbial degradation of synthetic biopolymers waste. *Polymers* **2019**, *11*, 1066. [CrossRef] [PubMed]
29. Shah, A.A.; Hasan, F.; Hameed, A.; Ahmed, S. Biological degradation of plastics: A comprehensive review. *Biotechnol. Adv.* **2008**, *26*, 246–265. [CrossRef] [PubMed]
30. Siracusa, V.; Rocculi, P.; Romani, S.; Rosa, M.D. Biodegradable polymers for food packaging: A review. *Trends Food Sci. Technol.* **2008**, *19*, 634–643. [CrossRef]
31. Chamas, A.; Moon, H.; Zheng, J.; Qiu, Y.; Tabassum, T.; Jang, J.H.; Abu-Omar, M.; Scott, S.L.; Suh, S. Degradation rates of plastics in the environment. *ACS Sustain. Chem. Eng.* **2020**, *8*, 3494–3511. [CrossRef]
32. Larché, J.F.; Bussiére, P.O.; Thérias, S.; Gardette, J.L. Photooxidation of polymers: Relating material properties to chemical changes. *Polym. Degrad. Stab.* **2012**, *97*, 25–34. [CrossRef]
33. Singh, B.; Sharma, N. Mechanistic implications of plastic degradation. *Polym. Degrad. Stab.* **2008**, *93*, 561–584. [CrossRef]
34. Banerjee, A.; Chatterjee, K.; Madras, G. Enzymatic degradation of polymers: A brief review. *Mater. Sci. Technol.* **2013**, *30*, 567–573. [CrossRef]
35. Lu, T.; Solis-Ramos, E.; Yi, Y.; Kumosa, M. UV degradation model for polymers and polimer matrix composites. *Polym. Degrad. Stab.* **2018**, *154*, 203–210. [CrossRef]
36. Jamshidian, M.; Tehrany, E.A.; Imran, M.; Jacquot, M.; Desobry, S. Poly-lactic acid: Production, applications, nanocomposites, and release studies. *Compr. Rev. Food Sci. Food Saf.* **2010**, *9*, 552–571. [CrossRef]
37. Kale, G.; Auras, R.; Singh, S.P.; Narayan, R. Biodegradability of polylactide bottles in real and simulated composting conditions. *Polym. Test.* **2007**, *26*, 1049–1061. [CrossRef]
38. Pastorelli, G.; Cucci, C.; Garcia, O.; Piantanida, G.; Elnaggar, A.; Cassar, M.; Strlič, M. Environmentally induced color change during natural degradation of selected polymers. *Polym. Degrad. Stab.* **2014**, *107*, 198–209. [CrossRef]
39. Sinyavsky, N.; Korneva, I. Study of optical properties of polymeric materials subjected to degradation. *J. Polym. Environ.* **2017**, *25*, 1280–1287. [CrossRef]
40. Bond, T.; Ferrandiz-Mas, V.; Felipe-Sotelo, M.; van Sebille, E. The occurrence and degradation of aquatic plastic litter based on polymer physicochemical properties: A review. *Crit. Rev. Environ. Sci. Technol.* **2018**, *48*, 685–722. [CrossRef]
41. Salomez, M.; George, M.; Fabre, P.; Touchaleaume, F.; Cesar, G.; Lajarrige, A.; Gastaldi, E. A comparative study of degradation mechanisms of PHBV and PBSA under laboratory-scale composting conditions. *Polym. Degrad. Stab.* **2019**, *167*, 102–113. [CrossRef]
42. Eubeler, J.P.; Bernhard, M.; Knepper, T.P. Environmental biodegradation of synthetic polymers II. Biodegradation of different polymer groups. *TrAC Trends Anal. Chem.* **2010**, *29*, 84–100. [CrossRef]
43. Emadian, S.M.; Onay, T.T.; Demirel, B. Biodegradation of bioplastics in natural environments. *Waste Manag.* **2017**, *59*, 526–536. [CrossRef] [PubMed]
44. Chiellini, E.; Corti, A.; D'Antone, S.; Solaro, R. Biodegradation of poly(vinyl alcohol) based materials. *Prog. Polym. Sci.* **2003**, *28*, 963–1014. [CrossRef]
45. Corti, A.; Solaro, R.; Chiellini, E. Biodegradation of poly(vinyl alcohol) in selected mixed microbial culture and relevant culture filtrate. *Polym. Degrad. Stab.* **2002**, *75*, 447–458. [CrossRef]
46. Tosin, M.; Weber, M.; Siotto, M.; Lott, C.; Innocenti, F.D. Laboratory test methods to determine the degradation of plastics in marine environment conditions. *Front. Microbiol.* **2012**, *3*, 225. [CrossRef] [PubMed]
47. Arcos-Hernandez, M.V.; Laycock, B.; Pratt, S.; Donose, B.C.; Nikolić, M.A.L.; Luckman, P.; Werker, A.; Lant, P.A. Biodegradation in a soil environment of activated sludge derived polyhydroxyalkanoate (PHBV). *Polym. Degrad. Stab.* **2012**, *97*, 2301–2312. [CrossRef]
48. Wu, C.S. Preparation and characterization of polyhydroxyalkanoate bioplastic-based green renewable composites from rice husk. *J. Polym. Environ.* **2014**, *22*, 384–392. [CrossRef]
49. Nakasaki, K.; Matsuura, H.; Tanaka, H.; Sakai, T. Synergy of two thermophiles enables decomposition of poly-ε-caprolactone under composting conditions. *FEMS Microbiol. Ecol.* **2006**, *58*, 373–383. [CrossRef] [PubMed]
50. Tabasi, R.Y.; Ajji, A. Selective degradation of biodegradable blends in simulated laboratory composting. *Polym. Degrad. Stab.* **2015**, *120*, 435–442. [CrossRef]
51. Mihai, M.; Legros, N.; Alemdar, A. Formulation-properties versatility of wood fiber biocomposites based on polylactide and polylactide/thermoplastic starch blends. *Polym. Eng. Sci.* **2014**, *54*, 1325–1340. [CrossRef]
52. Anstey, A.; Muniyasamy, S.; Reddy, M.M.; Misra, M.; Mohanty, A. Processability and biodegradability evaluation of composites from poly(butylene succinate) (PBS) bioplastic and biofuel co-products from Ontario. *J. Polym. Environ.* **2014**, *22*, 209–218. [CrossRef]
53. Kale, S.K.; Deshmukh, A.G.; Dudhare, M.S.; Patil, V.B. Microbial degradation of plastic: A review. *J. Biochem. Technol.* **2015**, *6*, 952–961.
54. Bishop, G.; Styles, D.; Lens, P.N.L. Environmental performance comparison of bioplastics and petrochemical plastics: A review of life cycle assessment (LCA) methodological decisions. *Resour. Conserv. Recycl.* **2021**, *168*, 105451. [CrossRef]

55. Spierling, S.; Venkatachalam, V.; Mudersbach, M.; Becker, N.; Herrmann, C.; Endres, H.J. End-of-life options for biobased plastics in a circular economy—Status quo and potential from a life cycle assessment perspective. *Resources* **2020**, *9*, 90. [CrossRef]
56. Hottle, T.A.; Bilec, M.M.; Landis, A.E. Biopolymer production and end of life comparisons using life cycle assessment. *Resour. Conserv. Recycl.* **2017**, *122*, 295–306. [CrossRef]
57. Lambert, S.; Sinclair, C.; Boxall, A. Occurrence, degradation, and effect of polymer-based materials in the environment. *Rev. Environ. Contam. Toxicol.* **2014**, *227*, 1–53. [PubMed]
58. Ramesh, P.; Vinodh, S. State of the art review on life cycle assessment of polymers. *Int. J. Sustain. Eng.* **2020**, *13*, 411–422. [CrossRef]
59. Lewis, H.; Verghese, K.; Fitzpatrick, L. Evaluating the sustainability impacts of packaging: The plastic carry bag dilemma. *Packag. Technol. Sci. Int. J.* **2010**, *23*, 145–160. [CrossRef]
60. Saling, P.; Gyuzeleva, L.; Wittstock, K.; Wessolowski, V.; Griesshammer, R. Life cycle impact assessment of microplastics as one component of marine plastic debris. *Int. J. Life Cycle Assess.* **2020**, *25*, 2008–2026. [CrossRef]
61. Agarwal, S. Biodegradable polymers: Present opportunities and challenges in providing a microplastic-free environment. *Macromol. Chem. Phys.* **2020**, *221*, 2000017. [CrossRef]
62. Puchalski, M.; Szparaga, G.; Biela, T.; Gutkowska, A.; Sztajnowski, S.; Krucińska, I. Molecular and supramolecular changes in polybutylene succinate (PBS) and polybutylene succinate adipate (PBSA) copolymer during degradation in various environmental conditions. *Polymers* **2018**, *10*, 251. [CrossRef]
63. Kim, H.S.; Yang, H.S.; Kim, H.J. Biodegradability and mechanical properties of agro-flour-filled polybutylene succinate biocomposites. *J. Appl. Polym. Sci.* **2005**, *97*, 1513–1521. [CrossRef]
64. Xu, J.; Guo, B.H. Poly(butylene succinate) and its copolymers: Research development and industrialization. *Biotechnol. J.* **2010**, *5*, 1149–1163. [CrossRef] [PubMed]
65. Shih, Y.F.; Lee, W.C.; Jeng, R.J.; Huang, M.H. Water bamboo husk-reinforced poly(butylenes succinate) biodegradable composites. *J. Appl. Polym. Sci.* **2006**, *99*, 188–190. [CrossRef]
66. Fujimaki, T. Processability and properties of aliphatic polyester BIONOLLE synthesized by polycondensation reaction. *Polym. Degrad. Stab.* **1998**, *59*, 209–214. [CrossRef]
67. Bechthold, I.; Bretz, K.; Kabasci, S.; Kopitzky, R.; Springer, A. Succinic acid: A new platform chemical for biobased polymers from renewable sources. *Chem. Eng. Technol.* **2008**, *31*, 647–654. [CrossRef]
68. McKinlay, J.B.; Vielle, C.; Zeikus, J.G. Prospects for bio-based succinate industry. *Appl. Microbiol. Biotechnol.* **2007**, *76*, 727–740. [CrossRef] [PubMed]
69. Cukalovic, A.; Stevens, C.V. Feasibility of production methods for succinic acid derivatives: A marriage of renewable resources and chemical technology. *Biofuels Bioprod. Biorefin.* **2008**, *2*, 505–529. [CrossRef]
70. Ioannidou, S.M.; Ladakis, D.; Moutousidi, E.; Dheskali, E.; Kookos, I.K.; Câmara-Salim, I.; Moreira, M.T.; Koutinas, A. Techno-economic risk assessment, life cycle analysis and life cycle costing for poly(butylene succinate) and poly(lactic acid) production using renewable resources. *Sci. Total Environ.* **2021**, *806*, 150594. [CrossRef] [PubMed]
71. Mousa, H.I.; Young, S.B. Polybutylene succinate life cycle assessment variations and variables. In Proceedings of the American Institute of Chemical Engineers 2012 Annual Meeting, Pittsburgh, PA, USA, 28 October–2 November 2012.
72. Zini, E.; Scandola, M. Green composites: An overview. *Polym. Compos.* **2011**, *32*, 1905–1915. [CrossRef]
73. Totaro, G.; Sisti, L.; Vannini, M.; Marchese, P.; Tassoni, A.; Lenucci, M.S.; Lamborghini, M.; Kalia, S.; Celli, A. A new route of valorization of rice endosperm by-product: Production of polymeric biocomposites. *Compos. Part B* **2018**, *139*, 195–202. [CrossRef]
74. Yap, S.Y.; Sreekantan, S.; Hassan, M.; Sudesh, K.; Ong, M.T. Characterization and biodegradability of rice husk-filled polymer composites. *Polymers* **2021**, *13*, 104. [CrossRef]
75. Sasimowski, E.; Majewski, Ł.; Grochowicz, M. Efficiency of twin-screw extrusion of biodegradable poly(butylene succinate)-wheat bran blend. *Materials* **2021**, *14*, 424. [CrossRef] [PubMed]
76. Rojas-Lema, S.; Arevalo, J.; Gomez-Caturla, J.; Garcia-Garcia, D.; Torres-Giner, S. Peroxide-induced synthesis of maleic anhydride grafted poly(butylene succinate) and its compatibilizing effect on poly(butylene succinate)/pistachio shell flour composites. *Molecules* **2021**, *26*, 5927. [CrossRef] [PubMed]
77. Hongsriphan, N.; Kamsantia, P.; Sillapasangloed, P.; Loychuen, S. Bio-based composite from poly(butylene succinate) and peanut shell waste adding maleinized linseed oil. *IOP Conf. Ser. Mater. Sci. Eng.* **2020**, *773*, 012046. [CrossRef]
78. Liminana, P.; Garcia-Sanoguera, D.; Quiles-Carrillo, L.; Balart, R.; Montanes, N. Optimization of maleinized linseed oil loading as a biobased compatibilizer in poly(butylene succinate) composites with almond shell flour. *Materials* **2019**, *12*, 685. [CrossRef] [PubMed]
79. Nanni, A.; Messori, M. Thermo-mechanical properties and creep modeling of wine lees filled polyamide 11 (PA11) and polybutylene succinate (PBS) bio-composites. *Compos. Sci. Technol.* **2020**, *188*, 107974. [CrossRef]
80. Picard, M.C.; Rodriguez-Uribe, A.; Thimmanagari, M.; Misra, M.; Mohanty, A.K. Sustainable biocomposites from poly(butylenes succinate) and apple pomace: A study on compatibilization performance. *Waste Biomass Valorizat.* **2020**, *11*, 3775–3787. [CrossRef]
81. Gowman, A.; Wang, T.; Rodriguez-Uribe, A.; Mohanty, A.K.; Misra, M. Bio-poly(butylene succinate) and its composites with grape pomace: Mechanical performance and thermal properties. *ACS Omega* **2018**, *3*, 15205–15216. [CrossRef] [PubMed]
82. Frollini, E.; Bartolucci, N.; Sisti, L.; Celli, A. Poly(butylene succinate) reinforced with different lignocellulosic fibers. *Ind. Crops Prod.* **2013**, *45*, 160–169. [CrossRef]

83. Feng, Y.H.; Zhang, D.W.; Qu, J.P.; He, H.Z.; Xu, B.P. Rheological properties of sisal fiber/poly(butylene succinate) composites. *Polym. Test.* **2011**, *30*, 124–130. [CrossRef]
84. Avérous, L.; Le Digabel, F. Properties of biocomposites based on lignocellulosic fillers. *Carbohydr. Polym.* **2006**, *66*, 480–493. [CrossRef]
85. Sirichalarmkul, A.; Kaewpirom, S. Enhanced biodegradation and processability of biodegradable package from poly(lactic acid)/poly(butylene succinate)/rice-husk green composites. *J. Appl. Polym. Sci.* **2021**, *138*, 50652. [CrossRef]
86. Laurichesse, S.; Avérous, L. Chemical modification of lignins: Towards biobased polymers. *Prog. Polym. Sci.* **2014**, *39*, 1266–1290. [CrossRef]
87. Sasimowski, E.; Majewski, Ł. Biodegradable polymer composition. Polish Patent No. PL239238B1, 13 October 2021.
88. *Plastics—Injection Moulding of Test Specimens of Thermoplastic Materials—Part 1: General Principles, and Moulding of Multipurpose and Bar Test Specimens*; ISO 294-1:2017; ISO: Geneva, Switzerland, 2017.
89. BioPBS FZ91PB—Technical Data Sheet. Available online: https://www.mcpp-global.com/en/mcpp-asia/products/product/biopbsTM-general-properties/ (accessed on 5 October 2021).
90. Greffeuille, V.; Abecassis, J.; Lapierre, C.; Lullien-Pellerin, V. Bran size distribution at milling and mechanical and biochemical characterization of common wheat grain outer layers: A relationship assessment. *Cereal Chem.* **2006**, *83*, 641–646. [CrossRef]
91. Kamal-Eldin, A.; Lærke, H.N.; Knudsen, K.E.B.; Lampi, A.M.; Piironen, V.; Adlercreutz, H.; Katina, K.; Poutanen, K.; Åman, P. Physical, microscopic and chemical characterization of industrial rye and wheat brans from the Nordic countries. *Food Nutr. Res.* **2009**, *53*, 1912. [CrossRef] [PubMed]
92. *Plastics—Methods of Exposure to Laboratory Light Sources—Part 2: Xenon-Arc Lamps*; ISO 4892-2:2013; ISO: Geneva, Switzerland, 2013.
93. *Plastics—Differential Scanning Calorimetry (DSC)—Part 1:General Principles*; ISO 11357-1:2016; ISO: Geneva, Switzerland, 2016.
94. Huang, Z.; Qian, L.; Yin, Q.; Yu, N.; Liu, T.; Tian, D. Biodegradability studies of poly(butylene succinate) composites filled with sugarcane rind fiber. *Polym. Test.* **2018**, *66*, 319–326. [CrossRef]
95. *Plastics—Injection Moulding of Test Specimens of Thermoplastic Materials—Part 4: Determination of Moulding Shrinkage*; ISO 294-4:2018; ISO: Geneva, Switzerland, 2018.
96. *Standard Practice for Computing the Color of Objects by Using the CIE System*; ASTM E308; ASTM International: West Conshohocken, PA, USA, 2018.
97. *Surface Texture (Surface Roughness, Waviness and Lay)*; ASME B46.1; American Society of Mechanical Engineers (ASME): London, UK, 2019.
98. *Paints and Varnishes—Determination of Gloss Value at 20 Degrees, 60 Degrees and 85 Degrees*; ISO 2813:2001; ISO: Geneva, Switzerland, 2001.
99. *Plastics—Thermoplastic Materials—Determination of Vicat Softening Temperature (VST)*; ISO 306:2013; ISO: Geneva, Switzerland, 2013.
100. *Plastics and Ebonite—Determination of Indentation Hardness by Means of a Durometer (Shore Hardness)*; ISO 868:2003; ISO: Geneva, Switzerland, 2001.
101. *Plastics—Methods of Test for the Determination of the Effects of Immersion in Liquid Chemicals*; ISO 175:2010; ISO: Geneva, Switzerland, 2010.
102. *Plastics—Determination of the Degree of Disintegration of Plastic Materials under Simulated Composting Conditions in a Laboratory-Scale Test*; ISO 20200:2015; ISO: Geneva, Switzerland, 2015.
103. Cai, L.; Qi, Z.; Xu, J.; Guo, B.; Huang, Z. Study on the photodegradation stability of poly(butylene succinate-co-butylene adipate)/TiO$_2$ nanocomposites. *J. Chem.* **2019**, *2019*, 5036019. [CrossRef]
104. Zhang, Y.; Xu, J.; Guo, B. Photodegradation behavior of poly(butylene succinate-co-butyleneadipate)/ZnO nanocomposites. *Colloids Surf. A Physicochem. Eng. Asp.* **2016**, *489*, 173–181. [CrossRef]
105. Carroccio, S.; Rizzarelli, P.; Puglisi, C.; Montaudo, G. MALDI investigation of photooxidation in aliphatic polyesters: Poly(butylene succinate). *Macromolecules* **2004**, *37*, 6576–6586. [CrossRef]
106. Yao, S.F.; Chen, X.T.; Ye, H.M. Investigation of structure and crystallization behavior of poly(butylene succinate) by fourier transform infrared spectroscopy. *J. Phys. Chem. B* **2017**, *121*, 9476–9485. [CrossRef]
107. Schick, C.; Wurm, A.; Mohamed, A. Vitrification and devitrification of the rigid amorphous fraction of semicrystalline polymers revealed from frequency-dependent heat capacity. *Colloid Polym. Sci.* **2001**, *279*, 800–806. [CrossRef]
108. Ratto, J.A.; Stenhouse, P.J.; Auerbach, M.; Mitchell, J.; Farrell, R. Processing, performance and biodegradability of a thermoplastic aliphatic polyester starch system. *Polymer* **1999**, *40*, 6777–6788. [CrossRef]
109. Sasimowski, E.; Majewski, Ł.; Grochowicz, M. Analysis of selected properties of injection moulded sustainable biocomposites from poly(butylene succinate) and wheat bran. *Materials* **2021**, *14*, 7049. [CrossRef] [PubMed]
110. Pegoretti, A.; Penati, A. Recycled poly(ethylene terephthalate) and its shortglass fibres composites: Effects of hygrothermal aging on thethermo-mechanical behavior. *Polymer* **2004**, *45*, 7995–8004. [CrossRef]
111. Kim, H.S.; Kim, H.J.; Lee, J.W.; Choi, I.G. Biodegradability of bio-flour filled biodegradable poly(butylene succinate) bio-composites in natural and compost soil. *Polym. Degrad. Stab.* **2006**, *91*, 1117–1127. [CrossRef]
112. Yasuniwa, M.; Satou, T. Multiple melting behavior of poly(butylene succinate). I. Thermal analysis of melt-crystallized samples. *J. Polym. Sci. Part B Polym. Phys.* **2002**, *40*, 2411–2420. [CrossRef]
113. Yoo, E.S.; Im, S.S. Melting behavior of poly(butylene succinate) during heating scan by DSC. *J. Polym. Sci. Part B Polym. Phys.* **1999**, *37*, 1357–1366. [CrossRef]

114. Qiu, Z.; Komura, M.; Ikehara, T.; Nishi, T. DSC and TMDSC study of melting behaviour of poly(butylene succinate) and poly(ethylene succinate). *Polymer* **2003**, *44*, 7781–7785. [CrossRef]
115. Signoria, F.; Pelagaggi, M.; Bronco, S.; Righetti, M.C. Amorphous/crystal and polymer/filler interphases in biocomposites from poly(butylene succinate). *Thermochim. Acta* **2012**, *543*, 74–81. [CrossRef]
116. Henke, L.; Zarrinbakhsh, N.; Endres, H.J.; Misra, M.; Mohanty, A.K. Biodegradable and bio-based green blends from carbon dioxide-derived bioplastic and poly(butylene succinate). *J. Polym. Environ.* **2017**, *25*, 499–509. [CrossRef]
117. Ou-Yang, Q.; Guo, B.; Xu, J. Preparation and characterization of poly(butylene succinate)/polylactide blends for fused deposition modeling 3D printing. *ACS Omega* **2018**, *3*, 14309–14317. [CrossRef]
118. Shahzad, A.; Isaac, D.H. Weathering of lignocellulosic polymer composites. In *Lignocellulosic Polymer Composites: Processing, Characterisation and Properties*; Thakur, V.K., Ed.; John Willey & Sons: Hoboken, NJ, USA, 2014.
119. Sasimowski, E.; Majewski, Ł.; Grochowicz, M. Analysis of selected properties of biocomposites based on polyethylene with a natural origin filler. *Materials* **2020**, *13*, 4182. [CrossRef] [PubMed]
120. Kim, H.S.; Lee, B.H.; Lee, S.; Kim, H.J.; Dorgan, J.R. Enhanced interfacial adhesion, mechanical, and thermal properties of natural flour-filled biodegradable polymer bio-composites. *J. Therm. Anal. Calorim.* **2011**, *104*, 331–338. [CrossRef]
121. Sahoo, S.; Misra, M.; Mohanty, A.K. Effect of compatinilizer and fillers on the properties of injection molded lignin-based hybrid green composites. *J. Appl. Polym. Sci.* **2013**, *127*, 4110–4121. [CrossRef]
122. National Institute of Standards and Technology. Chemistry WebBook—Succinic Acid. Available online: https://webbook.nist.gov/cgi/inchi?ID=C2338456&Mask=80 (accessed on 2 November 2021).
123. National Institute of Standards and Technology. Chemistry WebBook—1,4-Butanediol. Available online: https://webbook.nist.gov/cgi/cbook.cgi?ID=C110634&Mask=80 (accessed on 2 November 2021).
124. Robledo-Ortíz, J.R.; González-López, M.E.; Martín del Campo, A.S.; Pérez-Fonseca, A.A. Lignocellulosic materials as reinforcement of polyhydroxybutyrate and its copolymer with hydroxyvalerate: A review. *J. Polym. Environ.* **2021**, *29*, 1350–1364. [CrossRef]
125. Williams, R.S.; Knaebe, M.T.; Feist, W.C. Erosion rates of wood during natural weathering. Part II. Earlywood and latewood erosion rates. *Wood Fiber Sci.* **2001**, *33*, 43–49.
126. Stark, N.M. Effect of weathering cycle and manufacturing method on performance of wood flour and high-density polyethylene composites. *J. Appl. Polym. Sci.* **2006**, *100*, 3131–3140. [CrossRef]
127. Azwa, Z.N.; Yousif, B.F.; Manalo, A.C.; Karunasena, W. A review on the degradability of polymeric composites based on natural fibres. *Mater. Des.* **2013**, *47*, 424–442. [CrossRef]
128. Asif, M.; Liaqat, M.A.; Khan, M.A.; Ahmed, H.; Quddusi, M.; Hussain, Z.; Liaqat, U. Studying the effect of nHAP on the mechanical and surface properties of PBS matrix. *J. Polym. Res.* **2021**, *28*, 349. [CrossRef]
129. Patwary, F.; Matsko, N.; Mittal, V. Biodegradation properties of melt processed PBS/chitosan bio-nanocomposites with silica, silicate, and thermally reduced grapheme. *Polym. Compos.* **2018**, *39*, 386–397. [CrossRef]
130. Palai, B.; Mohanty, S.; Nayak, S.K. Synergistic effect of polylactic acid (PLA) and poly(butylene succinate-co-adipate) (PBSA) based sustainable, reactive, super toughened Eco-compoasite blown fims for flexible packaging applications. *Polym. Test.* **2020**, *83*, 106130. [CrossRef]
131. Quiles, L.G.; Vidal, J.; Luzi, F.; Dominici, F.; Cuello, Á.F.; Castell, P. Color fixation strategies on sustainable poly-butylene succinate using biobased itaconic acid. *Polymers* **2021**, *13*, 79. [CrossRef] [PubMed]
132. Matuana, L.M.; Jin, S.; Stark, N.M. Ultraviolet weathering of HDPE/wood-flour composites coextruded with clear HDPE cap layer. *Polym. Degrad. Stab.* **2011**, *96*, 97–106. [CrossRef]
133. Müller, U.; Rätzsch, M.; Schwanninger, M.; Steiner, M.; Zöbl, H. Yellowing and IR-changes of spruce wood as a result of UV-irradiation. *J. Photochem. Photobiol. B Biol.* **2003**, *69*, 97–105. [CrossRef]
134. Nordlund, E.; Aura, A.M.; Mattila, I.; Kössö, T.; Rouau, X.; Poutanen, K. Formation of phenolic microbial metabolites and short-chain fatty acids from rye, wheat, and oat bran and their fractions in the metabolical in vitro colon model. *J. Agric. Food Chem.* **2012**, *60*, 8134–8145. [CrossRef]
135. Mert, B.; Tekin, A.; Demirkesen, I.; Kocak, G. Production of microfluidized wheat bran fibers and evaluation as an ingredient in reduced flour bakery product. *Food Bioprocess Technol.* **2014**, *7*, 2889–2901. [CrossRef]
136. De Bondt, Y.; Rosa-Sibakov, N.; Liberloo, I.; Roye, C.; Van de Walle, D.; Dewettinck, K.; Goos, P.; Nordlund, E.; Courtin, C.M. Study into the effect of microfluidisation processing parameters on the physicochemical properties of wheat (*Triticum aestivum* L.) bran. *Food Chem.* **2020**, *305*, 125436. [CrossRef]
137. Thurber, H.; Curtzwiler, G.W. Suitability of poly(butylene succinate) as a coating for paperboard convenience food packaging. *Int. J. Biobased Plast.* **2020**, *2*, 1–12. [CrossRef]
138. Mizuno, S.; Maeda, T.; Kanemura, C.; Hotta, A. Biodegradability, reprocessability, and mechanical properties of polybutylene succinate (PBS) photografted by hydrophilic or hydrophobic membranes. *Polym. Degrad. Stab.* **2015**, *117*, 58–65. [CrossRef]
139. Kanemura, C.; Nakashima, S.; Hotta, A. Mechanical properties and chemical structures of biodegradable poly(butylene-succinate) for material reprocessing. *Polym. Degrad. Stab.* **2012**, *97*, 972–980. [CrossRef]
140. Jin, T.X.; Liu, C.; Zhou, M.; Chai, S.G.; Chen, F.; Fu, Q. Crystallization, mechanical performance and hydrolytic degradation of poly(butylene succinate)/graphene oxide nanocomposites obtained via in situ polymerization. *Compos. Part A Appl. Sci. Manuf.* **2015**, *68*, 193–201. [CrossRef]

141. Cho, K.; Lee, J.; Kwon, K. Hydrolytic degradation behavior of poly(butylene succinate)s with different crystalline morphologies. *J. Appl. Polym. Sci.* **2001**, *79*, 1025–1033. [CrossRef]
142. Muniyasamy, S.; Anstey, A.; Reddy, M.M.; Misra, M.; Mohanty, A. Biodegradability and compostability of lignocellulosic based composite materials. *J. Renew. Mater.* **2013**, *1*, 253–272. [CrossRef]
143. Da Silva, A.M.B.; Martins, A.B.; Santana, R.M.C. Biodegradability studiem of lignocellulosic fiber reinforced composites. In *Fiber Reinforced Composites*; Woodhead Publishing: Sawston, UK, 2021; pp. 241–271.
144. Soccio, M.; Dominici, F.; Quattrosoldi, S.; Luzi, F.; Munari, A.; Torre, L.; Lotti, N.; Puglia, D. PBS-based green copolymer as an efficient compatibilizer in thermoplastic inedible wheat flour/poly(butylene succinate) blends. *Biomacromolecules* **2020**, *21*, 3254–3269. [CrossRef]

Article

Laser Texturing as a Way of Influencing the Micromechanical and Biological Properties of the Poly(L-Lactide) Surface

Magdalena Tomanik [1], Magdalena Kobielarz [1,*], Jarosław Filipiak [1], Maria Szymonowicz [2], Agnieszka Rusak [3], Katarzyna Mroczkowska [4], Arkadiusz Antończak [4] and Celina Pezowicz [1]

1. Department of Mechanics, Materials and Biomedical Engineering, Wrocław University of Science and Technology, 50-370 Wrocław, Poland; magdalena.tomanik@pwr.edu.pl (M.T.); jaroslaw.filipiak@pwr.edu.pl (J.F.); celina.pezowicz@pwr.edu.pl (C.P.)
2. Department of Experimental Surgery and Biomaterials Research, Faculty of Dentistry, Wroclaw Medical University, 50-326 Wrocław, Poland; maria.szymonowicz@umed.wroc.pl
3. Division of Histology and Embryology, Department of Human Morphology and Embryology, Faculty of Medicine, Wroclaw Medical University, 50-368 Wrocław, Poland; agnieszka.rusak@umed.wroc.pl
4. Laser and Fiber Electronics Group, Faculty of Electronics, Wroclaw University of Science and Technology, 50-372 Wrocław, Poland; katarzyna.mroczkowska@pwr.edu.pl (K.M.); arkadiusz.antonczak@pwr.edu.pl (A.A.)
* Correspondence: magdalena.kobielarz@pwr.edu.pl

Received: 17 July 2020; Accepted: 24 August 2020; Published: 27 August 2020

Abstract: Laser-based technologies are extensively used for polymer surface patterning and/or texturing. Different micro- and nanostructures can be obtained thanks to a wide range of laser types and beam parameters. Cell behavior on various types of materials is an extensively investigated phenomenon in biomedical applications. Polymer topography such as height, diameter, and spacing of the patterning will cause different cell responses, which can also vary depending on the utilized cell types. Structurization can highly improve the biological performance of the material without any need for chemical modification. The aim of the study was to evaluate the effect of CO_2 laser irradiation of poly(L-lactide) (PLLA) thin films on the surface microhardness, roughness, wettability, and cytocompatibility. The conducted testing showed that CO_2 laser texturing of PLLA provides the ability to adjust the structural and physical properties of the PLLA surface to the requirements of the cells despite significant changes in the mechanical properties of the laser-treated surface polymer.

Keywords: poly(L-lactide); laser irradiation; surface enhancement; micromechanical properties; cytotoxicity

1. Introduction

Poly(L-lactide) (PLLA) is one of the more technologically advanced polymers. High product purity and high polymer biocompatibility are especially important for medical applications. Polymeric materials based on PLLA are often used for the production of, among others, bioresorbable screws, surgical sutures, vascular stents, and bone scaffolds [1–3]. Some medical devices, such as bone scaffolds, are complex structures of small geometrical dimensions, whose shaping by conventional methods, e.g., by injection or mechanical micromachining, is technologically challenging. The assumptions made at the design stage concerning the accuracy of execution of small-scale components cannot always be achieved using conventional processing methods.

The interaction of cells with the surfaces of biomaterials is a complex phenomenon that depends on many factors. The properties of an implant surface, including roughness, topography, and wettability directly affect tissue cells and must be adjusted to their mechanobiology [4–6]. The hydrophobic nature of PLLA [7–9] and its low surface roughness [10] adversely affect the adhesion of cells to the surface of the material [11], hence it is necessary to modify its surface properties. One of the possible ways of adjusting the material to cell preferences is to modify only its topography [4,5,12,13] without the need to interfere in the chemical bulk composition of the polymer. Technologies based on material processing with a laser beam are an excellent solution, enabling selective surface structuring of even very geometrically complex objects, regardless of their dimensions [12,14,15]. One way to structure the surfaces of biodegradable polymers is to irradiate the materials with a CO_2 laser beam [16–21]. It is known that a CO_2 laser surface irradiation causes formation of nano- and microstructures on the surfaces of the treated materials [10,21] which can alter their mechanical characteristics [21,22]. However, the use of a laser treatment process, due to the associated thermal effects, may adversely affect the material and change its properties [16,21,23–27]. The physicochemical properties of the irradiated material depend on the type of laser source and parameters of the irradiation process, such as, for example, the light wavelength, pulse duration, and number and energy of pulses.

A reliable assessment of the impact of laser irradiation on the degree of polymer degradation requires an analysis of a wide range of process parameters used during the processing of this material. In our previous work [16,27], an analysis of the influence of CO_2 laser energy density on the degree of PLLA degradation was initially conducted in the fluence range from 13 to 95 J/cm^2, i.e., from the minimum value that could be set on our laser system, to the value that ablated the material about half its thickness (for films with a thickness of about 250 µm). The analysis of all changes in the physicochemical properties was subordinated to the overriding goal of micro-processing of these polymers. The aim was to obtain information on how this polymer degrades (and what consequences it has) during the production of practically usable implants, i.e., stents [15,16,27]. Performing further research for such a large number of cases/fluences would be redundant, as certain regularities were observed for specific ranges of fluence. Therefore, we limited our considerations to three characteristic surface fluences: 24, 48, and 71 J/cm^2 [21,22]. The first fluence (24 J/cm^2) is the maximum value of the energy density at which neither surface layer melting nor ablation was observed. Despite this, significant changes were observed in both differential scanning calorimetry (DSC) and, above all, gel permeation chromatography (GPC) [16,27] (bimodal distribution and significant loss of molecular weight in the near-surface layer). In this case, however, Fourier transform infrared reflectance (FTIR) spectroscopy showed no changes. The erasure of the ageing peak in the DSC thermograms indicated that the polymer had crossed the glass transition in all its volume. The second fluence (48 J/cm^2) corresponded to the case of remelting the PLLA top layer. There were already visible changes (apart from those recorded for 24 J/cm^2 in GPC and DSC) in FTIR spectroscopy (appearance of new, characteristic peaks of vinyl groups, ketones, etc.). This fluence could also be considered the beginning of the polymer ablation threshold. The third characteristic fluence (71 J/cm^2) is the value at which intense ablation was noted. Here, the visible changes were similar but more intense compared to the values observed in GPC, FTIR, and DSC for 48 J/cm^2 [16,27]

Therefore, the aim of the study was to evaluate the effect of irradiation of PLLA with a CO_2 laser using different accumulated fluences, resulting in the formation of micro-groove patterns on the surface and affecting the micromechanical and cytotoxic properties of the material.

2. Material and Methods

2.1. Material and Surface Modification

The tests were carried out using PLLA (L210S, Evonik Industries AG, Essen, Germany), which underwent the process of formation described in detail in previous studies by our group [21,22]. Briefly, thin films of the material were produced by heating the granulate to a temperature of 200 °C, followed by melt processing. The specimens were excised from the obtained amorphous sheets with the crystallinity of $X_c \approx 2\%$ and average thickness of 350 µm. Biological testing was performed on discs with a diameter of 12 mm, while other tests used rectangular specimens measuring 20 × 5 mm.

All the specimens were irradiated using the Speedy 300 (Trotec Laser GmbH, Marchtrenk, Austria) engraving system, equipped with an air-cooled RF-excited (Radio Frequency-excited) pulsed CO_2 laser (10.6 ± 0.03 µm) (Series 48-2, Synrad Inc. Mukilteo, WA, USA) with the maximum average power of 25 W and pulse duration ranging from several dozen to several hundred µs depending on the power and the pulse repetition rate. The beam with the mode purity TEM_{00} at 95% (beam quality factor $M^2 < 1.2$) was focused on the surface of the material using a 38.1 mm (1.5″) focal length lens. The beam waist width was about $2w_0 = 125$ µm. The tests were conducted for the experimentally determined range of parameters (Table 1).

Table 1. Process parameters for CO_2 laser surface treatments of PLLA specimens.

Specimen	Optical Power P (W)	Scanning Speed V (cm/s)	Hatching (Line-to-Line Space) h (µm)	Pulse Repetition Rate (PRR) (Hz)	Pulse Energy E_i (mJ)	Accumulated Fluence F_A (J/cm^2)
Reference	-	-	-	-	-	-
F_1	0.447				15.5	24
F_2	0.867	7.1	25.4	2.8	31.0	48
F_3	1.286				45.8	71

The specimens intended for cytotoxicity testing were subjected to plasma sterilization. The PLLA specimens were sterilized in a Sterrad 100S plasma sterilizer (Advanced Sterilization Products, Irvine, CA, USA) at a temperature of 46 °C, with 2 exposures to hydrogen peroxide lasting 7 min each, under general sterilization conditions lasting 51 min.

2.2. Roughness and Wettability

In order to evaluate the surface properties of the PLLA thin films, both referenced (Ref) and micro-patterned (F_1–F_3) surfaces were examined. Surface roughness was determined based on height profiles using a Micro Combi Tester (CSM Instruments SA, Peseux, Switzerland) in the scratch test mode with a Rockwell indenter (CSM Instruments SA, Peseux, Switzerland) 10 µm in diameter. Unlike traditional profilometers, the optical module of the device with a magnification of the order of 50–200 times allowed for precise selection of the profile location and correlation of the profile curves with the recorded surface geometry of the specimens. One of the advanced scratch test (pre-scan) settings allowed us to register a profile so that after performing the proper scratch test it was possible to determine the actual penetration depth of the indenter into the material. This option was used to register surface profiles of the analyzed specimens. The contact force of the intender was set to 10 mN, so no scratches were observed on the surface of the material after the profile was made. The methodology used to determine the surface roughness was validated on surface roughness reference specimens with $R_z = 9.47$ µm, achieving a compliance level of 98%.

Six profile measurements were performed on the specimens from each measurement group in the direction perpendicular to the direction of laser beam propagation over a measuring length of 1.25 mm. The profiles obtained from a pre-scan allowed us to determine the R_z parameter understood as height of the roughness profile for the highest 5 and the lowest 5 points on the specimens. Based on the above, the theoretical R_a value (arithmetic mean of profile deviation) was calculated as 1/5 of the R_z value. Measurements were based on the ISO 4287 standard [28].

The effect of the developing roughness on the affinity of surfaces to the liquid(s) was examined by determining the wetting angle θ. Six drops with an average volume of 0.3 µl were applied on the examined surfaces; the tests were carried out at room temperature equal to 22 ± 1 °C using a Surftens Universal instrument (OEG Gesellschaft für Optik, Elektronik & Gerätetechnik mbH, Frankfurt (Oder), Germany).

2.3. In Vitro Cytotoxicity

The cytocompatibility effect of irradiation of the PLLA surface with a CO_2 laser using different accumulated fluences was evaluated on a culture Balb/3T3 of normal mouse fibroblast cell line (clone A31, American Type Culture Collection (ATCC CCL-163), Manassas, VA, USA). The cells were cultured in Dulbecco's Modified Eagle Medium (DMEM) with 4.5 g/L of glucose, 25 mM HEPES (Corning Inc., Corning, NY, USA), 1% L-glutamine with streptomycin and penicillin (Sigma-Aldrich, St. Louis, MO, USA), and 10% fetal bovine serum (Sigma-Aldrich, Saint Louis, Missouri, USA) under standard conditions of 37 °C and 5% CO_2, at constant humidity. Balb/3T3 fibroblasts were trypsinized (0.25% Trypsin-EDTA, Sigma-Aldrich, Saint Louis, Missouri, USA), suspended in culture medium, and seeded on a 6-well plate (Falcon, Corning Inc., Corning, NY, USA) at a density of 1.5×10^5 cells/well. After 24 h of cell incubation, discs of laser-treated and reference test material were applied to each well in such a way that the fibroblast monolayer was in contact with the polymer surface. The control group was cells without contact with the material. After 24 h, fibroblast morphology was assessed under the surface of the test material and at the edges of the specimens using a CKX-41 inverted phase contrast microscope (Olympus, Tokyo, Japan). To assess the cytotoxic effect [29], a grading scale was used where the changes in the culture over grade 2 (mild grade) were considered to be a cytotoxic effect [29]. On the other hand, a moderate degree of cytotoxicity was characterized by the presence of a zone of altered (degenerated or malformed) cells, which occurred only under the surface of the test material, with normal morphology of the cells around the material. Tests were also carried out on colonization of the test material by Balb/3T3 cells. Polymer specimens in the form of discs were placed in the wells of a 6-well plate (Falcon) and fibroblasts from the same culture were seeded on the polymer surface at a density of 1.5×10^5 cells/well. After 24 h, cell morphology and adhesion to the surfaces of the test materials were assessed using a CKX-41 inverted phase contrast microscope.

2.4. Micromechanical Properties

Tests of micromechanical properties were carried out on a Micro Combi Tester (CSM Instruments SA, Peseux, Switzerland) using a Vickers prismatic indenter with an apex angle of 136°. Specimens with an average thickness of 350 μm were mounted on an aluminum test rig of much higher stiffness using a very thin layer of crystal bond glue. The table was immobilized with adjustable clamps of the microhardness tester. An indentation load of 100 mN was used based on pre-tests, which allowed us both to avoid the effect of substrate on the specimen and enabled testing of the entire thickness of the surface layer. The resulting indentation depths were in the range of 3–4 μm to 10–12 μm for all the specimens. The measurement points, i.e., 20 indentations for a particular irradiated type of surface, were selected in the visual mode enabling a preview of the polymer surface under an optical microscope.

The tests for each indentation were run at a constant strain rate during the loading and unloading phases. As the indenter tip approached a maximum depth at 100 mN, the test was paused for 10 s and then the specimen was unloaded at the same strain rate. The unloading section of these data were analyzed to calculate the mechanical properties using the Oliver Pharr model [30]. Poisson's ratio (ν) of PLLA was set to 0.35, which is a representative value for polymers, while the parameters used for the diamond were E = 1140 GPa and ν = 0.07. The analysis focused on microhardness (H_{IT}, HV), Young's modulus (E_{IT}), maximum indentation depth (hm) as well as plastic strain energy (Wp) and elastic strain energy (We).

2.5. Statistical Analysis

The obtained results of the micromechanical properties and wetting angle were subjected to statistical analysis, starting with checking the normality of the obtained distributions (Shapiro–Wilk test), and were tested for homogeneity of variance (Levene's test). The results obtained within the study groups were averaged and standard deviations were determined. The statistical significance between individual groups was evaluated using one-way analysis of variance (one-way ANOVA) and post-hoc Tukey's tests; the statistical analysis was carried out for a significance level of $\alpha = 0.05$.

3. Results

3.1. Roughness and Wettability

The microscopic evaluation of the irradiated specimens' surfaces showed the presence of repetitive structures. As the laser irradiation parameters increased, the observed micro-patterns became larger (Figure 1).

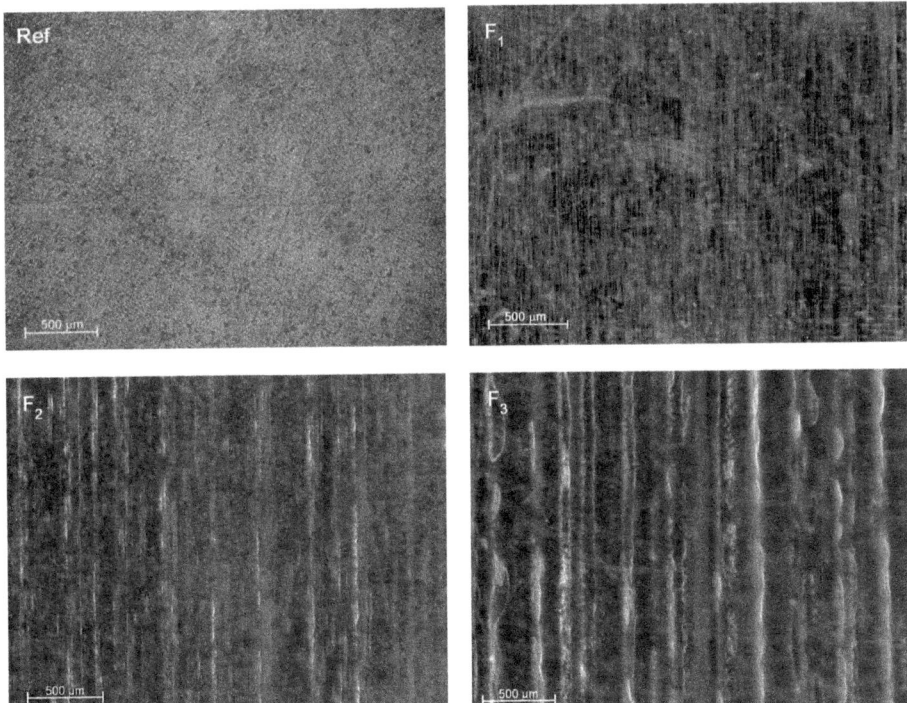

Figure 1. Microscopic images of the obtained surfaces (magnification 100×) recorded for reference material (Ref) and laser treated with accumulated fluences of 24 J/cm^2 (F_1), 48 J/cm^2 (F_2) and 71 J/cm^2 (F_3).

The determined profiles of surface roughness showed that increased use of accumulated fluence created micro-grooves, making the surface increasingly rough. Table 2 shows averaged R_z values as well as R_a values determined on their basis.

Table 2. Roughness parameters describing surface topography of micro-patterned PLLA.

Roughness Parameters	Reference	F_1	F_2	F_3
$R_z \left(\overline{X} \pm SD \right)$ (μm)	0.10 ± 0.04	0.21 ± 0.05	2.87 ± 0.31	6.67 ± 0.66
R_a * (μm)	0.02	0.04	0.57	1.33

* Values derived from the following correlation: $R_a = 1/5\ R_z$.

The conducted tests showed that surface structuring with the use of a CO_2 laser altered the wettability of poly(L-lactide) (Figure 2). The reference material was partially wettable, however, it was the most hydrophobic (θ = 72.3 ± 1.4°). Irradiation of the material with the lowest accumulated fluence (F_1) was almost unnoticeable as a micro-pattern (Figure 1) and did not significantly affect the surface wettability. The use of a higher fluence caused a significant change in the nature of the surface, not only compared to the reference material but also between individual groups. Irradiation of the surface decreased the angle of wettability θ relative to the reference material by 9.7% and 24.1% for accumulated fluences of, respectively, 48 J/cm^2 and 71 J/cm^2, indicating that the surface became more hydrophilic. The smallest angle (θ = 54.9 ± 2.4°) was noted for an accumulated fluence of 71 J/cm^2.

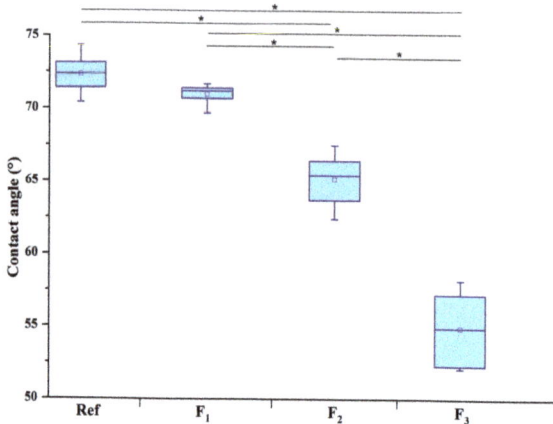

Figure 2. The wettability of the surface modified with different laser parameters; statistical significance between the study groups for water and PBS, obtained in one-way ANOVA and post-hoc Tukey's tests; * p < 0.001.

3.2. In Vitro Cytotoxicity

Reference material and material after laser treatment with the fluence of F_1 after application on fibroblast culture did not cause changes in the morphology of cells under the surface of the specimen, in the zone around the specimen, and in the zone outside of the specimen in the whole well, compared to the control cell culture (Figure 3). Materials after laser treatment F_2 and F_3 did not change the cell morphology compared to the control cell culture either in the zone around the laser-treated surface or in the zone outside of the specimen in the whole well. Changes in cell morphology, such as malformation or degeneration of cells, vacuolization, and also occasional lysis and growth inhibition were only observed for specimens that were laser treated with accumulated fluences F_2 and F_3 (Figure 3), which showed a mild degree (2) of cytotoxicity. According to the standard protocol stated in ISO 10993-5 for evaluation of in vitro cytotoxicity of medical devices, cytotoxicity effects can be considered when the reactivity grade of the cytotoxic effect is greater than 2 [29]. The conducted study showed that laser treatment changed the cytotoxicity of the materials against Balb/3T3 cells.

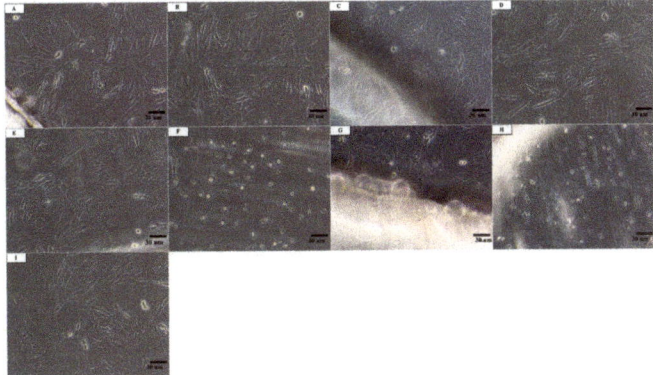

Figure 3. Culture of Balb/3T3 fibroblasts after 24-h contact with the test materials around and under the surface of the specimen: (**A**) and (**B**) Ref, (**C**) and (**D**) F_1, (**E**) and (**F**) F_2, (**G**) and (**H**) F_3, and (**I**) control cell culture without contact with materials. Altered cells occur only in a limited zone under the specimens. Magnification: ×100.

After 24 h after application of the culture, there was no observed cell adhesion to the surfaces of the test materials F_1, F_2, F_3, and Ref (Figure 4). The surfaces of all test materials showed cells with a rounded phenotype, which did not coat the test specimen. The cells in the zone around the specimens and in the zone outside the specimens in the whole well showed normal morphology (Figure 4).

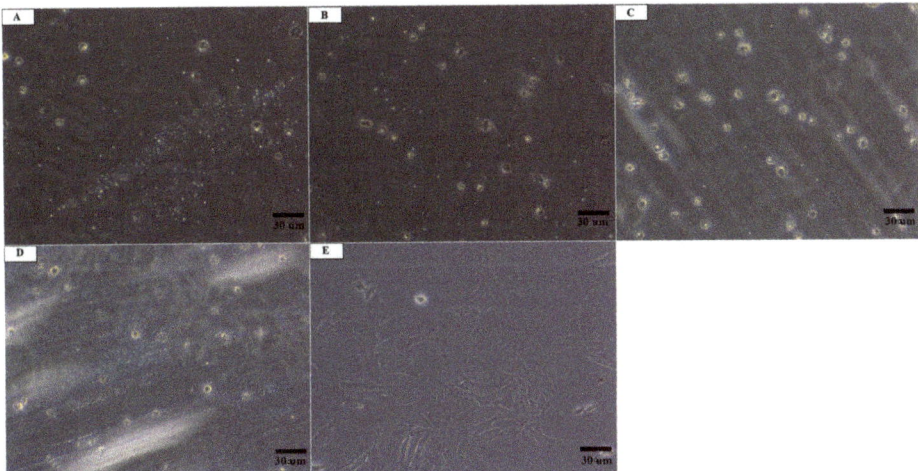

Figure 4. Culture of Balb/3T3 fibroblasts 24 h after application of the cells on the surfaces of the materials: (**A**) Ref, (**B**) F_1, (**C**) F_2, (**D**) F_3, and (**E**) control cell culture without contact with the material No cell colonization of the material surfaces was observed. Magnification: ×100.

3.3. Micromechanical Properties

Micromechanical properties were examined during a typical instrumented indentation test with constant parameters over several repetitions for each type of specimens. The obtained parameters were characteristic for all types of irradiated materials and indicated a specific relation to accumulated fluences (Table 3).

Table 3. Micromechanical properties presented as average and standard deviation ($\overline{X} \pm SD$) determined in the indentation test for references and irradiated specimens.

Specimen Type	E_{IT} (GPa)	H_{IT} (MPa)	HV	hm (μm)	We (nJ)	Wp (nJ)
Ref	2.1 ± 0.9	200.5 ± 53.2	18.9 ± 5.0	6.1 ± 0.9	116.0 ± 46.7	145.7 ± 27.3
F_1	3.4 ± 0.8	300.2 ± 54.9	28.3 ± 5.2	4.7 ± 0.5	80.6 ± 13.0	113.4 ± 5.6
F_2	4.1 ± 1.0	429.6 ± 205.8	40.5 ± 19.4	4.3 ± 0.7	73.6 ± 7.3	117.3 ± 9.3
F_3	0.4 ± 0.03	115.9 ± 17.7	10.9 ± 1.7	1.2 ± 0.4	376.6 ± 40.2	166.7 ± 25.0

Legend: E_{IT}—Young's modulus, H_{IT}—Instrumental microhardness, HV—Vickers microhardness, hm—Maximum indentation depth, We—Elastic strain energy, Wp—Plastic strain energy.

All mechanical parameters showed gradual changes with the trend maintained for all the obtained values, starting from the reference specimens to specimens irradiated with a medium accumulated fluence. However, at the fluence of 71 J/cm^2, all cases showed rapid (abrupt) changes of the trend. Irradiation of PLLA with the lowest and medium fluencies affected the material surface by increasing its hardness and stiffness. Plastic and elastic strain energies gradually decreased. Resistance of the material from the side of the irradiated surface increased. The highest used fluence strongly changed the mechanical behavior of the PLLA specimen. The trends in changes of the mechanical parameters were significantly reversed ($p < 0.05$). The maximal depth during the testing was almost twice higher compared with reference specimens and three times higher compared with specimens irradiated with 48 J/cm^2 fluence. PLLA became significantly more deformable and plastic than the reference material. This phenomenon is clearly visible in the relationships between the indentation force and depth (Figure 5). The mechanical curves obtained for the specimens irradiated with the fluences F_1 and F_2 were shifted as a result of indentation tests towards lower deformations with respect to the reference specimens while demonstrating the stiffer nature of the test surface. The relationships between the indentation force and depth for PLLA specimens irradiated with 71 J/cm^2 were shifted in the opposite direction to all other curves.

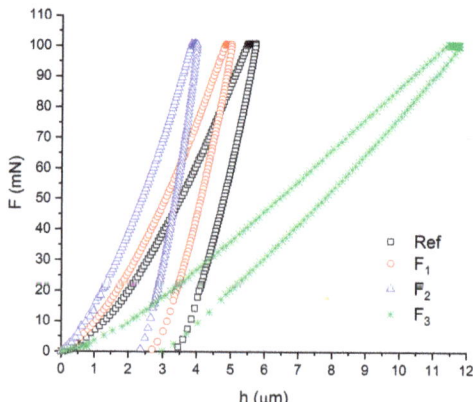

Figure 5. Typical loads vs. indentation depth curves for reference and irradiated PLLA specimens.

4. Discussion

A series of tests focused on assessing changes in the biomaterial surface allowed us to determine the potential of CO_2 laser texturing for biomedical applications. The properties of an implant surface, including roughness, topography, and wettability significantly determine the applicability of the implant and can be adjusted to the preferences of a given type of cell [4–6]. The almost completely smooth surface is not conducive to adhesion and proliferation of cells [10], requiring the use of additional material treatment. The biological potential of the surface can be increased by bulk modification of the material, mainly by altering the chemical composition [4,5,12,13] or by surface modification using, for example, a laser beam [10,12,14,15,31]. However, due to the action of the laser beam on the polymer material, photochemical degradation of the material may occur [24,25], significantly affecting the material properties of the polymer. Only a few studies have focused on assessing the impact of irradiation of biodegradable polymers with a CO_2 laser on their mechanical [15,22] or physicochemical properties [16,32]. The use of a low value of accumulated fluence strengthens and stiffens the material, which is visible in uniaxial tensile tests [22] and under hydrolytic degradation [21]. On the other hand, the highest applied value of accumulated fluence makes the surface of the material more pliable, so it strongly deforms and plasticizes (Figure 1).

The results of wetting angle tests indicate that the reference PLLA surface is the most hydrophobic, which adversely affects cell adhesion to the surface of the material [11]. The obtained wetting angle values for the reference material are consistent with the data presented by other authors [7–9]. In the process of surface laser structuring, an increase in the used accumulated fluence was accompanied by a decrease in the contact angle, resulting in better wetting (wettability) of the material and an increased surface roughness. In contrast to other studies [16,33] showing no relationship between CO_2 laser irradiation and wettability of the PLLA, our tests proved the significance of the process as there was an increase in hydrophilicity. A similar trend was also observed for other materials such as PEEK [34], LDPE [17] and nylon 6.6 [19], where CO_2 laser also improved the wettability of those materials. Surface wettability, in general, is determined by the functional groups (chemical composition) present on the surface and by surface roughness [35]. The observed decrease in the contact angle of water wetting increasing surface hydrophilicity is partly attributed due to the larger surface roughness [35] and changing relationship between polar and nonpolar functional groups, as well as between acidic or basic sites available at the modified surface region [36]. The chemical structure of the laser-irradiated PLLA surface was demonstrated in our previous studies [16,22,27]. The Fourier transform infrared spectroscopy (FTIR) confirmed both a decrease in the number of aliphatic ester segments and the defragmentation of the main chain following C-C and C-H bond scission [37,38]. The ATR/FTIR also proved the appearance of a few small but characteristic absorption bands for vinyl groups ($RCH=CH_2$), ketones ($RCOCH=CH_2$), and/or vinyl ethers ($ROCH=CH_2$) [22]. The appearance of these bands may indicate that the decomposition of PLLA by thermal CO_2 laser impact occurred, among others, by means of the cis elimination reactions, in which double carbon bonds were formed in the vinyl groups ($-CH=CH_2$) [16,22,27]. Admittedly, the use of X-ray photoelectron spectroscopy (XPS) did not demonstrate that processing with CO_2 laser irradiation caused no oxidation of the surface layer, despite the presence of oxygen in the process environment. The value of O/C remained constant [27]. The lack of photo-oxidation may be due to the fact that the degradation process was initiated by temperature and such PLLA decomposition generally does not generate free radicals.

Cytotoxicity tests were performed after sterilization of the PLLA in low-temperature plasma called "cold" plasma. The plasma treatment is a surface process commonly used to clean (sterilize) or etch polymers, alter their surface wettability, and improve the cell affinity to the polymer surface, resulting in better cytocompatibility [36]. Surface wettability increases as a result of the formation of the new chemical groups, mainly polar, introduced to the polymer surface (regardless of the gas used for modification) [39]. The newly formed functional groups enhance cell adhesion [40], owing to chemical interaction and higher surface energy increasing adhesive strengths on the polymer surface [41]. Plasma treatment modifies the surface morphology [35]. However, low-temperature plasma modifies only the upper layer of the polymer surface [42], assessed as approximately 10 nm in depth or less in depth (R_a < 10 nm) [36]. Therefore, it is generally accepted that plasma discharge can modify polymer surfaces, whereas the bulk properties remain unchanged. On the other hand, sterilization with hydrogen peroxide gas plasma is the recommended procedure for heat-sensitive poly(lactic acid) [43]. Hence, the process of plasma treatment enhances surface wettability, while having a negligible effect on the mechanical properties of the PLLA surface, where the maximum indentation depth (hm) is measured in micrometers (Table 3). Plasma sterilization is a compromise between surface modification mainly improving cytocompatibility of PLLA surface and physicochemical modification of surface layers at a shallow depth. Please note that all the tests were performed on non-sterilized PLLA specimens except for cytotoxic tests.

Proliferation and differentiation of cells depend on many factors, including mechanical conditions. The optimal conditions for stable maintenance of the continuity of the life cycle and performance of the functions are different for each type of cell. Cells present in load-bearing tissues are sensitive to mechanical stimulations through mechanoreceptors and, accordingly, to load parameters such as amount, duration, or even amplitude. Consequently, different metabolic pathways are initiated. However, cytocompatibility of materials is viewed mainly in terms of surface topography and chemicals interacting with the cells, including products of material degradation. The influence of the mechanical properties of the materials that are colonized by the cells on cytocompatibility of the materials is negligent. According to previous research [19,44], osteoblasts choose surfaces with greater roughness and their division occurs much faster on smooth surfaces; moreover, increased surface roughness also provides them with better adhesion [10,17,19,34]. The use of laser treatment to significantly increase surface roughness allows an increase in proliferation and differentiation of bone-forming cells. The increase in roughness is associated with the formation of nano- [45] or micropatterns [21] on aliphatic polymer surfaces following laser irradiation. The irradiated PLLA surfaces showed parallel and equidistant grooves and ridges. The pattern repeated itself and as the fluence increased, so did the distance between each two adjacent ridges (profile element width). To the best of our knowledge, this is the first presentation of results of roughness of PLLA surfaces irradiated by a CO_2 laser. Previously, R_a and R_z parameters were discussed mainly for PLLA combined with fillers in nano- and microscales, such as hydroxyapatite [46] or drug-intercalated fillers [47]. Roughness parameters of a mix of PLLA and particles (bulk modification) were comparable, or even higher, to those recorded here for PLLA irradiated with F_2 and F_3 fluences. Higher values of roughness parameters are justified if the bulk modification is processed by macromolecules incorporated into the polymer matrix and dependent on the concentration of the filler. Studies on commercially available implant materials, such as machined Ti6Al4V alloy, show R_a and R_z (0.48 ± 0.05 and 2.48 ± 0.29, respectively) strongly comparable to PLLA irradiated with 48 J/cm^2 fluence [48]. However, better osteoblast proliferation was found in the materials (etched or plasma-sprayed titanium) with higher surface roughness (Ra > 3.5 μm) which is unattainable for laser processing.

In the research conducted by Zheng et al. [20], pre-osteoblast adhesion and proliferation increased after laser treatment. An enhancement in the biocompatibility of the CO_2 laser-treated surfaces was also observed by Waugh et al. [19]. The conducted tests showed that surface modification with a CO_2 laser wavelength affected the cell viability and improved cell growth. Moreover, cell coverage was greater on textured than on non-textured surfaces. As infrared spectroscopy revealed the presence of a vinyl group in PLLA irradiated with higher laser fluences [22], there was a need to perform a biological test aimed at determining the cytotoxicity of the material. The performed biological assessment found no cytotoxic activity of the examined polymers, however, a mild cytotoxic effect was observed after laser treatment (F_2 and F_3). In accordance with the 10993-5 standard [29], changes in the culture over grade 2 are considered to be cytotoxic effect. Nevertheless, no cell colonization of the specimen surfaces was observed, which indicates that the examined surfaces of the materials do not have properties conducive to cell adhesion despite an improvement in the PLLA surface properties after irradiation with a CO_2 laser, i.e., increased roughness and surface wettability.

The micromechanical properties of unmodified PLA have been very rarely tested using the instrumented indentation method [49,50] and, to the best of our knowledge, so far no one has tested surface-modified PLA. Microhardness and elastic modulus of neat and extruded PLA were tested using a diamond Berkovich indenter with indentation load of 0.5 mN, resulting in the maximum indentation depths of, respectively, 1.5 μm and [49] 1 μm [50]. The elastic modulus of extruded PLA ranged from 3.9 ± 0.47 GPa on specimen edges to 4.1 ± 0.26 GPa in the middle of the specimen, while modulus and hardness of neat PLA were uniform and equal to 4.6 ± 0.47 GPa and 0.23 ± 0.034 GPa, respectively [50]. For thin films of pure PLA, Ajala et al. [49] obtained lower modulus and hardness values of 0.95 and 0.03 GPa, respectively. Modulus and hardness of unmodified PLLA presented here are equal to 2.1 ± 0.9 GPa and 0.2 ± 0.05 GPa, respectively. The mechanical parameters of laser-irradiated PLA with F_1 and F_2 accumulated fluencies determined in instrumented indentation tests increased, thus reinforcing the surface. For the highest accumulated fluence, the determined parameters sharply decreased, resulting in significantly greater indentation depth (up to almost 2.5–3.5% of the total thickness of the specimen crossing the boundary of the surface layer). Laser irradiation of the PLLA surface with accumulated fluence of 71 J/cm^2 led to a substantial weakness of both the surface and subsurface layers, with even lower parameters than those reported for non-modified PLA.

In the presented results, the mechanical parameters of the polymer were altered from stiff (groups F_1 and F_2) to compliant (F_3) while simultaneously changing the surface topography and wettability. The two features with the strongest impact on cell viability are chemical interactions and surface topography. The possibility of controlled structuring of the PLLA surface and the observed effect of irradiation on the polymer properties allow the design of its properties, selectively and locally. This way it is possible to adjust the structure of the implant surface so as to ensure the most favorable conditions of cooperation with the surrounding tissues, thus contributing to the reduction in tissue regeneration time.

Author Contributions: M.T.: Investigation, Writing—original draft, M.K.: Investigation, Formal analysis, Visualization, Writing—review and editing, J.F.: Writing—review and editing, M.S.: Investigation, Methodology, Writing—original draft, A.R.: Investigation, Writing—review and editing, K.M.: Investigation, A.A.: Methodology, Writing—review and editing, C.P.: Writing—review and editing, Supervision, Project administration, Funding acquisition. All authors have read and agreed to the published version of the manuscript.

Funding: This study was carried out with financial support from the National Science Centre on the basis of the decision number DEC-2013/09/B/ST8/02423.

Conflicts of Interest: The authors declare that they have no known competing financial interests or personal relationships that could have appeared to influence the work reported in this paper.

References

1. Santoro, M.; Shah, S.R.; Walker, J.L.; Mikos, A.G. Poly(lactic acid) nanofibrous scaffolds for tissue engineering. *Adv. Drug Deliv. Rev.* **2016**, *107*, 206–212. [CrossRef] [PubMed]
2. Lopes, M.S.; Jardini, A.L.; Filho, R.M. Poly (lactic acid) production for tissue engineering applications. *Procedia Eng.* **2012**, *42*, 1402–1413. [CrossRef]
3. Wiebe, J.; Nef, H.M.; Hamm, C.W. Current status of bioresorbable scaffolds in the treatment of coronary artery disease. *J. Am. Coll. Cardiol.* **2014**, *64*, 2541–2551. [CrossRef] [PubMed]
4. Altomare, L.; Gadegaard, N.; Visai, L.; Tanzi, M.C.; Farè, S. Biodegradable microgrooved polymeric surfaces obtained by photolithography for skeletal muscle cell orientation and myotube development. *Acta Biomater.* **2010**, *6*, 1948–1957. [CrossRef] [PubMed]
5. Ross, A.M.; Jiang, Z.; Bastmeyer, M.; Lahann, J. Physical aspects of cell culture substrates: Topography, roughness, and elasticity. *Small* **2012**, *8*, 336–355. [CrossRef]
6. Sartoretto, S.C.; Alves, A.T.N.N.; Resende, R.F.B.; Calasans-Maia, J.; Granjeiro, J.M.; Calasans-Maia, M.D. Early osseointegration driven by the surface chemistry and wettability of dental implants. *J. Appl. Oral Sci.* **2015**, *23*, 279–287. [CrossRef]
7. Bastekova, K.; Guselnikova, O.; Postnikov, P.; Elashnikov, R.; Kunes, M.; Kolska, Z.; Švorčík, V.; Lyutakov, O. Spatially selective modification of PLLA surface: From hydrophobic to hydrophilic or to repellent. *Appl. Surf. Sci.* **2017**, *397*, 226–234. [CrossRef]
8. Guo, C.; Cai, N.; Dong, Y. Duplex surface modification of porous poly (lactic acid) scaffold. *Mater. Lett.* **2013**, *94*, 11–14. [CrossRef]
9. Kiss, É.; Bertóti, I.; Vargha-Butler, E.I. XPS and wettability characterization of modified poly(lactic acid) and poly(lactic/glycolic acid) films. *J. Colloid Interface Sci.* **2002**, *245*, 91–98. [CrossRef]
10. Riveiro, A.; Maçon, A.L.B.; del Val, J.; Comesaña, R.; Pou, J. Laser surface texturing of polymers for biomedical applications. *Front. Phys.* **2018**, *6*. [CrossRef]

11. Fisher, J.P.; Reddi, A.H. Functional tissue engineering of bone: Signals and scaffolds. In *Topics in Tissue Engineering*; University of Oulu: Oulu, Finland, 2003.
12. Bhatla, A.; Yao, Y.L. Effect of laser surface modification on the crystallinity of poly(L-lactic acid). *J. Manuf. Sci. Eng. Trans. ASME* **2009**, *131*, 051004. [CrossRef]
13. Rytlewski, P.; Mróz, W.; Zenkiewicz, M.; Czwartos, J.; Budner, B. Laser induced surface modification of polylactide. *J. Mater. Process. Technol.* **2012**, *212*, 1700–1704. [CrossRef]
14. Kancharla, V.V.; Chen, S. Fabrication of biodegradable polymeric micro-devices using laser micromachining. *Biomed. Microdevices* **2002**, *4*, 105–109. [CrossRef]
15. Stepak, B.; Antończak, A.J.; Bartkowiak-Jowsa, M.; Filipiak, J.; Pezowicz, C.; Abramski, K.M. Fabrication of a polymer-based biodegradable stent using a CO_2 laser. *Arch. Civil Mech. Eng.* **2014**, *14*, 317–326. [CrossRef]
16. Antończak, A.J.; Stępak, B.D.; Szustakiewicz, K.; Wójcik, M.R.; Abramski, K.M. Degradation of poly(L-lactide) under CO_2 laser treatment above the ablation threshold. *Polym. Degrad. Stab.* **2014**, *109*, 97–105. [CrossRef]
17. Dadbin, S. Surface modification of LDPE film by CO_2 pulsed laser irradiation. *Eur. Polym. J.* **2002**, *38*, 2489–2495. [CrossRef]
18. Stępak, B.D.; Antończak, A.J.; Abramski, K.M. Rapid fabrication of microdevices by controlling the PDMS curing conditions during replication of a laser-prototyped mould. *J. Micromech. Microeng.* **2015**, *25*, 10. [CrossRef]
19. Waugh, D.G.; Lawrence, J.; Morgan, D.J.; Thomas, C.L. Interaction of CO_2 laser-modified nylon with osteoblast cells in relation to wettability. *Mater. Sci. Eng. C* **2009**, *29*, 2514–2524. [CrossRef]
20. Zheng, Y.; Xiong, C.; Wang, Z.; Li, X.; Zhang, L. A combination of CO_2 laser and plasma surface modification of poly(etheretherketone) to enhance osteoblast response. *Appl. Surf. Sci.* **2015**, *344*, 79–88. [CrossRef]
21. Kobielarz, M.; Tomanik, M.; Mroczkowska, K.; Szustakiewicz, K.; Oryszczak, M.; Mazur, A.; Antończak, A.; Filipiak, J. Laser-modified PLGA for implants: In vitro degradation and mechanical properties. *Acta Bioeng. Biomech.* **2020**, *22*, 179–197. [CrossRef]
22. Kobielarz, M.; Gazińska, M.; Tomanik, M.; Stępak, B.; Szustakiewicz, K.; Filipiak, J.; Antończak, A.; Pezowicz, C. Physicochemical and mechanical properties of CO_2 laser-modified biodegradable polymers for medical applications. *Polym. Degrad. Stab.* **2019**, *165*, 182–195. [CrossRef]
23. Fan, Y.; Nishida, H.; Shirai, Y.; Tokiwa, Y.; Endo, T. Thermal degradation behaviour of poly(lactic acid) stereocomplex. *Polym. Degrad. Stab.* **2004**, *86*, 197–208. [CrossRef]
24. Jia, W.; Luo, Y.; Yu, J.; Liu, B.; Hu, M.; Chai, L.; Wang, C. Effects of high-repetition-rate femtosecond laser micromachining on the physical and chemical properties of polylactide (PLA). *Opt. Express* **2015**, *23*, 26932–26939. [CrossRef] [PubMed]
25. Slepička, P.; Michaljaničová, I.; Sajdl, P.; Fitl, P.; Švorčík, V. Surface ablation of PLLA induced by KrF excimer laser. *Appl. Surf. Sci.* **2013**, *283*, 438–444. [CrossRef]
26. Michaljanicová, I.; Slepicka, P.; Heitz, J.; Barb, R.A.; Sajdlc, P.; Svorcíka, V. Comparison of KrF and ArF excimer laser treatment of biopolymer surface. *Appl. Surf. Sci.* **2015**, *339*, 144–150. [CrossRef]
27. Antończak, A.J.; Stępak, B.; Szustakiewicz, K.; Wójcik, M.; Kozioł, P.E.; Łazarek, Ł.; Abramski, K.M. Effect of CO_2 laser micromachining on physicochemical properties of poly(L-lactide). In Proceedings of the SPIE 2014 International Conference on Applications of Optics and Photonics, Aveiro, Portugal, 22 August 2014; Volume 9286, pp. 92860Z-1–92860Z-10.
28. ISO. *Surface Texture. Profile Method. Terms, Definitions and Surface Texture Parameters*; BS EN ISO 4287: Geometrical Product Specification (GPS); ISO: Geneva, Switzerland, 2000.
29. ISO. *ISO 10993-5: Biological Evaluation of Medical Devices—Part 3: Tests for Genotoxicity, Carcinogenicity and Reproductive Toxicity*; ISO: Geneva, Switzerland, 2003.
30. Oliver, W.C.; Pharr, G.M. Measurement of hardness and elastic modulus by instrumented indentation: Advances in understanding and refinements to methodology. *J. Mater. Res.* **2004**, *19*, 3–20. [CrossRef]
31. Castillejo, M.; Rebollar, E.; Oujja, M.; Sanz, M.; Selimis, A.; Sigletou, M.; Psycharakis, S.; Ranella, A.; Fotakis, C. Fabrication of porous biopolymer substrates for cell growth by UV laser: The role of pulse duration. *Appl. Surf. Sci.* **2012**, *258*, 8919–8927. [CrossRef]
32. Rodrigues, N.; Benning, M.; Ferreira, A.M.; Dixon, L.; Dalgarno, K. Manufacture and characterisation of porous PLA scaffolds. *Procedia CIRP* **2016**, *49*, 33–38. [CrossRef]

33. Kryszak, B.; Szustakiewicz, K.; Stępak, B.; Gazińska, M.; Antończak, A.J. Structural, thermal and mechanical changes in poly(L-lactide)/hydroxyapatite composite extruded foils modified by CO_2 laser irradiation. *Eur. Polym. J.* **2019**, *114*, 57–65. [CrossRef]
34. Riveiro, A.; Soto, R.; Comesaña, R.; Boutinguiza, M.; Del Val, J.; Quintero, F.; Lusquiños, F.; Pou, J. Laser surface modification of PEEK. *Appl. Surf. Sci.* **2012**, *258*, 9437–9442. [CrossRef]
35. Aflori, M.; Butnaru, M.; Tihauan, B.-M.; Doroftei, F. Eco-friendly method for tailoring biocompatible and antimicrobial surfaces of poly-L-lactic acid. *Nanomaterials* **2019**, *9*, 428. [CrossRef]
36. Slepickova-Kasalkova, N.; Slepicka, P.; Kolska, Z.; Svorcik, V. Wettability and other surface properties of modified polymers. *Wetting Wettability* **2015**. [CrossRef]
37. Correia, D.M.; Ribeiro, C.; Botelho, G.; Borges, J.; Lopes, C.; Vaz, F.; Carabineiro, S.A.C.; Machado, A.V.; Lanceros-Méndez, S. Superhydrophilic poly(L-lactic acid) electrospun membranes for biomedical applications obtained by argon and oxygen plasma treatment. *Appl. Surf. Sci.* **2016**, *371*, 74–82. [CrossRef]
38. Casalini, T.; Rossi, F.; Castrovinci, A.; Perale, G. A Perspective on polylactic acid-based polymers use for nanoparticles synthesis and applications. *Front. Bioeng. Biotechnol.* **2019**, *7*, 259–271. [CrossRef]
39. Slepičková Kasálková, N.; Slepička, P.; Bačáková, L.; Sajdl, P.; Švorčík, V. Biocompatibility of plasma nanostructured biopolymers. *Nucl. Instrum. Methods Phys. Res. B* **2013**, *307*, 642–646. [CrossRef]
40. Vesel, A.; Junkar, I.; Cvelbar, U.; Kovač, J.; Mozetič, M. Surface modification of polyester by oxygen- and nitrogen-plasma treatment. *Surf. Interface Anal.* **2008**, *40*, 1444–1453. [CrossRef]
41. Chan, C.M.; Ko, T.M.; Hiraoka, H. Polymer surface modification by plasmas and photons. *Surf. Sci. Rep.* **1996**, *24*, 1–54. [CrossRef]
42. Goddard, J.M.; Hotchkiss, J.H. Polymer surface modification for the attachment of bioactive compounds. *Prog. Polym. Sci.* **2007**, *32*, 698–725. [CrossRef]
43. Zhao, Y.; Zhu, B.; Wang, Y.; Liu, C.; Shen, C. Effect of different sterilization methods on the properties of commercial biodegradable polyesters for single-use, disposable medical devices. *Mater. Sci. Eng. C* **2019**, *105*, 110041–110049. [CrossRef]
44. Washburn, N.R.; Yamada, K.M.; Simon, C.G.; Kennedy, S.B.; Amis, E.J. High-throughput investigation of osteoblast response to polymer crystallinity: Influence of nanometer-scale roughness on proliferation. *Biomaterials* **2004**, *25*, 1215–1224. [CrossRef]
45. Slepička, P.; Siegel, J.; Lyutakov, O.; Slepičková-Kasálková, N.; Kolská, Z.; Bačáková, L.; Švorčík, V. Polymer nanostructures for bioapplications induced by laser treatment. *Biotechnol. Adv.* **2018**, *36*, 839–855. [CrossRef] [PubMed]
46. Terada, C.; Imamura, T.; Ohshima, T.; Maeda, N.; Tatehara, S.; Tokuyama-Toda, R.; Yamachika, S.; Toyoda, N.; Satomura, K. The effect of irradiation with a 405 nm blue-violet laser on the bacterial adhesion on the osteosynthetic biomaterials. *Int. J. Photoenergy* **2018**, *2018*, 1–10. [CrossRef]
47. Rapacz-Kmita, A.; Szaraniec, B.; Mikołajczyk, M.; Stodolak-Zych, E.; Dzierzkowska, E.; Gajek, M.; Dudek, P. Multifunctional biodegradable polymer/clay nanocomposites with antibacterial properties in drug delivery systems. *Acta Bioeng. Biomech.* **2020**, *22*, 1–19. [CrossRef]
48. Kubies, D.; Himmlová, L.; Riedel, T.; Chánová, E.; Balík, K.; Douděrová, M.; Bártová, J.; Pešáková, V. The interaction of osteoblasts with bone-implant materials: 1. The effect of physicochemical surface properties of implant materials. *Physiol. Res.* **2011**, *60*, 95–111. [CrossRef] [PubMed]
49. Ajala, O.; Werther, C.; Nikaeen, P.; Singh, R.P.; Depan, D. Influence of graphene nanoscrolls on the crystallization behavior and nano-mechanical properties of polylactic acid. *Polym. Adv. Technol.* **2019**, *30*, 1825–1835. [CrossRef]
50. Wright-Charlesworth, D.D.; Miller, D.M.; Miskioglu, I.; King, J.A. Nanoindentation of injection molded PLA and self-reinforced composite PLA after in vitro conditioning for three months. *J. Biomed. Mater. Res. Part A* **2005**, *74*, 388–396. [CrossRef]

© 2020 by the authors. Licensee MDPI, Basel, Switzerland. This article is an open access article distributed under the terms and conditions of the Creative Commons Attribution (CC BY) license (http://creativecommons.org/licenses/by/4.0/).

Article

Studies on the Uncrosslinked Fraction of PLA/PBAT Blends Modified by Electron Radiation

Rafał Malinowski [1,*], Krzysztof Moraczewski [2] and Aneta Raszkowska-Kaczor [1]

1   Łukasiewicz Research Network—Institute for Engineering of Polymer Materials and Dyes, 87100 Toruń, Poland; aneta.raszkowska-kaczor@impib.pl
2   Institute of Materials Engineering, Kazimierz Wielki University, 85064 Bydgoszcz, Poland; kmm@ukw.edu.pl
*   Correspondence: malinowskirafal@gmail.com

Received: 21 January 2020; Accepted: 25 February 2020; Published: 28 February 2020

Abstract: The results of studies on the uncrosslinked fraction of blends of polylactide and poly(butylene adipate-*co*-terephthalate) (PLA/PBAT) are presented. The blends were crosslinked by using the electron radiation and triallyl isocyanurate (TAIC) at a concentration of 3 wt %. Two kinds of samples to be investigated were prepared: one contained 80 wt % PLA and the other contained 80 wt % PBAT. Both blends were irradiated with the doses of 10, 40, or 90 kGy. The uncrosslinked fraction was separated from the crosslinked one. When dried, they were subjected to quantitative analysis, Fourier transform infrared spectroscopy (FTIR) measurements, an analysis of variations in the average molecular weight, and the determination of thermal properties. It was found that the electron radiation caused various effects in the studied samples, which depended on the magnitude of the radiation dose and the weight fractions of the components of the particular blends. This was evidenced by the occurrence of the uncrosslinked fractions of different amounts, a different molecular weight distribution, and the different thermal properties of the samples. It was also concluded that the observed effects were caused by the fact that the processes of crosslinking and degradation took place mostly in PLA, while PBAT appeared to be less susceptible to the influence of the electron radiation.

Keywords: polylactide; biodegradable blends; irradiation; crosslinking; degradation

1. Introduction

Studies on the effects of electron radiation on the properties of conventional polymers are already being carried out and have been for many years [1]. The development of knowledge in this field as well as learning the new phenomena and consequences of the interaction of this radiation with polymer materials resulted in an increase in the number of novel applications of polymers and put the use of the industrial radiation treatment into a new perspective. Changes occurring in the molecular structure of the polymers are the most common effects of the treatment with electron radiation. They are caused by the formation of various kinds of ions and radicals in the materials being radiation modified. This, in turn, leads to the occurrence of many processes, including synthesis, grafting, crosslinking, or radiation degradation [2]. As a result of these processes and due to specified conditions of the radiation treatment, novel polymer materials of modified properties and new potential applications are being produced.

Issues associated with the radiation treatment of polymers are complex, which results from the fact that various processes of different intensity may simultaneously occur in the polymers being modified this way. Therefore, the final effects of the irradiation, i.e., mostly crosslinking or degradation at different degrees, depend, e.g., on the yield of the individual processes [3]. Additionally, an active site (usually a radical one) being formed due to the irradiation, may move along a polymer chain. Thus, the processes induced by the absorption of the electron radiation energy by a polymer may occur in places of chain different with relation to location of the active site created originally. This results partially from the chemical structure of the macromolecules and partially from the conditions of the radiation treatment, mainly the radiation dose magnitude, dose rate, and the irradiation atmosphere. For example, polyethylene (PE) easily undergoes radiation crosslinking, whereas polypropylene (PP) solely undergoes radiation degradation, during which the molecular weight decreases [4]. This is caused by the presence in PP of lateral methyl groups that favor oxidation of the polymer and, thus, its degradation.

Various effects of the radiation treatment also occur in the biodegradable polymers, such as polylactide (PLA), poly(ε-caprolactone) (PCL), or poly(butylene adipate-co-terephthalate) (PBAT) [5–9]. They depend, for example, on the structure of macromolecules, the orderliness of carbon atoms in the main chain or the presence of aromatic groups. They are, however, much less recognized, which may result from the anxiety associated with a possible decrease in the susceptibility to the biodegradability of these polymers due to crosslinking. Nevertheless, not all the biodegradable polymers undergo crosslinking because of treatment with the electron radiation only. For example, PLA [10–12], when treated with the electron radiation, undergoes, first of all, degradation [13–15]. On the contrary, in PCL [16–18] or in PBAT [19–22], the radiation degradation processes generally are not dominant [23–25].

The influence of the electron radiation on the properties of blends of the biodegradable polymers (like PLA/PCL, PLA/PBAT or PHBV/PLA) [26–28] is even less recognized or totally unexplained. This statement does not apply to radiation sterilization processes that generally use low radiation doses that do not affect the properties and structure of polymers. In the case of using higher doses (above 10 kGy), the effects of the irradiation are much more complex. Such a situation results from different interactions of particular components of the blends with the radiation [29–34]. The low miscibility or immiscibility of the components are additional factors affecting complexity of the issue [35]. In order to increase the miscibility of the polymers, it is recommended to add to such blends low-molecular weight compounds that increase crosslinking and improve interfacial adhesion [31,36–41]. Triallyl isocyanurate (TAIC) is one such compound, which is being applied most often.

Knowledge about the influence of the electron radiation on polymers and polymer blends, including the effects associated with the crosslinking, may be acquired as a result of the investigation of a crosslinked fraction and phenomena occurring within that fraction. On the other hand, the results of the examination of the fraction that remains uncrosslinked in spite of irradiation are almost totally unknown. Most often, the amount of it is less than that of other fractions, which does not, however, mean that this fraction has to be degraded. Getting knowledge about its properties when irradiated may be an excellent supplement to the results of investigation of the crosslinked fraction and may also be a kind of reflection of the latter. In addition, this knowledge may give some information on the crosslinking of the studied materials. It may also contribute to the elucidation of the issues that have not been fully explained earlier. Therefore, the authors of the present article undertook studies aimed at determination of changes in some properties of the uncrosslinked fraction of the two kinds of blends of biodegradable polymers, occurring when the electron radiation doses of different magnitudes were applied. The studies presented in this work are a continuation of our previous research on the radiation treatment of PLA, PBAT, and its blends [13,25,32]. Therefore, the blends of PLA/PBAT type were chosen for studies.

2. Materials and Methods

2.1. Materials

Polylactide (PLA), type 2003D (NatureWorks®, Minnetonka, MN, USA), with a melt flow rate (MFR) equal to 2.8 g/10min (2.16 kg, 190 °C), a density of 1.24 g/cm^3, a melting temperature (T_m) of 155 °C, a number-average molecular weight of ca. 91 kDa, and a weight-average molecular weight of ca. 166 kDa, has been used in this work. This polymer contained 3.5% of D monomer units. The second polymer was poly(butylene adipate-co-terephthalate) (PBAT), type FBlend C1200 (BASF, Ludwigshafen, Germany) with a melt flow rate (MFR) equal to 10 g/10 min (2.16 kg, 190 °C), a density of 1.25 g/cm^3, a melting temperature (T$_m$) of 115 °C, a number-average molecular weight of ca. 35 kDa, and a weight-average molecular weight of ca. 73 kDa. Moreover, triallyl isocyanurate (TAIC) with a density equal to 1.16 g/cm^3 and a melting point of ca. 23–27 °C from Sigma-Aldrich GmbH (Munich, Germany) has also been applied. This compound was used in a liquid state as an agent to promote a crosslinking of the polymers upon the electron radiation. Dichloromethane (CH$_2$Cl$_2$) from Avantor Performance Materials Poland S.A. (Gliwice, Poland) was used during dissolution of the non-irradiated and irradiated samples in investigations of the uncrosslinked fraction, as well as in examination by gel permeation chromatography (GPC). Nitrogen (N$_2$) type Premier (Air Products, Warsaw, Poland) was used in the differential scanning calorimeter (DSC) examination.

2.2. Apparatus

The co-rotating twin screw extruder type BTSK 20/40D (Bühler, Braunschweig, Germany), equipped with the screws of a 40 L/D ratio and 20 mm diameter, and the three-opening die head was intended to be prepared from granulated samples of the PLA/PBAT blends. A linear accelerator of electrons, type Elektronika 10/10, was used in the irradiation of the obtained samples of the granulates. Attenuated total reflectance Fourier transform infrared (FTIR-ATR) spectrometer, type Cary 630 (Agilent Technologies, Santa Clara, CA, USA), was meant for the examination of the changes in the macromolecules' structure. In addition, gel permeation chromatograph (GPC), equipped with the set of two PLgel 5 µm MIXED-C columns, was designed for the investigation of the average molecular weights of the radiation modified samples. The last device applied in this work was a DSC, type DSC 1 STARe System (Mettler Toledo, Greifensee, Switzerland), designed for the determination of some thermal effects occurring in samples modified by electron radiation.

2.3. Sample Preparation

Samples of the uncrosslinked fractions were prepared in three stages. In the first stage, two kinds of granulated polymer blends of the PLA/PBAT type were prepared by using a co-rotating twin screw extruder type BTSK 20/40D (Bühler, Braunschweig, Germany) and a standard granulator. One kind of granule was the blend with the predominant content (80 wt %) of PLA, and the other one was the blend with the predominant content (80 wt %) of PBAT. TAIC in the amount of 3 wt % was added to each blend while being extruded. The extrusion was carried out at the temperatures of the particular barrel zones (I, II, III, and IV) equal to 180, 183, 186, and 190 °C, respectively, and at the die-head temperature of 190 °C [32]. Free degassing of possible gaseous products when released was applied at the screws with a length of L/D = 33. The screws rotational speed was constant (250 rpm). Before extrusion, the PLA was dried at 70 °C for 24 h, in order to avoid hydrolytic degradation. PBAT was not dried. The same temperature profile to extrude the two blends has been applied, because these were the lowest processing temperatures of PLA, which was present in both blends.

In the second stage, the prepared granulated blends were subjected to the radiation treatment with the use of a high-energy electron beam derived from accelerator type Elektronika 10/10 for the crosslinking of the studied materials. The applied radiation doses were 10, 40, or 90 kGy, the single dose not being larger than 20 kGy. This limitation was due to an increase in the temperature of the irradiated material, which might cause unwanted changes in the structure of the blends if larger doses were used. Therefore, some samples were irradiated several times. These doses were chosen because at the 10 kGy dose the crosslinking process began, at the 40 kGy dose the largest amount of gel fraction was obtained, and at the 90 kGy dose the degradation process was dominant. During the irradiation procedure, all the granulated samples were put on the belt conveyor moving at the speed of 0.3–1.2 m/min. The actual speed was related to the radiation dose absorbed by a polymer material being modified. The layer thickness of the irradiated samples was not larger than 20 mm. This ensured penetration of the irradiation beam on the all granules. Samples irradiation was carried out in the air.

In the third stage, the radiation crosslinked samples of the granulated blends were subjected to dissolution in CH$_2$Cl$_2$ in order to separate the uncrosslinked fraction from the crosslinked fraction.

This solvent was chosen because the studied samples dissolve well in it. The non-irradiated samples (L0 and B0, as indicated in Table 1) were also subjected to dissolution in that solvent. The samples were dissolving at room temperature for 24 h. The obtained solutions were filtered through medium flow quantitative filter papers. The crosslinked fractions that remained on the filter papers were used to determine the contents of the gel fractions. The uncrosslinked fractions present in the filtrates were used to prepare samples meant for the basic examinations. For this purpose, the filtrates were put on glass Petri dishes and left until the solvent had freely evaporated. The obtained specimens of the uncrosslinked fractions, designated with the symbols as shown in Table 1, were investigated by using techniques such as Fourier transform infrared spectroscopy (FTIR), gel permeation chromatography (GPC), and differential scanning calorimetry (DSC).

Table 1. Symbols of the studied samples.

Composition	Dose (kGy)			
	0	10	40	90
PLA/PBAT 80/20 (L type samples)	L0	L10	L40	L90
PLA/PBAT 20/80 (B type samples)	B0	B10	B40	B90

2.4. Methodology of Research

The content of the uncrosslinked fraction (N_g) of the studied samples was determined by a solvent extraction method, using CH_2Cl_2 as the solvent. All the specimens were subjected to dissolution at 20 ± 3 °C for 24 h. The mass ratios of the specimens and solvent were chosen so as to achieve a solution concentration (C_p) of 2%. The obtained solutions were filtered through medium flow quantitative filter papers. The gel fraction, which remained on the filter papers, was dried at 50 °C for 24 h. Using the mass (W_o) of a sample before it was subjected to dissolution and the mass (W_g) of that sample after it was subjected to dissolution and dried (W_g referred to the mass of the gel fraction), the content of the uncrosslinked fraction (N_g) in the particular samples was calculated according to the following relationship:

$$N_g = \left(1 - \frac{W_g}{W_o}\right) \times 100\%$$

The attenuated total reflectance Fourier transform infrared (ATR-FTIR) spectra of the studied samples were recorded in transmittance mode at a constant spectral resolution of 4 cm^{-1}, for the wavenumber ranged from 3700 to 500 cm^{-1}, after acquiring 8 scans.

The changes in the number-average molecular weight (M_n) and weight-average molecular weight (M_w) of all samples was examined by gel permeation chromatography (GPC) conducted in CH_2Cl_2 at room temperature with an eluent flow rate of 0.8 mL/min, using a set of two PLgel 5 µm MIXED-C columns. About 2 mg of each sample was applied as part of the preparation of the solution that was to be injected in the GPC columns. Polystyrene standards were used. GPC results have been shown as curves of the peak intensity (I) vs. the eluent volume (V).

DSC measurements were performed under nitrogen with rate flow of 50 ml/min. About 3 mg of the sample was placed on an aluminum pan for sampling. The samples were successively: heated from 20 to 180 °C at 10 °C/min, annealed at 180 °C for 3 min, cooled to 15 °C at 10 °C/min, and reheated to 180 °C at a rate of 10 °C/min. The second heating cycle was used in the analysis of the thermal properties of the studied samples. The glass transition temperature (T_g), cold crystallization temperature (T_{cc}), melting temperature (T_m), cold crystallization enthalpy (H_{cc}), and melting enthalpy (H_m) were determined.

3. Results

3.1. Quantitative Determination of the Uncrosslinked Fraction

Dependences of the content of the uncrosslinked fraction (N_g) of the samples L and B on the magnitude of the radiation dose are shown in Figure 1. As can be seen, the N_g values of the B samples are on the average a dozen or so percent larger than those of the L samples. This indicates a higher susceptibility of PLA to the radiation crosslinking in the presence of TAIC than that of PBAT. Larger N_g values, corresponding to smaller contents of the gel fraction of PBAT, result from the chemical structure of the macromolecules of that polymer, mostly from the presence of aromatic groupings that enable the absorption of a part of the electron radiation and the dissipation of it in the form of heat. This is connected to the so-called protective effect and to making at least some of the active sites, which may move along polymer chains, inactive [42,43]. Such a situation is the reason for the creation of a smaller number of crosslinking bonds and, thus, for the formation of a smaller amount of the gel fraction while, at the same time, a larger amount of the uncrosslinked fraction is formed.

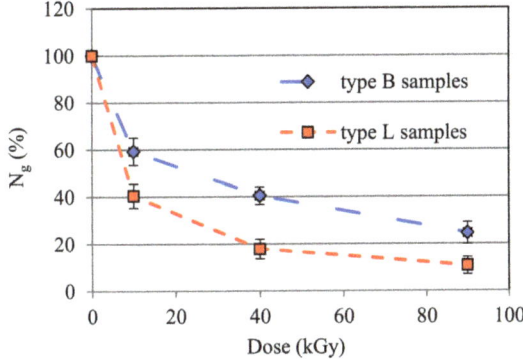

Figure 1. Results of the content of the uncrosslinked fraction (N_g) of the samples type L and B.

From Figure 1, it also follows that the N_g values of the samples of both types rapidly decrease as the radiation dose increases. Already, the smallest radiation dose causes a reduction in the N_g values down to ca. 40% (sample L10) and 59% (sample B10). The largest dose, in turn, causes a decrease in the values of N_g to ca. 10% (sample L90) and 25% (sample B90). Considering the values of the confidence intervals of the particular results, it has to be stated that essential variations in the values of Ng occurred for all irradiated samples except for samples L40 and L90. A reduction in the contents of the uncrosslinked fraction in the particular samples upon the increase in the radiation dose points out the fact that the crosslinking becomes more effective. However, the radiation modified blends do not undergo the complete crosslinking since the N_g values do not fall down to zero. This may indicate that the applied amount of TAIC could be insufficient for the complete crosslinking of the studied blends. Besides, the amount of an uncrosslinked fraction at the level of 10% or 25% is relatively large anyway.

The data shown in Figure 1 also indicate that the gel fraction does not form in the non-irradiated samples when solely TAIC was used. This is proved by the maximum N_g values of these samples, which are associated with the easy dissolution of those materials in the applied solvent. It is an important piece of information because not all the low molecular weight multifunctional compounds show a similar effect [44]. Some of them, e.g., trimethylopropane triacrylate (TMPTA), may cause the formation of small amounts of the gel fraction in the case of several polymers already at the stage of processing these materials and without the application of the electron radiation.

3.2. FTIR Spectroscopy

The uncrosslinked fractions of the studied samples may exhibit various properties, which may result from different macromolecular structures of PLA and/or PBAT after irradiation. In addition, the blends of both types when irradiated may show slightly different weight percentages of PLA and PBAT compared to the original blends produced at the stage of extrusion, which may result from the crosslinking of both polymers at different degrees. This is proved to some extent by the N_g data. In the extreme case, the uncrosslinked fraction might be composed of solely one polymer if the other one would be completely crosslinked and adhesion at the interface between these two polymers would be insufficient. The FTIR spectra of the L type samples (L0, L10, L40, and L90) and those of the B type samples (B0, B10, B40, and B90) are shown in Figures 2 and 3, respectively. The results of the FTIR measurements enable us to determine compositions of the studied samples.

Considering the FTIR spectra, only the bands characteristic of PLA or PBAT were taken into account [45–47]. In the case of PLA, there were analyzed bands assigned to (i) the asymmetric and symmetric stretching vibrations of the CH_3 group present in the saturated hydrocarbons (2994 and 2946 cm^{-1}, respectively), (ii) the stretching vibrations of the C=O group (1748 cm^{-1}), (iii) asymmetric bending vibrations of the CH_3 group (1452 cm^{-1}), (iv) the deformation and symmetric bending vibrations of the CH group (1382 and 1360 cm^{-1}, respectively), and (v) the stretching vibrations of the C–O–C group (1180, 1079, and 1042 cm^{-1}) because C–O can form a bond with different atoms and groups, so the vibration absorptions are more complex. Moreover, the band at 869 cm^{-1} can be assigned to the amorphous phase and the one at 757 cm^{-1} to the crystalline phase of PLA [48]. However, other authors report that the band at 869 cm^{-1} can be assigned to the absorption of the (O–CH–CH_3) ester and the one at 757 cm^{-1} can be assigned to the rocking vibration absorption of α-methyl [49]. The bond at 697 cm^{-1} is assigned to the vibrations of the carbonyl group (C=O). In the case of PBAT, in turn, the bands ascribed to (i) the asymmetric stretching vibrations of the CH_2 group (2948 cm^{-1}), (ii) the stretching vibrations of the C=O group of the ester bond (1711 cm^{-1}), (iii) the skeletal vibrations of the aromatic ring (1504 cm^{-1}), (iv) the in-plane bending vibrations of the CH_2 group (1409 cm^{-1}), (v) the symmetric stretching vibrations of the C–O group (1267 cm^{-1}), (vi) the left–right symmetric stretching vibration absorption of the C–O group (1104 cm^{-1}), (vii) the vibrations of the hydrogen atom of the aromatic ring (1018 cm^{-1}), (viii) the symmetric stretching vibration of the trans C–O group (935 cm^{-1}), and (ix) the bending vibration absorption of CH-plane of the benzene ring (727 cm^{-1}) were taken into account.

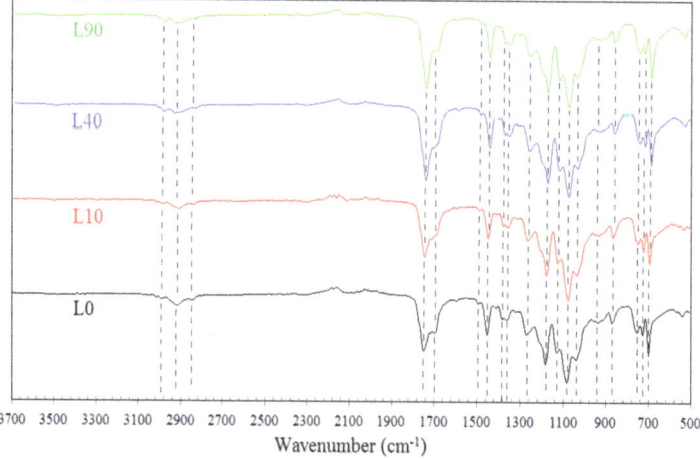

Figure 2. Results of the Fourier transform infrared spectroscopy (FTIR) measurements of the samples type L.

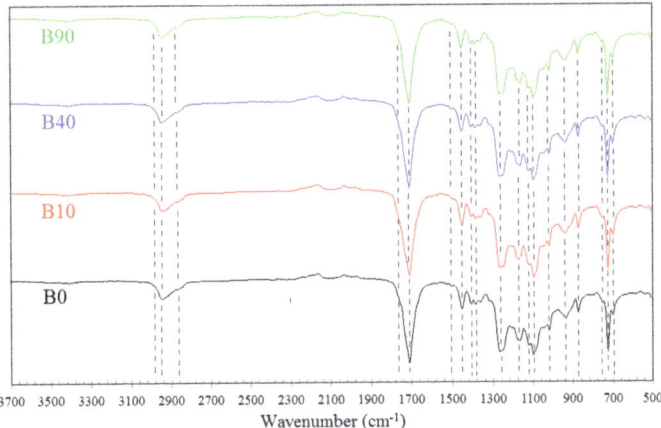

Figure 3. Results of the FTIR measurements of the samples type B.

The bands characteristic of both PLA and PBAT appear in each spectrum, which indicates that all the studied samples contain macromolecules of the two polymers. The bands of the L and B samples that have been irradiated using the same radiation dose differ from one another in intensity only, which results from the predominant content of one of the polymers. It is also important that the positions of the characteristic bands in the spectra corresponding to the irradiated samples did not change, or changed only slightly, in respect of those relating to the non-irradiated samples. This points out that the studied samples contained no crosslinked macromolecules of both polymers.

3.3. Molecular Weight

Treatment with the electron radiation essentially influences the average molecular weights of the polymers being modified. This is connected with the degradation of the polymers, the change in the structure of macromolecules (e.g., branching and lengthening), or crosslinking (partial or complete). GPC data, illustrated in Figures 4 and 5 and summarized in Table 2, indicate that effects occurring in the samples with the predominant content of PLA are somewhat different in comparison to those in the samples with the prevailing content of PBAT.

Two types of molecular weight distributions, i.e., monomodal and bimodal, were found in the case of the L samples (Figure 4). The non-irradiated blend and that irradiated with the dose of 10 kGy exhibit the monomodal distribution and molecular weights close to one another. These samples differ in the degree of polydispersion only, which is slightly higher in the irradiated sample. The distribution peak of the latter sample is lower and its full width at half maximum is larger compared to the non-irradiated sample. The higher degree of the polydispersion of the irradiated sample results from the partial radiation degradation of polylactide, which easily degrades when exposed to the electron radiation. The difference in the degree of polydispersion may also be caused by the fact that the discussed sample contains PLA (being susceptible to the radiation degradation) in the predominant amount. The monomodal distribution of the molecular weight of the non-irradiated sample (L0) is due to the fact that macromolecules of both polymers (PLA and PBAT) may probably have different hydrodynamic volumes. This, in turn, may result in the retention times of the macromolecules of different lengths being similar. Thus, despite the different molecular weights of the individual components of the blend, the molecular weight distribution of L0 sample can be a monomodal.

The bimodal distribution occurs in the sample irradiated with the dose of 40 kGy, which is more obvious when the derivative of the GPC curve is taken into account. Such a distribution is even more evident in the case of the sample irradiated with the dose of 90 kGy. The occurrence of the bimodal distribution indicates that the short macromolecules (degraded or non-degraded oligomeric) as well as the macromolecules of the molecular weight larger than that of the macromolecules of the non-irradiated sample are present in the samples L40 and L90. This may result from the formation of branched, elongated, or partially crosslinked structures. Furthermore, it can also be seen in Figure 4 that a shift in the positions of particular peaks in the direction of larger or smaller eluent volumes (V) appears as the radiation dose increases. The former direction corresponds to the samples containing macromolecules of the diminishing molecular weight, whereas the latter one corresponds to the samples L40 and L90 only, in which the molecular weight of a part of the macromolecules increases as the radiation dose rises. From Table 2, it also follows that the number average molecular weight of one of the fractions of the L samples (the fraction that contains macromolecules with smaller molecular weights, peak 2) decreased by a factor of almost six when the maximum radiation dose was applied. On the other hand, if the fraction containing macromolecules with larger molecular weights is concerned (peak 1), the larger molecular weight of sample L40, as compared to that of sample L90, may result from the more effective combining of some PLA macromolecules into more complex structures than in the sample irradiated with the dose of 90 kGy, which may facilitate the degradation of the PLA macromolecules.

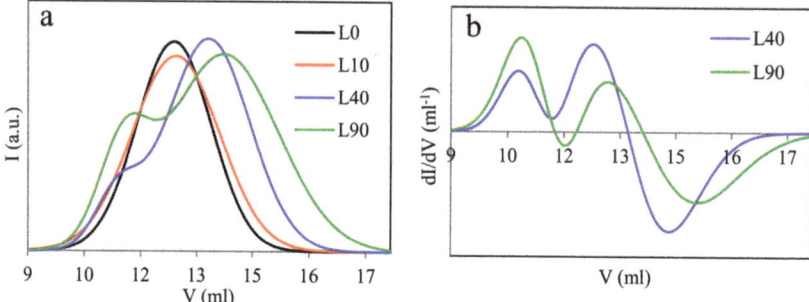

Figure 4. Results of the gel permeation chromatography (GPC) measurements for the samples with the predominant content of polylactide (PLA) (**a**—GPC curves for the samples type L; **b**—derivatives (*dI/dV*) of the GPC curves for the samples L40 and L90).

The bimodal distribution does not practically occur in the B samples (Figure 5). It can only slightly be seen in the case of the sample irradiated with the 90 kGy dose. An insignificant peak of the position clearly shifted to the left in respect of the remaining peaks, which is more distinct on the derivative of the GPC curve, indicates a contribution to that sample of the macromolecules with a larger molecular weight. However, these macromolecules should not be completely crosslinked, but at most be elongated or partially crosslinked only. The monomodal distribution of the molecular weight of the B0 sample before radiation treatment is due to the same reason as for the non-irradiated L0 sample.

The shift in the positions of the highest peaks towards larger eluent volumes proves the decrease in the average molecular weights of the B samples, occurring with the increasing radiation dose. It can be due to the shortening of the macromolecules (degradation) or the reduction in the number of the longer macromolecules that undergo the crosslinking while forming the gel fraction. Thus, only the shorter macromolecules would remain in the studied sample, which had not undergone crosslinking nor degradation. Besides, while considering the results of the N_g determination (Figure 1), one can state that the number of the shorter macromolecules in the B samples is relatively high. The data summarized in Table 2 also point out that the number average molecular weights of the B samples decrease by the factor of three when they get irradiated with the maximum dose (peak 2). Thus, the reduction in the molecular weights of the B samples is smaller by the factor of two in relation to that of the L samples. This proves again that PLA is more susceptible to the radiation degradation compared to PBAT.

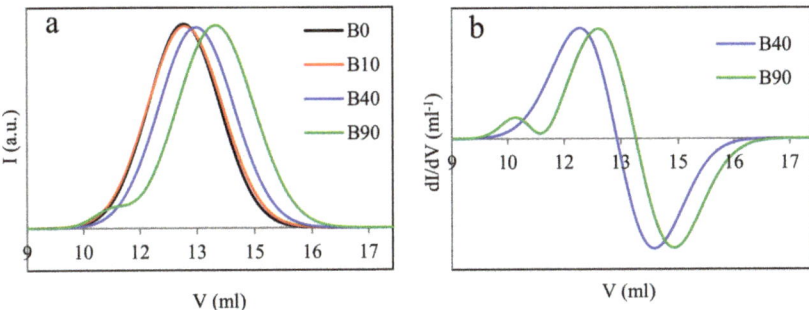

Figure 5. Results of GPC measurements for the samples with the predominant content of poly(butylene adipate-*co*-terephthalate) (PBAT) (**a**—GPC curves for the samples type B; **b**—derivatives (*dI/dV*) of the GPC curves for the samples B40 and B90).

When considering the results of determination of both the content of the uncrosslinked fraction (N_g) and the molecular weight, one may state that the occurrence of the bimodal distribution is closely dependent on both the weight fraction of PLA and the radiation dose magnitude. This distribution becomes more evident when the PLA weight fraction and the radiation dose increase. This may indicate that properties of the studied samples will be determined by the PLA phase. On the other hand, the PLA fraction of the B samples is relatively small. Thus, the properties of those samples should mainly be determined by the radiation dose magnitude, which was confirmed by the results of the DSC measurements (Section 3.4, Figure 7).

Table 2. Number-average molecular weight (M_n), weight-average molecular weight (M_w), and polydispersion degree (PD) of the samples type L and B.

Sample	M_n (kDa)		M_w (kDa)		PD (M_w/M_n)	
	Peak 1	Peak 2	Peak 1	Peak 2	Peak1	Peak 2
L0	–	71.9	–	157.4	2.2	–
L10	–	51.4	–	159.9	3.1	–
L40	356.3	22.6	440.3	55.2	2.5	1.2
L90	253.0	13.0	354.4	33.3	2.6	1.4
B0	–	53.5	–	115.4	2.2	–
B10	–	50.3	–	158.9	3.2	–
B40	–	38.7	–	99.2	2.6	–
B90	393.2	21.2	458.5	50.3	2.4	1.2

3.4. DSC Results

The DSC second heating curves recorded for the samples L and B are shown in Figures 6 and 7, respectively. The data concerning particular phase transitions are summarized in Table 3. They include the glass transition temperature (T_g) of the PLA phase, the cold crystallization temperature (T_{cc}) of the PLA phase, the cold crystallization enthalpy (H_{cc}) of the PLA phase, the melting temperature of the crystalline phases of PBAT (T_m (PBAT)) and PLA (T_m (PLA)), and the enthalpies of the melting of the crystalline phases of PBAT (H_m (PBAT)) and PLA (H_m (PLA)). Detailed discussions on the observed phenomena is presented for the samples L and B separately.

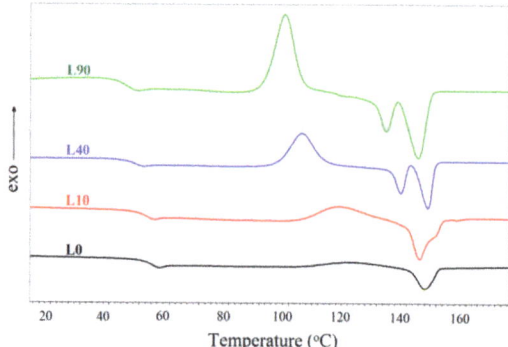

Figure 6. DSC curves recorded for the samples type L.

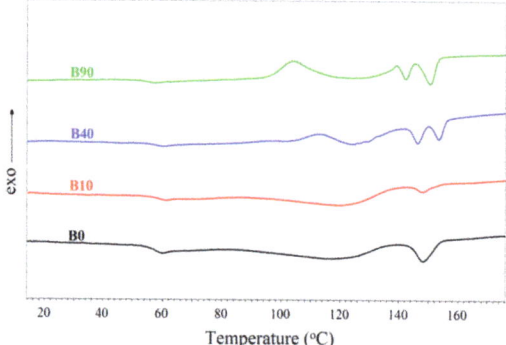

Figure 7. DSC curves recorded for the samples type B.

Table 3. Data derived from the differential scanning calorimeter (DSC) curves recorded for the samples type L (Figure 6) and B (Figure 7).

Sample	T_g (°C)	T_{cc} (°C)	H_{cc} (J/g)	T_m (°C) (PBAT)	T_m (°C)(PLA)		H_m (J/g) (PBAT)	H_m*(J/g) (PLA)
					Peak 1	Peak 2		
L0	57.0	124.0	7.2	–	149.4	–	–	7.2
L10	54.6	120.7	12.9	–	147.5	153.3	–	13.0
L40	51.5	107.9	18.2	–	141.0	150.2	–	20.3
L90	48.2	102.2	17.5	–	136.3	147.1	–	21.6
B0	57.5	–	–	120.4	148.5	–	7.2	2.4
B10	56.2	–	–	120.2	148.3	–	8.8	0.8
B40	56.1	113.2	2.1	–	146.4	153.5	–	2.8
B90	54.1	103.9	4.2	–	142.2	150.5	–	4.2

* $H_m = H_{m, peak1} + H_{m, peak2}$.

3.4.1. Analysis of the DSC Results for Samples L

The lowering of the glass transition temperature by ca. 9 °C occurred in the region of the glass transition of PLA as the radiation dose was increased (Figure 6). This could be caused by the presence in the studied samples of the short macromolecules or oligomeric structures, acting as plasticizer and being, e.g., products of the radiation degradation of PLA or PBAT. The glass transition temperature of PBAT could not be determined since it lies in the region far below 0 °C (beyond the studied range).

The cold crystallization occurs in all the studied samples over the temperature range of ca. 102–125 °C. It appears only in the temperature region characteristic of the cold crystallization of PLA, which indicates that the polylactide phase of the samples was initially amorphous. A lack of the peak corresponding to the cold crystallization of PBAT indicates that the phase of this polymer exhibited maximum crystallinity. Besides, the PBAT fraction in the L samples was small and the effect connected with the cold crystallization of that polymer could hardly be noticed. From Figure 6, it also follows that the cold crystallization temperature of the PLA phase decreases and the cold crystallization enthalpy of that phase increases as the radiation dose rises. Such a relationship points out that the rate of ordering of the macromolecules increases as the radiation dose increases. This can occur when the macromolecules become shorter or adhesion at the interface improves (in the case of the polymer blends) and the transport barrier to crystallization becomes reduced. The latter effect seems to be confirmed by the reduction of the full width at half the maximum of the cold crystallization peaks with the rising radiation dose. The increase in the cold crystallization of PLA upon the rising radiation dose may also be caused by the nucleation effect of the shorter macromolecules of PBAT [50].

Slight changes in crystallinity of the studied samples could be observed as well. The crystallinity was determined from both the cold crystallization enthalpies and the enthalpies of melting of the crystalline phase. The obtained data indicated that the L0 and L10 samples were amorphous because the values of the enthalpies of cold crystallization and the melting of these samples were close to each other. The L40 and L90 samples already exhibited some crystallinity that increased as the radiation dose rose. This might confirm the conclusion that more extensive and quicker ordering of the macromolecules occurs in the samples irradiated with the maximum doses, i.e., these samples exhibit a greater ability to crystallize.

An analysis of the melting peaks of the crystalline phase gives information on the particularly interesting effects. Figure 6 illustrates melting of only the polylactide phase. The peak assigned to the melting of the PBAT phase cannot be seen because they occur in the region of the cold crystallization of PLA. The weight fraction of polylactide predominates in the discussed samples and the cold crystallization heat of PLA may be larger than the heat of the melting of the PBAT crystalline phase. The peaks ascribed to the melting of the PLA crystalline phase are single or double. The single peak concerns the non-irradiated sample (L0) only. In the case of the irradiated samples, the phase transition may be reflected by a single peak with two apexes or two peaks that are almost entirely separated, which can especially be seen in the curves of the L40 and L90 samples.

A single peak with two apexes, assigned to the melting of the crystalline phase, can often be observed in the case of PLA and its blends with other polymers [50–52]. It reflects the melting and recrystallization of original crystalline structures into more stable forms (in this case the melting peak occurs at lower temperatures) and the melting of the more stable crystalline structures (the melting peak occurs at higher temperatures). From Figure 6, it follows that the recrystallization into the more stable crystalline structures occurs more easily in the samples irradiated with larger doses. This is indicated by the decreasing ratio of the melting enthalpy corresponding to the lower temperature peak to that relating to the higher temperature peak.

In the case of two separated peaks that correspond to two different phase transitions, another effect can also occur. It concerns melting of two crystalline phases with different molecular weights, which can appear in the radiation modified samples. This would confirm the GPC results indicating that a bimodal distribution of molecular weights occurs in the samples irradiated with the largest doses. Two clearly separated peaks on the DSC curves, ascribed to the melting of the crystalline phases, correspond to the bimodal distribution of molecular weights in the same samples (irradiated with the doses of 40 or 90 kGy), which was observed on the GPC curves. Two melting peaks can also be seen in the case of the sample irradiated with the 10 kGy dose. However, one of them (occurring at a higher range of temperatures) is very small and the relevant bimodal distribution was not observed on the GPC curves. Nevertheless, the presence of this tiny peak is closely connected with a much larger degree of polydispersion of the L10 sample with respect to the remaining samples (Table 2). This fact may indicate that already the smallest radiation dose induces essential radiation processes in the studied sample. However, the discussed effect, which is associated with the melting of the crystalline phases of two fractions with different molecular weights, is less probable or does not predominate. This conclusion may be justified by the fact that only one peak assigned to the cold crystallization occurs on the DSC curves of all samples, which points rather out the melting of the crystalline phase and its further recrystallization.

The DSC results also show that the melting temperature of the crystalline phase, corresponding to both peaks, decreases as the radiation dose increases. This is especially visible when these peaks occur in the range of lower temperatures: the melting temperature drops from ca. 149 (sample L0) to ca 136 °C (sample L90). This fact is associated with a notable reduction in the molecular weight of the relevant samples, which can be confirmed by the GPC results. In the range of higher temperatures, the melting temperature of the crystalline phase of the L40 sample is by ca. 3 °C higher than that of the L90 sample, which results from a larger molecular weight of one of the fractions of the L40 sample as compared to the analogous fraction of the L90 sample (Table 2).

3.4.2. Analysis of the DSC Results for Samples B

Several thermal effects of the phase transitions, occurring in the B samples (Figure 7), somewhat differ from those observed in the samples with the predominating PLA fraction. The direction of the change in the glass transition temperature upon the increasing radiation dose is similar to that occurring in the L samples. However, the reduction in the values of this temperature in the B samples irradiated with the maximum dose is smaller than that of the L samples and is equal to ca. 3 °C. This is due to the effect connected with the plasticization of the samples by a part of the degraded macromolecules of PLA or PBAT. However, the reduction in the glass transition temperature of the B samples, due to the plasticization of those materials by the degraded macromolecules of PLA or PBAT, is smaller compared to the L samples because the number average molecular weight of the degraded macromolecules in the B samples is larger than that in the relevant L samples, which was shown above (GPC results, Table 2). Thus, the ability of the plasticizer to reduce the glass transition temperature of the relevant samples decreases as the molecular weight of the plasticizer increases.

From Figure 7, it also follows that the DSC curves of the samples non-irradiated or irradiated with the 10-kGy dose exhibit the presence of two peaks corresponding to the melting of the crystalline phases of PBAT (at ca. 120 °C) and PLA (at ca. 150 °C). These curves do not include the cold crystallization peaks, contrary to the DSC curves of the L samples, in which the peaks assigned to the cold crystallization of PLA were observed. A lack of the peak of the cold crystallization of PBAT indicates that this phase already exhibited some crystallinity. On the other hand, the peak of the cold crystallization of PLA is invisible because this peak and the peak corresponding to the melting of the crystalline phase of PBAT occur over the same temperature range and one may assume that the thermal effect of the melting of the crystalline phase of PBAT is larger than that of the cold crystallization of PLA. The larger thermal effect of the melting of PBAT might result from the fact that the PBAT phase was initially crystalline, the PBAT fraction predominates in the B samples and PLA hardly crystallizes. The last statement would agree with the literature data that point out a very slow crystallization of PLA, especially under conditions of the actual experiment, when the cooling rate was 10 °C/min [50].

The cold crystallization of the PLA phase is still observed in the samples irradiated with the dose of at least 40 kGy. This is due to the reduction in the average molecular weight of the PBAT macromolecules that, because of essential degradation of them, may act as a nucleation agent and contribute to the cold crystallization of the PLA phase. At the same time, the two peaks assigned to the melting of the crystalline phase of PLA occur. The peak corresponding to the melting of the PBAT phase is invisible because, in the mentioned samples, the thermal effect of the cold crystallization of the PLA is larger than the thermal effect connected with the melting of the crystalline phase of PBAT, contrary to the effects observed in the samples irradiated with the smaller doses. The remaining effects, associated with the crystallization and melting of the B40 and B90 samples, are analogous to those of the L40 and L90 samples. Therefore, they are not discussed here.

4. Conclusions

The results of the investigation of the uncrosslinked fraction of the radiation modified PLA/PBAT blends discussed in the present article may contribute to a better understanding of some phenomena occurring during the radiation treatment of the biodegradable polymers and their blends. The applied research techniques enabled us to determine the most important changes occurring in the studied samples.

It was found that the electron radiation causes the occurrence of various effects in the biodegradable polymers and blends of them, which mostly depend on the composition of the materials being irradiated, radiation dose magnitude, and the presence of low molecular weight multifunctional compounds, such as TAIC. The investigation of the uncrosslinked fraction of the irradiated materials revealed an occurrence of effects, such as a different amount of that fraction, changes in the molecular weights of the macromolecules, the formation of new macromolecular structures, and thermal changes associated with, e.g., the cold crystallization or melting of crystalline phases. All the investigated samples contained the macromolecules of both PLA and PBAT, which indicated that any phase of both kinds of the blends was not completely crosslinked, independently of whether a given phase predominated or not. Much larger amounts of the uncrosslinked fractions occur in the samples of the predominating content of PBAT, which points out that this polymer is less susceptible than PLA to the radiation crosslinking in the presence of TAIC. This is mainly due to the presence in the PBAT structure of the aromatic groupings and the so-called protective effect. Samples with the predominating content of PLA more clearly exhibit the presence of the bimodal distribution of molecular weights and a larger reduction in the molecular weight magnitudes than the samples with the predominating content of PBAT. This shows that the intensity of the radiation degradation processes occurring in the studied blends increases as the weight fraction of PLA rises. The macromolecules present in the samples of both kinds of the uncrosslinked fractions exhibit different structures. In addition to the short macromolecules formed due to the radiation degradation, the short macromolecules being only an oligomeric fraction of the studied samples, which underwent neither crosslinking nor degradation, may be present. The longer macromolecules, in turn, are formed rather due to the lengthening by mutual binding at the chain ends or the formation of branched structures. There are no macromolecules crosslinked completely, which results from the occurrence of the cold crystallization and melting of the crystalline phases. The molecular weight of the macromolecules in the samples with the predominating content of PBAT decreases upon the irradiation more slowly compared to that in the samples with the predominating content of PLA. Therefore, the plasticization effect in the former samples is less evident than that in the latter samples. As a result, the glass transition temperature of the samples with the predominating content of PBAT decreases upon the irradiation more slowly than that of the samples with the predominating content of PLA. The shortening of the macromolecules in the PBAT phase, occurring with the increase in the electron radiation dose, may beneficially influence the process of the nucleation of the crystallization of the PLA phase, which usually is hindered and runs slowly. An increase in the electron radiation dose facilitates the recrystallization of the original crystalline structures into more stable forms, which is especially visible in the samples with the predominating content of PLA.

The results of the investigation of the uncrosslinked fraction of the radiation modified blends of PLA and PBAT presented in this article, constitute only a small fragment of wider considerations about the radiation treatment of the biodegradable polymers. Apart from that, there are many issues connected with the explanation and verification of some phenomena and hypotheses, which requires carrying out a larger number of additional studies. These concerns, e.g., the influence of the electron radiation on susceptibility to the biodegradation of the biodegradable polymers and their blends, the verification of the possibility of the occurrence of the specified mechanisms of crosslinking or degradation, and the specification of post-radiation effects. These issues should be the subjects of the next studies carried out by the scientists. Results presented in this article relate to issues of the radiation treatment of biodegradable materials. The modification of the properties of these materials by radiation treatment is poorly understood, and the modified materials in this way may be more widely used, e.g., in medicine, tissue engineering, or the packaging sector.

Author Contributions: Conceptualization and methodology, R.M.; investigation, R.M., K.M. and A.R.-K.; resources, A.R.-K.; writing—original draft preparation, R.M., K.M. and A.R.-K.; writing—review and editing, R.M. and K.M. All authors have read and agreed to the published version of the manuscript.

Funding: This research has partially been funded by the National Science Centre, grant number DEC-2012/07/D/ST8/02773.

Acknowledgments: The authors of this work would like to express their special thanks to the Institute of Nuclear Chemistry and Technology in Warsaw, Poland, for technical support during the radiation treatment of studied samples.

Conflicts of Interest: The authors declare no conflict of interest.

References

1. Charlesby, A. *Atomic Radiation and Polymers*, 1st ed.; Pergamon Press: New York, NY, USA, 1960.
2. Clough, R.L.; Shalaby, S.W. *Radiation Effects on Polymers*, 1st ed.; American Chemical Society: Washington, DC, USA, 1991.
3. Zenkiewicz, M. Some problems of polymeric materials modification with high energy electron radiation. *Polimery* **2005**, *50*, 4–9. [CrossRef]
4. Czuprynska, J. The effect of high-energy electron beam radiation on polymer properties. *Polimery* **2002**, *47*, 8–14. [CrossRef]
5. Shivam, P. Recent Developments on biodegradable polymers and their future trends. *Int. Res. J. Sci. Eng.* **2016**, *4*, 17–26.
6. Doppalapudi, S.; Jain, A.; Khan, W.; Domb, A.J. Biodegradable polymers – an overview. *Polym. Adv. Techn.* **2014**, *25*, 427–435. [CrossRef]
7. Luckachan, G.E.; Pillai, C.K.S. Biodegradable Polymers – A review on recent trends and emerging perspectives. *J. Polym. Environ.* **2011**, *19*, 637–676. [CrossRef]
8. Scaffaro, R.; Maio, A.; Sutera, F.; Gulino, E.F.; Morreale, M. Degradation and Recycling of Films Based on Biodegradable Polymers: A Short Review. *Polimery* **2019**, *11*, 651. [CrossRef]
9. Vroman, I.; Tighzert, L. Biodegradable Polymers. *Materials* **2009**, *2*, 307–344. [CrossRef]
10. Castro-Aguirre, E.; Iñiguez-Franco, F.; Samsudin, H.; Fang, X.; Auras, R. Poly(lactic acid)—Mass production, processing, industrial applications, and end of life. *Adv. Drug Deliv. Rev.* **2016**, *107*, 333–366. [CrossRef]
11. Lim, L.-T.; Auras, R.; Rubino, M. Processing technologies for poly(lactic acid). *Prog. Polym. Sci.* **2008**, *33*, 820–852. [CrossRef]
12. Sonchaeng, U.; Iñiguez-Franco, F.; Auras, R.; Selke, S.; Rubino, M.; Lim, L.-T. Poly(lactic acid) mass transfer properties. *Prog. Polym. Sci.* **2018**, *86*, 85–121. [CrossRef]
13. Malinowski, R. Effect of high energy b-radiation and addition of triallyl isocyanurate on the selected properties of polylactide. *Nucl. Instrum. Methods Phys. Res. Sect. B* **2016**, *377*, 59–66. [CrossRef]
14. Mansouri, M.; Berrayah, A.; Beyens, C.; Rosenauer, C.; Jama, C.; Maschke, U. Effects of electron beam irradiation on thermal and mechanical properties of poly(lactic acid) films. *Polym. Degrad. Stab.* **2016**, *133*, 293–302. [CrossRef]
15. Loo, J.; Ooi, C.; Boey, F. Degradation of poly(lactide-co-glycolide) (PLGA) and poly(l-lactide) (PLLA) by electron beam radiation. *Biomaterials* **2005**, *26*, 1359–1367. [CrossRef] [PubMed]
16. Moraczewski, K. Characterization of multi-injected poly(ε-caprolactone). *Polym. Test.* **2014**, *33*, 116–120. [CrossRef]
17. Mohamed, R.M.; Yusoh, K. A Review on the Recent Research of Polycaprolactone (PCL). *Adv. Mater. Res.* **2015**, *1134*, 249–255. [CrossRef]
18. Labet, M.; Thielemans, W. Synthesis of polycaprolactone: A review. *Chem. Soc. Rev.* **2009**, *38*, 3484. [CrossRef]
19. Gan, Z.; Kuwabara, K.; Yamamoto, M.; Abe, H.; Doi, Y. Solid-state structures and thermal properties of aliphatic–aromatic poly(butylene adipate-co-butylene terephthalate) copolyesters. *Polym. Degrad. Stab.* **2004**, *83*, 289–300. [CrossRef]
20. Jian, J.; Xiangbin, Z.; Xianbo, H. An overview on synthesis, properties and applications of poly(butylene-adipate-co-terephthalate)–PBAT. *Adv. Ind. Eng. Polym. Res.* **2020**, *3*, 19–26. [CrossRef]

21. Herrera, R.; Franco, L.; Rodríguez-Galán, A.; Puiggalí, J. Characterization and degradation behavior of poly(butylene adipate-co-terephthalate)s. *J. Polym. Sci. Part A Polym. Chem.* **2002**, *40*, 4141–4157. [CrossRef]
22. Costa, A.R.M.; Almeida, T.; Silva, S.M.; De Carvalho, L.H.; Canedo, E.L. Chain extension in poly(butylene-adipate-terephthalate). Inline testing in a laboratory internal mixer. *Polym. Test.* **2015**, *42*, 115–121. [CrossRef]
23. Malinowski, R. Effect of electron radiation and triallyl isocyanurate on the average molecular weight and crosslinking of poly(ε-caprolactone). *Polym. Adv. Technol.* **2015**, *27*, 125–130. [CrossRef]
24. Hwang, I.-T.; Jung, C.-H.; Kuk, I.-S.; Choi, J.-H.; Nho, Y.-C. Electron beam-induced crosslinking of poly(butylene adipate-co-terephthalate). *Nucl. Instruments Methods Phys. Res. Sect. B Beam Interactions Mater. Atoms* **2010**, *268*, 3386–3389. [CrossRef]
25. Malinowski, R. Application of the electron radiation and triallyl isocyanurate for production of aliphatic-aromatic co-polyester of modified properties. *Int. J. Adv. Manuf. Technol.* **2016**, *87*, 3307–3314. [CrossRef]
26. Zembouai, I.; Kaci, M.; Bruzaud, S.; Pillin, I.; Audic, J.-L.; Shayanfar, S.; Pillai, S.D. Electron beam radiation effects on properties and ecotoxicity of PHBV/PLA blends in presence of organo-modified montmorillonite. *Polym. Degrad. Stab.* **2016**, *132*, 117–126. [CrossRef]
27. Shin, B.Y.; Han, D.H. Compatibilization of immiscible poly(lactic acid)/poly(ε-caprolactone) blend through electron-beam irradiation with the addition of a compatibilizing agent. *Rad. Phys. Chem.* **2013**, *83*, 98–104. [CrossRef]
28. Kumara, P.H.S.; Nagasawa, N.; Yagi, T.; Tamada, M. Radiation-induced crosslinking and mechanical properties of blends of poly(lactic acid) and poly(butylene terephthalate-co-adipate). *J. Appl. Polym. Sci.* **2008**, *109*, 3321–3328. [CrossRef]
29. Leonard, D.J.; Pick, L.; Farrar, D.; Dickson, G.R.; Orr, J.F.; Buchanan, F. The modification of PLA and PLGA using electron-beam radiation. *J. Biomed. Mater. Res. Part A* **2009**, *89*, 567–574. [CrossRef]
30. Yoshii, F.; Darwis, D.; Mitomo, H.; Makuuchi, K. Crosslinking of poly(ε-caprolactone) by radiation technique and its biodegradability. *Radiat. Phys. Chem.* **2000**, *57*, 417–420. [CrossRef]
31. Nagasawa, N.; Kaneda, A.; Kanazawa, S.; Yagi, T.; Mitomo, H.; Yoshii, F.; Tamada, M. Application of poly(lactic acid) modified by radiation crosslinking. *Nucl. Instruments Methods Phys. Res. Sect. B Beam Interactions Mater. Atoms* **2005**, *236*, 611–616. [CrossRef]
32. Malinowski, R.; Janczak, K.; Moraczewski, K.; Raszkowska-Kaczor, A. Analysis of swelling degree and gel fraction of polylactide/poly(butylene adipate-co-terephthalate) blends crosslinked by radiation. *Polimery* **2018**, *63*, 25–30. [CrossRef]
33. Suhartini, M.; Mitomo, H.; Nagasawa, N.; Yoshii, F.; Kume, T. Radiation crosslinking of poly(butylene succinate) in the presence of low concentrations of trimethallyl isocyanurate and its properties. *J. Appl. Polym. Sci.* **2003**, *88*, 2238–2246. [CrossRef]
34. Malinowski, R. Some effects of radiation treatment of biodegradable PCL/PLA blends. *J. Polym. Eng.* **2018**, *38*, 635–640. [CrossRef]
35. Imre, B.; Pukánszky, B. Compatibilization in bio-based and biodegradable polymer blends. *Eur. Polym. J.* **2013**, *49*, 1215–1233. [CrossRef]
36. Malinowski, R.; Rytlewski, P.; Janczak, K.; Raszkowska-Kaczor, A.; Moraczewski, K.; Stepczyńska, M.; Żuk, T. Studies on functional properties of PCL films modified by electron radiation and TAIC additive. *Polym. Test.* **2015**, *48*, 169–174. [CrossRef]
37. Ng, H.-M.; Bee, S.; Ratnam, C.T.; Sin, L.T.; Phang, Y.-Y.; Tee, T.-T.; Rahmat, A.R. Effectiveness of trimethylopropane trimethacrylate for the electron-beam-irradiation-induced cross-linking of polylactic acid. *Nucl. Instruments Methods Phys. Res. Sect. B Beam Interactions Mater. Atoms* **2014**, *319*, 62–70. [CrossRef]
38. Han, C.; Ran, X.; Zhang, K.; Zhuang, Y.; Dong, L. Thermal and mechanical properties of poly(e-caprolactone) crosslinked with γ radiation in the presence of triallyl isocyanurate. *J. Appl. Polym. Sci.* **2007**, *103*, 2676–2681. [CrossRef]
39. Quynh, T.M.; Mitomo, H.; Nagasawa, N.; Wada, Y.; Yoshii, F.; Tamada, M. Properties of crosslinked polylactides (PLLA & PDLA) by radiation and its biodegradability. *Eur. Polym. J.* **2007**, *43*, 1779–1785.
40. Mitomo, H.; Kaneda, A.; Quynh, T.M.; Nagasawa, N.; Yoshii, F. Improvement of heat stability of poly(l-lactic acid) by radiation-induced crosslinking. *Polimery* **2005**, *46*, 4695–4703. [CrossRef]

41. Abdel-Rehim, H.A.; Yoshii, F.; Kume, T. Modification of polycaprolactone in the presence of polyfunctional monomers by irradiation and its biodegradability. *Polym. Degrad. Stab.* **2004**, *85*, 689–695. [CrossRef]
42. Reinholds, I.; Kalkis, V.; Zicans, J.; Meri, R.M.; Elksnite, I. Thermomechanical and deformation properties of electron beam modified polypropylene copolymer grafted with acrylic monomer. *Mater. Sci. Appl. Chem.* **2012**, *25*, 16–21.
43. Dawes, K.; Glover, L.C.; Vroom, D.A. The Effects of Electron Beam and g-Irradiation on Polymeric Materials. In *Physical Properties of Polymers Handbook*; Springer Science and Business Media LLC: Berlin, Germany, 2007; pp. 867–887.
44. Malinowski, R.; Żenkiewicz, M.; Richert, A. Effect of some crosslinking factors on gelation and swelling degree of polylactide. *Przem. Chem.* **2012**, *91*, 1596–1599.
45. Wang, L.F.; Rhim, J.W.; Hong, S.I. Preparation of poly(lactide)/poly(butylene adipate-co-terephthalate) blend films using a solvent casting method and their food packaging application. *LWT Food Sci. Technol.* **2016**, *68*, 454–461. [CrossRef]
46. Al-Itry, R.; Lamnawar, K.; Maazouz, A. Improvement of thermal stability, rheological and mechanical properties of PLA, PBAT and their blends by reactive extrusion with functionalized epoxy. *Polym. Degrad. Stab.* **2012**, *97*, 1898–1914. [CrossRef]
47. Sirisinha, K.; Somboon, W. Melt characteristics, mechanical, and thermal properties of blown film from modified blends of poly(butylene adipate-co-terephthalate) and poly(lactide). *J. Appl. Polym. Sci.* **2011**, *124*, 4986–4992. [CrossRef]
48. Auras, R.; Harte, B.R.; Selke, S. An Overview of Polylactides as Packaging Materials. *Macromol. Biosci.* **2004**, *4*, 835–864. [CrossRef]
49. Weng, Y.-X.; Jin, Y.; Meng, Q.-Y.; Wang, L.; Zhang, M.; Wang, Y.-Z. Biodegradation behavior of poly(butylene adipate-co-terephthalate) (PBAT), poly(lactic acid) (PLA), and their blend under soil conditions. *Polym. Test.* **2013**, *32*, 918–926. [CrossRef]
50. Jiang, L.; Wolcott, M.; Zhang, J. Study of Biodegradable Polylactide/Poly(butylene adipate-co-terephthalate) Blends. *Biomacromolecules* **2006**, *7*, 199–207. [CrossRef]
51. Sarasua, J.-R.; Prud'Homme, R.E.; Wisniewski, M.; Le Borgne, A.; Spassky, N. Crystallization and Melting Behavior of Polylactides. *Macromolecules* **1998**, *31*, 3895–3905. [CrossRef]
52. Arruda, L.C.; Magaton, M.; Bretas, R.E.S.; Ueki, M.M. Influence of chain extender on mechanical, thermal and morphological properties of blown films of PLA/PBAT blends. *Polym. Test.* **2015**, *43*, 27–37. [CrossRef]

© 2020 by the authors. Licensee MDPI, Basel, Switzerland. This article is an open access article distributed under the terms and conditions of the Creative Commons Attribution (CC BY) license (http://creativecommons.org/licenses/by/4.0/).

Article

Characterization of Biodegradable Food Contact Materials under Gamma-Radiation Treatment

Karolina Wiszumirska [1], Dorota Czarnecka-Komorowska [2,*], Wojciech Kozak [1], Marta Biegańska [1], Patrycja Wojciechowska [1], Maciej Jarzębski [3], and Katarzyna Pawlak-Lemańska [4]

[1] Department of Industrial Products and Packaging Quality, Institute of Quality Science, Poznan University of Economics and Business, Al. Niepodległosci 10, 61-875 Poznan, Poland
[2] Polymer Processing Division, Institute of Materials Technology, Faculty of Mechanical Engineering, Poznan University of Technology, Piotrowo 3, 61-138 Poznan, Poland
[3] Department of Physics and Biophysics, Faculty of Food Science and Nutrition, Poznan University of Life Sciences, Wojska Polskiego 38/42, 60-637 Poznan, Poland
[4] Department of Technology and Instrumental Analysis, Institute of Quality Science, Poznan University of Economics and Business, Al. Niepodległosci 10, 61-875 Poznan, Poland
* Correspondence: dorota.czarnecka-komorowska@put.poznan.pl

Abstract: Radiation is an example of one of the techniques used for pasteurization and sterilization in various packaging systems. There is a high demand for the evaluation of the possible degradation of new composites, especially based on natural raw materials. The results of experimental research that evaluated the impact of radiation technology on biodegradable and compostable packaging materials up to 40 kGy have been presented. Two commercially available flexible composite films based on aliphatic–aromatic copolyesters (AA) were selected for the study, including one film with chitosan and starch (AA-CH-S) and the other with thermoplastic starch (AA-S). The materials were subjected to the influence of ionizing radiation from 10 to 40 kGy and then tests were carried out to check their usability as packaging material for the food industry. The results showed that the mechanical properties of AA-S films improved due to the radiation-induced cross-linking processes, while in the case of AA-CH-S films, a considerable decrease in the elongation at break was observed. The results also showed a decrease in the WVTR in the case of AA-S and no changes in barrier properties in the case of AA-CH-S. Both materials revealed no changes in the odor analyzed by sensory analysis. In the case of the AA-S films, the higher the radiation dose, the faster the biodegradation rate. In the case of the AA-CH-S film, the radiation did not affect biodegradation. The performed research enables the evaluation of the materials intended for direct contact with food. AA-CH-S was associated with unsatisfactory parameters (exceeding the overall migration limit and revealing color change during storage) while AA-S showed compliance at the level of tests carried out. The study showed that the AA-CH-S composite did not show a synergistic effect due to the presence of chitosan.

Keywords: biodegradable polymer; packaging materials; gamma radiation; quality; safety; food contact materials

1. Introduction

The bioplastic market is developing an alternative to the conventional plastics used in the packaging industry [1–4]. Global bioplastics production capacities in 2021 reached 2.42 million tons. Currently, bioplastics constitute less than one percent of the more than 367 million tons of plastic produced annually [5,6]. However, with the increase in demand and the emergence of increasingly sophisticated biopolymers, applications, and products, this is a constantly growing market [7]. According to the latest market data developed by European Bioplastics, in cooperation with the nova-Institute research institute, the global production capacity of bioplastics will increase from around 2.11 million tons in 2018 to around 5.22 million tons in 2023 [5,6].

Bioplastics that are biobased, biodegradable, or both have similar, or even the same, properties as conventional plastics and offer additional environmental benefits, such as a reduced carbon footprint or additional waste management options, such as composting [8,9]. With the presence and availability of the bioplastics market, modified and improved properties, such as flexibility, durability, printability, transparency, barrier properties, heat resistance, gloss, and many more, have been significantly enhanced. Owing to these changes, bioplastics are becoming increasingly attractive to the demanding packaging industry [10–14].

Bioplastics are a large family of materials that can be divided into the following three main groups [15–17]: (i) biobased or partially biobased non-biodegradable plastics, such as biobased PE (polyethylene), PP (polypropylene), or PET (polyethylene terephthalate) (so-called drop-ins) and biobased technical performance polymers, such as PTT (polytrimethylene terephthalate); (ii) plastics that are both biobased and biodegradable, such as PLA (polylactide) and PHA (polyhydroxyalkanoates) or PBS (polybutylene succinate); (iii) plastics that are based on fossil resources and are biodegradable, such as PBAT (poly(butylene adipate-co-terephthalate) and PCL (polycaprolactone). Biobased or partially biobased durable plastics, such as biobased PE and PET, have the same properties as polymers produced from petroleum, this is why can be mechanically recycled in the existing recycling streams. In addition, materials such as PLA and PHA (produced from raw materials such as starch) have good barrier properties, which are important for the packaging industry, and are biodegradable and compostable (if confirmed by the appropriate certificate) [5,6,18–20].

Biodegradable shopping and waste collection bags are available on the consumer market. A new market potential (niche) is emerging for rigid bioplastics, including food packaging and take-out meals, disposable cutlery, and stationery. The portfolio of biodegradable products for the automotive and agriculture sectors is also expanding to the consumer electronics [21–23].

The components that appear in the tested materials in the context of the impact of gamma radiation are briefly discussed below. Aliphatic–aromatic copolyesters (AAC) are a hybrid combination characterized by good strength properties and resistance to the thermal degradation of aromatic polyesters and biodegradability, due to the presence of aliphatic polyesters [24,25]. Aromatic polyesters exhibit water hydrolytic resistance due to the hydrophobic benzene rings in their chemical structure, while aliphatic polyesters undergo spontaneous hydrolytic degradation under the influence of moisture because they contain short methylene chains separated by ester bonds. They are also biodegradable in the presence of enzymes [26]. AAC degradation carried out in compost under elevated temperature conditions is significantly different from degradation in aquatic environments. In liquid media, the AAC degradation process is usually much slower, which is mainly due to the lower temperature and composition of the microflora [27]. The aliphatic–aromatic polyesters include PBTA copolyesters (based on terephthalic acid, adipic acid, and 1,4-butanediol) and PBTS copolyesters (based on terephthalic acid, succinic acid, and 1,4-butanediol). The relatively low price of AAC compared to PHA and PCL, as well as its good utility and processing properties, favor the development of this group of plastics. Based on the efficiency tests of the radiolytic emission of hydrogen $G(H_2)$ during irradiation, it can be assumed that AAC is radiation-resistant in the scope of low-dose applications [28].

Starch is a polysaccharide that consists of amylose and amylopectin. The ratio of these polysaccharides depends on the source of starch (potato, rice, corn, etc.). The effect of ionizing radiation on starch is demonstrated through the ways in which glycosidic bonds break in the polymer and modification of its crystalline structure [29]. Materials based on starch are brittle and hydrophilic, which limits their processing and use. To overcome this problem, starch is mixed with various synthetic and natural polymers. Mixtures of starch with various compostable polymeric materials show insolubility or exhibit enhanced strength and other advantageous features. Usually, starch-based mixtures exhibit favorable strength, processing, and performance characteristics, e.g., they have greater water resistance [30,31].

Chitosan (poly [β-(1,4)-2-ammonium-2-deoxy-D-glucopyranose) is a pseudo-natural, non-toxic biopolymer [32–34]. It is made of N-acetyl-D-glucosamine and D-glucosamine residues connected by β-1,4-glycosidic bonds. It is industrially obtained by the chemical N-deacetylation of chitin, which is the main component of the cell wall of fungi of the class *Zygomycetes*, e.g., *Absidia, Mucor* and *Rhizopus*. It is also found in the outer skeletal structures of numerous invertebrates, including crustaceans and insects [35]. Its hydrophilic nature and sensitivity to changes in the pH of the environment mean that the stability of chitosan is much worse than that of chitin. It dissolves well in aqueous acid solutions. Crosslinking causes an increase in the space between the chains, which results in the partial degradation of its crystalline structure and a decrease in its solubility. Chitosan is a bioactive, biocompatible, biodegradable, non-toxic polymer with high adhesion, which is why it is used in many fields, including agriculture, environmental protection, food, and the cosmetics industry, as well in biomedicine [36–39]. Due to its antimicrobial properties, chitosan can be used in food products to extend their shelf life and as an ingredient in packaging films. Chitosan and its derivatives have strong biocidal activity against various groups of G(+) and G(−) bacteria, fungi, and viruses. The properties of chitosan depend on its molecular weight, degree of deacetylation, concentration, pH, and the composition of the environment in which it is located. Chitosan can be made into films that exhibit high gas barriers. However, their fragility requires the use of plasticizers such as polyols (glycerine, sorbitol and polyethylene glycol) and fatty acids (stearic and palmitic acid) [40]. Mathew and Abraham reported that the physical modification of starch–chitosan mixtures with gamma radiation or ultrasound can modify them through cross-linking, thereby improving the functionality of the materials [41].

When analyzing the possibility of using packaging materials in radiation technologies, the risk concerning the potential packaging–product interaction should be considered [42,43]. Gamma irradiation is not a surface treatment, because the photon energy is high enough to penetrate the materials [44]. The impact of ionizing radiation on the polymer may lead to the formation of free radicals, and the formation of reactive intermediates. These reactions may occur during direct contact and may be deferred in time and their effects may be delayed [45,46]. There is a risk leading to unsatisfactory barrier properties, or exceeding the limit of overall and specific migration, as stipulated by Regulation (EC) No. 1935/2004 of the European Parliament [47], Commission Regulation (EU) No. 10/2011 [48] and the requirements of the Commission Regulation (EC) No. 2023/2006 [49] for materials and articles intended to come into contact with food.

Ionizing radiation is an alternative method to conventional methods of product preservation and sterilization, guaranteeing a high degree of hygienization during storage and use. Various food and industrial products can be subjected to radiation sterilization, for example surgical implants, medical utensils, special purpose food, or herbs and spices. The originality of the use of ionizing radiation to combat pathogens involves the sterilization of the packaging before filling or sterilization of the packaged product. Maintaining the specific properties of the product in the irradiation process and protection against secondary contamination requires the use of appropriate packaging, which must additionally meet the requirements of radiation treatment.

Research on the impact of ionizing radiation on packaging made of various materials, including plastic, has been carried out for many years, but there are few reports on biodegradable materials [50,51]. A prerequisite for considering the application potential of new solutions in radiation technologies is the identification of the possible changes caused by ionizing radiation. This is important because it can lead to the process of cross-linking or degradation of the material, and the chemical substances formed as a result of radiolysis can cause negative phenomena (e.g., sensory changes), which may only become apparent after a certain period.

The widespread use of biodegradable materials, as well as other packaging materials, requires verification of the compliance with the requirements in at least two aspects, which are as follows: the technological conditions in radiation pasteurization/sterilization and safety of use, which is reflected in the applicable law. The aim of the research was to determine whether the increasing dose of ionizing radiation up to 40 kGy affects the structure and properties of biodegradable films and flexible films for food packaging, including an aliphatic–aromatic copolyester with thermoplastic starch (AA-S) and an aliphatic–aromatic copolyester with chitosan and thermoplastic starch (AA-CH-S), and whether it is advisable to conduct further research with these materials and their composites.

2. Materials and Methods

2.1. Materials

Tests were conducted on biodegradable and compostable flexible films for food packaging, including an aliphatic–aromatic copolyester with thermoplastic starch (AA-S) and an aliphatic–aromatic copolyester with chitosan and thermoplastic starch (AA-CH-S) as presented in Table 1. Both materials are commercially available and were purchased directly from the manufacturers.

Table 1. Samples' description.

Sample	Characteristics	Symbol
Aliphatic–aromatic copolyester with chitosan and thermoplastic starch film	Flexible film, milky, translucent, food contact material; industrially compostable (about 14 days)	AA-CH-S
Aliphatic–aromatic copolyester with thermoplastic starch film	Flexible film, creamy yellow, food contact material; industrially compostable	AA-S

2.2. Mechanical Properties

The tensile properties of the films in the machine direction (MD) were tested according to the standard ISO 527-1 and ISO 523-3 [52,53], using a universal testing machine Zwick/Roell Z005 (Ulm, Germany) equipped with testXpert II V3.7 software. The tensile characteristics were measured at room temperature with a crosshead speed of 200 mm/min. Film tensile strength (σ_M) and elongation at break (ε_B) were evaluated. The reported data represent the average of 10 measurements. The samples before testing were climatized at 23 ± 2 °C, with an air humidity of $50 \pm 5\%$.

2.3. Overall Migration

Migration tests were performed based on the appropriate relative standards, which specify the type of contact between the packaging material and a simulant, and the time and temperature in which the test is carried out. The conditions of the test reflect risks that are higher than the real risks, while the principle of the "worst case scenario" ensures that the application of food contact materials (FCMs) is safe in their real conditions of use. Under EU/10/2011, all packaging materials should (which is understood as "must") meet the requirements concerning overall migration (OML, overall migration limit) and specific migration (SML, specific migration limit). The OML means the total of the non-volatile substances that are transferred from the packaging material or article into a simulant (a liquid imitating food). The residue is expressed in milligrams and square decimetres (mg/dm^2) on the sample's surface, which is in direct contact with the simulant. OML cannot exceed 10 mg/dm^2 (or 60 mg/kg of food stuff in the case of products intended for infants and young children) of the non-volatile substances that diffuse from the surface of the material, which comes into direct contact with food. In the case of packaging that is suitable for all kinds of foodstuff and food simulants, A (or water), B (acetic acid 3% (w/v)), and D2 (vegetable oil) are used for the migration test. The selection of test conditions and the methodology were decided based on the standard EN 1186-1, EN 1186-3 [54,55]. The method of total immersion was used, using two food simulants, water and food simulant D1 (ethanol 50% (v/v)), for 10 days at a temperature 60 °C. The contact surface of the material with the model fluid was 2.0 dm^2, according to the guidelines of the above-mentioned standards. The film edge surface was omitted. The test was performed in five replicates for each of the model fluids (water and ethyl alcohol).

2.4. Sensory Analysis of Packaging Materials

Organoleptic (olfactometric) evaluation is a test that is complementary to migration tests. It is a non-instrumental analysis carried out with the help of the human senses (sight, taste, and smell). The test allows the verification of whether the material releases volatile compounds that cause odor formation in the packaging material. Fragrances can arise from the degradation products of plastics or processing additives (dyes and plasticizers), solvents or printing inks. Sensory analysis can be extended with instrumental identification methods. Confirming the sensory indifference of the packaging material in direct contact with food is a key aspect of the packaging material's compliance with EC/1935/2004 (Art. 3) [47]. Sensory evaluation was carried out following the standardized methods described in the standards. The test was conducted by an experienced team of six specialists in the field of sensory analysis. The samples were prepared based on the methodology of Regulation 10/2011 from 14 January 2011 on plastic materials and products intended to come into contact with food, which specifies the requirements and tests of global migration. Distilled water was chosen as the model fluid. The test was carried out following the methodology of DIN 10955: 2004 [56]. A five-grade scale of flavor or aroma intensity was used, where 0 means that there is no noticeable aroma transfer and no flavors; 1—the transfer of smell/unpleasant aftertaste is palpable (still difficult to define); 2—moderate odor/unpleasant aftertaste transfer (the assessor can generally determine the origin of the smell, but no specific substance can be identified); 3—moderately strong odor/unpleasant aftertaste; 4—strong transfer of smell/unpleasant aftertaste (the assessor can identify the responsible substance) [56]. It was assumed that the median of 3 or more results is inconsistent with the provisions of EC/1935/2004, art. 3 [47].

2.5. Oxygen Transmission Rate

The oxygen transmission rate (OTR) was determined by ASTM F3136 [57], using the OxyPerm system that consists of the Oxysense 325 oxygen analyzer and dedicated test chambers for oxygen permeation measurements. Periodic measurement of oxygen concentration within the test chambers during the permeation process was based on the fluorescence quenching time measurements obtained by ASTM F2714-08 [58].

The test samples were cut in the shape of squares with a side of 6.5 cm. Then, the samples were placed in the test chambers, which were flushed with nitrogen (5.0 purity) to obtain an oxygen-free atmosphere. Then, the OTR measurements began, consisting of the periodic measurement of the oxygen concentration inside the measuring chambers. It was assumed that the end of the study was the moment in which the correlation coefficient between the individual measurements was greater than 0.95. The OTR measurements (cc/m^2/24 h) were performed in controlled conditions at a temperature of 23 ± 2 °C and a relative humidity of 60 ± 10%, respectively. The test was performed in three replicates for all variants of the investigated materials.

2.6. Water Vapor Transmission Rate

The water vapor transmission rate (WVTR) was determined according to ASTM E96/E96M-16 [59], using the desiccant method and the EZ-Cup Vapometer (Thwing-Albert Instrument Company, West Berlin, NJ, USA). The Vapometer consists of an aluminium cup and threaded flanged ring with two neoprene gaskets and with an additional Teflon seal that holds the specimen in place. Specimens were cut into a circular shape with a diameter of 74 mm. The measuring cups were first filled up to $\frac{3}{4}$ of their internal height with silica gel (previously dried), which was used as a water absorption agent (desiccant).

The specimens were then placed between the two mentioned rubber gaskets, fitted onto the cup collar and pressed down using the threaded aluminium ring. The effective diameter of the specimen exposed to the environment was 63.5 mm. Next, the cups were placed in a controlled climatic test chamber that operated at a temperature of 23 ± 2 °C and a relative humidity of 60 ± 10%, respectively. The cups were weighed periodically using a laboratory balance until 10 data points were collected. The water-vapor transmission rate (WVTR) was calculated as follows [60]:

$$WVTR = \frac{\Delta W}{t \cdot A} \quad (1)$$

where $WVTR$ represents the water vapor transmission rate (g·m^{-2}·d^{-1}); ΔW is the weight change (g); t indicates the time of the experiment (h); A is the test area (m^2).

2.7. Color Measurements

The color of the AA-CH-S and AA-S films was analyzed by using the tristimulus colorimeter MINOLTA CR310. Standard CIE conditions with illumination were used. The configuration included the illuminant D65 and an angle of 10. The readings were taken using the CIELAB system (L^*, a^*, b^*), and presented as the L^* value (color brightness). The results are expressed as the mean value of five measurements. Color was evaluated for the bioplastics AA-CH-S and AA-S before and after ionization irradiation after the 12-month storage period. Based on the obtained results, the total color difference (ΔE) of the irradiated film samples stored for 12 months concerning the non-irradiated material (sample 0) was calculated. Color change (ΔE) can be calculated from the following equation:

$$\Delta E = \sqrt{(\Delta L)^2 + (\Delta a)^2 + (\Delta b)^2} \quad (2)$$

where L is the lightness (also referred to as luminance) parameter (maximum value of 100 represents a perfectly reflecting diffuser; the minimum value of zero represents the color black); a is the axis of the red–green character ($+a$ = redder; $-a$ = greener); b is the axis of the yellow–blue character ($+b$ = yellower; $-b$ = bluer) [61].

For the interpretation of data, the criterion of the International Commission on Illumination CIE regarding the acceptability of color was adopted. When $0 < \Delta E < 1$, the observer does not notice the color difference; when $1 < \Delta E < 2$, only an experienced observer notices the color difference; when $2 < \Delta E < 3.5$, the color difference is noticeable for an inexperienced observer; when $3.5 < \Delta E < 5$, the color difference is noticeable; when $5 < \Delta E$, the observer notices two different colors.

2.8. Structure Analysis by Fourier Transform Infrared (FTIR) Spectroscopy

Spectral profiles of the non-irradiated and irradiated biofilms were recorded in the reflection mode in the range 4000–400 cm^{-1} with a resolution of 1 cm^{-1} by an attenuated total reflection Fourier transform infrared spectroscope (Perkin-Elmer Spectrum 100 IR, ATR, Waltham, MA, USA). Spectra of the biofilms were recorded at room temperature directly on the diamond crystal. Each spectrum recorded was an average of 16 successive scans. In addition, ten acquisitions were performed for each experiment by the manual rotation of biofilm samples. The ten replicates for each sample of non-irradiated and irradiated biofilms were averaged before further data processing and the mean spectra for each were used in subsequent analysis. Principal component analysis (PCA) was performed on the IR transmission spectra in the range of 4000–400 cm^{-1}. The cross-validation method was used to check the models. Multivariate data analysis was performed using the Unscrambler 9.7 software (CAMO, Oslo, Norway).

2.9. Surface Morphology Using SEM Microscopy

The surface microstructure of the films was studied using a scanning electron microscope (SEM; Zeiss EVO 40 with the electron accelerating voltage of 17 kV). Before the analysis, the surfaces of the films were coated with gold (Sputter Coater SCD 050). The images were taken at the magnification of 500×.

2.10. Biodegradation in the Activated Sludge Environment

The biodegradation test was carried out under active sludge conditions. The test proceeded for 70 days at a temperature of 55 °C. Periodic weight loss measurements were applied using the weight method with an accuracy of 0.0001 g, every 14 days. Due to the equipment capabilities and long analysis time, samples irradiated under air conditions were used with two doses of 10 and 40 kGy; all the samples were run in triplicate. The presented results are a preliminary attempt to determine the biodegradability of the tested materials in a liquid environment.

3. Results and Discussion

3.1. Materials

The radiation process was conducted on a laboratory scale using a cobalt (^{60}Co) gamma radiation chamber GC 5000 with a dose rate of 8.5 kGy/h. The following radiation doses were used: 10, 20, and 40 kGy. The samples were conditioned before radiation process under the following standardized conditions: 22 ± 2 °C and a relative air humidity of 65 ± 2%.

3.2. Mechanical Properties

In the case of the AA-CH-S film that contained three components (an aliphatic–aromatic copolymer, starch, and chitosan), the changes in the tensile strength caused by the radiation were small and rather accidental (Figure 1). A further increase in the radiation dose (from 10 to 40 kGy) did not cause a significant change. On the other hand, a significant decrease in the elongation at break was observed (Figure 2) from about 400% for the control sample to 50–80% in the case of the irradiated samples. This is possibly due to the phenomenon of creating additional bonds, and thus reducing chain mobility. It is noteworthy that the impact on the elongation at break values remained at a low level regardless, of the radiation dose applied (10, 20, or 40 kGy).

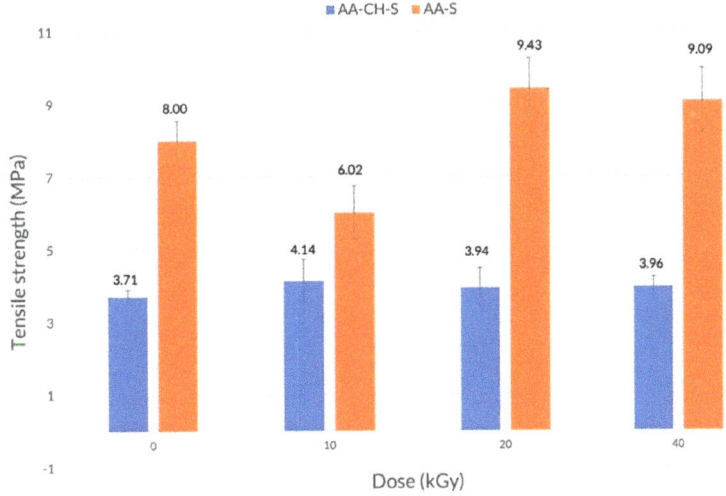

Figure 1. Tensile strength of AA-CH-S (blue) and AA-S (orange) films irradiated with doses of 10, 20, and 40 kGy in the machine direction.

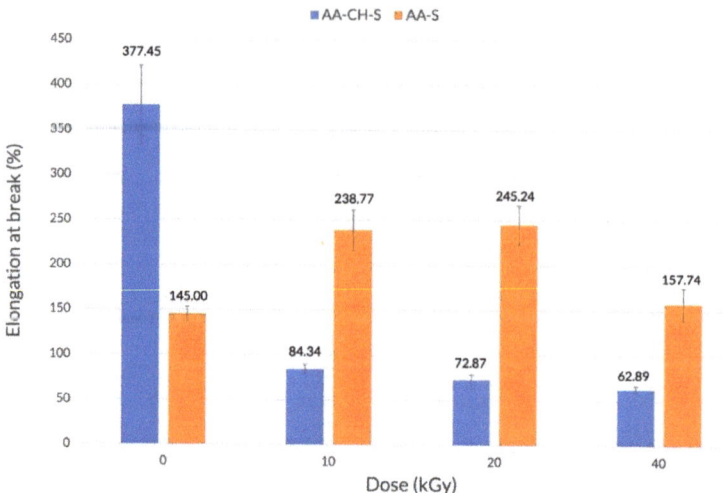

Figure 2. Elongation at break for AA-CH-S (blue) and AA-S (orange) films irradiated with doses of 10, 20, and 40 kGy in the machine direction.

The AA-S film was characterized by higher mechanical strength than the AA-CH-S film (Figure 1). Interestingly, an increase in the radiation dose (from 10 to 20 and 40 kGy) caused a slight increase in the tensile strength values from 8 MPa for the control sample to 9.4 and 9.1 MPa, respectively. An increase in the elongation at break values, from 145% for the reference sample to 239% and 245% for the samples treated with the radiation doses of 10 and 20 kGy, respectively, was also observed (Figure 2). A further increase in the radiation dose to 40 kGy resulted in a decrease (157%) in the elongation at break value.

3.3. Overall Migration

The overall migration analysis revealed that in the case of the AA-CH-S sample, the migration size exceeds the permissible limit of 10 mg/dm^2, according to the Commission Regulation (EU) No. 10/2011 [48], regardless of radiation dose; therefore, this material cannot be allowed to come into contact with hydrated food and oil in water emulsions (Figure 3). Interestingly, the allowed limit (Figure 3, red line) was also exceeded in the case of a reference sample. In the case of AA-S, the values did not exceed the acceptable level of overall migration, regardless of the radiation dose.

3.4. Sensory Analysis of Packaging Materials

The study allows for the assessment of the intensity of free volatile compounds, the presence of which may affect the sensory neutrality of the packaging. Packaging materials comply with FCM requirements only if they meet the limits for global, specific migration and sensory inertness. Failure to meet the requirements of even one parameter disqualifies the material as safe for FCMs. In some cases, the human senses, although burdened with many restrictions regarding the reception of stimuli, can detect a foreign flavor note, which remains on the verge of detectability of the used instrumental methods, which is why this research has become popular in the packaging industry in recent years. The results of the sensory analysis are complemented by an analysis of volatile organic compounds (VOCs), which can be derived from packaging components, printing inks, and other contaminants. Examples of such VOCs are acetone, cyclohexanone, ethanol, methyl ethyl ketone, ethyl acetate, toluene, and others.

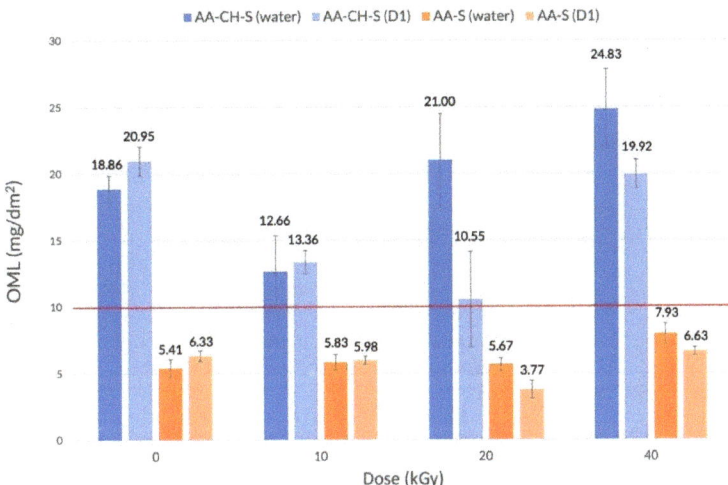

Figure 3. Global migration of AA-CH-S (blue) and AA-S (orange) films in two model solutions (water; D1).

Some food products, especially those containing significant amounts of fat, are particularly sensitive to foreign smells, which can be detected by consumers. These products, e.g., butter cookies, chocolate, almonds, and mineral water, are used as indicator products in the sensory testing of packaging materials. Sensory evaluation can be carried out according to the standardized methodologies described in the standards. The test is performed by a team trained in sensory evaluation. In the case of an analogous food evaluation team, the methods and reference substances are well-described and known. In the case of sensory analysis of packaging materials, there are no flavor patterns to train and check the sensitivity of the team. On the one hand, basic knowledge in the field of the sensory evaluation of food is used, but the key is the sensitivity to flavors and aromas associated with the production process of a given packaging material.

The tested samples did not have any foreign taste or smell (Table 2). The taste was defined as neutral or close to the neutral taste of water. The smell was imperceptible or slightly palpable. It was observed that with the increase in radiation dose, a slightly higher level of foreign taste and smell was noticeable; however, the results were at an acceptable, low level. On this basis, it can be concluded that radiation does not cause unacceptable changes in the taste or smell of the packaging materials tested.

Table 2. Sensory analysis of AA-CH-S and AA-S films.

Dose (kGy)	AA-CH-S Flavor/Odor Median	AA-S Flavor/Odor Median
0	0.0/0.0	0.0/0.0
10	0.0/0.0	0.0/0.0
20	0.5/1.0	0.0/0.5
40	0.5/1.0	0.0/1.0

3.5. Oxygen Transmission Rate

Oxygen permeability is one of the most commonly studied properties of packaging films. The tested biodegradable plastics were characterized by considerable gas permeability, i.e., low barrier properties. The OTR results for AA-S and AA-CH-S films presented in Table 3 revealed that radiation does not alter oxygen permeability. No changes in the barrier properties of the tested samples were observed, regardless of the radiation dose.

Table 3. OTR (cc/m^2/24 h) results of AA-S and AA-CH-S films.

Dose (kGy)	AA-CH-S	AA-S
0	85 ± 10	120 ± 15
10	80 ± 5	115 ± 5
20	85 ± 8	120 ± 10
40	90 ± 10	115 ± 10

3.6. Water Vapor Transmission Rate

The water vapor transmission rate (WVTR) data for the AA-S and AA-CH-S films are presented in Table 4. The results showed that in the case of the AA-S sample, the radiation caused a two-fold increase in the permeability of water vapor as a result of the radiation for all the analyzed doses. Interestingly, no significant changes in water vapor permeability were observed in the case of the AA-CH-S films.

Table 4. WVTR (g·m^{-2}·d^{-1}) results of AA-S and AA-CH-S films.

Dose (kGy)	AA-CH-S	AA-S
0	300 ± 30	110 ± 15
10	340 ± 30	210 ± 10
20	300 ± 40	220 ± 5
40	280 ± 10	210 ± 10

3.7. Color Measurements

The total color difference (ΔE) parameter determined for the analyzed films is presented in Table 5. The results showed that in the case of both tested materials, the magnitude of the changes correlated with the radiation dose used. However, in the case of the AA-S sample, the color changes remained at an acceptable level of $\Delta E < 1$, since the observer did not notice the color difference. In the case of the AA-CH-S sample, the changes in color were more pronounced ($\Delta E < 1$) only for the radiation dose of 10 kGy. For the radiation dose of 20 kGy, the changes in color can be noticed by an experienced observer, and for the radiation dose of 40 kGy, the color difference is noticeable.

Table 5. The average total color difference (ΔE) results for the analyzed films.

Dose (kGy)	AA-CH-S	AA-S
10	0.77	0.21
20	1.62	0.28
40	3.79	0.36

3.8. Structure Analysis by Fourier Transform Infrared (FTIR) Spectroscopy

The aim of the study was to characterize the changes in the structural properties of the biofilms (AA-S and AA-CH-S) during ionization irradiation with different doses. For spectral characteristics, infrared spectroscopy was used. The FT-IR spectra of the non-irradiated AA-CH-S sample are presented in Figure 4.

The following bands were recognized in these spectra according to the literature data [62–64]: aromatic ring stretching vibrations of the C-H group (3321 cm^{-1}); aliphatic chain stretching vibrations of the C-H groups (2956 cm^{-1}); stretching vibrations from the carbonyl groups C=O (1723 cm^{-1}); aromatic ring stretching vibrations of the C=C group (1541 cm^{-1}). In the tested material, the presence of chitosan is most likely indicated by the band that is visible at the wavelength 2916 cm^{-1}. The intensive band at 1723 cm^{-1} that is characteristic of C=O binding may be caused by the degradation of chitosan, while the medium intensity band at 1386 cm^{-1} may originate from the C-H deformation vibrations in the chitosan methylene groups. By comparing the band characteristics of polysaccharides (1200–1030 cm^{-1}), pyranose rings (785–730 cm^{-1}), and the spectra of a material that contained corn starch, the presence of starch in the AA-CH-S material was confirmed [62–64].

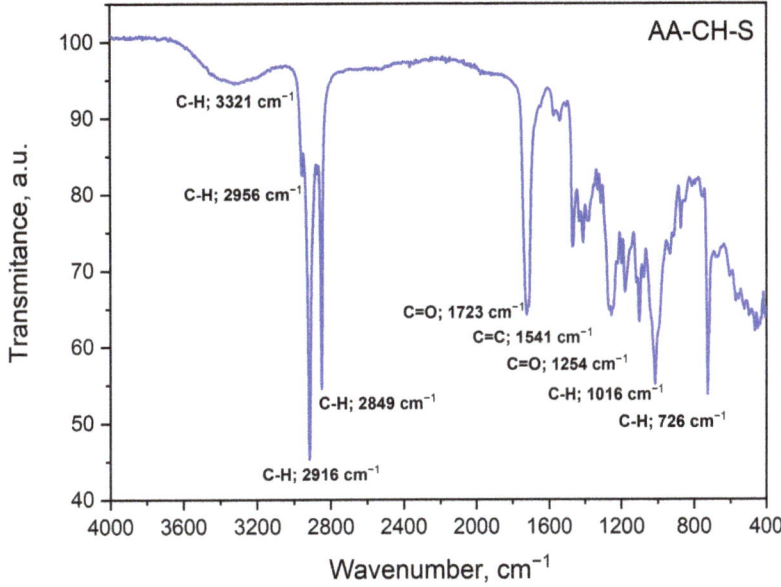

Figure 4. ATR FT-IR spectra of the AA-CH-S film.

For the comparison, the mid-infrared spectra were registered and analyzed for the non-irradiated AA-S film (Figure 5).

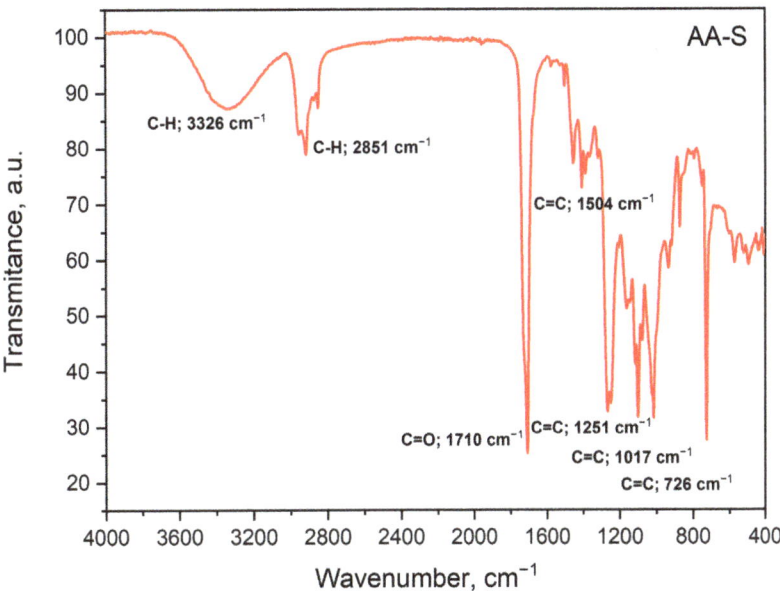

Figure 5. ATR FT-IR spectra of the AA-S film.

The following bands were recognized based on the literature data (Figure 5) [62]: aromatic ring stretching vibrations of the C-H group (3326 cm^{-1}); aliphatic chain stretching vibrations of the C-H groups (2851 cm^{-1}) confirmed by the presence of strong bands in the region of ester stretching vibrations of C-O (1300–1100 cm^{-1}); stretching vibrations from the C=O carbonyl groups (1710 cm^{-1}); aromatic ring stretching vibrations of the C=C group (1504 cm^{-1}); a group of bands in the range of 785–730 cm^{-1} also found in materials that contained starch, derived from the pyranose ring of polysaccharides.

Principal component analysis (PCA) was performed on the recorded infrared spectra to examine the main differences in the spectra of the examined materials irradiated by different doses of ionization radiation. Only analyses of pure spectra, without chemometrics, show no significant differences between the samples exposed to ionizing radiation. Infrared spectra can be used to distinguish reference material from radiation-treated samples, but further distinction due to the dose used is no longer possible. As a result of the PCA analysis, we could discriminate between the samples with different doses of ionization irradiation for the AA-CH-S and AA-S materials, respectively (Figures 6 and 7).

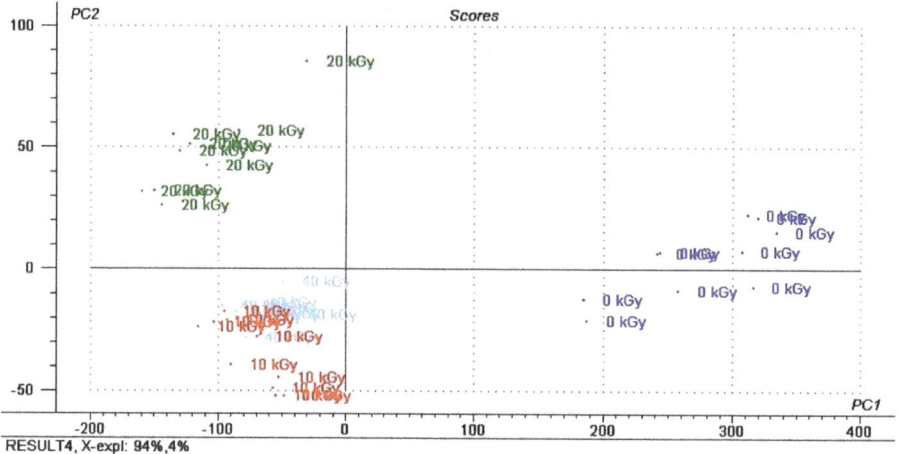

Figure 6. PCA results of IR spectra of the AA-CH-S film and the score plot for two significant principal components, PC1 vs. PC2 (non-irradiated (0 kGy) and irradiated with different doses (10, 20 and 40 kGy)).

The IR spectra of the AA-CH-S film irradiated by different ionization radiation doses analyzed using a PCA score plot revealed the clear separation between the samples (Figure 6). The first (PC1) and second (PC2) principal components accounted for 94.66% and 3.48% of the variance in the experimental data, respectively. The first three PCs described over 99% of the data. PC1 was strongly positively correlated with general irradiation. PC2 is dependent on the quantitative dose. Different levels of separation of the samples irradiated by 20 kGy doses in comparison to the samples irradiated by 40 kGy doses may mean irreversible changes in the structure of the material detected by IR spectroscopy.

Figure 6 presents the PCA results of the IR spectra of the AA-S material and the score plot for two significant principal components, PC1 vs. PC2. The material was non-irradiated (0 kGy) and irradiated with different doses. The IR spectra of the AA-S material irradiated by different ionization radiation doses analyzed using the PCA score plot revealed the blurred separation between the samples (Figure 7). The first (PC1) and second (PC2) principal components accounted for 85.56% and 11.58% of the variance in the experimental data, respectively. The first six principal components (PCs) described more than 99% of the data for this sample (data series). PC1 was strongly negatively correlated with the quantitative irradiation dose.

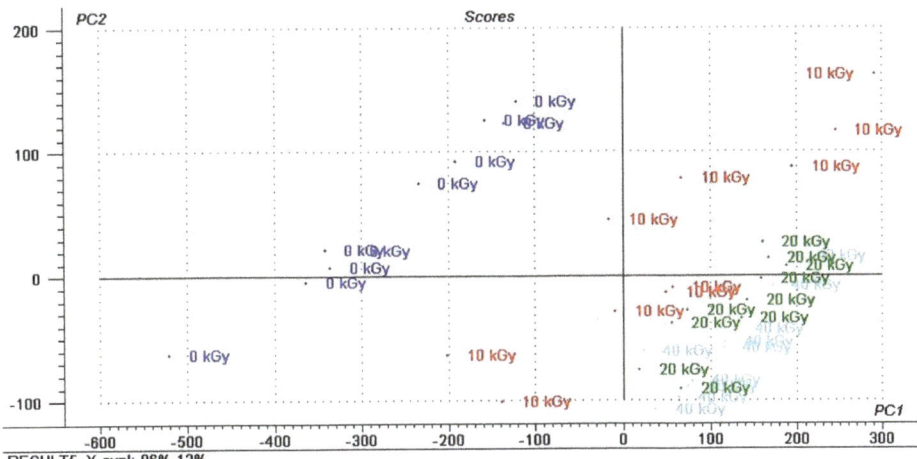

Figure 7. PCA results of IR spectra of the AA-S film and score plot for two significant principal components, PC1 vs. PC2 (non-irradiated (0 kGy) and irradiated with different doses (10, 20 and 40 kGy)).

3.9. Surface Morphology Using SEM Microscopy

SEM microscopic images of biobased composites after exposure to radiation was made for the surface change evaluation. The morphology of the samples' surface is presented in Figures 8 and 9. SEM images of the AA-CH-S film (Figure 8) revealed visible long structures with sharp edges, which originate from the filler chitosan. More SEM microphotographs of the AA-CH-S samples are presented in Supplementary Data in Figures S1–S18. There are no changes in the surface morphology between the reference sample (AA-CH-S 0 kGy) and the irradiated films.

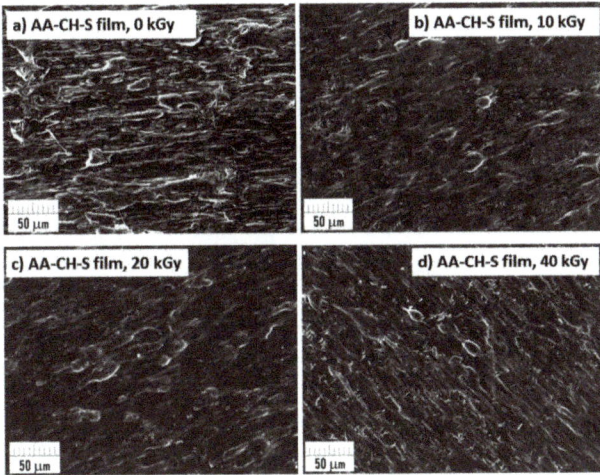

Figure 8. SEM micrographs of AA-CH-S films (0, 10, 20; 40 kGy) taken at 500× magnification.

The AA-S film contains starch, but its grains are not recognizable in the microscopic image and a homogenous structure can be observed. There are no visible changes or destruction in the morphology of the reference sample (AA-S 0 kGy) or the irradiated sample (Figure 9). More SEM microphotographs of the AA-S samples are presented in Supplementary Data in Figures S19–S30.

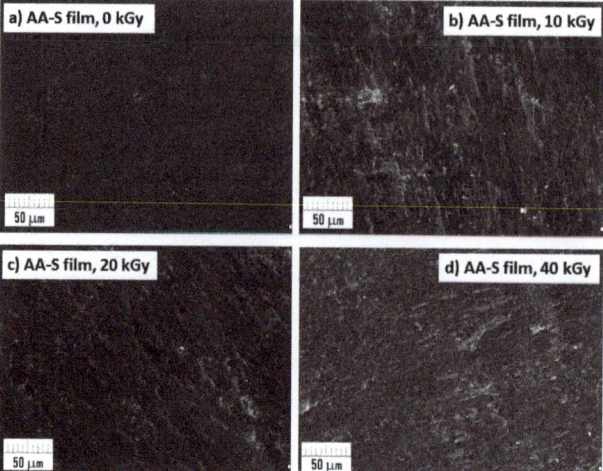

Figure 9. SEM micrographs of AA-S films (0, 10, 20; 40 kGy) taken at 500× magnification.

Regardless of the dose used, neither structural changes nor physical damage during the irradiation process were observed.

3.10. Biodegradation in Activated Sludge Environment

Biodegradation is a specific process that takes place in stages and depends on many external factors, including the characteristics of the material, pH, and humidity of the environment, temperature, and types of microorganisms. This work involved pilot studies on biodegradation in activated sludge conditions for the samples irradiated with a radiation dose of 10 and 40 kGy, and a reference material (Figure 10). The results showed that in the case of the AA-S samples, the higher the radiation dose, the faster the biodegradation rate. In the case of the AA-CH-S material, the radiation did not affect the biodegradation process.

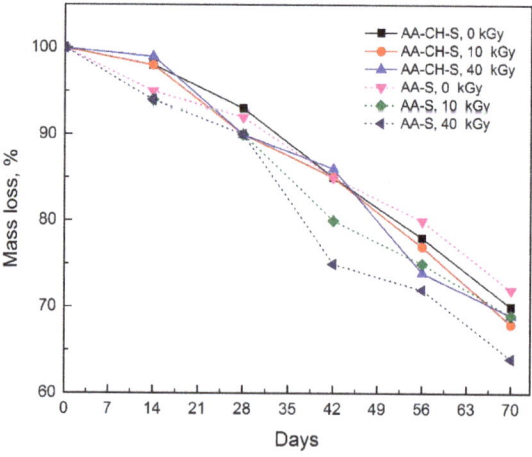

Figure 10. The degree of biodegradability in activated sludge was measured by the mass loss of the AA-S and AA-CH-S films exposed to 10 and 40 kGy radiation doses.

4. Conclusions

This paper presents the results of the tests that were carried out to determine the changes in the biodegradable materials AA-CH-S and AA-S caused by gamma radiation (10, 20, and 40 kGy exposure doses). The results showed that the mechanical properties of AA-S were improved due to the radiation-induced cross-linking processes, while in the case of AA-CH-S, a considerable decrease in the elongation at break was observed. The results also showed a decrease in the WVTR in the case of AA-S and no changes in barrier properties in the case of AA-CH-S. Both materials revealed no changes in the odor analyzed by the sensory analysis.

Our tests attempt to determine the biodegradability of the tested materials in a liquid environment. It has been shown that ionizing radiation, in the range of the radiation doses applied, affected only the AA-S material. Furthermore, an important aspect of the research was also the evaluation of the packaging–product interaction through the study of overall migration, which determines the suitability of the packaging material for contact with the packaged products, particularly food. The results revealed the unacceptable overall migration of AA-CH-S. Our study showed that for new packing system development based on natural material composites, evaluation of the radiation impact should be intensively studied before final product implementation.

Supplementary Materials: The following are available online at https://www.mdpi.com/article/10.3390/ma16020859/s1, Figures S1–S18: selected SEM micrographs of AA-CH-S films taken at different magnification; Figures S19–S30 selected SEM micrographs of AA-S films taken at different magnification.

Author Contributions: Conceptualization, K.W., D.C.-K. and W.K., methodology; K.W., D.C.-K., W.K. and K.P.-L.; investigation, K.W., D.C.-K., W.K., P.W., M.J. and K.P.-L.; writing—original draft preparation, K.W., D.C.-K., W.K., M.B., P.W., M.J. and K.P.-L.; manuscript revision, K.W., D.C.-K., W.K., M.B., P.W., M.J. and K.P.-L.; supervision, K.W., D.C.-K. and W.K.; project administration, K.W., D.C.-K. and W.K. All authors have read and agreed to the published version of the manuscript.

Funding: This research received no external funding.

Institutional Review Board Statement: Not applicable.

Informed Consent Statement: Not applicable.

Data Availability Statement: Not applicable.

Conflicts of Interest: The authors declare no conflict of interest.

References

1. Tsang, Y.F.; Kumar, V.; Samadar, P.; Yang, Y.; Lee, J.; Ok, Y.S.; Song, H.; Kim, K.H.; Kwon, E.E.; Jeon, Y.J. Production of Bioplastic through Food Waste Valorization. *Environ. Int.* **2019**, *127*, 625–644. [CrossRef] [PubMed]
2. Delgado, M.; Felix, M.; Bengoechea, C. Development of Bioplastic Materials: From Rapeseed Oil Industry by Products to Added-Value Biodegradable Biocomposite Materials. *Ind. Crops Prod.* **2018**, *125*, 401–407. [CrossRef]
3. Bilo, F.; Pandini, S.; Sartore, L.; Depero, L.E.; Gargiulo, G.; Bonassi, A.; Federici, S.; Bontempi, E. A Sustainable Bioplastic Obtained from Rice Straw. *J. Clean. Prod.* **2018**, *200*, 357–368. [CrossRef]
4. Rajesh Banu, J.; Kavitha, S.; Yukesh Kannah, R.; Poornima Devi, T.; Gunasekaran, M.; Kim, S.H.; Kumar, G. A Review on Biopolymer Production via Lignin Valorization. *Bioresour. Technol.* **2019**, *290*, 121790. [CrossRef] [PubMed]
5. European Bioplastics Materials—European Bioplastics e.V. Available online: https://www.european-bioplastics.org/bioplastics/materials/ (accessed on 27 November 2022).
6. European Bioplstics Bioplastic Facts and Figures. Available online: https://docs.european-bioplastics.org/publications/EUBP_Facts_and_figures.pdf (accessed on 27 November 2022).
7. Yu, L.; Dean, K.; Li, L. Polymer Blends and Composites from Renewable Resources. *Prog. Polym. Sci.* **2006**, *31*, 576–602. [CrossRef]
8. Assman, K. Polimery Biodegradowalne - Przykłady Zastosowań. In *Nowoczesne Materiały Polimerowe I Ich Przetwórstwo*; Tomasz, K., Ed.; Politechnika Lubelska: Lublin, Poland, 2017; Volume 3, pp. 51–70. ISBN 978-83-7947-300-7.
9. Czarnecka-Komorowska, D.; Tomasik, M.; Thakur, V.K.; Kostecka, E.; Rydzkowski, T.; Jursa-Kulesza, J.; Bryll, K.; Mysłowski, J.; Gawdzińska, K. Biocomposite Composting Based on the Sugar-Protein Condensation Theory. *Ind. Crops Prod.* **2022**, *183*. [CrossRef]

10. Kraśniewska, K.; Pobiega, K.; Gniewosz, M. Pullulan-Biopolymer with Potential for Use as Food Packaging. *Int. J. Food Eng.* **2019**, *15*, 20190030. [CrossRef]
11. Acha, C.; Blanchard, R.; Brodsky, J.; Ding, L.; Fox, A.; Grosvenor, E.; Gibson, K.; Hoy, A.; Hughes, J.; Lee, K.; et al. On the Mechanism of Electron Beam Radiation-Induced Modification of Poly(Lactic Acid) for Applications in Biodegradable Food Packaging. *Appl. Sci.* **2022**, *12*, 1819. [CrossRef]
12. Abramowska, A.; Cieśla, K.A. The Influence of Electron and Gamma Irradiation on the Properties of Starch:PVA Films—The Effect of Irradiation Dose. *Nukleonika* **2021**, *66*, 3–9. [CrossRef]
13. Madera-Santana, T.J.; Meléndrez, R.; González-García, G.; Quintana-Owen, P.; Pillai, S.D. Effect of Gamma Irradiation on Physicochemical Properties of Commercial Poly(Lactic Acid) Clamshell for Food Packaging. *Radiat. Phys. Chem.* **2016**, *123*, 6–13. [CrossRef]
14. Li, L.; Chen, H.; Wang, M.; Lv, X.; Zhao, Y.; Xia, L. Development and Characterization of Irradiated-Corn-Starch Films. *Carbohydr. Polym.* **2018**, *194*, 395–400. [CrossRef]
15. Díez-Pascual, A.M. Synthesis and Applications of Biopolymer Composites. *Int. J. Mol. Sci.* **2019**, *20*, 2321. [CrossRef] [PubMed]
16. Salapare, H.S.; Amigoni, S.; Guittard, F. Bioinspired and Biobased Materials. *Macromol. Chem. Phys.* **2019**, *220*, 1900241. [CrossRef]
17. van den Oever, M.; Molenveld, K. Replacing Fossil Based Plastic Performance Products by Bio-Based Plastic Products-Technical Feasibility. *New Biotechnol.* **2017**, *37*, 48–59. [CrossRef]
18. Habel, C.; Schöttle, M.; Daab, M.; Eichstaedt, N.J.; Wagner, D.; Bakhshi, H.; Agarwal, S.; Horn, M.A.; Breu, J.; Habel, C.; et al. High-Barrier, Biodegradable Food Packaging. *Macromol. Mater. Eng.* **2018**, *303*, 1800333. [CrossRef]
19. Sangroniz, A.; Sangroniz, L.; Gonzalez, A.; Santamaria, A.; del Rio, J.; Iriarte, M.; Etxeberria, A. Improving the Barrier Properties of a Biodegradable Polyester for Packaging Applications. *Eur. Polym. J.* **2019**, *115*, 76–85. [CrossRef]
20. Aniśko, J.; Barczewski, M. Polylactide: From Synthesis and Modification to Final Properties. *Adv. Sci. Technol. Res. J.* **2021**, *15*, 9–29. [CrossRef]
21. Peelman, N.; Ragaert, P.; de Meulenaer, B.; Adons, D.; Peeters, R.; Cardon, L.; van Impe, F.; Devlieghere, F. Application of Bioplastics for Food Packaging. *Trends Food Sci. Technol.* **2013**, *32*, 128–141. [CrossRef]
22. Ruggero, F.; Gori, R.; Lubello, C. Methodologies to Assess Biodegradation of Bioplastics during Aerobic Composting and Anaerobic Digestion: A Review. *Waste Manag. Res.* **2019**, *37*, 959–975. [CrossRef]
23. Abramowska, A.; Cieśla, K.A.; Buczkowski, M.J.; Nowicki, A.; Głuszewski, W. The Influence of Ionizing Radiation on the Properties of Starch-PVA Films. *Nukleonika* **2015**, *60*, 669–677. [CrossRef]
24. Flores, I.; Martínez De Ilarduya, A.; Sardon, H.; Müller, A.J.; Muñoz-Guerra, S. Synthesis of Aromatic-Aliphatic Polyesters by Enzymatic Ring Opening Polymerization of Cyclic Oligoesters and Their Cyclodepolymerization for a Circular Economy. *ACS Appl. Polym. Mater.* **2019**, *1*, 321–325. [CrossRef]
25. Ukielski, R.; Kondratowicz, F.; Kotowski, D. Production, Properties and Trends in Development of Biodegradable Polyesters with Particular Respect to Aliphatic-Aromatic Copolymers/Produkcja, Wlasciwosci i Kierunki Rozwoju Biodegradowalnych Poliestrow Ze Szczegolnym Uwzglednieniem Kopolimerow Alifatyczno-Aromatycznych. *Polimery* **2013**, *58*, 167–177.
26. Chandra, R.; Rustgi, R. Biodegradable Polymers. *Prog. Polym. Sci.* **1998**, *23*, 1273–1335. [CrossRef]
27. Witt, U.; Müller, R.J.; Deckwer, W.D. Biodegradation Behavior and Material Properties of Aliphatic/Aromatic Polyesters of Commercial Importance. *J. Environ. Polym. Degrad.* **1997**, *5*, 81–89. [CrossRef]
28. Kubera, H.; Assman, K.; Czaja-Jagielska, N.; Melski, K.; Głuszewski, W.; Migdał, W.; Zimek, Z. Impact of Ionizing Radiation on the Properties of a Hydrobiodegradable Aliphatic-Aromatic Copolyester. *Nukleonika* **2012**, *57*, 621–626.
29. Khan, B.; Bilal Khan Niazi, M.; Samin, G.; Jahan, Z. Thermoplastic Starch: A Possible Biodegradable Food Packaging Material—A Review. *J. Food Process. Eng.* **2017**, *40*, e12447. [CrossRef]
30. Knitter, M.; Czarnecka-Komorowska, D.; Czaja-Jagielska, N.; Szymanowska-Powałowska, D. Manufacturing and Properties of Biodegradable Composites Based on Thermoplastic Starch/Polyethylene-Vinyl Alcohol and Silver Particles. In *International Scientific-Technical Conference MANUFACTURING*; Springer: Cham, Switzerland, 2019; pp. 610–624. [CrossRef]
31. Nadia, N.; Othman, S.A. Gamma Radiation Effects on Biodegradable Starch Based Blend With Different Polyester: A Review. *J. Adv. Res. Dyn. Control Syst.* **2019**, *62*, 244–249.
32. Guo, Y.; Wang, H. Preparation and Properties of Edible Packaging Films Based on Chitosan with Microcrystalline Cellulose from Tomato Peel Pomace. *J. Biobased Mater. Bioenergy* **2019**, *14*, 1–8. [CrossRef]
33. Ahmed, K.B.M.; Khan, M.M.A.; Siddiqui, H.; Jahan, A. Chitosan and Its Oligosaccharides, a Promising Option for Sustainable Crop Production—A Review. *Carbohydr. Polym.* **2020**, *227*, 115331. [CrossRef]
34. Dong, Z.; Cui, H.; Wang, Y.; Wang, C.; Li, Y.; Wang, C. Biocompatible AIE Material from Natural Resources: Chitosan and Its Multifunctional Applications. *Carbohydr. Polym.* **2020**, *227*, 115338. [CrossRef]
35. Malinowska-Pańczyk, E.; Sztuka, K.; Kołodziejska, I. Substancje o Działaniu Przeciwdrobnoustrojowym Jako Składniki Biodegradowalnych Folii z Polimerów Naturalnych. *Polimery* **2010**, *55*, 627–633.
36. Rinaudo, M. Chitin and Chitosan: Properties and Applications. *Prog. Polym. Sci.* **2006**, *31*, 603–632. [CrossRef]
37. Michieletto, A.; Lorandi, F.; de Bon, F.; Isse, A.A.; Gennaro, A. Biocompatible Polymers via Aqueous Electrochemically Mediated Atom Transfer Radical Polymerization. *J. Polym. Sci.* **2020**, *58*, 114–123. [CrossRef]
38. Madni, A.; Kousar, R.; Naeem, N.; Wahid, F. Recent Advancements in Applications of Chitosan-Based Biomaterials for Skin Tissue Engineering. *J. Bioresour. Bioprod.* **2021**, *6*, 11–25. [CrossRef]

39. Saad, E.M.; Elshaarawy, R.F.; Mahmoud, S.A.; El-Moselhy, K.M. New Ulva Lactuca Algae Based Chitosan Bio-Composites for Bioremediation of Cd(II) Ions. *J. Bioresour. Bioprod.* **2021**, *6*, 223–242. [CrossRef]
40. Srinivasa, P.C.; Ramesh, M.N.; Tharanathan, R.N. Effect of Plasticizers and Fatty Acids on Mechanical and Permeability Characteristics of Chitosan Films. *Food Hydrocoll.* **2007**, *21*, 1113–1122. [CrossRef]
41. Sindhu, M.; Abraham, T. Emilia Characterisation of Ferulic Acid Incorporated Starch–Chitosan Blend Films. *Food Hydrocoll.* **2008**, *22*, 826–835. [CrossRef]
42. Komolprasert, V. Packaging Food for Radiation Processing. *Radiat. Phys. Chem.* **2016**, *129*, 35–38. [CrossRef]
43. Eyssa, H.M.; Sawires, S.G.; Senna, M.M. Gamma Irradiation of Polyethylene Nanocomposites for Food Packaging Applications against Stored-Product Insect Pests. *J. Vinyl Addit. Technol.* **2019**, *25*, E120–E129. [CrossRef]
44. Silvestre, C.; Cimmino, S.; Stoleru, E.; Vasile, C. Application of Radiation Technology to Food Packaging. In *Applications of Ionizing Radiation in Materials Processing*; Yongxia, S., Andrzej, G.C., Eds.; Institute of Nuclear Chemistry and Technology: Warszawa, Poland, 2017; pp. 461–484.
45. Jedson, A.; Brant, C.; Naime, N.; Lugão, A.B.; Ponce, P. Influence of Ionizing Radiation on Biodegradable Foam Trays for Food Packaging Obtained from Irradiated Cassava Starch. *Braz. Arch. Biol. Technol.* **2018**, *61*, 1–16. [CrossRef]
46. Negrin, M.; Macerata, E.; Consolati, G.; Quasso, F.; Genovese, L.; Soccio, M.; Giola, M.; Lotti, N.; Munari, A.; Mariani, M. Gamma Radiation Effects on Random Copolymers Based on Poly(Butylene Succinate) for Packaging Applications. *Radiat. Phys. Chem.* **2018**, *142*, 34–43. [CrossRef]
47. *(EC) No 1935/2004*; Regulation (EC) No 1935/2004 of the European Parliament and of the Council of 27 October 2004 on Materials and Articles Intended to Come into Contact with Food and Repealing Directives 80/590/EEC and 89/109/EEC. The European Parliament: Strasbourg, France; The Council of The European Union: Brussels, Belgium, 2004.
48. *(EU) No 10/2011*; Commission Regulation (EU) No 10/2011 of 14 January 2011 on Plastic Materials and Articles Intended to Come into Contact with Food. The European Commission: Brussels, Belgium, 2011.
49. *(EC) No 2023/2006*; Commission Regulation (EC) No 2023/2006 of 22 December 2006 on Good Manufacturing Practice for Materials and Articles Intended to Come into Contact with Food. The Commission of the European Communities: Brussels, Belgium, 2006.
50. Drobny, J.G. Ionizing Radiation and Polymers: Principles, Technology, and Applications. In *Ionizing Radiation and Polymers: Principles, Technology, and Applications*; Elsevier Inc.: Waltham, MA, USA, 2012; pp. 1–298. [CrossRef]
51. Hara, M. Effects of Ionizing Radiation on Biopolymers for Applications as Biomaterials. *Biomed. Mater. Devices* **2022**, *1*, 1–18. [CrossRef]
52. *ISO 527-1:2019*; Plastics—Determination of Tensile Properties—Part 1: General Principles. International Organization for Standardization: Geneva, Switzerland, 2019. Available online: https://www.iso.org/standard/75824.html (accessed on 24 November 2022).
53. *ISO 527-3:2018*; Plastics—Determination of Tensile Properties—Part 3: Test Conditions for Films and Sheets. International Organization for Standardization: Geneva, Switzerland, 2018. Available online: https://www.iso.org/standard/70307.html (accessed on 29 November 2022).
54. *EN 1186-1:2002*; Materials and Articles in Contact with Foodstuffs—Plastics—Part 1: Guide to The Selection of Conditions and Test Methods for Overall Migration. Technical Committee; European Committee for Standardization: Brussels, Belgium, 2002. Available online: https://standards.iteh.ai/catalog/standards/cen/9b16242d-0ec6-4df8-90eb-38fb76e4cc10/en-1186-1-2002 (accessed on 24 November 2022).
55. *EN 1186-3:2022*; Materials and Articles in Contact with Foodstuffs—Plastics—Part 3: Test Methods for Overall Migration in Evaporable Simulants. Technical Committee; European Committee for Standardization: Brussels, Belgium, 2002. Available online: https://standards.iteh.ai/catalog/standards/cen/9e66297d-3b68-4d07-ae09-b8b37057b6ce/en-1186-3-2022 (accessed on 24 November 2022).
56. *DIN 10955*; Sensory Analysis—Testing of Packaging Materials and Packaging Materials and Packages for Food Products. European Committee for Standardization: Brussels, Belgium, 2004. Available online: https://www.en-standard.eu/din-10955-sensory-analysis-testing-of-packaging-materials-and-packages-for-food-products/ (accessed on 27 November 2022).
57. *ASTM F3136-15*; Standard Test Method for Oxygen Gas Transmission Rate through Plastic Film and Sheeting Using a Dynamic Accumulation Method. ASTM International: West Conshohocken, PA, USA, 2022. Available online: https://www.astm.org/f3136-15.html (accessed on 27 November 2022).
58. *ASTM F2714-08*; Standard Test Method for Oxygen Headspace Analysis of Packages Using Fluorescent Decay. ASTM International: West Conshohocken, PA, USA, 2021. Available online: https://www.astm.org/f2714-08r21.html (accessed on 27 November 2022).
59. *ASTM E96/E96M-22*; Standard Test Methods for Gravimetric Determination of Water Vapor Transmission Rate of Materials. ASTM International: West Conshohocken, PA, USA, 2022. Available online: https://www.astm.org/e0096_e0096m-22.html (accessed on 27 November 2022).
60. Shi, A.M.; Wang, L.J.; Li, D.; Adhikari, B. Characterization of Starch Films Containing Starch Nanoparticles: Part 1: Physical and Mechanical Properties. *Carbohydr. Polym.* **2013**, *96*, 593–601. [CrossRef] [PubMed]
61. Chorobiński, M.; Skowroński, Ł.; Bieliński, M. Methodology for Determining Selected Characteristics of Polyethylene Dyeing Using CIELab System. *Polimery* **2019**, *64*, 690–696. [CrossRef]

62. Silverstein, R.M.; Webster, F.X.; Kiemle, D.J.; Bryce, D.L. *Spectrometric Identification of Organic Compounds*; Wiley: New York, NY, USA, 2014; ISBN 0470616377.
63. Paluszkiewicz, C.; Stodolak, E.; Hasik, M.; Blazewicz, M. FT-IR Study of Montmorillonite-Chitosan Nanocomposite Materials. *Spectrochim. Acta Part A Mol. Biomol. Spectrosc.* **2011**, *79*, 784–788. [CrossRef]
64. Silva, S.M.L.; Braga, C.R.C.; Fook, M.V.L.; Raposo, C.M.O.; Carvalho, L.H.; Canedo, E.L. Application of Infrared Spectroscopy to Analysis of Chitosan/Clay Nanocomposites. In *Infrared Spectroscopy: Materials Science, Engineering and Technology*; Theophile, T., Ed.; InTech: Rijeka, Croatia, 2012; pp. 43–62. ISBN 978-953-51-0537-4.

Disclaimer/Publisher's Note: The statements, opinions and data contained in all publications are solely those of the individual author(s) and contributor(s) and not of MDPI and/or the editor(s). MDPI and/or the editor(s) disclaim responsibility for any injury to people or property resulting from any ideas, methods, instructions or products referred to in the content.

MDPI
St. Alban-Anlage 66
4052 Basel
Switzerland
Tel. +41 61 683 77 34
Fax +41 61 302 89 18
www.mdpi.com

Materials Editorial Office
E-mail: materials@mdpi.com
www.mdpi.com/journal/materials

www.ingramcontent.com/pod-product-compliance
Lightning Source LLC
LaVergne TN
LVHW070211100526
838202LV00015B/2033